The Advanced Part
of a Treatise on the
Dynamics of a System
of Rigid Bodies

EDWARD JOHN ROUTH

CAMBRIDGE
UNIVERSITY PRESS

CAMBRIDGE UNIVERSITY PRESS

Cambridge, New York, Melbourne, Madrid, Cape Town,
Singapore, São Paolo, Delhi, Mexico City

Published in the United States of America by Cambridge University Press, New York

www.cambridge.org
Information on this title: www.cambridge.org/9781108050357

© in this compilation Cambridge University Press 2013

This edition first published 1892
This digitally printed version 2013

ISBN 978-1-108-05035-7 Paperback

THE ADVANCED PART

OF A TREATISE ON THE

DYNAMICS OF A SYSTEM OF RIGID BODIES

BEING PART II. OF A TREATISE ON THE WHOLE SUBJECT.

THE ADVANCED PART

OF A TREATISE ON THE

DYNAMICS OF A SYSTEM OF RIGID BODIES.

BEING PART II. OF A TREATISE ON THE WHOLE
SUBJECT.

𝔚𝔦𝔱𝔥 numerous 𝔈𝔵𝔞𝔪𝔭𝔩𝔢𝔰.

BY

EDWARD JOHN ROUTH, Sc.D., LL.D., F.R.S., &c.

HON. FELLOW OF PETERHOUSE, CAMBRIDGE ;
FELLOW OF THE SENATE OF THE UNIVERSITY OF LONDON.

FIFTH EDITION, REVISED AND ENLARGED.

London:

MACMILLAN AND CO.

AND NEW YORK.

1892

First Edition, 1860. *Second Edition*, 1868.

Third Edition, 1877. *Fourth Edition*, 1882. *Fifth Edition*, 1892.

PREFACE.

In this edition many additions and improvements have been made, particularly in the last half of the book. Many parts have been re-written in the hope of making the explanations clearer and briefer. A few sections have been omitted to make room for more important matter. New subjects, not discussed in the former editions, have been introduced in order to make the treatise as complete as possible. Though more than a year has elapsed since the publication of the first volume of this edition, I have not found the time at my disposal during the interval too long for these changes.

Following the same plan as in Vol. I., the several chapters have been made as independent as possible. The object in view is that the reader should be able to select his own order of study. Historical notices and references have been given throughout the book. I have endeavoured to join to every theorem or problem the name of the writer who, as far as I know, was the first to enunciate or solve it.

Numerous examples have been given throughout the book. Some of these are intended to be merely simple exercises, but many are important as illustrating and completing the theories given in the text. Sometimes when the principles of a theory had been explained numerous applications seemed to arise. Instead of loading the text with these it appeared preferable to put them into the form of examples and to give such hints as would make the solution easy. Everywhere the results have been given, and care has been taken to secure their accuracy; but amongst so many theorems, it cannot be expected that no errors have escaped detection.

I wish to express my thanks to Mr J. M. Dodds of Peterhouse for his kind assistance in correcting so many of the proof sheets.

EDWARD J. ROUTH.

PETERHOUSE,
February, 1892.

CONTENTS.

CHAPTER I.

MOVING AXES AND RELATIVE MOTION.

CHAPTER II.

OSCILLATIONS ABOUT EQUILIBRIUM.

CHAPTER III.

OSCILLATIONS ABOUT A STATE OF MOTION.

CHAPTER IV.

MOTION OF A BODY UNDER NO FORCES.

CHAPTER V.

MOTION OF A BODY UNDER ANY FORCES.

CHAPTER VI.

NATURE OF THE MOTION GIVEN BY LINEAR EQUATIONS AND THE CONDITIONS OF STABILITY.

CHAPTER VII.

FREE AND FORCED OSCILLATIONS.

CHAPTER VIII.

DETERMINATION OF THE CONSTANTS OF INTEGRATION IN TERMS OF THE INITIAL CONDITIONS.

CHAPTER IX.

CALCULUS OF FINITE DIFFERENCES.

CHAPTER X.

CALCULUS OF VARIATIONS.

CHAPTER XI.

PRECESSION AND NUTATION.

CHAPTER XII.

MOTION OF THE MOON ABOUT ITS CENTRE.

CHAPTER XIII.

MOTION OF A STRING OR CHAIN.

CHAPTER XIV.

MOTION OF A MEMBRANE.

NOTES.

In order that the plan of the book may be understood the following short summary is given of the subjects treated of in Part I.

Chap. 1. Theory of moments of inertia and the ellipsoids of inertia.

Chap. 2. D'Alembert's Principle and other fundamental theorems.

Chap. 3. Theory of motion about a fixed axis with applications to the pendulum, the numerical value of g, the watch balance, the ballistic pendulum, the anemometer.

Chap. 4. General principles of motion in two dimensions. Special consideration of stress, friction, impulses and relative motion.

Chap. 5. Geometry of motion in three dimensions, with Euler's equations.

Chap. 6. On Momentum, with the discussion of sudden changes of motion.

Chap. 7. On Vis Viva and Work, with some general theorems by Carnot, Bertrand, Thomson and Gauss.

Chap. 8. Lagrange's equations. Theory of reciprocation, the Hamiltonian transformation and the Modified function.

Chap. 9. Small oscillations. Several methods described. Lagrange's method, the energy test of stability and the Cavendish experiment.

Chap. 10. Some special problems. Oscillations of rolling bodies, and Lagrange's rule with regard to large tautochronous motions.

DYNAMICS.

CHAPTER I.

Moving Axes.

1. In many problems in dynamics it is found that the axes of reference suitable to the initial state of the motion are not well adapted to follow the body under consideration during its whole course of motion. It is therefore sometimes convenient to use axes which themselves move in space so that they always keep those positions which are most appropriate to the instantaneous position of the body. Thus, to take a simple case, in dynamics of a particle we sometimes resolve our forces along the tangent and normal to the path. This is practically the same as using a set of Cartesian axes which move so as to be always parallel to the tangent and normal. This theory has been generalised in Vol. I. Chap. IV. where the motion is referred to any two lines whatever which move in one plane. We now propose to extend the theory still further. We shall discuss the general equations of motion of a particle and then those of a rigid body referred to any rectangular axes which move as we may find convenient.

2. If we make the axes to which we refer the body move, it is clear that we must have some means of determining the position and motion of these axes in space. This might be effected by having another set of axes which are themselves fixed in space and to which in turn we might refer the moving axes. This is the course adopted by Euler; thus in the equations usually called after his name (Vol. I. Chap. V.) he uses two sets of axes. The advantage of giving motion to the axes is however greatly

diminished if we must also use a set of fixed axes throughout the motion. For this reason *we shall now determine the motion of the moving axes by angular velocities* θ_1, θ_2, θ_3 *about themselves.* In other words, we regard the axes as if they were a material system of three straight lines at right angles whose motion at any instant was given by three coexistent angular velocities about axes which instantaneously coincided with them. In this way we do not use any fixed axes except at the beginning or end of the solution, and only in such a manner as we may find convenient.

3. In order to understand how the motion of a body is referred to moving axes let us first suppose that the body is turning about a fixed point. Taking this point as origin we determine the motion of the body by three angular velocities ω_1, ω_2, ω_3 about the axes in the same manner as if the axes were fixed in space. The position of the body at the time $t + dt$ may be constructed from that at the time t by turning the body through the angles $\omega_1 dt$, $\omega_2 dt$, $\omega_3 dt$ successively round the instantaneous positions of the axes. But it must be remembered that $\omega_3 dt$ does not now give the angle the body has been turned through relatively to the plane xz, but relatively to some plane fixed in space passing through the instantaneous position of the axis of z. The angle turned through relatively to the plane of xz is $(\omega_3 - \theta_3)\, dt$.

If there be no fixed point we use the construction explained in Vol. I. Chap. V. We represent the motion of the body by the six components u, v, w; ω_1, ω_2, ω_3 referred to any origin, the axes being treated as if they were fixed for the moment. Here u, v, w are the resolved parts in the directions of the axes of the velocity of the origin or base point, and ω_1, ω_2, ω_3 are the resolved parts about the same axes of the angular velocity of the body. In the same way the motion of the axes is given by the components of motion p, q, r; θ_1, θ_2, θ_3, the moving axes being themselves the instantaneous axes of reference.

In most cases however the axes will be made to turn round some point which either is fixed or may be treated as fixed. Their directions in space are made to vary in a manner suitable to the purpose we have in hand. We then have p, q, r all zero. Since any point may be reduced to rest by the method explained in Vol. I. Chap. IV. this supposition, which will be generally made, does not really limit our choice of axes.

4. **Fundamental Theorem.** *A system of rectangular axes moves in any manner about a fixed point* O, *it is required to establish the kinematical relations between these axes and a system of axes fixed in space and coincident with them at any time* t.

Let Ox, Oy, Oz be the positions of the moving axes at the time t; after an interval dt these assume new positions, which we represent by Ox', Oy', Oz'. The change of position may be represented by a rotation θdt about some instantaneous axis, which we may represent by OI. Let θ_1, θ_2, θ_3 be the components of the angular velocity θ, so that the axes are moved from their positions Ox, Oy, Oz at the time t into their positions Ox', Oy', Oz' at the time $t + dt$ by the three rotations $\theta_1 dt$, $\theta_2 dt$, $\theta_3 dt$ about Ox, Oy, Oz performed in any order.

Let us represent by the symbol R any directed quantity or vector, such as a force, a velocity, the moment of a couple about its axis, or an angular momentum. Let us suppose that the vector may be resolved and compounded according to the "parallelogram law." Let us represent its components parallel to the three axes Ox, Oy, Oz by the symbols U, V, W. In the time dt the vector R has changed its magnitude and direction; in the same time the axes have also changed. The components of the vector at the time $t + dt$ in the then direction of the axes of reference, i.e. in the directions Ox', Oy', Oz' are $U + dU$, $V + dV$, $W + dW$.

We wish to find the increase in the time dt of the component in the direction of the axis Ox *supposed fixed in space*. Describe a sphere of unit radius whose centre is at O and let the axes cut the sphere in the points x, y, z, x', y', z'. Thus we have two spherical triangles xyz, $x'y'z'$, all whose sides are right angles. The resolved part of the vector at the time $t + dt$ along the axis Ox is

$$(U + dU)\cos xx' + (V + dV)\cos xy' + (W + dW)\cos xz'.$$

The rotations about Ox and Oy cannot alter the arc xy, but the rotation about Oz will move y' away from x by the arc $\theta_3 dt$. In the same way the rotations about Ox and Oz cannot alter the arc xz but the rotation about Oy will move z' towards x by the arc $\theta_2 dt$. Therefore

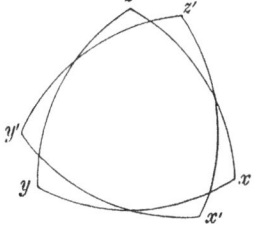

$$xy' = xy + \theta_3 dt,$$
$$xz' = xz - \theta_2 dt.$$

Also the cosine of the arc xx' differs from unity by the square of a small quantity. Substituting λ, we find that at the time $t + dt$ the component of the vector along Ox is

$$U + dU - V\theta_3 dt + W\theta_2 dt.$$

The rate of increase of the component of the vector in the direction Ox is

$$U_1 = \frac{dU}{dt} - V\theta_3 + W\theta_2.$$

In the same way the rates of increase of the components in the directions Oy, Oz are

$$V_1 = \frac{dV}{dt} - W\theta_1 + U\theta_3,$$

$$W_1 = \frac{dW}{dt} - U\theta_2 + V\theta_1.$$

We have here practically used two sets of axes. One set Ox, Oy, Oz moves about the fixed origin according to the law determined by the angular velocities θ_1, θ_2, θ_3, these are the axes of reference. Another set coincides with Ox, Oy, Oz at the time t, but is fixed in space and is therefore left behind by the axes of reference as they move in the time dt. The symbols U, V, W represent the resolved parts of the vector along either set of axes at the time t. The symbols $U + dU$, $V + dV$, $W + dW$ represent the components along the *moving axes* at the time $t + dt$; and $U + U_1 dt$, $V + V_1 dt$, $W + W_1 dt$, represent the components along the *fixed axes* at the same time $t + dt$.

5. **Important Applications.** We may now apply this general theorem to a variety of vectors*.

(1) Let the vector R be the radius vector of a moving point P. Then U, V, W represent the co-ordinates x, y, z; while U_1, V_1, W_1 represent the component velocities in space. These we now represent by u, v, w. Therefore

$$u = \frac{dx}{dt} - y\theta_3 + z\theta_2,$$

$$v = \frac{dy}{dt} - z\theta_1 + x\theta_3,$$

$$w = \frac{dz}{dt} - x\theta_2 + y\theta_1.$$

(2) Let the vector R be the velocity of a moving point P. Then U, V, W represent the component velocities u, v, w; while U_1, V_1, W_1 represent the accelerations. These we represent by X, Y, Z. Therefore

---

* The sets of equations (1) (2) (3) were given in this form by the late Prof. Slesser (*Cambridge Quarterly Journal*, Vol. II., 1858) to whom the two special cases given further on in Art. 12 had been previously shown by the author, together with their application to the motion of spheres. Other proofs were given of them in the following number of the *Quarterly Journal* by Rev. P. Frost. All four sets of equations were given by R. B. Hayward in Vol. x. of the *Cambridge Transactions*, 1856. Similar results were also given in *Liouville's Journal*, 1858.

$$X = \frac{du}{dt} - v\theta_3 + w\theta_2,$$

$$Y = \frac{dv}{dt} - w\theta_1 + u\theta_3,$$

$$Z = \frac{dw}{dt} - u\theta_2 + v\theta_1.$$

(3) Let the vector R be the angular velocity Ω of a body. Then $U,\ V,\ W$ are the components of ω about the moving axes, let us call these $\omega_1,\ \omega_2,\ \omega_3$. Let $\omega_x,\ \omega_y,\ \omega_z$ be the components about the fixed axes. Then we have

$$\frac{d\omega_x}{dt} = \frac{d\omega_1}{dt} - \omega_2\theta_3 + \omega_3\theta_2,$$

$$\frac{d\omega_y}{dt} = \frac{d\omega_2}{dt} - \omega_3\theta_1 + \omega_1\theta_3,$$

$$\frac{d\omega_z}{dt} = \frac{d\omega_3}{dt} - \omega_1\theta_2 + \omega_2\theta_1.$$

(4) Let the vector R be the angular momentum of a body. Let $h_1,\ h_2,\ h_3$ be its components about the moving axes; $h_x,\ h_y,\ h_z$ the components about fixed axes. Then

$$\frac{dh_x}{dt} = \frac{dh_1}{dt} - h_2\theta_3 + h_3\theta_2,$$

$$\frac{dh_y}{dt} = \frac{dh_2}{dt} - h_3\theta_1 + h_1\theta_3,$$

$$\frac{dh_z}{dt} = \frac{dh_3}{dt} - h_1\theta_2 + h_2\theta_1.$$

If the origin of co-ordinates is also in motion, these equations require some modifications. Let $(p,\ q,\ r)$ be the resolved parts of the velocity of the origin in the directions of the axes. If $(u,\ v,\ w)$ represent the resolved velocities of the centre of gravity in space i.e. referred to axes fixed in space we must add $p,\ q,\ r$ respectively to the expressions for $u,\ v,\ w$ given by (1). Supposing $(u,\ v,\ w)$ to continue to represent the velocities referred to axes fixed in space, the expressions (2) will be unaltered. On the same supposition we must add $m(-vr+wq)$, $m(-wp+ur)$, $m(-uq+vp)$ respectively to the expressions for dh_x/dt &c. given by (4), where m is the mass of the body.

To prove this let us determine the parts of dh_x and dh_1 due to the translational and rotational motion of the axes separately. Those of the latter are given by the formulae (4); to find those of the former, let $H_x,\ H_y,\ H_z$ be the angular momenta about parallel axes through the centre of gravity. Then, by Vol. I. Chap. I.,

$$h_x = h_1 = H_x - mvz + mwy.$$

The differential coefficient dh_x/dt is obtained from this on the supposition that we write $r + dz/dt$, $q + dy/dt$ for dz/dt and dy/dt, because these are the resolved velocities in space of the centre of gravity. The differential coefficient dh_1/dt is obtained without the addition of r and q. We therefore have

$$dh_x/dt = dh_1/dt - mvr + mwq.$$

We may notice that, if the moving set of axes be fixed in the body and move with it, $\theta_1 = \omega_1$, $\theta_2 = \omega_2$, $\theta_3 = \omega_3$. The third set of equations then show that

$$\frac{d\omega_x}{dt} = \frac{d\omega_1}{dt}, \quad \frac{d\omega_y}{dt} = \frac{d\omega_2}{dt}, \quad \frac{d\omega_z}{dt} = \frac{d\omega_3}{dt}.$$

These simplified forms are the ones used by Euler in obtaining his equations of motion of a rigid body about a fixed point. See Vol. I. Chap. V.

6. The above results may be obtained in other ways, but there is an obvious advantage in deducing them all by one method.

The equations connecting (u, v, w) with the co-ordinates (x, y, z) may be obtained as follows. The resolved velocities in space of a point P are not given by dx/dt, dy/dt, dz/dt. These are the resolved velocities *relatively* to the moving axes. To find the motion *in space* we must add to these the resolved velocities due to the motion of the axes. If we supposed the particle to be rigidly connected with the axes, its velocities would be expressed by the forms $\theta_2 z - \theta_3 y$, &c. given in Vol. I. Chap. V. By adding the parts together the actual resolved velocities of the particle are found to be those given above.

Since acceleration is the rate of increase of velocity, just as velocity is the rate of increase of space, it is clear that the relations which hold between accelerations and velocities must be the same as those which hold between velocities and spaces. Thus the relations (2) between (X, Y, Z) and (u, v, w) follow at once from those between (u, v, w) and (x, y, z).

7. • Ex. 1. Let the motion be referred to *oblique* moving axes so that the sides of the spherical triangle xyz are a, b, c, and the angles A, B, C. Let the equal quantities $\sin a \sin b \sin C$, $\sin b \sin c \sin A$, $\sin c \sin a \sin B$ be called μ. Prove that, if the velocity be represented by the three *components* u, v, w parallel to these axes, then the *resultant* acceleration parallel to the axis of z is

$$Z = \frac{dw}{dt} + \frac{du}{dt}\cos b + \frac{dv}{dt}\cos a - u\theta_2\mu + v\theta_1\mu,$$

with similar expressions for X and Y.

This may be done by the use of the spherical triangles xyz, $x'y'z'$, by first proving $zx' = b + \theta_2 dt \sin c \sin A$, $zy' = a - \theta_1 dt \sin c \sin B$, and then substituting as before.

Ex. 2. Prove in the same way that, if x, y, z be the co-ordinates referred to oblique axes moving about a fixed origin, and u', v', w' the *resultant* velocities parallel to the axes, $w' = \frac{dz}{dt} + \frac{dx}{dt}\cos b + \frac{dy}{dt}\cos a - x\theta_2\mu + y\theta_1\mu$, with similar expressions for u' and v'.

Ex. 3.   Prove also that the equations connecting the components, u, v, w with the co-ordinates x, y, z referred to axes with a fixed origin are

$$w = \frac{dz}{dt} + \begin{vmatrix} \mu^{-1}\sin^2 c & -\cot B & -\cot A \\ \theta_3 & \theta_1 & \theta_2 \\ z & x & y \end{vmatrix}$$

with two similar expressions for u and v.

Since w' is the component parallel to z of $(u, v, w,)$, we have $u\cos b + v\cos a + w = w'$, with similar expressions for u' and v'.   By solving these we get the required values of u, v, w.

Ex. 4.   If the whole acceleration be represented by the three components X, Y, Z parallel to the axes, prove that the expressions for these in terms of u, v, w may be obtained from those given in the last example by changing x, y, z into u, v, w and u, v, w into X, Y, Z.

8.   *To explain another general method of obtaining the kine-matical relation between fixed and moving axes.*

Let U, V, W be, as before, the components of a vector R.   Let OL be any straight line *fixed in space* making with the moving axes the angles α, β, γ.   Let R_1 be the resolved part of the vector along OL.   Then

$$R_1 = U\cos\alpha + V\cos\beta + W\cos\gamma,$$

$$\therefore \frac{dR_1}{dt} = \frac{dU}{dt}\cos\alpha + \frac{dV}{dt}\cos\beta + \frac{dW}{dt}\cos\gamma$$

$$- U\sin\alpha\,\frac{d\alpha}{dt} - V\sin\beta\,\frac{d\beta}{dt} - W\sin\gamma\,\frac{d\gamma}{dt}.$$

Since OL is any fixed line in space, let it be so chosen that the moving axis of z coincides with it at the time t.   Then $\alpha = \frac{1}{2}\pi$, $\beta = \frac{1}{2}\pi$, $\gamma = 0$, also $dR_1/dt = W_1$.   Since α is the angle OL makes with the moving axis of x, $d\alpha/dt$ expresses the rate at which the axis of x is separating from a fixed straight line coincident with the axis of z and this is clearly θ_2.   Similarly $d\beta/dt = -\theta_1$, hence

$$W_1 = \frac{dW}{dt} - U\theta_2 + V\theta_1$$

where W_1 expresses the rate of increase of the component W along the fixed axis of z.   The other two equations follow in the same way.   The principle of this method is due to the late Prof. Slesser.

We may obtain the relations between the second and higher differential coefficients in the same way, though the expressions become more complicated.   Since U_1, V_1, W_1 follow the parallelogram law, we have

$$\frac{dR_1}{dt} = \left(\frac{dU}{dt} - V\theta_3 + W\theta_2\right)\cos\alpha + \left(\frac{dV}{dt} - W\theta_1 + U\theta_3\right)\cos\beta + \left(\frac{dW}{dt} - U\theta_2 + V\theta_1\right)\cos\gamma.$$

Repeating the same reasoning, we finally obtain

$$\frac{dW_1}{dt} = \frac{d}{dt}\left(\frac{dW}{dt} - U\theta_2 + V\theta_1\right) - \theta_2\left(\frac{dU}{dt} - V\theta_3 + W\theta_2\right) + \theta_1\left(\frac{dV}{dt} - W\theta_1 + U\theta_3\right).$$

9. We have now obtained a method of transforming the equations of motion with regard to fixed axes into those with regard to axes moving about a fixed origin.

Let any general equation true for *all* fixed axes having a given origin be

$$\psi \{\omega_x, \, d\omega_x/dt, \, \&c....\} = 0,$$

where ω_x, ω_y, ω_z are the angular velocities about the fixed axes.

Since the fixed axes are *arbitrary in position*, let them be so chosen that the three moving axes are passing through them at the moment under consideration; thus at that instant the two sets are coincident. The equations relative to the moving axes may then be deduced by replacing ω_x, ω_y, ω_z in the general equation $\psi = 0$ by the corresponding quantities ω_1, ω_2, ω_3 for the moving axes; and $d\omega_x/dt$, &c. by the equivalents written above in Art. 5.

The same remarks apply if, instead of ω_x, ω_y, ω_z, the components of any other vector entered into the equation.

10. **General equations of Motion.** *To state the general equations of motion of a system of moving bodies referred to any rectangular axes moving about a fixed origin.*

Let m be the mass of any one body of the system. Let the impressed forces on the body be represented by the three forces mX, mY, mZ acting at its centre of gravity and the three couples L, M, N. We suppose that the unknown reactions of the other bodies of the system are included in these expressions.

Let (u, v, w) be the resolved velocities in space of the centre of gravity of the body. The equations of motion for fixed axes are $u = dx/dt$, $X = du/dt$, &c. When the axes move, these become

$$u = \frac{dx}{dt} - \theta_3 y + \theta_2 z \quad\ldots\ldots\ldots\ldots\ldots\ldots(1),$$

$$X = \frac{du}{dt} - \theta_3 v + \theta_2 w \quad\ldots\ldots\ldots\ldots\ldots\ldots(2),$$

with corresponding expressions for the other coordinate axes.

Let (h_1, h_2, h_3) be the angular momenta of the body about parallels to the co-ordinate axes drawn through the centre of gravity.

The equations of moments for fixed axes are $dh_x/dt = L$, &c., Vol. I. Chap. II. When the axes are in motion these become

$$L = \frac{dh_1}{dt} - h_2\theta_3 + h_3\theta_2 \quad\ldots\ldots\ldots\ldots\ldots(4),$$

with similar expressions for M and N.

The expressions for (h_1, h_2, h_3) in terms of the angular velocities of the body are given in Vol. I. Chap. V. If ω_1, ω_2, ω_3 be the angular velocities of the body about the parallels to the axes through the centre of gravity, and A, F, &c. the moments and products of inertia, the fundamental relation is

$$h_1 = A\omega_1 - F\omega_2 - E\omega_3$$

with similar expressions for h_2 and h_3. But there are many others which cannot be repeated here.

Besides the dynamical equations there will be the geometrical equations which express the connections of the system. As every such forced connection is accompanied by some reaction, the number of geometrical equations will be the same as the number of unknown reactions. Thus we have sufficient equations to determine the motion.

Ex. A heavy rigid body is spitted on a smooth circularly-cylindrical rod, on which it can slide, and which passes through its centre of gravity, and the rod is made to rotate uniformly with angular velocity ω in a right circular cone, semi-vertical angle α, about a vertical axis. If C is the moment of inertia about the rod, A and B about two lines fixed in the body perpendicular to the rod, one of which is inclined at an angle ϕ to the plane through the vertical axis and the rod, and if D, E, F are the products of inertia; prove that

$$Cd^2\phi/dt^2 = \omega^2 \sin^2\alpha \{(B-A)\sin\phi\cos\phi + F\cos 2\phi\} - \omega^2\sin\alpha\cos\alpha (E\sin\phi + D\cos\phi).$$

By resolving the angular velocity ω we find $\omega_1 = -\omega\sin\alpha\cos\phi$, $\omega_2 = \omega\sin\alpha\sin\phi$, $\omega_3 = \dot\phi + \omega\cos\alpha$. Substituting these in the expressions for $h_1 h_2 h_3$ given in Art. 10, and equating to zero the moment of the effective forces about the vertical, the result follows at once.   [Math. Tripos, 1885.

11. The motion of the moving axes has been supposed to be determined by the three angular velocities θ_1, θ_2, θ_3. To find their actual position in space we use the Eulerian geometrical equations already given in Vol. I. Chap. V. Let θ, ψ, ϕ be the Eulerian angular coordinates of the moving axes referred to any axes fixed in space. We then have

$$\theta_1 = \frac{d\theta}{dt}\sin\phi - \frac{d\psi}{dt}\sin\theta\cos\phi,$$

$$\theta_2 = \frac{d\theta}{dt}\cos\phi + \frac{d\psi}{dt}\sin\theta\sin\phi,$$

$$\theta_3 = \frac{d\phi}{dt} + \frac{d\psi}{dt}\cos\theta.$$

These geometrical equations determine θ, ϕ, ψ when θ_1, θ_2, θ_3 are known.

12. **Two important special cases**. There are two cases in which the equations of motion just found admit of great simplification. As these often occur, it is worth while to discuss them separately.

In the first case we suppose the body to be turning round some point O fixed in space and to be such that *two of the principal moments of inertia at the fixed point are equal*.

Let OC be the axis of unequal moment of inertia and let us take this as the moving axis of z. Let us choose as the other axes of reference two other axes OA, OB which turn round OC in any manner we please. To fix this let χ be the angle the plane COA makes with some plane OCF fixed in the body and passing through OC. Then we have $\theta_1 = \omega_1$, $\theta_2 = \omega_2$, and $\theta_3 = \omega_3 + d\chi/dt$. Also $h_1 = A\omega_1$, $h_2 = B\omega_2$, $h_3 = C\omega_3$. The equations of moments, Art. 10, are now

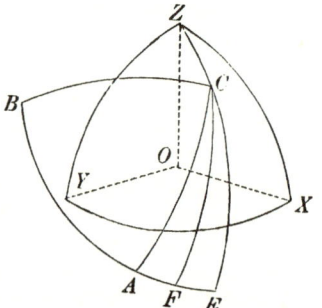

$$A\left(\frac{d\omega_1}{dt} - \omega_2\frac{d\chi}{dt}\right) - (A - C)\,\omega_2\omega_3 = L$$

$$A\left(\frac{d\omega_2}{dt} + \omega_1\frac{d\chi}{dt}\right) + (A - C)\,\omega_3\omega_1 = M$$

$$C\frac{d\omega_3}{dt} \qquad\qquad = N$$

In this case the most convenient geometrical equations to express the relations of these moving axes to axes OX, OY, OZ fixed in space are those usually called Euler's geometrical equations. They are given at length in the last article, where ω_1, ω_2 and $\omega_3 + d\chi/dt$ must of course be written on the left-hand sides for θ_1, θ_2, θ_3. In the figure $ZC = \theta$, $XZC = \psi$, $ECA = \phi$.

13. Since $d\chi/dt$ is arbitrary, it may be chosen to simplify either the dynamical equations or the geometrical equations.

I. If we put $d\chi/dt = -\omega_3$, the moving axes of reference move round the axis of OC with an angular velocity relatively to the body equal and opposite to that of the body, so that if the axis OC were fixed in space the axes of reference would be also fixed in space. The dynamical equations then become

$$A\frac{d\omega_1}{dt} + C\omega_2\omega_3 = L$$

$$A\frac{d\omega_2}{dt} - C\omega_1\omega_3 = M$$

$$C\frac{d\omega_3}{dt} \qquad\qquad = N$$

The geometrical equations however are not much simplified.

We may also choose $d\chi/dt = -\omega_3(A - C)/A$ when the dynamical equations take the simple forms

$$A\frac{d\omega_1}{dt} = L, \qquad A\frac{d\omega_2}{dt} = M, \qquad C\frac{d\omega_3}{dt} = N.$$

II. We may so choose $d\chi/dt$ that $\phi = 0$. In this case the plane COA always passes through a straight line OZ fixed in space. Euler's geometrical equations then become

$$\frac{d\theta}{dt} = \omega_2, \qquad -\frac{d\psi}{dt}\sin\theta = \omega_1, \qquad -\frac{d\chi}{dt} + \frac{d\psi}{dt}\cos\theta = \omega_3.$$

If we substitute these values in the equations of Art. 12, they take the form

$$\left.
\begin{aligned}
-\frac{A}{\sin\theta}\frac{d}{dt}\left(\sin^2\theta\,\frac{d\psi}{dt}\right) + C\omega_3\frac{d\theta}{dt} &= L \\
A\left\{\frac{d^2\theta}{dt^2} - \sin\theta\cos\theta\left(\frac{d\psi}{dt}\right)^2\right\} + C\sin\theta\,\omega_3\frac{d\psi}{dt} &= M \\
C\frac{d\omega_3}{dt} &= N
\end{aligned}
\right\}$$

14. **Second special case.** In the second special case we suppose as before that the body is turning about a fixed point, but *that all the moments of inertia at the fixed point are equal.* In this case there are three sets of axes which may be chosen with advantage.

Firstly. We may choose axes fixed in space. Since every axis is a principal axis in the body, the general equations of motion become

$$\frac{d\omega_1}{dt} = \frac{L}{A}, \quad \frac{d\omega_2}{dt} = \frac{M}{A}, \quad \frac{d\omega_3}{dt} = \frac{N}{A}.$$

Secondly. We may choose one axis, as that of OC, fixed in space and let the other two move round it in any manner, when, as in the first special case, the equations of motion become

$$\left.
\begin{aligned}
\frac{d\omega_1}{dt} - \omega_2\frac{d\chi}{dt} &= \frac{L}{A} \\
\frac{d\omega_2}{dt} + \omega_1\frac{d\chi}{dt} &= \frac{M}{A} \\
\frac{d\omega_3}{dt} &= \frac{N}{A}
\end{aligned}
\right\}$$

Thirdly. We can take as axes any three straight lines at right angles moving in space in any proposed manner. The equations of motion may be deduced from the first set just written down by the help of the general rule for changing from fixed to moving axes. We have therefore

$$\frac{d\omega_1}{dt} - \omega_2\theta_3 + \omega_3\theta_2 = \frac{L}{A},$$

$$\frac{d\omega_2}{dt} - \omega_3\theta_1 + \omega_1\theta_3 = \frac{M}{A},$$

$$\frac{d\omega_3}{dt} - \omega_1\theta_2 + \omega_2\theta_1 = \frac{N}{A}.$$

The geometrical equations may be conveniently expressed in the forms given to them in Art. 18.

15. Numerous examples showing the utility of the above forms of dynamical equations will be found in the following chapters of this work, and especially in that *on motion under any forces*. The following is an instance of their application to a problem on small oscillations, which includes many cases of frequent occurrence*.

A body, which can turn freely about a fixed point O, rotates with uniform angular velocity about one of the principal axes at O, and is under the action of given forces. A small disturbance being given, it is required to find the small oscillations.

Let OC be the principal axis about which the body rotates and n the constant angular velocity. After disturbance OC makes small oscillations about a straight line OZ fixed in space. Describe a sphere with centre O and radius unity; let the principal axes OA, OB, OC intersect the sphere in the points A, B, C and let OZ cut it in the point Z. Draw perpendiculars ZM, ZN, on the arcs CB, CA and let $p = ZM$, $q = ZN$. Then $(p, q, 1)$ are the direction cosines of OZ referred to the principal axes. Also p and q are the coordinates of C referred to axes OX, OY moving round OZ with an angular velocity n. Hence the velocities of C resolved parallel to MC and CN are respectively equal to $q' + pn$ and $-p' + qn$, (see Vol. I. Art. 211) where accents denote differentiations with regard to the time. But these velocities are ω_1 and ω_2. We have therefore

$$\omega_1 = q' + pn, \quad \omega_2 = -p' + qn \ldots\ldots(1).$$

Substituting these in Euler's equations, we find

$$\left.\begin{array}{l} Aq'' + (A + B - C)\, np' - (B - C)n^2q = L \\ -Bp'' + (A + B - C)\, nq' + (A - C)n^2p = M \end{array}\right\} (2).$$

---

* A more detailed account of the equations discussed in this article was given in the first edition of this book. As however it is generally easier to repeat the process of deriving these equations from general principles than to quote them from memory this brief account has been thought sufficient.

The moments of the forces L and M are zero in the undisturbed position and must be expressed in terms of p, q by the geometry peculiar to the problem. Since the squares of p and q are neglected, we have

$$L = a_1 p + a_2 q, \quad M = b_1 p + b_2 q \dots\dots\dots\dots\dots\dots\dots\dots(3),$$

where a_1, a_2, b_1, b_2 are constants.

If θ, ϕ, ψ be the Eulerian angles when estimated positively in the manner described in Vol. I. Art. 256, we see at once by the figure in that article that

$$p = -\sin\theta\cos\phi, \quad q = \sin\theta\sin\phi, \quad r = \cos\theta.$$

We may also notice that the *approximate* equations (1) follow immediately from the *accurate* equations

$$\omega_1 \cos\theta = q' + p\omega_3, \quad \omega_2 \cos\theta = -p' + q\omega_3,$$

which are given under the heading *geometry of moving axes*, Art. 18.

The quantities L, M, N are strictly the moments of the impressed forces about the axes OA, OB, OC respectively. In determining their values in any particular problem, it will be found useful to notice that, since these moments are small, they are to the first approximation equal to the moments about axes OX, OY, and OZ, the two former of which revolve round the latter with a uniform angular velocity equal to n.

We may also notice that, if $(p', q', 1)$ are the direction cosines of any straight line OP near OC referred to the axes OA, OB, OC, its direction cosines referred to OX, OY, OZ are $(p'-p, q'-q, 1)$. As a corollary we infer that the direction cosines of OC referred to OX, OY, OZ are $(-p, -q, 1)$.

In this way the *determination of the motion can be made to depend on the solution of two linear differential equations with constant coefficients.*

When the body is uniaxal, so that $A = B$ we may sometimes with advantage use one of the systems of axes described in Art. 13. For example if we take as the axes of OA, OB, OC the set in which $d\chi/dt = -n$, these axes are very nearly fixed in space. Let OX, OY, OZ be their mean positions; let $(P, Q, 1)$ be the direction cosines of OC referred to these fixed axes, so that $P = \sin\theta\cos\psi$, $Q = \sin\theta\sin\psi$. Or, if we construct as before a sphere of radius unity, and draw CM, CN perpendicular to the arcs YZ and XZ, $CM = P$ and $CN = Q$. We therefore have

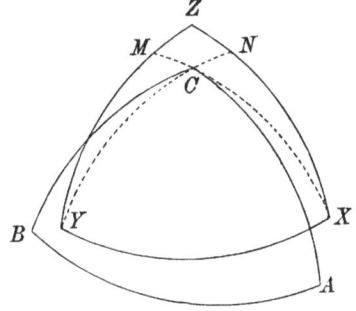

$$\omega_1 = -Q', \quad \omega_2 = P' \dots\dots\dots\dots\dots\dots\dots\dots\dots\dots (4).$$

Substituting, the equations of motion take the form

$$\left. \begin{array}{r} -AQ'' + CnP' = L \\ AP'' + CnQ' = M \end{array} \right\} \quad \dots\dots\dots\dots\dots\dots\dots\dots (5).$$

As before we notice that L, M, N are strictly the moments of the forces about the oscillating axes OA, OB, OC but, since they are small quantities, we may replace them by the moments of the forces about the fixed axes OX, OY, OZ. This property will often enable us to find the moments without difficulty and to express L and M in the linear forms

$$L = a_1 P + a_2 Q, \quad M = b_1 P + b_2 Q.$$

16. **The Geometry of Moving Axes.** In order to use moving axes it is necessary to be able to express with respect to these axes any conditions which may exist with regard to straight lines or points which move independently in space. We have therefore placed together in the following articles a few of the more important conditions.

17. *To express the geometrical conditions that a point whose co-ordinates are* (x, y, z) *is fixed in space.*

This may be done by equating to zero the resolved velocities of the point as given in Art. 4. We thus obtain the conditions

$$p + \frac{dx}{dt} - y\theta_3 + z\theta_2 = 0,$$

with two similar equations.

18. *To express the geometrical conditions that a straight line whose direction cosines are* (l, m, n) *moves parallel to itself in space, or that its direction is fixed in space.*

Let a straight line OL of unit length be drawn from any point O fixed in space parallel to the given straight line. The co-ordinates of L referred to axes which turn round O as an origin so as to be always parallel to the moving axes will be l, m, n. Since OL is fixed in space, the resolved velocities of L are zero. The required geometrical conditions are therefore

$$\frac{dl}{dt} - m\theta_3 + n\theta_2 = 0,$$

with two similar equations. Since $l^2 + m^2 + n^2 = 1$, these three equations are equivalent to two independent conditions.

It is sometimes necessary to express the direction of the straight line by the Eulerian angles θ, ϕ, ψ, as explained in Vol. I. Chap. V. The moving axes are there called OA, OB, OC, and the straight line whose direction is to be fixed in space is represented by OZ. We see that the equations just written down are equivalent to two of those usually called *Euler's geometrical equations*, but expressed in a symmetrical form. The third of Euler's equations follows from Art. 19.

19. Sometimes, while using moving axes, we require to refer the motion of some straight line OM connected with the moving axes to an axis of reference fixed in space. The object of the following theorem is to show how this may be done.

Let the direction cosines of a straight line OM fixed relatively to the moving axes be (λ, μ, ν), and let it be required to refer the motion of OM to some straight line OL fixed in space whose direction cosines at the time t are (l, m, n). Let the angle LOM be θ, and let ψ be the angle which the plane LOM makes with any plane fixed in space passing through OL. Then it may be shown that

$$\cos \theta = l\lambda + m\mu + n\nu,$$

$$\left.\sin^2\theta \frac{d\psi}{dt} = \theta_1 (l - \lambda \cos \theta) + \theta_2 (m - \mu \cos \theta) + \theta_3 (n - \nu \cos \theta)\right\}.$$

If θ_l, θ_m, be the resolved parts of the angular velocities about OL, OM respectively, the last equation may be written in the form

$$\sin^2\theta \frac{d\psi}{dt} = \theta_l - \theta_m \cos \theta.$$

If the straight line OM be not fixed relatively to the axis, then (λ, μ, ν) will be variable, and we must add to the right-hand side of the second equation the determinant

$$\left(\lambda \frac{d\mu}{dt} - \mu \frac{d\lambda}{dt}\right) n + \left(\mu \frac{d\nu}{dt} - \nu \frac{d\mu}{dt}\right) l + \left(\nu \frac{d\lambda}{dt} - \lambda \frac{d\nu}{dt}\right) m.$$

In this determinant we may replace λ, μ, ν, by any quantities $\lambda\kappa$, $\mu\kappa$, $\nu\kappa$ proportional to them (whether κ be variable or not), provided we divide the determinant by κ^2.

The mode of proof may be indicated as follows. Let P be a point in OM at a distance unity from O, and let P move about with OM. Draw PQ perpendicular to OL. First, let OM be fixed to the system of axes. Let the angular velocity of the system about its instantaneous axis be resolved into three components viz., θ_l about OL, θ_x about a perpendicular to OL in the plane LOM, and θ_y about a perpendicular to the plane LOM. The velocity of P is 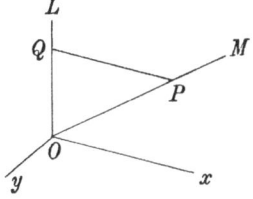 $\theta_l.PQ - \theta_x.OQ$. Since the velocity of P is also $PQ.d\psi/dt$, we have $\sin \theta d\psi/dt = \theta_l \sin \theta - \theta_x \cos \theta$. Now $\theta_l \cos \theta + \theta_x \sin \theta = \theta_m$, whence substituting for θ_x we have the result in the question.

The additional term due to the motion of OM relative to the system may be easily found by treating the system as if it were at rest. The quantities in brackets in the determinant are the moments about the axes of the velocity of P. Resolving these about OL, the determinant follows at once.

20. The motion of a body being given when referred to axes fixed in the body by the angular velocities $(\omega_1, \omega_2, \omega_3)$, it is sometimes necessary to find the motion of the instantaneous axis in space. This is clearly only a case of the theorem in Art. 21. Let OM be the instantaneous axis, OL, as before, the fixed line in space, then $\theta_l = \theta_m \cos \theta$. The expression for $\sin^2 \theta d\psi/dt$ is reduced therefore to the determinant above. The following examples are obtained by combining Arts. 18 and 21, accents denoting differentiations with regard to the time.

Ex. 1. If Ω be the angular velocity about the instantaneous axis, prove that

$$\Omega^2 \frac{d\psi}{dt} = \omega_1' l' + \omega_2' m' + \omega_3' n'. \qquad n \frac{d\psi}{dt} = \omega_3 + \frac{l'm'' - l''m'}{l'^2 + m'^2 + n'^2}.$$

Ex. 2. Show that $\qquad d\psi/dt = \theta_l + D/(l'^2 + m'^2 + n'^2)$,

where, as before, θ_l is the angular velocity of the body about OL and D is the following determinant $l\,(m'n'' - m''n') + n\,(n'l'' - n''l') + n\,(l'm'' - l''m')$.

Ex. 3. Show that $\Omega^2 - \theta_l^2 = l'^2 + m'^2 + n'^2$.

Ex. 4. Show that the equation to the plane LOM referred to the axes fixed in the body is
$$l'x + m'y + n'z = 0.$$

21. Use of Moving axes in Solid Geometry. As we have sometimes to displace the axes of coordinates independently of the motion of the body, and even to change the axes without altering the time, it is convenient to have the fundamental principle of Art. 4 expressed without reference to dynamical ideas. This is effected in the following proposition.

Let a system of moving axes be screwed from one position Ox, Oy, Oz to a consecutive position Ox', Oy', Oz' by the small rotations $d\phi_1$, $d\phi_2$, $d\phi_3$ about their instantaneous positions. Let U, V, W be the projections or components of a straight line or vector OL on Ox, Oy, Oz, where U, V, W may be either constant or variable. Let $U + dU$, $V + dV$, $W + dW$ be the projections of the consecutive position OL' of the straight line on Ox', Oy', Oz'; and $U + \delta U$, $V + \delta V$, $W + \delta W$ the projections of OL' on Ox, Oy, Oz. Then

$$\left. \begin{aligned} \delta U &= dU - V d\phi_3 + W d\phi_2 \\ \delta V &= dV - W d\phi_1 + U d\phi_3 \\ \delta W &= dW - U d\phi_2 + V d\phi_1 \end{aligned} \right\}.$$

These follow from Art. (4) by writing $\theta_1 dt = d\phi_1$, $\theta_2 dt = d\phi_2$, $\theta_3 dt = d\phi_3$.

If the length OL is taken equal to unity, the projections U, V, W become the direction cosines of the line. These equations then tell us at once the changes in space of the direction cosines when the changes relative to the moving axes are known.

Thus if $\delta\chi$ be the angle between two consecutive positions of a line OL, whose direction cosines referred to the moving axes are U, V, W, we have

$$(\delta\chi)^2 = (\delta U)^2 + (\delta V)^2 + (\delta W)^2.$$

Also the direction cosines of the plane through two consecutive positions of OL are proportional to $V\delta W - W\delta V$, $W\delta U - U\delta W$, $U\delta V - V\delta U$.

It is not our object here to show the utility of moving axes in Solid Geometry further than to prove those theorems which are required in Dynamics. It will be found however that both curves and surfaces are sometimes most easily treated by referring them to a set of moving axes in which the origin travels along the curve or surface and the directions of the axes are such tangents and normals as may be suitable to the property under discussion. We may refer the reader to a paper by the author in the *Cambridge Mathematical Journal* (Vol. VII. 1866), where the application of moving axes to the curvature of curves is illustrated by several examples. The following examples though of no immediate importance will be found useful further on.

Ex. 1.   The principal axes at any point P of a curve are the radius of curvature, the tangent and the binormal. If these be respectively taken as the axes of x, y, z, prove that the components of motion by which the axes are screwed along the curve through an arc dy are $p=0$, $q=dy$, $r=0$; $d\phi_1=0$, $d\phi_2=-d\tau$, $d\phi_3=-d\epsilon$, where $d\tau$ and $d\epsilon$ are the angles of torsion and contingence.

Ex. 2.   The principal axes at any point O of a surface are the tangents to the lines of curvature and the normal to the surface. Let these be called the axes of x, y, z. Let it be required to move the axes from O into the position of the principal axes at a neighbouring point O' on the axis of x. If $OO'=dx$ the six components of motion for the base point O are given by

$$p=dx,\ q=0,\ r=0;\ d\phi_1=0,\ d\phi_2=-\frac{dx}{\rho},\ \left(\frac{1}{\rho}-\frac{1}{\rho'}\right)d\phi_3=\frac{d}{dy}\left(\frac{1}{\rho}\right)dx,$$

where ρ, ρ' are the principal radii of curvature for the sections xz, yz respectively. By combining this with a corresponding motion along the axis of y, we can move the axes from O into the positions of the principal axes at any neighbouring point O' on the surface.

Ex. 3.   Show that the equation to a surface referred to the principal axes at any point O is

$$z=\tfrac{1}{2}\left\{\frac{x^2}{\rho}+\frac{y^2}{\rho'}\right\}+\tfrac{1}{6}\left\{x^3\frac{d}{dx}\left(\frac{1}{\rho}\right)+3x^2y\frac{d}{dy}\left(\frac{1}{\rho}\right)+3xy^2\frac{d}{dx}\left(\frac{1}{\rho'}\right)+y^3\frac{d}{dy}\left(\frac{1}{\rho'}\right)\right\}+\&c.$$

22.   **Equations of Motion of a changing body.**   It may be noticed that the three general equations of motion whose type is

$$dh_1/dt-\theta_3h_2+\theta_2h_3=L \quad\dotfill\ (1)$$

are not restricted to a rigid body. They hold even when the system is a collection of particles moving amongst themselves. We may therefore apply them to find the motion of a body which is changing its shape by transference of heat, or by some other cause, and is also turning freely in space about its centre of gravity as a fixed point*.

---

* The equations of motion of a changing body were given by Liouville in 1858 in the third volume of his Journal in the form shown in equations (6) of the text. The equations marked (9) agree with those given by Prof. Darwin in the *Phil. Trans.* 1876, *On the influence of geological changes on the earth's axis of rotation.* The equations (10) are in substance the same as those of Sir W. Thomson in the Appendix C of Prof. Darwin's paper. These are also to be found in the *Mécanique Céleste* of Tisserand, 1891. The first use of mean axes is ascribed by Tisserand to Gyldén, *Société Royale d'Upsala*, 1871.

In many cases the mode in which the body is changing its shape is given, so that, if we find the motion in space of any three rectangular axes connected in a known manner with the changing body, the motion of every part of the body is known. The axes thus chosen to define the motion in space of the body may for shortness be called *axes of the body*.

There is one method of choosing these axes which has the advantage of simplifying the equations of motion. Let a system of axes $O\xi$, $O\eta$, $O\zeta$ move about the centre of gravity as origin with such angular velocities that, if at any instant the changing body were suddenly to become rigid the motion of the axes in the time dt would be the same as if they were fixed in the body. These axes possess the property that the angular momentum of the changing body about any one of them is the same as that of an ideal rigid body which is attached to the axes, and has the same instantaneous moments and products of inertia as the changing body. The angular momenta can therefore be expressed by the usual formulæ for a rigid body, viz. $h_1 = A\Omega_1 - F\Omega_2 - E\Omega_3$, &c.

To make this point clear; let U, V, W be the resolved velocities in space of a particle of mass m, Ω_1, Ω_2, Ω_3 the angular velocities of the axes. Then as in Art. 5

$$U = \xi' - \eta\Omega_3 + \zeta\Omega_2, \quad V = \eta' - \zeta\Omega_1 + \xi\Omega_3, \text{ &c.}$$

Let $\qquad A_\xi = \Sigma m\,(\eta\zeta' - \zeta\eta'), \quad A_\eta = \Sigma m\,(\zeta\xi' - \xi\zeta'), \quad A_\zeta = \Sigma m\,(\xi\eta' - \eta\xi'),$

so that A_ξ, A_η, A_ζ are the angular momenta of that part of the motion of the particles of the body which is *relative to the axes* ξ, η, ζ. Let h_ξ, h_η, h_ζ be the whole angular momenta about the axes, we then find by substitution

$$h_\zeta = \Sigma m\,(\xi V - \eta U) = A_\zeta + C\Omega_3 - E\Omega_1 - D\Omega_2 \dots\dots\dots \dots\dots (2),$$

and two similar equations, where A, B, C, D, E, F are the moments and products of inertia of the body about the axes of ξ, η, ζ.

The choice of axes we have described makes them such that

$$A_\xi = 0, \quad A_\eta = 0, \quad A_\zeta = 0 \dots\dots\dots\dots\dots\dots\dots\dots(3).$$

Such axes have been called *mean axes* by Tisserand in his *Mécanique Céleste*. He remarks that they are characterised by the property that the changes in the body do not take the form of currents round them.

We may notice that the positions of the axes are not strictly defined by the property that $A_\xi = 0$, $A_\eta = 0$, $A_\zeta = 0$. These equations only determine the motion when their initial positions have been chosen. To take a single instance, let the body be initially at rest and let *internal* changes beginning at any instant alter its shape and structure. It is evident that at the beginning and throughout all these changes, the angular momentum about any axis fixed in space is zero. It follows that any rectangular axes fixed in space form a mean system. The angular momenta A_ξ, A_η, A_ζ depend on the motion relative to the axes of ξ, η, ζ, and are independent of Ω_1, Ω_2, Ω_3. We may therefore now superimpose on the body and the axes any the same state of motion, and the axes will continue to be a mean system.

It sometimes happens that the changes under consideration are so slow, though long continued, that the body presents the appearance of being unaltered and rigid when viewed for any short time. It is evident that in such cases the mean axes will also be sensibly fixed in the body.

23. Let $O\xi$, $O\eta$, $O\zeta$ be the axes of the body, whether mean axes or not. Let Ox, Oy, Oz be any other set of axes to which we wish to refer the motion. Let

ω_1, ω_2, ω_3 be the angular velocities of the axes of the body about the axes of reference; θ_1, θ_2, θ_3 those of the axes of reference themselves. The angular momenta of the body about the axes of reference are then

$$\left.\begin{aligned} h_x &= A_x + A\omega_1 - F\omega_2 - E\omega_3 \\ h_y &= A_y + B\omega_2 - D\omega_3 - F\omega_1 \\ h_z &= A_z + C\omega_3 - E\omega_1 - D\omega_2 \end{aligned}\right\} \dots\dots\dots\dots\dots\dots(4)$$

where A_x, A_y, A_z are the components of A_ξ, A_η, A_ζ about the axes of reference, and are zero if the axes of the body are mean axes. Also A, B, C, D, E, F are here the moments and products of inertia of the body about the axes of reference.

To obtain the equations of motion we substitute these values of h_x, h_y, h_z in the equations

$$h'_x - \theta_3 h_y + \theta_2 h_z = L, \text{ &c., &c.} \dots\dots\dots\dots\dots\dots \quad (5).$$

If the axes of the body are chosen as the axes of reference, we have $\theta_1 = \omega_1$, $\theta_2 = \omega_2$, $\theta_3 = \omega_3$, &c. The equations, after substitution for h_x &c., take the form

$$\left.\begin{aligned} \frac{d}{dt}\{A\omega_1 - F\omega_2 - E\omega_3 + A_x\} + (C - B)\,\omega_2\omega_3 \\ + D(\omega_3{}^2 - \omega_2{}^2) + F\omega_1\omega_3 - E\omega_1\omega_2 + \omega_2 A_z - \omega_3 A_y = L \end{aligned}\right\} \dots\dots\dots\dots(6),$$

with two other equations.

In these equations A, B, C, D, E, F are the moments and products of inertia about the axes of the body, while the angular velocities ω_1, &c., and the moments L, M, N are referred to these as axes of coordinates.

If the instantaneous positions of the principal axes are taken as the axes of reference, the expressions (4) for h_x, h_y, h_z assume very simple forms. The equations of motion (5) now become

$$\frac{d}{dt}(A\omega_1 + A_x) - \theta_3(B\omega_2 + A_y) + \theta_2(C\omega_3 + A_z) = L \dots\dots\dots\dots(7),$$

with two similar equations. In these we write

$$\theta_1 = \omega_1 + a_1, \qquad \theta_2 = \omega_2 + a_2, \qquad \theta_3 = \omega_3 + a_3 \dots\dots\dots\dots(8),$$

so that a_1, a_2, a_3 are the angular velocities with which the principal axes are separating from axes of the body. This substitution is made because in most cases a_1, a_2, a_3 are very small. The equations now take the form

$$\frac{d}{dt}(A\omega_1) - (B - C)\,\omega_2\omega_3 - B\omega_2 a_3 + C\omega_3 a_2 + \frac{d}{dt}A_x - A_y(\omega_3 + a_3) + A_z(\omega_2 + a_2) = L \dots(9),$$

with two similar equations. In these equations A, B, C are the instantaneous values of the principal moments of inertia of the body, and the angular velocities ω_1, &c., a_1, &c., are referred to the principal axes as axes of coordinates.

24. These equations admit of simplification when the instantaneous axis of rotation is nearly coincident with one principal axis. Taking this axis as that of z, both ω_1 and ω_2 are then small quantities. If also the internal changes are small and periodic, or slow and limited, so that the principal axes do not wander much in the body, the angular velocities ω_1 and ω_2 will remain small throughout the motion. In such cases we may sometimes be able to neglect the angular momenta A_x, A_y, A_z due to these internal changes. Taking a set of axes in the body such that the principal axes do not deviate far from them, the angular velocities a_1, a_2, a_3 will also be small. We shall also suppose that $N = 0$ and that L, M are small.

The third of equations (9) then shows that $d\,(C\omega_3)/dt$ differs from zero by the squares of small quantities. We may therefore write in the small terms of the two first equations $C\omega_3 = G$, where G is a constant. We thus obtain

$$\left.\begin{array}{l} \dfrac{d}{dt}(A\omega_1) - \left(\dfrac{B-C}{BC}\,G - a_3\right)B\omega_2 + Ga_2 = L \\[3mm] \dfrac{d}{dt}(B\omega_2) - \left(\dfrac{C-A}{CA}\,G - a_3\right)A\omega_1 - Ga_1 = M \end{array}\right\} \quad\dots\dots\dots\dots(10).$$

When the body is uniaxal and remains so throughout all changes, $A = B$. Since we may now take any axes in the equator of the body as principal axes, we may further simplify the equations by so choosing these axes that $a_3 = 0$.

In using these equations the internal changes of the body relatively to the axes of the body are supposed to be given, so that a_1, a_2, a_3 and A, B, C are known functions of t. These differential equations when solved will then determine ω_1 and ω_2. The motion in the body of the instantaneous axis follows at once. If required, θ_1, θ_2, θ_3 also may be found, and the motion in space of the principal axes may be deduced from Euler's equations.

Taking the case of a uniaxal body, let us suppose that the motion in the body of the axis of figure Oz is given by its angular co-ordinates $(\xi, \eta, 1)$ referred to the axes $O\xi$, $O\eta$, $O\zeta$; then ξ, η are known functions of t. Since (a_1, a_2, a_3) are the angular velocities with which the principal axes are moving relatively to axes in the body, we have $a_1 = -d\eta/dt$ and $a_2 = d\xi/dt$.

If we also suppose that the changes in the body are such that, though the positions in the body of the principal axes are sensibly altered, yet the changes in magnitude of A, A, C are so small that we may neglect their variations when multiplied by ω_1, ω_2, the equations become

$$\left.\begin{array}{l} \dfrac{d\omega_1}{dt} + \mu\omega_2 + \nu\,\dfrac{d\xi}{dt} = L \\[3mm] \dfrac{d\omega_2}{dt} - \mu\omega_1 + \nu\,\dfrac{d\eta}{dt} = M \end{array}\right\} \quad\dots\dots\dots\dots\dots\dots(11),$$

where $\mu = G\,(C-A)/AC$ and $\nu = G/A$.

In other problems the positions of the principal axes may be fixed in the body while the changes in the moments of inertia are given. In such cases we put $a_1 = 0$, $a_2 = 0$, $a_3 = 0$ and regard A, B, C as known functions of the time.

Ex. 1. Let the earth be regarded as a uniaxal body, having all its principal moments of inertia nearly equal, and rotating about its axis of figure with an angular velocity n. If the internal changes of the earth are such that the pole of the axis of figure has a small annual motion round its mean place so that its coordinates are $\xi = p\cos mt$, $\eta = q\sin mt$, the magnitudes of the principal moments of inertia remaining sensibly unaltered, prove that the co-ordinates of the pole of the instantaneous axis of rotation are

$$\xi_1 = \frac{\mu^2 p + \mu mq}{\mu^2 - m^2}\cos mt + H\cos(\mu t + K), \qquad \eta_1 = \frac{\mu^2 q + \mu mp}{\mu^2 - m^2}\sin mt + H\sin(\mu t + K),$$

where H, K are two arbitrary constants. Helmert's problem. *Astron. Nachr.* Vol. cxxvi.

To prove these results we put $L = 0$, $M = 0$ in the equations (11) and substitute for ξ, η their given values; then $\xi_1 = \xi + \omega_1/n$ and $\eta_1 = \eta + \omega_2/n$.

In the actual case of the earth $2\pi/\mu$ is equal to ten months nearly, and $2\pi/m$ is equal to a year.

Ex. 2. *An ellipsoid, whose centre O is fixed, contracts by cooling, and being set in motion in any manner is under the action of no forces. Find the motion.*

The principal diameters are principal axes at O throughout the motion. Let us take them as axes of reference. The expressions for the angular moments about the axes are $h_1 = A\omega_1$, $h_2 = B\omega_2$, $h_3 = C\omega_3$. The equations (6) then become

$$\frac{d}{dt}(A\omega_1) - (B-C)\,\omega_2\omega_3 = 0$$

and two similar equations.

Multiplying these equations by $A\omega_1$, $B\omega_2$, $C\omega_3$, adding, and integrating we see that $A^2\omega_1^2 + B^2\omega_2^2 + C^2\omega_3^2$ is constant throughout the motion. To obtain another integral, let $A = A_0 f(t)$, $B = B_0 f(t)$, $C = C_0 f(t)$ where $f(t)$ expresses the law of cooling which has been supposed such that the body changes its form very slowly. Let $\omega_1 f(t) = \Omega_1$, $\omega_2 f(t) = \Omega_2$, $\omega_3 f(t) = \Omega_3$, and put $dt/dt' = f(t)$, then the equations become

$$A_0 \frac{d\Omega_1}{dt'} - (B_0 - C_0)\Omega_2\Omega_3 = 0,$$

and two similar equations. These may be treated as in the chapter on the motion of a body under no forces. *Liouville's Journal*, 1858.

On relative motion.

25. **Clairaut's Theorem***. The theory of relative motion is best understood by viewing it in as many aspects as possible. We shall therefore now consider a method of determining the motion which is more elementary, and does not in the result make an exclusive use of Cartesian co-ordinates.

Let it be required to refer the motion of a particle P to any given set of moving axes. Let P_0 be the position of P at any time t and let P_0 be attached to the axes and move with them during any short interval. Let f represent the acceleration of P_0 in direction and magnitude at the time t. The particle P will of course separate from P_0, but as is explained in dynamics of a particle the actual acceleration of P in space is the resultant of its acceleration relative to P_0 treated as a fixed point and the acceleration f of P_0. The acceleration of P_0 is called the "*acceleration of the moving space.*"

Let x, y, z be the co-ordinates of the particle P referred to the

---

* This method of determining the relative motion of a particle was first given by Clairaut in 1742, and afterwards the same rule was demonstrated in a different manner by Coriolis. The arguments of the former were criticized and improved by M. Bertrand in the nineteenth volume of the *Journal Polytechnique*. The mode of proof of the latter is altogether independent of all co-ordinates. Another demonstration by the use of polar co-ordinates was given in Vol. XII. of the *Quarterly Journal of Mathematics* by the Rev. H. W. Watson.

moving axes, and let X, Y, Z be the impressed forces on the particle resolved parallel to the axes. Let p, q, r be the resolved velocities of the origin; adding these to the right-hand sides of equation (1) in Art. 5 and substituting in (2) we have

$$\frac{X}{m} = \frac{d^2x}{dt^2} - 2\frac{dy}{dt}\theta_3 + 2\frac{dz}{dt}\theta_2 + Ax + By + Cz + D,$$

with similar expressions for Y and Z. Here A, B, C, D are functions of θ_1, θ_2, θ_3, p, q, r and their differential coefficients with regard to t which it is unnecessary to write down. If x, y, z were constants, all the terms of X would disappear except the last four. These then, with the corresponding terms in Y and Z, express the acceleration f of a point P_0, rigidly attached to the axes, but occupying the instantaneous position of P.

We have now to examine the effect of the remaining terms. The motion of the axes of reference during any interval dt may be constructed by a screw motion along and round some central axis OI. Let Udt be the translation along and Ωdt the rotation round OI. Let V represent the velocity of P relative to these axes, and let θ be the angle made by the direction of V with OI. Consider now the second and third terms of X taken together, and the corresponding terms of Y and Z, neglecting for the moment all the other terms. If we multiply the expressions for X, Y, Z by θ_1, θ_2, θ_3 respectively the sum of these terms is zero. The resultant of the accelerations is therefore perpendicular to OI. Again, if we multiply the expressions for X, Y, Z by dx/dt, dy/dt, dz/dt respectively the sum of the terms is again zero. The resultant of the accelerations is therefore perpendicular to the direction of the relative velocity V. Finally, by adding up the squares of the terms, we find that the magnitude of the resultant acceleration is $2\Omega V \sin \theta$.

To determine the manner in which these forces should be applied, we must transpose the terms which represent them to the other sides of the equations. The first equation then becomes

$$m\frac{d^2x}{dt^2} = X + 2m\left(\frac{dy}{dt}\theta_3 - \frac{dz}{dt}\theta_2\right) - m(Ax + By + Cz + D),$$

and the other two take similar forms. These are the equations of motion of a particle referred to fixed axes, moving under the same impressed forces as before, but with two additional forces. These are, first, a force equal and opposite to that represented by mf, where f is the acceleration of the point of moving space occupied by the particle; and secondly, a force whose magnitude has been shown to be $2mV\Omega \sin \theta$. To determine the direction of this force, let the axis of z be taken along the axis OI, and let the plane of yz be parallel to the direction of motion of the particle, then $\theta_1 = 0$, $\theta_2 = 0$, and $dx/dt = 0$. We then easily see that this force disappears from the equations giving md^2y/dt^2 and md^2z/dt^2; while in that giving md^2x/dt^2, we have the single term $2m\theta_3 dy/dt$. The magnitude of this force is obviously $2mV\Omega \sin \theta$, and it acts along the positive direction of the axis of x. This is the left-hand side when the receding particle is viewed from the central axis OI.

When these equations have been integrated, the arbitrary constants are to be determined from the initial values of x, y, z, dx/dt, dy/dt, dz/dt. These differential coefficients are clearly the components of the initial velocity of the particle, taken relatively to the moving axes.

26. **Relative motion of a particle.** *We may express these conclusions in the following rule.*

In finding the motion of a particle of mass m with reference to any moving axes we may treat the axes as if they were fixed in space, provided that we regard the particle as acted on, in addition to the impressed forces, by two other forces:

(1) a force equal and opposite to mf, where f represents in direction and magnitude the acceleration of the point of moving space occupied by the particle. The force mf is called the "*force of moving space;*"

(2) a force perpendicular to the direction of relative motion of the particle, and also to the central axis or axis of rotation of the moving axes. This force is measured by $2mV\Omega \sin \theta$ where V is the relative velocity of the particle, Ω the resultant angular velocity of the moving axes, and θ is the angle between the direction of the velocity and the central axis. This force is called *the compound centrifugal force.*

To find the direction in which the force is to be applied; stand with the back along the central axis so that the rotation appears to be in the direction of the hands of a watch; then viewing the particle receding from the central axis the force acts to the left-hand. The central axis may be conveniently called the *axis of the centrifugal forces.*

27. Ex. If the particle be constrained to move along a curve which is itself moving in any manner, the compound centrifugal force, being perpendicular to the direction of the relative velocity of the particle, may be included in the reaction of the curve. The only force which it is necessary to impress on the particle is the force of the moving space. If the curve be turning about a fixed axis with an angular velocity Ω, the components of the accelerating force of moving space are clearly $\Omega^2 r$ tending directly from the axis of rotation, and $rd\Omega/dt$ perpendicular to the plane containing the particle and the axis, where r is the distance of the particle from the axis. This agrees with the result obtained in the section on relative motion in Vol. I. Chap. IV. Art. 213.

28. In finding the compound centrifugal force it is useful to remember, that we may resolve the angular velocity Ω or the linear velocity V in any manner that we please, and find the forces due to each of the components separately. Though we have thus more than two forces which must be applied to the particle, yet, by making a proper resolution, some of these may either produce no effect, and may therefore be omitted, or may produce an effect which is easily taken account of.

29. **Relative motion of a Rigid body.** When we wish to apply Clairaut's theorem to the motion of a rigid body, we must consider each particle to be acted on by the two forces which depend on the position and velocity of that particle. To find the resultant of all these forces, we generally have to effect an integration throughout the body. This integration though not difficult is sometimes troublesome. Methods of abbreviating the process have been formulated but they are omitted here because such problems are generally more easily solved by using the methods described in Art. 10.

30. **Principle of Vis Viva applied to moving axes.** *Suppose the system at any instant to become fixed to the set of moving axes relative to which the motion is required, and calculate what would then be the effective forces on the system. These have been called in Art. 25 the forces of moving space. If we apply them as additional impressed forces on the system, but reversed in direction, we may use the equation of Vis Viva to determine the relative motion as if the axes were fixed in space.* This theorem is due to Coriolis, *Journal Polytech.* 1831.

If we follow the notation of Art. 24 the accelerations of any point P resolved parallel to the rectangular moving axes are

$$\frac{d^2x}{dt^2} - 2\frac{dy}{dt}\theta_3 + 2\frac{dz}{dt}\theta_2 + Ax + By + Cz + D$$

with two similar expressions for the axes of y and z. The last four terms, with the corresponding terms in the other expressions, are the resolved accelerations of a point P_0 rigidly attached to the axes, but occupying the instantaneous position of P. Let us call these X_0, Y_0, Z_0.

Let us now recur to the proof of the principle of Vis Viva given in Vol. I. Chap. VII. Art. 350. To adapt that proof to our present case we have merely to substitute the above expressions for d^2x/dt^2, &c. in the general equation of virtual moments. After substituting for the displacements δx, δy, δz their values dx, dy, dz, it is clear that the terms containing dx/dt, dy/dt, dz/dt disappear,.. The equation after integration becomes

$$\Sigma m\left\{\left(\frac{dx}{dt}\right)^2 + \left(\frac{dy}{dt}\right)^2 + \left(\frac{dz}{dt}\right)^2\right\} = 2\Sigma m \int\{(X-X_0)dx + (Y-Y_0)dy + (Z-Z_0)dz\} + C.$$

31. *Another proof.* This theorem of Coriolis also follows at once from that given in Art. 25 for all kinds of relative motion. The mode of proof just given has the advantage of recurring to first principles.

It is clear that when we use the principle of virtual velocities any force whose line of action is perpendicular to the displacement given to its point of application must disappear from the equation. Now in the principle of Vis Viva the displacement given to every point is the elementary arc described by that point in the time dt relative to the axes. The compound centrifugal force acts perpendicularly to this arc, and therefore disappears from the equation. But the virtual moments of the forces of moving space are not zero, and must be allowed for in the equation.

32. *Ex. A sphere rolls on a perfectly rough plane, which turns with a uniform angular velocity* n *about a horizontal axis in its own plane. Supposing the motion of the sphere to take place in a vertical plane perpendicular to the axis of rotation, find the motion of the sphere relatively to the plane.*

Let Ox be the trace described by the sphere as it rolls on the plane, and let Oy be drawn through the axis of rotation perpendicular to Ox in the plane of motion of the sphere. Let nt be the angle which Ox makes with a horizontal plane through the axis of rotation. Let ϕ be the angle that the radius of the sphere which was initially perpendicular to the plane makes with the axis of y. Let x, y be the co-ordinates of P the centre of the sphere, and Mk^2 the moment of inertia of the sphere about a diameter.

If the sphere were fixed relatively to the plane its effective forces would be Mn^2x and Mn^2y acting at the centre of gravity, and a couple $Mk^2 dn/dt = 0$ round the centre of gravity. See Vol. I. Chap. IV., note to Art. 450. Also the impressed force, viz., gravity, is equivalent to $g \sin nt$ and $-g \cos nt$ parallel to the moving axes. The equation of Vis Viva for relative motion is therefore

$$\tfrac{1}{2} \frac{d}{dt} \left\{ \left(\frac{dx}{dt}\right)^2 + \left(\frac{dy}{dt}\right)^2 + k^2 \left(\frac{d\phi}{dt}\right)^2 \right\} = n^2 x \frac{dx}{dt} + n^2 y \frac{dy}{dt} + g \sin nt \frac{dx}{dt} - g \cos nt \frac{dy}{dt}.$$

Here $dx/dt = a\, d\phi/dt$ and $dy/dt = 0$. We have therefore $\left(1 + \dfrac{k^2}{a^2}\right) \dfrac{d^2x}{dt^2} = n^2 x + g \sin nt$.

This equation might also have been derived from the formulæ for moving axes given in Vol. I. Chap. IV. Art 211.

If $k^2 = \tfrac{2}{5} a^2$ this equation leads to $x = -\dfrac{5}{12} \dfrac{g}{n^2} \sin nt + A e^{n\sqrt{\frac{5}{7}}t} + B e^{-n\sqrt{\frac{5}{7}}t}$

where A and B are two constants which depend on the initial conditions of the sphere.

On Motion relative to the Earth.

33.   The motion of a body on the surface of the earth is not exactly the same as if the earth were at rest. As an illustration of the use of the equations of this chapter, we shall proceed to determine the equations of motion of a particle referred to axes of co-ordinates fixed in the earth and moving with it.

Let O be any point on the surface of the earth whose latitude is λ. Thus λ is the angle which the normal to the surface of still water at O makes with the plane of the equator. Let the axis of z be the vertical at O, measured positively in the direction opposite to gravity. Let the axes of x and y be respectively a tangent to the meridian and a perpendicular to it, their positive directions being respectively south and west. In the figure the axis of y is dotted to indicate that it is perpendicular to the plane of the paper. Let

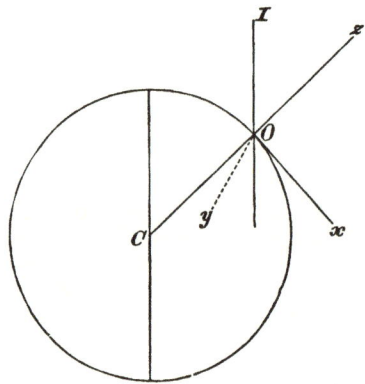

ω be the angular velocity of the earth, b the distance of the point O from the axis of rotation.

We may reduce the point O to rest by applying to every point under consideration an acceleration equal and opposite to that of O, and therefore equal to $\omega^2 b$ and tending from the axis of rotation. We must also apply a velocity equal and opposite to the initial velocity of O. This velocity is ωb. The whole figure will then be turning about an axis OI, parallel to the axis of rotation of the earth, with an angular velocity ω.

When the particle has been projected from the earth it is acted on by the attraction of the earth and the applied acceleration $\omega^2 b$. The attraction of the earth is not what we call gravity. Gravity is the resultant of the attraction of the earth and the centrifugal force, and the earth is of such a form that this resultant acts perpendicular to the surface of still water. If it were not so, particles resting on the earth would tend to slide along the surface. It appears, therefore, that the force on a particle at O, *after O has been reduced to rest*, is equal to gravity. Let this be represented by g.

The equations of motion are much simplified if we neglect such small quantities as the difference between the attractions of the earth at different points near O. If a is the equatorial radius of the earth, the attraction at a height z above O is nearly equal to $g(1 - z/a)$. Since a is 20926629 feet and $2\pi/\omega$ is 24 hours, we easily find that the centrifugal force at the equator, $\omega^2 a$, is equal to $g/289$. Hence if we neglect the small term gz/a we must also neglect $\omega^2 z$ at all points near O. The term $\omega^2 b$

is not neglected, because at places near the equator b is nearly as large as the radius of the earth.

Since the earth is turning round OI with angular velocity ω, the resolved part about Oz is $\omega \sin \lambda$, since the angle IOz is the complement of λ; since the rotation is from west to east, the resolved angular velocity is from y to x, which is the negative direction, hence $\theta_3 = -\omega \sin \lambda$. The resolved angular velocity round Ox is $\omega \cos \lambda$, and is from y to z, which is the positive direction, hence $\theta_1 = \omega \cos \lambda$. Also since OI is perpendicular to Oy, $\theta_2 = 0$. Hence, by Art. 4, the actual velocities of any particle whose co-ordinates are (x, y, z) are

$$\left.\begin{aligned} u &= \frac{dx}{dt} + \omega \sin \lambda y \\[2mm] v &= \frac{dy}{dt} - \omega \cos \lambda z - \omega \sin \lambda x \\[2mm] w &= \frac{dz}{dt} + \omega \cos \lambda y. \end{aligned}\right\}.$$

To find the equations of motion it is only necessary to substitute these values in the equations of Art. 5.

We thus have

$$\left.\begin{aligned} \frac{d^2x}{dt^2} + 2\omega \sin \lambda \frac{dy}{dt} &= X \\[2mm] \frac{d^2y}{dt^2} - 2\omega \cos \lambda \frac{dz}{dt} - 2\omega \sin \lambda \frac{dx}{dt} &= Y \\[2mm] \frac{d^2z}{dt^2} + 2\omega \cos \lambda \frac{dy}{dt} &= -g + Z \end{aligned}\right\},$$

where the terms (X, Y, Z) include all the accelerating forces, except gravity, which act on the particle. These equations agree with those given by Poisson, *Journal Polytechnique*, 1838.

34. If we retain the terms containing ω^2, and include the difference between the attractions at (x, y, z) and O in the forces X, Y, Z, the equations of motion are

$$\frac{d^2x}{dt^2} + 2\omega \sin \lambda \frac{dy}{dt} - \omega^2 \sin^2 \lambda x - \omega^2 \sin \lambda \cos \lambda z = X,$$

$$\frac{d^2y}{dt^2} - 2\omega \cos \lambda \frac{dz}{dt} - 2\omega \sin \lambda \frac{dx}{dt} - \omega^2 y = Y,$$

$$\frac{d^2z}{dt^2} + 2\omega \cos \lambda \frac{dy}{dt} - \omega^2 \cos^2 \lambda z - \omega^2 \sin \lambda \cos \lambda x = -g + Z.$$

35. **Ex. 1.**   As an example, let us consider the case of a particle dropped from a height h. The initial conditions are therefore x, y, dx/dt, dy/dt, dz/dt all zero, and $z = h$. As a first approximation, neglect all the terms containing the small factor ω. Then we have $x = 0$, $y = 0$, $z = h - \frac{1}{2}gt^2$.

For a second approximation, we may substitute these values of (x, y, z) in the small terms. We have after integration

$$x = 0, \quad y = -\tfrac{1}{3}\omega \cos \lambda g t^3, \quad z = h - \tfrac{1}{2}gt^2.$$

Thus there will be a small deviation towards the east, proportional to the cube of the time of descent. There will be no southerly deviation, and the vertical motion will be the same as if the earth were at rest.

An elementary demonstration of this result will make the whole argument clearer. Let the particle be dropped from a height h vertically over O. Then, O being reduced to rest, the particle is really projected eastwards with a velocity $\omega h \cos \lambda$. Hence, if the direction of gravity did not alter owing to the rotation of the earth about OI, the particle would describe a parabola, and the easterly deviation would be $(\omega h \cos \lambda)\, t$, where t is the time of falling. Since $h = \frac{1}{2}gt^2$, this deviation is $\frac{1}{2}\omega \cos \lambda g t^3$. The rotation ω about OI is equivalent to $\omega \sin \lambda$ about Oz and $\omega \cos \lambda$ about Ox. The former does not alter the position of OC the normal to the surface of the earth, which is the direction of gravity. The latter turns OC in any time t through an angle $\omega \cos \lambda t$. Thus gravity gradually changes its direction as the particle falls. The particle is therefore acted on by a westerly component $= g \sin (\omega \cos \lambda t)$, which, since ωt is small, is nearly equal to $g\omega t \cos \lambda t$. Let y' be the distance of the particle from the position of the plane xz in space at the moment when the particle began to fall, and let y' be measured positively to the west. The equation of motion of the particle in space is therefore $d^2 y'/dt^2 = g\omega t \cos \lambda$. Integrating this and remembering that, as explained above, $dy'/dt = -\omega h \cos \lambda$ when $t = 0$, we get $y' = -\omega h t \cos \lambda + \frac{1}{6}g\omega t^3 \cos \lambda$. When the particle reaches the ground we have $y' = y$ very nearly, and $h = \frac{1}{2}gt^2$, thus the deviation westwards is $-\frac{1}{3}\omega g t^3 \cos \lambda$, which is the same as before. If it be not evident that $y' = y$, it may be shown thus. In the time t, Oy, Oz have turned through a very small angle $\theta = \omega \cos \lambda t$, hence, as in transformation of axes, $y' = y \cos \theta - z \sin \theta$, which gives $y' = y$ when we reject the squares of θ.

Ex. 2.   A particle is projected vertically upwards in vacuo with a velocity V. Show that on reaching the ground again there is no deviation to the south but the deviation to the west is $4\omega \cos \lambda V^3/3g^2$.   [Laplace, iv. p. 341.]

Ex. 3.   A particle is dropped from a height h and falls to the earth. If the resistance of the air be kv^n, where v is the relative velocity of the particle and air, show that the deviation to the south is still zero, but the deviation to the east is $\frac{1}{3}\omega \cos \lambda\, g t^3 \left\{ 1 - \dfrac{3 k g^{n-1} t^n}{(n+1)(n+2)} \right\}$ where t is the time of descent and the squares of k are neglected. Laplace gives the expansion for several powers of k when the resistance varies as the square of the relative velocity. [*Mec. Celeste*, iv. p. 337.]

36.   In many cases it will be found convenient to refer the motion to axes more generally placed. Let O be the origin, and let the axes be fixed relatively to the earth, but in any directions at right angles to each other. Let θ_1, θ_2, θ_3 be the resolved

parts of ω about these axes, then θ_1, θ_2, θ_3 are known constants. After substituting from Art. 4 in the equations of motion given in Art. 5 we get

$$\frac{d^2x}{dt^2} - 2\frac{dy}{dt}\,\theta_3 + 2\frac{dz}{dt}\,\theta_2 = X\,,$$

$$\frac{d^2y}{dt^2} - 2\frac{dz}{dt}\,\theta_1 + 2\frac{dx}{dt}\,\theta_3 = Y\,,$$

$$\frac{d^2z}{dt^2} - 2\frac{dx}{dt}\,\theta_2 + 2\frac{dy}{dt}\,\theta_1 = -g + Z\,.$$

For example, if we wish to determine the motion of a projectile, it is convenient to take the axis of z vertical and the plane of xz to be the plane of projection. Let the axis of x make an angle β with the meridian, the angle being measured from the south towards the west. Then

$$\theta_1 = \omega \cos \lambda \cos \beta, \quad \theta_2 = -\omega \cos \lambda \sin \beta, \quad \theta_3 = -\omega \sin \lambda.$$

These equations may be solved in any particular case by the method of continued approximation. If we neglect the small terms we get a first approximation to the values of (x, y, z). To find a second approximation we may substitute these values in the terms containing ω, and integrate the resulting equations. As the equations are only true on the supposition that ω^2 may be neglected, we cannot proceed to a third approximation.

37. Ex. 1. A particle is projected with a velocity V in a direction making an angle α with the horizontal plane, and such that the vertical plane through the direction of projection makes an angle β with the plane of the meridian, the angle β being measured from the south towards the west. If x be measured horizontally in the plane of projection, y be measured horizontally in a direction making an angle $\beta + \frac{1}{2}\pi$ with the meridian, and z vertically upwards from the point of projection, prove that
$$x = V \cos \alpha t + (V \sin \alpha t^2 - \tfrac{1}{3}gt^3)\,\omega \cos \lambda \sin \beta,$$

$$y = (V \sin \alpha t^2 - \tfrac{1}{3}gt^3)\,\omega \cos \lambda \cos \beta + V \cos \alpha t^2 \omega \sin \lambda,$$

$$z = V \sin \alpha t - \tfrac{1}{2}gt^2 - V \cos \alpha t^2 \omega \cos \lambda \sin \beta,$$

where λ is the latitude of the place, and ω the angular velocity of the earth about its axis of figure.

Show also that the increase of range on the horizontal plane through the point of projection is
$$4\omega \sin \beta \cos \lambda \sin \alpha\, (\tfrac{1}{3} \sin^2 \alpha - \cos^2 \alpha)\, V^3/g^2,$$
and that the deviation to the right of the plane of projection is
$$4\omega \sin^2 \alpha\, (\tfrac{1}{3} \cos \lambda \cos \beta \sin \alpha + \sin \lambda \cos \alpha)\, V^3/g^2.$$

Ex. 2. A bullet is projected from a gun nearly horizontally with great velocity so that the trajectory is nearly flat, prove that the deviation is nearly equal to $Rt\omega \sin \lambda$, where R is the range, and the other letters have the same meaning as in the last question. The deviation is always to the right of the plane of firing in the Northern hemisphere, and to the left in the Southern hemisphere. It is asserted (*Comptes Rendus*, 1866) that the deviation due to the earth's rotation as calculated by this formula is as much as half the actual deviation in Whitworth's gun.

We may arrive at this result in an elementary way. The bullet after it leaves the earth describes its path in space, while the axes of reference turn with the earth round the vertical at the point of projection with an angular velocity $\omega \sin \lambda$. The bullet is therefore left behind by the axes, and after a time t we have $y = xt\omega \sin \lambda$. As explained in Art. 35, the effect of the resolved angular velocity about a horizontal line at the point of projection is here neglected. The solution is therefore only approximately true when the trajectory is flat and the error increases with the time of flight.

The terms containing the factor ω are so small that it is unusual to make any allowance for them in aiming a gun at a target, except when the velocity of projection V is very great. In such cases it is enough to retain the terms which contain V as a factor. If however the trajectory is flat, the vertical velocity $V \sin a$ is small and there is no reason for retaining the term. Taking only the principal terms, we see from the results of the last example that

$$x = V \cos at \qquad y = V \cos at^2 \omega \sin \lambda$$

$$z = V \sin at - \tfrac{1}{2}gt^2 - V \cos at^2 \omega \cos \lambda \sin \beta.$$

It follows that throughout the motion $y = x\omega t \sin \lambda$. It appears also that the time the bullet takes to reach the target is (on these suppositions) independent of the motion of the earth. The vertical deviation of the bullet from its parabolic path at the moment of reaching the target is $-xt\omega \cos \lambda \sin \beta$. This is to be measured upwards when positive ; the deviation is therefore upwards or downwards according as the target is on the east or the west side of the meridian.

It may be objected that in obtaining these results we have neglected the resistance of the air, whose effects in altering the parabolic path are much greater than those of the rotation of the earth. So long however as we reject the squares both of ω and of the constant of resistance, the deviations due to ω from an unresisted are the same as those from a resisted path.

38. Disturbance of a Pendulum. Let us apply the equations of Art. 36 to determine the effect of the rotation of the earth on the motion of a pendulum. In this, as in some other cases, it is found advantageous to refer the motion to axes not fixed in the earth but moving in some known manner. Let the axis of z be vertical as before, and let the axes of x and y move slowly round the vertical with angular velocity $\omega \sin \lambda$ in the direction from the south towards the west. In this case we have

$$\theta_1 = \omega \cos \lambda \cos \beta, \quad \theta_2 = -\omega \cos \lambda \sin \beta,$$

and $\qquad \theta_3 = -\omega \sin \lambda + \omega \sin \lambda = 0,$

where β is the angle the axis of x makes with the tangent to the meridian, so that $d\beta/dt = \omega \sin \lambda$. If, as before, we neglect quantities which contain the square of ω as a factor, the terms which contain $d\theta_1/dt$ and $d\theta_2/dt$ must be omitted. Hence the required equations may be obtained from those of Art. 36 by putting $\theta_3 = 0$.

If m be the mass of the particle, l the length of the string, and T the tension ; the equations are

$$\frac{d^2x}{dt^2} - 2\omega \cos \lambda \sin \beta \frac{dz}{dt} = -\frac{T}{m}\frac{x}{l}$$

$$\frac{d^2y}{dt^2} - 2\omega \cos \lambda \cos \beta \frac{dz}{dt} = -\frac{T}{m}\frac{y}{l}$$

$$\frac{d^2z}{dt^2} + 2\omega \cos \lambda \sin \beta \frac{dx}{dt} + 2\omega \cos \lambda \cos \beta \frac{dy}{dt} = -g + \frac{T}{m}\frac{l-z}{l}$$

the origin being taken at the lowest point of the arc of oscillation.

If the oscillation be sufficiently small z differs from zero by small quantities of the order of α^2, where α is the semi-angle of oscillation. The last equation then shows that T differs from mg by quantities of the order of $\omega\alpha$ at least. If then we neglect terms of the order of $\omega\alpha^2$ and α^3, we may put mg for T in the two first equations, and neglect the terms containing $\omega dz/dt$. The equations of motion thus become the same as for a pendulum attached to a fixed point. If $ln^2 = g$ the solutions of the equations are clearly

$$x = A \cos (nt + C), \qquad y = B \sin (nt + D).$$

The small oscillations of a pendulum on the earth referred to axes turning round the vertical with angular velocity $\omega \sin \lambda$ are therefore the same as those of an imaginary pendulum suspended from an absolutely fixed point.

Let us then suppose the pendulum to be drawn aside so as to make with the vertical a small angle α and then let go. Relatively therefore to the axes moving round the vertical with angular velocity $\omega \sin \lambda$ we must suppose the particle to be projected with a velocity $l \sin \alpha\omega \sin \lambda$ perpendicular to the initial plane of displacement. We have then when $t = 0$, $x = l\alpha$, $y = 0$, $dx/dt = 0$, $dy/dt = -l\alpha\omega \sin \lambda$. It is then easy to see that in the above values of x and y, C and D are both zero, and that the particle describes an ellipse, the ratio of the axes being $\omega \sin \lambda (l/g)^{\frac{1}{2}}$. The effect of the rotation of the earth is to make this ellipse turn round the vertical with uniform angular velocity $\omega \sin \lambda$ in a direction from south to west. If the angle α be not so small that its square may be neglected, it is known by dynamics of a particle that, independently of all considerations of the rotation of the earth, there will be a progression of the apsides of the ellipse. It is therefore necessary for the success of the experiment that the length l of the pendulum should be very great. This motion of the apsides depending on the magnitude of α is in the opposite direction to that caused by the rotation of the earth.

It also appears that the time of oscillation is unaffected by the rotation of the earth, provided the arc of oscillation be so small that the effects of forces whose magnitude contains the factor $\omega\alpha^2$ may be neglected.

39. **Ex. 1.** In Foucault's experiment, a long pendulum is suspended from a point over the centre of a circular table, and the arc of oscillation is seen to pass from one diameter to another. Show that the arc of the circular rim of the table described by the plane of oscillation in one day is equal to the difference in length between two parallels of latitude one through the centre and the other through the northern or southern point of the rim. This theorem is due to Prof. J. R. Young.

Ex. 2. A heavy particle is suspended from a fixed point of support by a string of length a, and the effect of the rotation of the earth is neglected. In the two following cases the path of the particle is very nearly an ellipse whose apses advance in each complete revolution of the particle through an angle $\beta \cdot 2\pi$. If b and c be the major and minor semi-axes of the ellipse, prove (1) that when b and c are small compared with a, $\beta = \frac{3}{8}bc/a^2$, and (2) that when b and c are not small compared with a, but are very nearly equal, $(\beta+1)^{-2} = 1 - \frac{3}{4}b^2/a^2$.

Ex. 3. A pendulum, at rest relatively to the earth, is started in any direction with a small angular velocity, show that the oscillations will take place in a vertical plane turning uniformly round the vertical so that the pendulum becomes vertical once in each half oscillation.

Ex. 4. Let θ be the angle which a pendulum of length l makes with the vertical, and ϕ the angle which the vertical plane containing the pendulum makes with a vertical plane which turns round the vertical with uniform angular velocity $\omega \sin \lambda$ in a direction from south to west. Prove that, when terms depending on ω^2 are neglected, the equations of motion become

$$\left(\frac{d\theta}{dt}\right)^2 + \sin^2\theta \left(\frac{d\phi}{dt}\right)^2 = \frac{2g}{l}\cos\theta + A, \quad \frac{d}{dt}\left(\sin^2\theta \frac{d\phi}{dt}\right) = 2\sin^2\theta \cos(\phi+\beta)\,\omega \cos\lambda \frac{d\theta}{dt},$$

where A is an arbitrary constant, and the other letters have the meanings given to them in Art. 36. See M. Quet in *Liouville's Journal*, 1853.

These equations will be found convenient in treating the motion of a pendulum. They may be easily obtained by transforming those given in Art. 38 to polar co-ordinates.

40. **Disturbances of motion in one plane.** In the first volume of this treatise a chapter was devoted to the discussion of the motion of a body or a system of bodies constrained to remain in a fixed plane. This plane was treated as if it were really fixed in space. But since no plane can be found which does not move with the earth, it is important to determine what effect the rotation of the earth will have on the motion of these bodies. Let us treat this as an example of the method of Clairaut and Coriolis given in Art. 25.

Let the plane make an angle λ with the axis of the earth. Let a point O in this plane be on the surface of the earth, and let it be reduced to rest. Then, as proved in Art. 33, the moving bodies while in the neighbourhood of O are acted on by their weights in a direction normal to the surface of the earth. The earth is now turning round an axis through O parallel to the axis of figure with a constant angular velocity ω. Let this angular velocity be resolved into two, viz., $-\omega \sin \lambda$ about an axis perpendicular to the plane, and $\omega \cos \lambda$ about an axis in the plane. Now the square of ω is to be rejected, hence, by the principle of the superposition of small motions, we may determine the whole effect of these two rotations by adding together the effects produced by each separately.

It is a known theorem that if a particle be constrained to move in a plane which turns round any axis in that plane with a constant angular velocity $\omega \cos \lambda$, the motion may be found by regarding the plane as fixed and impressing an acceleration $\omega^2 r \cos^2 \lambda$ on the particle, where r is the distance of the particle from the axis. This may be deduced, as in Art. 26, from the theorem of Clairaut. This impressed acceleration is to be neglected because it depends on the square of ω. The angular velocity $\omega \cos \lambda$ has therefore no sensible effect.

If the bodies be free to move in the plane, the effect of the rotation $-\omega \sin \lambda$ is to turn the axes of reference round the normal to the plane drawn through the point O. If then we calculate the motion without regard to the rotation of the earth, taking the initial conditions relative to fixed space, the effect of the rotation of the earth may be allowed for by referring this motion to axes turning round the normal with angular velocity $-\omega \sin \lambda$. For example, if the body be a heavy particle suspended by a long string from a point O fixed relatively to the earth, it is really constrained to move in a horizontal plane, and the reasoning given above shows that the plane of oscillation will appear to a spectator on the earth to revolve with angular velocity $-\omega \sin \lambda$ round the vertical.

If the bodies be constrained to revolve with the plane, it will be required to find the motion relatively to that plane. We must therefore apply to each particle the force of moving space and the compound centrifugal force. If r be the distance of any particle of mass m from O, the former is $mr\omega^2 \sin^2 \lambda$. This is to be neglected because it depends on the square of ω. The latter is therefore the only force to be considered. Let us replace it by a resultant force acting at the centre of gravity of the body and a couple. We notice that, by Art. 24, the components of the compound centrifugal force on any particle are algebraic functions of dx/dt, dy/dt, dz/dt of the first degree. By Vol. I. Art. 14, their moment about the centre of gravity is equal to that of the compound centrifugal forces after the centre of gravity has been reduced to rest. Since each particle of the body is then moving in the plane of constraint perpendicular to its radius vector drawn from the centre of gravity as origin, the compound centrifugal force on it acts along the radius vector, and has therefore no moment about the centre of gravity. The couple therefore is zero. Again, the resultant force at the centre of gravity is the same as if all the mass were collected at that point, and is therefore equal to $-2MV\omega \sin \lambda$, where M is the mass of the body and V the velocity of the centre of gravity.

The effect of the rotation of the earth may therefore be allowed for by treating the earth as fixed and applying this force at the centre of gravity of the body. The ratio of this force to gravity for a particle moving 32 feet per second, is at most $4\pi/24.60.60$, which is less than a five thousandth. This is so small that, except under special circumstances, its effect is imperceptible.

41. Disturbance of the motion of a rigid body. Hitherto we have considered chiefly the motion of a single particle. The effect of the rotation of the earth on the motion of a rigid body will be more easily understood when the methods to be described in the following chapters have been read. If, for example, a body be set in rotation about its centre of gravity, it will not be difficult to determine its motion as viewed by a spectator on the earth, when we know its motion in space. It seems, therefore, sufficient here to consider the peculiarities which these problems present, and to seek illustrations which do not require any extended use of the equations of motion.

42. The effect of the rotation of the earth is in general so small compared with that of gravity, that it is necessary to fix the centre of gravity in order that the effects of the former may be perceptible. Even when this is done, the friction on the points of support and the other resistances, cannot be wholly done away with. If, however, the apparatus be made with such care that these resistances are small, the effects of the rotation of the earth may be made to accumulate, and after some time to become sufficiently great to be clearly perceptible.

If a body be placed at rest relatively to the earth and free to turn about its centre of gravity as a fixed point, it is actually in rotation about an axis parallel to the axis of the earth. Unless this axis be a principal axis, the body does not continue to rotate about it, and thus a change takes place in its state of motion. By referring to Euler's equations, we see that the change in the position of the axis of rotation is due to the terms $(A - B)\omega_1\omega_2$, $(B - C)\omega_2\omega_3$, $(C - A)\omega_3\omega_1$. The body having been placed apparently at rest, ω_1, ω_2, ω_3 are small quantities of the same order as the angular velocity of the earth; these terms are, therefore, of the order of the squares of small quantities. Whether they are great enough to produce any visible effect or not depends on their ratio to the frictional forces which could be called into play.   But, since these frictional forces are sufficient to prevent any relative motion, these terms will in general be just cancelled by the frictional couples introduced into the right-hand sides of Euler's equations. The body, therefore, continues at rest relatively to the earth.

In order that some visible effect may be produced, it is usual to impress on the body a very great angular velocity about some axis. If this be the axis of ω_3, the terms in Euler's equations, which are due to the centrifugal forces, and which contain ω_3 as a factor, become greater than when ω_3 had no such initial value. The greater this initial angular velocity, the greater these terms will be, and the more visible we may expect their effects on the body to be.

If the angular velocity thus communicated to the body be sufficient to turn it only once in a second, it is still $24 \times 60 \times 60$ times as great as the angular velocity of the earth. In such problems, therefore, we may regard the angular velocity of the earth as so small, compared with the existing angular velocities of the body, that the *square* of the ratio may be neglected.

As an example* of the application of these principles, we have selected one case of the gyroscope, which admits of an elementary solution. More general cases are considered further on.

43. **Ex.** *The centre of gravity of a solid of revolution is fixed, while the axis of figure is constrained to remain in a plane fixed relatively to the earth. The solid being set in rotation about its axis of figure, it is required to find the motion.*

Let us refer the motion to moving axes. Let the centre of gravity be the origin, the plane of yz the plane fixed relatively to the earth. Let the axis of figure be the axis of z, and let it make an angle χ with the projection of the axis of rotation of the earth on the plane of yz. Let this projection, for the sake of brevity, be called the axis of χ. Let p be the angular velocity of the earth about its axis, a the angle which the normal to the plane of yz makes with the axis of the earth. We suppose p to be reckoned positive when the rotation is in the standard direction usually taken as positive, so that when viewed from the positive extremity of the axis, the rotation appears to be in the direction of the hands of a watch. Since the earth turns from west by south to east, it follows, if the angle a be measured from the northern extremity P of the axis, that p is really negative and is represented in Art. 33 by $-\omega$. The motion of the moving axes is given by

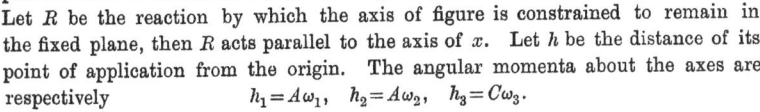

$$\theta_1 = p \cos a + d\chi/dt,$$
$$\theta_2 = p \sin a \sin \chi,$$
$$\theta_3 = p \sin a \cos \chi.$$

Let ω_1, ω_2, ω_3 be the angular velocities of the body about the moving axes; A, A, C the principal moments of inertia at the centre of gravity. Let R be the reaction by which the axis of figure is constrained to remain in the fixed plane, then R acts parallel to the axis of x. Let h be the distance of its point of application from the origin. The angular momenta about the axes are respectively $h_1 = A\omega_1,$ $h_2 = A\omega_2,$ $h_3 = C\omega_3.$

---

* M. Quet has published in *Liouville's Journal*, 1853, a memoir on relative motion and the application to the pendulum and several forms of the gyroscope. The problem considered in Art. 43 is one of those solved by him, though in a different manner.

The application of Lagrange's equations to relative motion has been discussed by Ed. Bour in a memoir presented to the French Academy in 1856 and afterwards published in *Liouville's Journal*, 1863. He forms an expression for the vis viva similar to that given in Art. 44, equation (1), and applies it to various problems. The principal object of his memoir is to show by the solution of some problems a little more complicated than those usually given in treatises on mechanics the advantages which result by using the canonical forms of Hamilton and Jacobi. He therefore continually uses the principal function of Hamilton to obtain the solutions of his problems. Lagrange's equations have also been used by Lottner in *Crelle's Journal*, 1857. His processes are somewhat complicated, but they have been abbreviated by Prof. Gilbert of Louvain, who supplied a "compte rendu" to the *Association Française* in 1878 and another to the Academy in 1882, Tome xciv. In both of these he continually refers to a memoir published by him, which however the author has not seen.

Substituting in Art. 10, the equations of motion are

$$
\left.
\begin{aligned}
A \frac{d\omega_1}{dt} - A\omega_2\theta_3 + C\omega_3\theta_2 &= 0 \\[2mm]
A \frac{d\omega_2}{dt} - C\omega_3\theta_1 + A\omega_1\theta_3 &= Rh \\[2mm]
C \frac{d\omega_3}{dt} - A\omega_1\theta_2 + A\omega_2\theta_1 &= 0
\end{aligned}
\right\}.
$$

Since the axis of z is fixed in the body, we see by Art. 3, that $\omega_1 = \theta_1$, $\omega_2 = \theta_2$. *The last equation of motion, therefore, shows that ω_3 is constant.* It should however be remembered that ω_3 is not the apparent angular velocity of the body as viewed by a spectator on the earth. If Ω_3 be the angular velocity relatively to the moving axes, we have by Art. 3, $\Omega_3 = \omega_3 - \theta_3$, so that

$$
\Omega_3 + p \sin a \cos \chi = \text{constant.}
$$

Thus the body, if so small a difference could be perceived, would appear to rotate slower or quicker the nearer its axis approached one extremity or the other of the projection of the axis of the earth's rotation on the fixed plane.

The first equation of motion, after substitution for ω_1, ω_2, θ_2, θ_3, their values in terms of χ, becomes

$$
A \frac{d^2\chi}{dt^2} - Ap^2 \sin^2 a \sin \chi \cos \chi + Cnp \sin a \sin \chi = 0,
$$

where n has been written for ω_3. The second term should be rejected as compared with the third, since it depends on the square of the small quantity p, Art. 33. We have, therefore,

$$
\frac{d^2\chi}{dt^2} = - \frac{C}{A} np \sin a \sin \chi.
$$

This is the equation of motion of a pendulum under the action of a force constant in magnitude, and whose direction is along the axis of χ, *i.e.* the projection of the axis of rotation of the earth on the fixed plane. The body being set in rotation about its axis of figure, we see that that axis immediately begins to approach one extremity or the other of the axis of χ with a continually increasing angular velocity. When the axis of figure reaches the axis of χ, its angular velocity begins to decrease, and it comes to rest when it makes an angle on the other side of the axis of χ equal to its initial value. The oscillation will then be repeated continually.

The axis of figure oscillates about that extremity of the axis of χ, which, when χ is measured from it, makes the coefficient on the right-hand side of the last equation negative. This extremity is such that, when the axis of figure is passing through it, the rotation n of the body is in the same direction as the resolved rotation p of the earth.

If we compare bodies of different form, we see that the time of oscillation depends only on the ratio of C to A. It is otherwise independent of the structure or form of the body. The greater this ratio the quicker will the oscillation be. For a solid of revolution the ratio is greatest when $\Sigma mz^2 = 0$. In this case the ratio is equal to 2, and the body is a circular disc or ring.

If we compare the different planes in which the axis may be constrained to remain, we see that the motion is the same for all planes making the same angle with the axis of the earth. It is therefore independent of the inclination of the plane to the horizon at the place of observation. The time of oscillation is least, and the motion of the axis most perceptible, when $\alpha = \frac{1}{2}\pi$, *i.e.* when the plane is parallel to the axis of rotation of the earth. If the plane be perpendicular to the axis of the earth, the axis of figure does not oscillate, but if the initial value of $d\chi/dt$ is zero, it remains at rest in whatever position it may be placed.

44. Application of Lagrange's equations. Let the body be referred to a system of axes with a fixed origin O, whose angular motions about themselves are given by θ_1, θ_2, θ_3. If λ, μ, ν are the direction cosines of their instantaneous axis OI, and θ the angular velocity about it, then $\theta_1 = \lambda\theta$, $\theta_2 = \mu\theta$, $\theta_3 = \nu\theta$. The vis viva of the body is

$$2T = \Sigma m \{(x' - y\theta_3 + z\theta_2)^2 + (y' - z\theta_1 + x\theta_3)^2 + (z' - x\theta_2 + y\theta_1)^2\}$$

where accents denote differential coefficients with regard to the time. Let $2R$ be the vis viva of the motion relative to the moving axes, then

$$2R = \Sigma m (x'^2 + y'^2 + z'^2).$$

We find by expansion

$$T = R + N\theta + \tfrac{1}{2}I\theta^2 \quad\dots\dots\dots\dots\dots(1).$$

where $N = \lambda\Sigma m (yz' - zy') + \mu\Sigma m (zx' - xz') + \nu\Sigma m (xy' - yx')$

$$I = \nu^2\Sigma m (x^2 + y^2) + \&c. - 2\lambda\mu\Sigma mxy - \&c.$$

so that N *is the angular momentum of the relative motion, and* I *is the moment of inertia of the body about the instantaneous axis* OI *of the axes of reference.*

We may verify this result for the case of a rigid body turning about the origin as a fixed point by noticing that its Vis Viva is

$$A (\Omega_1 + \theta\lambda)^2 + B (\Omega_2 + \theta\mu)^2 + C (\Omega_3 + \theta\nu)^2,$$

where Ω_1, Ω_2, Ω_3 are the relative angular velocities of the body. Expanding this we arrive at equation (1).

If the origin O of the moving axes is not fixed, let α, β, γ be its components of acceleration in space along the axes of reference. To reduce O to rest we apply these with reversed signs to every point of the system, Art. 33. The resultant of each of these systems of parallel forces is a single force acting at the centre of gravity of the body. These may be included in the force function by adding to U the term

$$K = - M (\alpha\bar{x} + \beta\bar{y} + \gamma\bar{z}) \quad\dots\dots\dots\dots(2),$$

where $\bar{x}, \bar{y}, \bar{z}$ are the co-ordinates of the centre of gravity, and M is the mass of the body.

The Lagrangian function is therefore

$$L = R + N\theta + \tfrac{1}{2}I\theta^2 + U + K \quad\dots\dots\dots\dots(3),$$

and if q be any one of the independent variables on which the position of the body is made to depend, we have the typical equation

$$\frac{d}{dt}\frac{dL}{dq'} - \frac{dL}{dq} = 0 \quad \dots\dots\dots\dots\dots(4).$$

In applying these equations to find the motion of a body relative to the earth we neglect as explained above the term $\frac{1}{2}I\theta^2$. For the reasons given in Art. 42, the centre of gravity of the body is usually fixed relatively to the earth, so the terms in the force function due to gravity do not appear. The term represented by K may also be omitted for the same reason. If, then, gravity is the only acting force the Lagrangian function reduces to

$$L = R + N\theta \quad \dots\dots\dots\dots\dots\dots(5).$$

One integral of the equations (4) can be found by the principle of Vis Viva. The method of treating the equations used in Vol. I. Chap. VIII. Art. 407, does not apply here because T is not a homogeneous function of the velocities. For the sake of increased generality, let us suppose that $L = L_0 + L_1 + \&c. + L_n$, where L_n is a homogeneous function of n dimensions of the velocities of the co-ordinates. When L has the value (3) this expression reduces to the first three terms. Multiplying each of the equations comprised in the typical form (4) by the corresponding q' and adding the results we have

$$\Sigma \left\{ \frac{d}{dt}\left(q'\frac{dL}{dq'} \right) - q''\frac{dL}{dq'} \right\} - \Sigma q'\frac{dL}{dq} = 0, \quad \dots\dots\dots\dots\dots(6),$$

where Σ implies summation for all the variables. Now

$$\Sigma q'\frac{dL_n}{dq'} = nL_n, \qquad\qquad \Sigma \left(q''\frac{dL_n}{dq'} + q'\frac{dL_n}{dq} \right) = \frac{dL_n}{dt},$$

since L_n does not contain t explicitly. It immediately follows by integrating (6) that

$$(n-1)L_n + (n-2)L_{n-1} + \&c. + L_2 - L_0 = h, \quad \dots\dots\dots\dots\dots(7),$$

where the term L_1 is absent and h is an arbitrary constant. When the expression for L contains only three terms, this reduces to $L_2 - L_0 = h$ or

$$R - \tfrac{1}{2}I\theta^2 - U - K = h. \quad \dots\dots\dots\dots\dots\dots(8).$$

In applying this equation to motion relative to the earth when θ^2 is rejected and the centre of gravity is fixed we have

$$R = h. \quad \dots\dots\dots\dots\dots\dots\dots(9).$$

Ex. 1. *As an example of the use of these equations, let us consider the problem already solved in Art. 43.*

To find R and N we notice that Oz separates from $O\chi$ in a fixed plane with angular velocity χ', the relative motion may therefore be constructed by the angular velocities $\Omega_1 = \chi'$, $\Omega_2 = 0$, $\Omega_3 = \phi'$ where ϕ is the angle a plane through Oz fixed in the body makes with the plane χOz. We therefore have

$$2R = A\chi'^2 + C\phi'^2, \qquad N = A\chi'\cos\alpha + C\phi'\sin\alpha\cos\chi,$$
$$T = \tfrac{1}{2}(A\chi'^2 + C\phi'^2) + p\,(A\chi'\cos\alpha + C\phi'\sin\alpha\cos\chi),$$
$$+ \tfrac{1}{2}p^2\{A(\cos^2\alpha + \sin^2\alpha\sin^2\chi) + C\sin^2\alpha\cos^2\chi\},$$

where the notation of Art. 43 has been followed. Using this value of T as the Lagrangian function and taking q to be ϕ and χ in turn, we have

$$\phi' + p \sin a \cos \chi = n,$$

$$A\chi'' + pC\phi' \sin a \sin \chi - p^2 \sin^2 a \, (A - C) \sin \chi \cos \chi = 0.$$

Eliminating ϕ' from the second equation we obtain the same results as in Art. 43.

Ex. 2. If gravity at each point of a body is regarded as the resultant of the terrestrial attraction and the centrifugal force at that point, prove that the force function U differs from that due to gravity by $-\frac{1}{2}I\theta^2 - \frac{1}{2}Mb^2\theta^2$, where b is the distance of the centre of gravity from the axis of the earth. It will be observed that in forming the Lagrangian function L the terms $\frac{1}{2}I\theta^2$ in U and T cancel each other, so that when the centre of gravity of the body is fixed and the force function due to gravity is treated as a constant the expression $L = R + N\theta$ is correct including the square of θ.   [Gilbert's Theorem.]

45. Ex. A very general form of the gyroscope is that in which the axis of the gyrating body is free to move in all directions about the centre of gravity, which is fixed relatively to the earth. One construction by which this freedom may be obtained is as follows.

A uniaxal body can turn freely about its axis of figure $C'OC$, which is pivoted on the inside of a metal ring $CY_1C'Y_1'$ so that $C'OC$ is a diameter, the point O being the centre of gravity of the body and the centre of the ring. The external extremities of that diameter $Y_1'OY_1$ of this ring which is perpendicular to $C'OC$ are pivoted at two points Y_2, Y_2' on the inside of a second ring external to the former, having $Y_2'OY_2$ for a diameter, and O for its centre. This external ring is free to

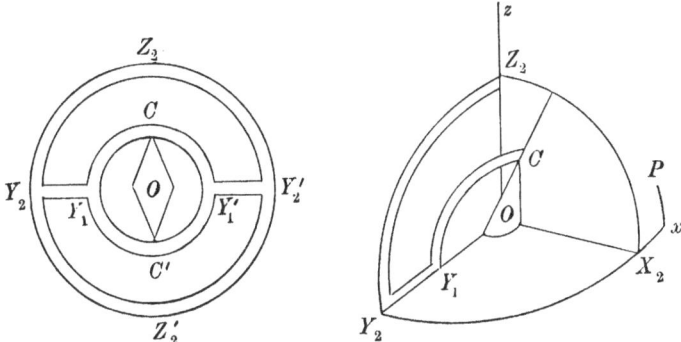

move about a diameter $Z_2'OZ_2$ perpendicular to $Y_2'OY_2$. The diameter OZ_2 is fixed relatively to the earth and will be taken as an axis of z, the plane of xz is also fixed relatively to the earth and will be taken to contain the straight line OP, drawn parallel to the northern direction of the axis of rotation of the earth.

In the first diagram the internal and external rings are shown folded into the plane of Y_2Z_2. In the second diagram all that portion of the figure is represented which lies in the positive octant of the axes $X_2Y_2Z_2$. The inner ring has been turned round its axis Y_2Y_2' through an angle θ. The axis Ox which is fixed relatively to the earth and lies in the plane X_2Y_2 has also been sketched.

Let the angle xOX_2 which defines the position in space of the external ring be ψ; let the angle Z_2OC defining the position of the internal ring be θ; and let the angle made by a plane passing through OC and fixed in the uniaxal body with X_2Z_2 be ϕ. These angles are the co-ordinates of the gyroscope, which has therefore three degrees of freedom. With two rings only we notice that the angles θ, ψ, ϕ are the Eulerian angular co-ordinates of the uniaxal body.

By increasing the number of rings we could increase the degrees of freedom and generalize the instrument. On the other hand we can reduce the number of independent co-ordinates by introducing any restrictions we please. Thus in the example discussed in Art. 43 where the axis OC is restricted to lie in one plane, we have ψ equal to a constant.

Let (A, A, C), (A_1, A_1, C_1), (A_2, A_2, C_2) be the principal moments of inertia at O of the uniaxal body, the internal and external rings. We then have

$$2R = A\,(\theta'^2 + \sin^2\theta\,\psi'^2) + C\,(\phi' + \psi'\cos\theta)^2 + A_1\,(\theta'^2 + \cos^2\theta\,\psi'^2) + C_1\psi'^2\sin^2\theta + A_2\psi'^2.$$

The first two terms represent the vis viva of the uniaxal body, the third and fourth terms represent that of the internal ring. These are obtained from the first two by putting $\phi'=0$, interchanging A and C in the coefficients of ψ'^2, and adding the suffixes.

Let λ, μ, ν be the direction cosines of OP referred to OC as axis of Z_1, OY_1Y_2 as axis of Y_1 and an axis OX_1 perpendicular to both; let i be the angle zOP. The angular momentum about OP is then

$$N = -A\sin\theta\,\psi'\lambda + A\theta'\mu + C\,(\phi' + \psi'\cos\theta)\,\nu$$
$$- C_1\sin\theta\,\psi'\lambda + A_1\theta'\mu + A_1\psi'\cos\theta\,\nu + A_2\psi'\cos i.$$

We also have the geometrical relations

$$\lambda = -\cos i\sin\theta + \sin i\cos\theta\cos\psi,$$
$$\mu = -\sin i\sin\psi,$$
$$\nu = \cos i\cos\theta + \sin i\sin\theta\cos\psi.$$

Representing the angular velocity of the earth by p measured positively in the direction X_2Y_2, and putting

$$P = A_1 + A_2 + (A - A_1 + C_1)\sin^2\theta,$$

the Lagrangian function becomes, when the square of p is rejected,

$$L = \tfrac{1}{2}(A + A_1)\theta'^2 + \tfrac{1}{2}P\psi'^2 + \tfrac{1}{2}C\,(\phi' + \psi'\cos\theta)^2$$
$$+ p\cos i\,[P\psi' + C\,(\phi' + \psi'\cos\theta)\cos\theta]$$
$$+ p\sin i\,[\{C\phi' + (C + A_1 - A - C_1)\psi'\cos\theta\}\sin\theta\cos\psi - (A + A_1)\theta'\sin\psi].$$

The equation corresponding to vis viva becomes

$$(A + A_1)\theta'^2 + P\psi'^2 + C\,(\phi' + \psi'\cos\theta)^2 = a.$$

Putting q in equation (4) equal to ϕ and ψ in turn we have

$$\phi' + (\psi' + p\cos i)\cos\theta + p\sin i\sin\theta\cos\psi = \beta,$$

$$\frac{d}{dt}\Big[P\,(\psi' + p\cos i) + C\,\{\phi' + (\psi' + p\cos i)\cos\theta\}\cos\theta\Big]$$
$$+ p\sin i\,[C\phi'\sin\theta\sin\psi + (2A_1 + C - C_1)\cos\psi\theta' + 2(A - A_1 - C + C_1)\sin^2\theta\cos\psi\theta'] = 0,$$

where a and β are arbitrary constants.

When the fixed axis Oz is parallel to the axis of the earth, $i=0$. The last equation is then a perfect differential, and we thus have a third integral.

46. Ex. 1. Show that a person furnished with the particular form of the gyroscope described in Art. 43, could, without any Astronomical observations, determine the latitude of the place, the direction. of the rotation of the earth, and the length of the sidereal day. This remark is due to M. Quet.

Ex. 2. If the body be a rod, and its centre of gravity supported without friction, prove that it could rest in relative equilibrium either parallel or perpendicular to the projection of the earth's axis on the plane of constraint. If it be placed in any other position, its motion will be very slow, depending on p^2, but it will oscillate about a mean position perpendicular to the projection of the earth's axis.

Ex. 3. If the axis of figure be acted on by a frictional force producing a retarding couple, whose moment about the axis of x bears a constant ratio μ to the moment of the reactional couple about the axis of y, and if the fixed plane be parallel to the axis of the earth, find the small oscillations about the position of equilibrium. Show that the position at any time t is given by

$$\chi = Le^{-\lambda t} \cos\left[(Cnp/A - \lambda^2)^{\frac{1}{2}} t + M\right],$$

where $2A\lambda = \mu(Cn - 2Ap)$, and L and M are two constants depending on the initial conditions.

Ex. 4. The centre of gravity of a solid of revolution is fixed, while the axis of figure is constrained to remain in the surface of a smooth right cone fixed relatively to the earth. Show that the axis of figure will oscillate about the projection of the axis of rotation of the earth on the surface of the cone, and that the time of a complete small oscillation about the mean position will be $2\pi (A \sin \epsilon / Cpn \sin \beta)^{\frac{1}{2}}$, where ϵ is the semi-angle of the cone, β the inclination of its axis to the axis of the earth, and the other letters have the same meaning as before. This problem is discussed both by Quet and Bour.

Ex. 5. The fixed axis OZ_2 of the external ring of a two ringed gyroscope is placed parallel to the axis of revolution of the earth, prove that

$$(A + A_1)\,\theta'^2 + \frac{(E - Cn \cos \theta)^2}{(A + C_1)\sin^2 \theta + A_1 \cos^2 \theta + A_2} = F,$$

where n, E and F are arbitrary constants.   [Lottner's Problem.]

Ex. 6. Two equal heavy rods CA, CB are connected by a hinge at C with a spring so that they tend to make a known angle with each other. The free ends A and B are then tied together and the whole is suspended by a string OC attached to the hinge. The system is left to itself until it is at rest relatively to the earth. If the string which fastens A and B be now cut, the arms separate from each other. Show that the system will immediately have an apparent angular velocity round the vertical equal to $p \sin \lambda (I' - I)/I'$, where I, I' are the moments of inertia of the system about the vertical OC respectively before and after the string joining A and B was cut, p is the angular velocity of the earth about its axis, and λ is the latitude of the place. In which direction will the system turn? This apparatus was devised by M. Poinsot, who considered that the experiment would be so effective that the latitude of the place could be deduced from the observed angular velocity. See *Comptes Rendus*, 1851, Tome XXXII. page 206.

Ex. 7. If a river is flowing due north, prove that the pressure on the eastern bank at a depth z is increased by the change of latitude of the running water in the ratio $gz + bv\omega \sin l : gz$, where b is the breadth of the stream, v its velocity, l the latitude and ω the angular velocity of the earth about its axis. [Math. Tripos, 1875.]

Ex. 8. A wave like the Tide-wave travels along a river with its crest at right angles to the banks. Deduce from Clairaut's rule (Art. 25) that the tide is higher on one bank than on the other, and shew that the height of the tide decreases in geometrical progression for equal increments of distance from one bank.

The general line of argument is as follows. Since the motion of the water is very nearly in a horizontal plane we may (by Art. 40) disregard the rotation of the earth provided we apply to every particle an acceleration $2\omega v \sin \lambda$ perpendicular to its direction of motion, *i.e.* perpendicular to the direction of the river. Hence the river must be so much higher on one side than the other that the pressure due by gravity to the difference of level is equal to that due to the applied acceleration. If ζ be the altitude of the tide above the mean level at a distance y from that side of the river at which the tide is highest, we have $-gd\zeta = 2\omega v \sin \lambda dy$. But in the theory of tides as undisturbed by the rotation it is proved that $v = \zeta\sqrt{g/h}$. By integration we find $\zeta = Ce^{-2\omega y \sin \lambda/\sqrt{gh}}$.

CHAPTER II.

Lagrange's Method with indeterminate multipliers.

47. In the first volume of this treatise Lagrange's method of finding the small oscillations of a system about a position of equilibrium has been explained. It is our object, not to repeat those explanations, but rather to examine how that theory is modified by the use of indeterminate multipliers. In a dynamical problem it generally happens that we want to know how some particular quantities change with the time. Now it is one of the chief advantages of Lagrange's method that it gives a large choice of quantities which may be taken as co-ordinates. The quantities we most wish to find are therefore usually chosen for the independent co-ordinates and their variations can then be found from Lagrange's equations. But sometimes we find that this introduces a great complication of symbols. Perhaps we lose thereby some principle of symmetry which would have abbreviated and simplified the whole process. We now propose to consider what modifications must be introduced into the equations when those particular quantities whose values we most require cannot be conveniently introduced as independent co-ordinates. For this purpose the method of indeterminate multipliers may be used with great advantage.

48. Let the system be referred to any co-ordinates θ, ϕ, &c. which are so small that we may reject all powers of them except the lowest which occur. They should therefore be so chosen that they vanish in the position of equilibrium. Let n be the number of those co-ordinates. Assuming that the geometrical equations do not contain the time explicitly, the vis viva $2T$ will be a quadratic function of the velocities, and may therefore be expanded in a series of the form

$$2T = A_{11}\theta'^2 + 2A_{12}\theta'\phi' + A_{22}\phi'^2 + \&\text{c}.$$

Here the coefficients A_{11}, &c. are all functions of θ, ϕ, &c. and we may suppose them to be expanded in a series of some powers of these co-ordinates. Since the oscillations are so small that we may reject all powers of the small quantities except the lowest which occur, we may reject all except the constant terms of these series. We shall therefore regard the coefficients A_{11}, &c. as constants.

We must now make an expansion for the force function U in a series of powers of θ, ϕ, &c. If the co-ordinates θ, ϕ, &c. were all independent, the terms containing the first powers would vanish, because by the principle of virtual velocities $dU/d\theta$, $dU/d\phi$, &c. are zero in the position of equilibrium for all variations of θ, ϕ, &c. which are consistent with the geometrical conditions. But as this does not necessarily occur when θ, ϕ, &c. are connected by geometrical relations, we take as our expansion

$$U - U_0 = C_1\theta + C_2\phi + \&c. + \tfrac{1}{2}C_{11}\theta^2 + C_{12}\theta\phi + C_{22}\phi^2 + \&c.,$$

where U_0 is a constant which is easily seen to be the value of U in the position of equilibrium. We may notice that the coefficients C_1, C_2, &c. are not unrestricted. They must be such that the equations of equilibrium are all satisfied.

Since the co-ordinates θ, ϕ, &c. are not independent there will be some geometrical relations which connect them. To simplify matters, let us suppose that there are but two such relations. Let these be $f(\theta, \phi, \&c.) = 0$, $F(\theta, \phi, \&c.) = 0$. We may also expand these in powers of the co-ordinates in the following manner:

$$f = G_1\theta + G_2\phi + \&c. + \tfrac{1}{2}G_{11}\theta^2 + G_{12}\,\theta\phi + \tfrac{1}{2}G_{22}\phi^2 + \&c.$$

$$F = H_1\theta + H_2\phi + \&c. + \tfrac{1}{2}H_{11}\theta^2 + H_{12}\,\theta\phi + \tfrac{1}{2}H_{22}\phi^2 + \&c.$$

The constant terms of these series are omitted because the geometrical equations are to be satisfied when the system is in equilibrium, i.e. when $\theta = 0$, $\phi = 0$, &c.

We have now to substitute these series in the Lagrangian equations. Referring to Chap. VIII. of Vol. I. these are represented by the type

$$\frac{d}{dt}\frac{dT}{d\theta'} - \frac{dT}{d\theta} = \frac{dU}{d\theta} + \lambda\frac{df}{d\theta} + \mu\frac{dF}{d\theta},$$

with similar equations for ϕ, ψ, &c. Here λ, μ are indeterminate multipliers whose values have to be found from the equations thus written down. The results of these substitutions are obviously

$$A_{11}\theta'' + \&c. = C_1 + C_{11}\theta + \&c. + \lambda(G_1 + \&c.) + \mu(H_1 + \&c.),$$

$$A_{12}\theta'' + \&c. = C_2 + C_{12}\theta + \&c. + \lambda(G_2 + \&c.) + \mu(H_2 + \&c.),$$

$$\&c. = \&c.$$

49. Since the system has been disturbed from a position of equilibrium these equations are all satisfied by $\theta = 0$, $\phi = 0$, &c. We thus obtain the equilibrium values of λ, μ. Let these be λ_0, μ_0. Then

$$\left.\begin{array}{l} 0 = C_1 + \lambda_0\, G_1 + \mu_0 H_1 \\ 0 = C_2 + \lambda_0\, G_2 + \mu_0 H_2 \\ 0 = \&c. \end{array}\right\}.$$

These are the equations of equilibrium already alluded to. The force function U being a known function of the co-ordinates, the co-efficients C_1, C_2, &c. are all known; and thus any two of these equations will determine λ_0, μ_0. The remaining equations will then be identically satisfied, because the quantities C_1, C_2, &c. are not unrestricted, but are such that the equations of equilibrium are all satisfied.

Let the dynamical values of λ and μ be $\lambda = \lambda_0 + \lambda_1$, $\mu = \mu_0 + \mu_1$. Then λ_1 and μ_1 are small quantities whose squares can be rejected. The equations of oscillation then become

$$A_{11}\theta'' + A_{12}\phi'' + \ldots = C_{11}\theta + C_{12}\phi + \ldots$$
$$+ \lambda_0\,(G_{11}\theta + G_{12}\phi + \ldots) + \lambda_1 G_1$$
$$+ \mu_0\,(H_{11}\theta + H_{12}\phi + \ldots) + \mu_1 H_1$$

$$A_{12}\theta'' + A_{22}\phi'' + \ldots = C_{12}\theta + C_{22}\phi + \ldots$$
$$+ \lambda_0\,(G_{12}\theta + G_{22}\phi + \ldots) + \lambda_1 G_2$$
$$+ \mu_0\,(H_{12}\theta + H_{22}\phi + \ldots) + \mu_1 H_2$$
$$\&c. = \&c.$$

We have here as many equations as there are co-ordinates. Besides these we have as many geometrical equations as indeterminate multipliers. These are

$$\left.\begin{array}{l} G_1\theta + G_2\phi + \ldots = 0 \\ H_1\theta + H_2\phi + \ldots = 0 \end{array}\right\}.$$

Thus we have on the whole sufficient equations to find all the unknown quantities θ, $\phi \ldots \lambda_1$, μ_1.

50. To solve these we proceed exactly as in the corresponding method described in Vol. I., where the co-ordinates θ, ϕ, &c. are all independent, except that we now include λ_1, μ_1 amongst the variables to be determined. We take as our typical solution

$$\theta = M\sin(pt + \alpha), \quad \phi = N\sin(pt + \alpha),\ \&c.$$
$$\lambda_1 = D\sin(pt + \alpha), \quad \mu_1 = E\sin(pt + \alpha).$$

Substituting these in the equations we see that $\sin(pt + \alpha)$ can be divided out from every equation. Writing

$$\left.\begin{array}{l} \bar{C}_{11} = C_{11} + \lambda_0 G_{11} + \mu_0 H_{11} \\ \bar{C}_{12} = C_{12} + \lambda_0 G_{12} + \mu_0 H_{12} \\ \&c. = \&c. \end{array}\right\},$$

we thus obtain

$$\left.\begin{array}{l} (A_{11}p^2 + \bar{C}_{11}) M + (A_{12}p^2 + \bar{C}_{12}) N + \ldots + G_1 D + H_1 E = 0 \\ (A_{12}p^2 + \bar{C}_{12}) M + (A_{22}p^2 + \bar{C}_{22}) N + \ldots + G_2 D + H_2 E = 0 \\ \qquad\qquad\qquad\qquad\qquad\qquad \&c. = 0 \\ \quad G_1 M + G_2 N + \ldots \qquad\qquad\qquad = 0 \\ \quad H_1 M + H_1 N + \ldots \qquad\qquad\qquad = 0 \end{array}\right\}.$$

Eliminating the ratios M, N, &c. D, E, we have the determinantal equation

$$\begin{vmatrix} A_{11}p^2 + \bar{C}_{11}, & A_{12}p^2 + \bar{C}_{12}, \ldots & G_1, & H_1 \\ A_{12}p^2 + \bar{C}_{12}, & A_{22}p^2 + \bar{C}_{22}, \ldots & G_2, & H_2 \\ \&c., & \&c., & \&c., \&c. \\ G_1, & G_2, & 0, & 0 \\ H_1, & H_2, & 0, & 0 \end{vmatrix} = 0.$$

If there be n co-ordinates, this is an equation of the nth degree to find p^2. Taking any root positive or negative, the preceding equations determine the corresponding ratios of M, N, &c. Taking all the roots in turn and adding together these partial solutions we have a solution complete with its $2n$ constants. These constants have to be determined from the initial values of the co-ordinates and their velocities.

51. This determinant differs from that used when there are no indeterminate multipliers in two respects. (1) There is a change in the quantities C_{11}, C_{12}, &c. represented by the insertion of the bar over the letters, (2) the determinant is bordered by the coefficients G_1, H_1, &c. of the *first* powers of the co-ordinates in the geometrical equations.

We notice that there is a very great simplification of the process when the *force function is such that the coefficients of the first powers of the co-ordinates in its expansion are all zero.* In this case C_1, C_2, &c. are zero, hence from the equations of equilibrium $\lambda_0 = 0$, $\mu_0 = 0$. Thus $\bar{C}_{11} = C_{11}$, $\bar{C}_{12} = C_{12}$, &c. = &c. It immediately follows that it is unnecessary to calculate the terms of the second order in the geometrical equations, for these disappear from the equations of motion. This of course is an important simplification. Further, the final determinant only differs from that used when there are no indeterminate multipliers by being bordered by the coefficients G_1, &c. H_1, &c.

This simplification occurs when the *position about which the system oscillates is a position of equilibrium for all variations of the co-ordinates, although the constraints compel the system to oscillate in a given limited manner.*

52. Brief Summary. In order to indicate the method of proceeding in any particular case we shall now sum up the general line of argument.

Expand the semi vis viva T and the force function U in powers of the co-ordinates θ, ϕ, &c. and their differential coefficients θ', ϕ', &c., all powers above the second being rejected. Multiply the geometrical relations $f = 0$, $F = 0$ by $\lambda = \lambda_0 + \lambda_1$ and $\mu = \mu_0 + \mu_1$ where λ_1 and μ_1 are small quantities of the same order as the co-ordinates θ, ϕ, &c. and expand these products, all powers of the small quantities above the second being rejected. *First,* taking the expression $U + \lambda f + \mu F$, equate to zero the coefficient of the first power of each co-ordinate, we thus have equations to find λ_0, μ_0. *Secondly,* omitting the accents in the expression for T and also the constant terms in U, form the discriminant of

$$Tp^2 + U + \lambda f + \mu F$$

with regard to the co-ordinates and the subsidiary variables λ_1, μ_1. Equating this determinant to zero, we have an equation to find the values of p.

53. On Principal Oscillations. The equations which determine the constants M, N, &c. D, E are shown above. Solving these we see that their ratios are equal to the ratios of the minors of the constituents of any row we please in the determinantal equation. If we represent these minors by $I_{11}(p^2)$, $I_{12}(p^2)$, &c. the oscillations of the system are represented by

$$\theta = L_1 I_{11}(p_1^2) \sin(p_1 t + \alpha_1) + L_2 I_{11}(p_2^2) \sin(p_2 t + \alpha_2) + \&c.,$$

$$\phi = L_1 I_{12}(p_1^2) \sin(p_1 t + \alpha_1) + L_2 I_{12}(p_2^2) \sin(p_2 t + \alpha_2) + \&c.,$$

$$\&c. = \&c.,$$

where L_1, L_2 &c. are constants which depend on the initial conditions.

When the initial co-ordinates are such that all the constants L_1, L_2, &c. vanish except one, the expressions for θ, ϕ ... λ, μ are reduced to the trigonometrical expressions in some one column. The co-ordinates θ, ϕ, &c. then bear to each other *ratios which are constant throughout the motion.* It follows also that the *values of the co-ordinates* θ, ϕ, &c. *repeat at a constant interval,* viz. the period of the trigonometrical expression in the one column preserved. Referring to Vol. I. we see that the characteristics of a principal oscillation are satisfied.

54. The system being referred to any co-ordinates θ, ϕ, &c. it may *be required to find how it should be disturbed from its position of equilibrium that it may describe any proposed principal oscillation.* We see that the system must be so displaced that its co-ordinates θ, ϕ, &c. have the ratios of the minors of any row of the determinantal equation. It is also necessary that the initial velocities θ', ϕ', &c. have the same ratio. These conditions are necessary and sufficient.

55. Putting this into algebraical language, we say that when a system is performing a principal oscillation of the type $\sin(p_1 t + a_1)$, then

$$\frac{\theta}{I_{11}(p_1{}^2)} = \frac{\phi}{I_{12}(p_1{}^2)} = \&c. = L_1 \sin(p_1 t + a_1).$$

We also infer from these equations that throughout the motion $\theta'' = -p_1{}^2\theta$, $\phi'' = -p_1{}^2\phi$, &c.

56. **Principal Co-ordinates.** *It may be required to find formulæ of transformation by which we may change any co-ordinates θ, ϕ, &c. into principal co-ordinates.* According to the definitions laid down in Vol. I. a system is referred to principal co-ordinates ξ, η, &c. when the vis viva $2T$ and the force function U are expressed in the forms
$$\left. \begin{array}{l} 2T = \xi'^2 + \eta'^2 + \zeta'^2 + \dots \\ 2(U - U_0) = c_{11}\xi^2 + c_{22}\eta^2 + c_{33}\zeta^2 + \dots \end{array} \right\}.$$

Lagrange's equations then take the form $\xi'' - c_{11}\xi = 0$, $\eta'' - c_{22}\eta = 0$, &c., so that the whole motion is given by $\xi = E \sin(p_1 t + a_1)$, $\eta = F \sin(p_2 t + a_2)$, &c., where E, F, &c. are the constants of integration and $p_1{}^2 = -c_{11}$, $p_2{}^2 = -c_{22}$, &c.

When the initial conditions are such that all the constants E, F, &c. are zero *except one* the system is said to be performing a principal oscillation. If then we write $x = \sin(p_1 t + a_1)$, $y = \sin(p_2 t + a_2)$, x will be a multiple of ξ, y a multiple of η, and so on. The expressions for θ, ϕ, &c. given in Art. 53, now reduce to

$$\theta = L_1 I_{11}(p_1{}^2) x + L_2 I_{11}(p_2{}^2) y + \dots$$
$$\phi = L_1 I_{12}(p_1{}^2) x + L_2 I_{12}(p_2{}^2) y + \dots$$
$$\&c. = \&c.$$

These formulæ will enable us to change any co-ordinates θ, ϕ, &c. into others x, y, &c. which make T and U assume the forms
$$\left. \begin{array}{l} 2T = a_{11}x'^2 + a_{22}y'^2 + \dots \\ 2(U - U_0) = c_{11}x^2 + c_{12}y^2 + \dots \end{array} \right\}.$$

The n constants L_1, L_2, &c., are arbitrary multipliers of x, y, &c., and may, if we please, be so chosen as to make a_{11}, a_{22}, &c. each equal to unity.

On Lagrange's Determinant.

57.   On examining Lagrange's method of finding the oscillations of a system we see that the whole process depends on the solution of a certain determinantal equation.   Even the stability or in-stability of the equilibrium depends on the nature of its roots.   If this equation can be solved, the character of the motion and the periods of oscillation (if the motion be oscillatory) are immediately apparent.   If the equation cannot be solved, we may expand the determinant and discuss its roots by the methods given in the theory of equations.   But without expanding the determinant we may sometimes accomplish the same purpose by the following theorem.   We shall begin with the determinant in its simplest form as it is obtained in Vol. I. Chap. IX.; we shall then consider the modifications introduced by bordering it with any quantities.

58.   **Separation of Roots.**   Let the determinantal equation be written in the form*

$$\Delta = \begin{vmatrix} A_{11}p^2 + C_{11}, & A_{12}p^2 + C_{12}, & \&c. \\ A_{12}p^2 + C_{12}, & A_{22}p^2 + C_{22}, & \&c. \\ \&c. & \&c. \end{vmatrix} = 0.$$

Let us form from this determinant a minor by erasing the first row and the first column.   We may then form from this minor a second minor, and so on.   Thus we have a series of functions of p^2 whose degrees regularly diminish from the nth to the first.   Let us call the successive determinants thus formed Δ, Δ_1, Δ_2, &c.   The de-terminant Δ is not altered if we border it with a column of zeros on the right-hand side and a row of zeros at the bottom, provided we put unity in the vacant corner.   We may therefore consider that $\Delta_n = 1$.

By a theorem in determinants, if I_{11}, I_{12}, &c. be the minors of the several constituents of Δ, we have $\Delta\Delta_2 = I_{11}I_{22} - I_{12}^2$, and we notice that $I_{11} = \Delta_1$.   Let us suppose p^2 to increase gradually from $p^2 = -\infty$ to $p^2 = +\infty$, then when p^2 passes through a value which makes $\Delta_1 = 0$ we see that Δ and Δ_2 must have opposite signs.   The same argument applies to every one of the series

---

* The proposition that the roots of Lagrange's determinant, when written in this general form, are all real is due to Sir W. Thomson.   It is the extension of a corresponding theorem for that particular form of the equation which occurs when the vis viva is expressed as the sum of the squares of the velocities of the co-or-dinates.   Several proofs of this latter theorem will be found in Lesson VI. of Dr Salmon's *Higher Algebra*.   The simplest of these is the one given by Dr Salmon himself.   He also proves that the roots are separated by those of the leading minors.   The proof in the text is an extension of his line of argument to Lagrange's determinant in its general form.   Another line of argument is indicated in the examples in Art. 71.

Δ, Δ_1, Δ_2, &c., whenever any one of them vanishes the determinants on each side have opposite signs*.

Using these determinants like Sturm's functions we see that a variation of sign can be lost or gained only at one end of the series. It can be lost at the end Δ only when p^2 passes through a root of the equation $\Delta = 0$, and it will be regained again as p^2 passes through the next root in order of magnitude, *unless a root of the equation* $\Delta_1 = 0$ *lies between these two*.

If then we can prove that n variations of sign are lost as p^2 passes from $p^2 = -\infty$ to $p^2 = +\infty$ it is clear that the equation $\Delta = 0$ must have n real roots and these roots will be separated by the roots of the equation $\Delta_1 = 0$.

Now the coefficient of the highest power of p^2 in the determinant Δ is the discriminant of T, and is therefore positive. The coefficient of the highest power of p^2 in Δ_1 is the discriminant of T after θ' has been put zero, and this also is positive. Thus the coefficients of the highest powers of p^2 in every one of the determinants Δ, Δ_1, Δ_2, &c. are positive. If then we substitute $-\infty$ for p^2, these determinants are alternately positive and negative, if we substitute $+\infty$ for p^2 the determinants are all positive. It follows that n variations of sign are lost as p^2 passes from $p^2 = -\infty$ to $p^2 = +\infty$.

Summing up we see that *the roots of each determinant of the series* Δ, Δ_1, Δ_2, &c. *are all real and the roots of each separate or lie between the roots of the determinant next before it in the series*.

59. Resuming our line of argument we see that as p^2 increases from $p^2 = -\infty$ to $p^2 = +\infty$ a variation of sign in the series Δ, Δ_1, &c. is lost when p^2 passes through a root of $\Delta = 0$, and once lost this

---

* In this reasoning we have for the sake of brevity omitted the case in which two or more successive determinants in the series Δ, Δ_1, Δ_2, &c. vanish for the same value of p^2. But this omission is of no real importance, for we may change these determinants into others whose constituents are slightly different from those of the given determinants but are such that no successive two of the series have a common root. In the limit, therefore, when these arbitrary changes of the constituents are indefinitely small, the roots of the series of determinants will still be real and the roots of each will separate, *or coincide with*, the roots of the next before it in the series.

To show that these changes are possible, let Δ, Δ_1, Δ_2 be any three consecutive members of the series. Let us suppose that Δ_2 does not vanish while the two members (and perhaps others) just before it are zero. Then from the equation in the text, we have $I_{12} = 0$. Let us add to each of the constituents of which I_{12} is the minor the small quantity a. The determinant Δ_1 is unaltered and remains equal to zero. The determinant Δ undergoes a slight alteration, so that in its new form the equation just quoted becomes $\Delta\Delta_2 = -a^2\Delta_2^2$. Thus Δ is no longer zero. In this way whenever any two consecutive members of the series of determinants vanish, one may be rendered finite.

variation cannot be regained. It immediately follows *that as* p^2 *passes from* $p^2 = \alpha$ *to* $p^2 = \beta$ *if* κ *variations of sign are lost there are exactly* κ *roots of the equation* $\Delta = 0$ *between these limits.*

60. It will be noticed that in this line of argument no assumption has been made about the functions

$$T = \tfrac{1}{2} A_{11} \theta'^2 + A_{12} \theta' \phi' + \tfrac{1}{2} A_{22} \phi'^2 + \ldots \Big\}$$
$$U - U_0 = \tfrac{1}{2} C_{11} \theta^2 + C_{12} \theta \phi + \tfrac{1}{2} C_{22} \phi^2 + \ldots\ldots \Big\} ,$$

except that the successive discriminants of the former are all positive. This may be expressed by saying that T is a one-signed positive function, i.e. a function which keeps the positive sign for all values of the variables and never vanishes except when all the variables are zero. That the vis viva is a one-signed positive function is of course evident. The necessary and sufficient conditions that a quadric function should be a one-signed positive function are given in Williamson's *Differential Calculus*. They may be briefly summed up by saying that the successive discriminants are all positive. A short proof is given in a note at the end of the volume.

61. **Equal Roots.** Since the roots of any one of the leading minors I_{11}, I_{22}, &c. separate the roots of Lagrange's determinant, it follows that when the latter has r roots each equal to p_1, each of the former must have $r - 1$ roots each equal to p_1. For the same reason any leading second minor such as Δ_2 must have $r - 2$ roots each equal to p_1.

Consider next any other minor of the determinant. By proper changes of rows and columns we may represent this by I_{12}. Since $\Delta \Delta_2 = I_{11} I_{22} - I_{12}^2$, it follows that I_{12} must also have $r - 1$ roots equal to p_1.

On the whole we conclude that *if Lagrange's determinant have* r *equal roots, then every first minor has* r − 1 *roots equal to each of these.* In the same way it follows from this, that *every second minor has* r − 2 *roots equal to each of these, and so on.*

62. This theorem will often enable us to detect the presence of equal roots in Lagrange's determinant. We equate any minor to zero and thus obtain an equation to find p^2, which is sometimes of a very simple form.

Suppose for example the system had two co-ordinates, so that (Art. 60)

$$2T = A_{11} \theta'^2 + 2 A_{12} \theta' \phi' + A_{22} \phi'^2 \Big\}$$
$$2 (U - U_0) = C_{11} \theta^2 + 2 C_{12} \theta \phi + C_{22} \phi^2 \Big\} .$$

If we form Lagrange's determinant, we see that the minors cannot be zero unless $C_{11}/A_{11} = C_{12}/A_{12} = C_{22}/A_{22}$, each of these ratios being equal to $-p^2$. Unless therefore these conditions are satisfied there cannot be two equal roots.

63. The equation used in solid geometry to determine the lengths of the axes of a conicoid is an equation of Lagrange's form. As a consequence of this theorem, the usual conditions for a surface of revolution follow at once by equating each of the minors to zero.

64. The Bordered Determinant. Let us now border Lagrange's determinant with any arbitrary quantities f, g, h, &c., so that we obtain the determinantal equation.

$$\Delta' = \begin{vmatrix} A_{11}p^2 + C_{11}, & A_{12}p^2 + C_{12}...f \\ A_{12}p^2 + C_{12}, & A_{22}p^2 + C_{22}...g \\ \hdotsfor{2} \\ f & g \qquad 0 \end{vmatrix} = 0.$$

Regarding this as a function of p^2, we see that its degree is one less than that of Δ. We shall now consider how the roots of this equation are connected with those of Lagrange's.

If we remove the zero in the corner of Δ' and write $ap^2 + c$ in its place, where a and c are any quantities however small, we obtain another equation which is of Lagrange's form but one degree higher than Δ. The expression for $2T$ from which this new equation is derived is the same as the former with the addition of the term ax'^2 where x is some new variable. If then a be positive, we may apply the theorem proved in Art. 58 to this new determinant. Call this new determinant D', then the roots of D' are all real and are separated by those of the first minor of any constituent in the leading diagonal. But the determinant Δ is the minor of the last constituent in that diagonal. The roots of D' are therefore all real and are separated by those of Δ. If we put a and c both infinitely small, two roots of the equation $D' = 0$ are each infinite, and the other roots may be made to approximate as closely as we please to those of $\Delta' = 0$. Hence we infer that *whatever the quantities* f, g, &c. *may be, the roots of the determinantal equation* $\Delta' = 0$ *are real and separate or lie between those of* $\Delta = 0$.

65. The original determinant Δ has n columns and n rows. The determinant Δ' has been derived from Δ by bordering it with n arbitrary quantities forming a new column and a new row with zero in the corner. In the same way we may border the determinant Δ' with a new set of n arbitrary quantities f', g', &c., filling up the vacant spaces near the corner with zeros. Thus we obtain a new determinant with *four* zeros in the corner, which we may call Δ''. This determinant is of one degree less than Δ' and its roots are all real and separate those of Δ'.

66. Lastly let us form the series of $n+1$ determinants Δ, Δ', Δ'' &c., terminating with a constant. Each determinant is derived from the one before by bordering it with n arbitrary quantities with zeros near the corner, so that the determinants are all symmetrical. Proceeding as in Art. 64, we may regard this set of determinants as the limiting cases of other determinants which are all of Lagrange's form, but of degrees successively higher than Δ. The last of these, being in the limit a constant, will have all its roots infinitely great. Prefixing to this second set of determinants the set formed (as described in Art. 58) by cutting off rows and columns, we have a complete series of determinants separated into two sets by the determinant Δ. They begin with unity and terminate with a determinant whose roots (in the limit) are all infinitely large. It follows by the theorem in Art. 58 that in passing from $p^2 = \alpha$ to $p^2 = \beta$ no variation of sign can be lost in the complete series because no root of the last determinant can lie between the finite quantities α and β. But if κ roots of the determinant Δ lie between these limits, κ variations of sign must be lost in the first set of determinants. Hence as many variations of sign are gained in the second set of determinants as are lost in the first set. Summing up we infer that *as* p^2 *passes from* $p^2 = \alpha$ *to* $p^2 = \beta$, *if* κ *variations of sign are gained in the series* Δ, Δ', Δ'', &c. *there are exactly* κ *roots of the equation* $\Delta = 0$ *between these limits*.

67. **Ex. 1.** In the theorem of Art. 64 show without putting $a = 0$ that the roots of Δ' separate or lie between those of Δ.

Ex. 2. In the theorem of Art. 66 show that if variations of sign are lost as p^2 passes from $p^2 = a$ to $p^2 = \beta$, then a is greater than β.

Ex. 3. If the system be referred to principal co-ordinates, show that the determinantal equations $\Delta' = 0$, $\Delta'' = 0$ may be written in the form

$$\frac{f^2}{A_{11}p^2 + C_{11}} + \frac{g^2}{A_{22}p_2 + C_{22}} + \ldots = 0,$$

$$\frac{(fg' - f'g)^2}{(A_{11}p^2 + C_{11})(A_{22}p^2 + C_{22})} + \frac{(gh' - g'h)^2}{(A_{22}p^2 + C_{22})(A_{33}p^2 + C_{33})} + \ldots = 0.$$

68. **Invariants of the System.** In order to determine the values of p^2 it will often be necessary to expand the determinant. When there are only a few co-ordinates this can be done without difficulty. In other cases we may use Taylor's theorem. Let Δ be the discriminant of T and let Π represent the operation

$$\Pi = C_{11} \frac{d}{dA_{11}} + C_{12} \frac{d}{dA_{12}} + C_{23} \frac{d}{dA_{23}} + \ldots$$

Then Lagrange's determinant becomes when expanded

$$\Delta p^{2n} + \Pi (\Delta) p^{2n-2} + \Pi^2 (\Delta) \frac{p^{2n-4}}{1 \cdot 2} + \ldots = 0.$$

If Δ' be the discriminant of U and Π' represent the operation Π when the letters A and C are interchanged, we may write the equation in the form

$$\Delta' + \Pi' (\Delta') p^2 + \Pi'^2 (\Delta') \frac{p^4}{1 \cdot 2} + \ldots = 0.$$

When there are only three co-ordinates we may adopt the notation used in the chapter on Invariants in Dr Salmon's Conics.

69. It is sometimes convenient to change the co-ordinates from θ, ϕ, &c. to others x, y, &c. connected by linear relations. Let these be

$$\left.\begin{array}{l} \theta = l_1 x + l_2 y + l_3 z + \ldots \\ \phi = m_1 x + m_2 y + m_3 z + \ldots \\ \&c. = \&c. \end{array}\right\}.$$

In whatever manner this is done it is clear that the equation giving the times of oscillation must be the same. The ratios of the coefficients of the several powers of p^2 are therefore invariable. Let μ be the determinant of transformation, i.e. the determinant whose rows are the coefficients of x, y, z, &c. in the equations of transformation just written down. Then by a known theorem in determinants the discriminant Δ is changed into $\mu^2 \Delta$. Hence all the other coefficients are altered in the same ratio. *The coefficients Δ, Π (Δ), &c. are therefore called the invariants of the system. The sign of each of these, and the ratio of any two, are unaltered by any transformation of co-ordinates.*

70. **Ex. 1.** If a system be in equilibrium, show that the equilibrium will be stable if $-\Pi (\Delta)$, $\Pi^2 (\Delta)$, $-\Pi^3 (\Delta)$, &c. be all positive.

We notice (1) that Δ is necessarily positive, (2) since the roots of Lagrange's equation are all real, these are the conditions given by Descartes' theorem that the roots should be all positive.

Ex. 2. The same dynamical system can oscillate about the same position of equilibrium under two different sets of forces. If ρ_1, ρ_2, &c. σ_1, σ_2, &c. be the periods of oscillation when the two sets act separately, R_1, R_2, &c. the periods when they act together, prove that $\Sigma \dfrac{1}{\rho^2} + \Sigma \dfrac{1}{\sigma^2} = \Sigma \dfrac{1}{R^2}$.

This follows from the fact that $\Pi (\Delta)$ contains C_{11}, &c. only in their first powers.

Ex. 3. Two different systems of bodies when acted on by the same set of forces oscillate in periods ρ_1, ρ_2, &c., σ_1, σ_2, &c. If R_1, R_2, &c. be the periods when they are both acted on by this set of forces, prove that $\Sigma\rho^2 + \Sigma\sigma^2 = \Sigma R^2$.

71. Ex. 1. Let T and U be given in their simplest forms, i.e. referred to principal co-ordinates, and let these be

$$2T = a_1 \theta'^2 + a_2 \phi'^2 + \ldots$$
$$2(U - U_0) = c_1 \theta^2 + c_2 \phi^2 + \ldots$$

It is required to transform these to general co-ordinates by using the formulæ of Art. 69, and thence to construct the general form of Lagrange's determinant. For the sake of brevity let $B_1 = a_1 p^2 + c_1$, $B_2 = a_2 p^2 + c_2$, &c., let there be κ of these. Also let $I(l_1)$, $I(l_2)$, &c. be the minors of l_1, l_2, &c. in the determinant of transformation, called μ in Art. 69. Then show (1) that Lagrange's determinant is equal to $\mu^2 B_1 B_2 \ldots B_\kappa$, (2) that the minor of the leading constituent of Lagrange's determinant is equal to $\{I(l_1)\}^2 B_2 B_3 \ldots B_\kappa + \{I(m_1)\}^2 B_1 B_3 \ldots B_\kappa + \ldots$, (3) that Lagrange's determinant when bordered with f, g, h, &c. with zero in the vacant corner is equal to

$$- \begin{vmatrix} f & g & h & \ldots \\ m_1 m_2 m_3 \ldots \\ n_1 n_2 n_3 \ldots \\ \ldots\ldots\ldots\ldots \end{vmatrix}^2 B_2 B_3 \ldots B_\kappa - \begin{vmatrix} l_1 l_2 l_3 & \ldots \\ f & g & h & \ldots \\ n_1 n_2 n_3 \ldots \\ \ldots\ldots\ldots \end{vmatrix}^2 B_1 B_3 \ldots B_\kappa - \ldots$$

Ex. 2. Deduce from the analytical results of the last article that if T and U be any expressions which can be derived by a *real* linear transformation from the forms

$$2T = a_1 \theta'^2 + a_2 \phi'^2 + \ldots$$
$$2(U - U_0) = c_1 \theta^2 + c_2 \phi^2 + \ldots$$

where the a's and the c's have any signs, then (1) the roots of Lagrange's determinant are all real, (2) that they will be separated by those of any leading minor, and (3) that they will also be separated by those of the bordered determinant.

Energy of an Oscillating System.

72. *A system is referred to its principal co-ordinates, it is required to find its kinetic and potential energies.*

Let the co-ordinates be ξ, η, &c. so that the vis viva $2T$ and force function U are given by

$$\left. \begin{array}{l} 2T = \xi'^2 + \eta'^2 + \ldots \\ 2(U - U_0) = -p_1^2 \xi^2 - p_2^2 \eta^2 - \ldots \end{array} \right\}.$$

Then by Lagrange's equations Art. 56, we have.

$$\xi = E \sin(p_1 t + \alpha_1), \quad \eta = F \sin(p_2 t + \alpha_2), \text{ &c.}$$

Substituting these in the expressions for T and U just written down, we find

$$2T = p_1^2 E^2 \cos^2(p_1 t + \alpha_1) + p_2^2 F^2 \cos^2(p_2 t + \alpha_2) + \text{&c.},$$
$$2(U_0 - U) = p_1^2 E^2 \sin^2(p_1 t + \alpha_1) + p_2^2 F^2 \sin^2(p_2 t + \alpha_2) + \text{&c.}$$

Here T is the kinetic energy of the system, and when the position of equilibrium is the position of reference, $U_0 - U$ is the potential energy.

From these expressions we infer that the *whole energy of a system oscillating about a position of equilibrium is the sum of the energies of its principal oscillations.*

73. **Mean kinetic and Potential energies.** The mean value of $E^2 \cos^2(pt + \alpha)$ with regard to time from $t = 0$ to $t = t$ is $\dfrac{E^2}{t} \displaystyle\int_0^t \cos^2(pt + \alpha)\, dt$, which after integration reduces to $\frac{1}{2}E^2$ when t is very great. The mean value of $E^2 \sin^2(pt + \alpha)$ is of course the same. We therefore infer that the *mean kinetic energy of a system oscillating about a position of equilibrium is equal to the mean potential energy, the mean being taken for a long period* and the position of equilibrium being the position of reference. Thus the energy of the system is on the whole equally distributed into kinetic and potential energies. Sometimes one has an excess and sometimes the other, but in any long time their shares are equal.

74. **Energy of any system.** *To find the energy of a system oscillating about a position of equilibrium referred to any co-ordinates.*

Let the general co-ordinates be θ, ϕ, &c. so that the kinetic energy T and the potential energy $U_0 - U$ are given by

$$\left. \begin{aligned} 2T &= A_{11}\theta'^2 + 2A_{12}\theta'\phi' + \dots \\ 2(U - U_0) &= C_{11}\theta^2 + 2C_{12}\theta\phi + \dots \end{aligned} \right\} .$$

We have just proved that the whole energy is the sum of the energies of the principal oscillations. Let us therefore find the whole energy of that principal oscillation whose type (Art. 55) is

$$\frac{\theta}{M_1} = \frac{\phi}{N_1} = \text{&c.} = \sin(p_1 t + \alpha_1),$$

where $\qquad M_1 = L_1 I_{11}(p_1^2), \quad N_1 = L_1 I_{12}(p_1^2)$ &c.

Substituting in the expression for T we find

$$2T = [A_{11}M_1^2 + 2A_{12}M_1N_1 + \dots]\, p_1^2 \cos^2(p_1 t + \alpha_1).$$

Let us indicate by the symbol T_1 the result of substituting for θ', ϕ', &c. in T the coefficients M_1, N_1, &c. of the column in Art. 53 which represents the principal oscillation whose type is $\sin(p_1 t + \alpha_1)$. Then T_2 will indicate the result of substituting M_2, N_2, &c. and so on. We see therefore that the whole *kinetic energy* of the system is

$$T_1 p_1^2 \cos^2(p_1 t + \alpha_1) + T_2 p_2^2 \cos^2(p_2 t + \alpha_2) + \text{&c.}$$

If U_1, U_2, &c. indicate the results of the same substitutions in $U - U_0$, we find that the *potential energy* of the system is

$$= -U_1 \sin^2(p_1 t + \alpha_1) - U_2 \sin^2(p_2 t + \alpha_2) - \text{&c.}$$

If we compare the expressions for the kinetic and potential energies of a principal oscillation obtained in Art. 72, we see that the coefficients of the trigonometrical terms are equal. We therefore infer that

$$T_1 p_1^2 + U_1 = 0, \quad T_2 p_2^2 + U_2 = 0, \&c. = 0.$$

Adding together the two expressions for the kinetic and potential energies we find that the *whole energy is represented by*

$$T_1 p_1^2 + T_2 p_2^2 + \ldots\ldots$$

75. We may also deduce the equation $T_1 p_1^2 + U_1 = 0$ from the equations given in Art. 50 to find M, N, &c. If we multiply these by M, N, &c. respectively (omitting the two last) and add the results, we obviously have, since λ and μ are here absent,

$$(A_{11} M^2 + 2 A_{12} MN + \ldots) p^2 + (C_{11} M^2 + 2 C_{12} MN + \ldots) = 0,$$

which is the result to be proved when written at length.

Effect of changes in the system.

76. **Effect of an increase of inertia.** Supposing the system to be oscillating about its position of equilibrium under a given set of forces, it is required to find the effect of increasing the inertia of any part of the system without altering the forces.

Let
$$\left. \begin{aligned} 2T &= A_{11} \theta'^2 + 2 A_{12} \theta' \phi' + \ldots \\ 2(U - U_0) &= C_{11} \theta^2 + 2 C_{12} \theta \phi + \ldots \end{aligned} \right\} \quad \ldots\ldots\ldots\ldots\ldots\ldots\ldots(1)$$

where the A's and C's are all given by the conditions of the question. Suppose we add on to $2T$ the quantity

$$\mu \left(\theta' + b\phi' + \&c. \right)^2,$$

it is required to find the change in the periods of oscillation.

Let us change the co-ordinate θ by writing $\theta_1 = \theta + b\phi + \&c.$, then eliminating θ we find that T and U take the forms

$$\left. \begin{aligned} 2T &= (A_{11} + \mu) \theta_1'^2 + 2 A'_{12} \theta_1' \phi' + \ldots \\ 2(U - U_0) &= \qquad C_{11} \theta_1^2 + 2 C'_{12} \theta_1 \phi + \ldots \end{aligned} \right\} \quad \ldots\ldots\ldots\ldots\ldots\ldots(2),$$

where A'_{12} &c., C'_{12} &c. are the coefficients as altered by the change of variables. The periods are now given by the determinant

$$\begin{vmatrix} (A_{11} + \mu) p^2 + C_{11}, & A'_{12} p^2 + C'_{12}, & \&c. \\ A'_{12} p^2 + C'_{12}, & & \&c. \end{vmatrix} = 0.$$

If we put $\mu = 0$, this equation gives the periods before the increase of inertia. We write this in the form $f(p^2) = 0$. Let I be the minor of the leading constituent in the determinant. Then the equation to find the altered periods is

$$u = f(p^2) + \mu p^2 I = 0.$$

We notice that I is independent of μ so that μ enters into the equation only in the first power. The coefficients of the highest powers of p^2 in $f(p^2)$ and I are the first and second discriminants of T and are therefore both positive, Art. 60.

Let the roots of $f(p^2) = 0$ be p_1^2, p_2^2, &c., and the roots of $I = 0$ be q_1^2, q_2^2, &c., both series being arranged in descending order of magnitude.' The roots of $I = 0$ separate those of $f(p^2) = 0$ by Art. 58, hence the terms of the series $p_1^2, q_1^2, p_2^2, q_2^2,$ &c. are arranged in descending order. The case in which some of these quantities are equal may be regarded as the limit of the case in which they are all different, however small those differences may be. Since all the oscillations of the system are real the values of $p_1{}^2$, $p_2{}^2$ &c. are positive.

In order to discover how the roots of the equation $u = 0$ have been altered by the introduction of μ, we put p^2 in succession equal to $p_1{}^2$, $p_2{}^2$, &c. We see that u takes the sign of I and is therefore alternately positive and negative, beginning with a positive value. Thus u now vanishes for values of p^2, the greatest of which lies between $p_1{}^2$ and $p_2{}^2$, the next greatest between $p_2{}^2$ and $p_3{}^2$ and so on. *Thus all the roots have been decreased* *.

But putting p^2 in succession equal to $q_1{}^2$, $q_2{}^2$, &c., we see that u takes the sign of $f(p^2)$ which is independent of μ. These signs are therefore the same as before the introduction of μ. It appears therefore that no value of μ can so decrease the root $p_1{}^2$ that it becomes less than $q_1{}^2$, or so decrease the root $p_2{}^2$ that it becomes less than $q_2{}^2$ and so on. *Thus the roots continue to be separated by the roots $I = 0$.*

Now I is the minor of the leading constituent in Lagrange's determinant, that is $I = 0$ is the equation which gives the periods when we introduce into the system the constraint $\theta_1 = 0$. Hence we infer that though *all the values of p^2 are decreased by an increase μ to the inertia of any part of the system, yet no increase however great can so reduce them that any one passes the corresponding value obtained by absolutely fixing the part whose inertia was increased.*

It immediately follows that if any of the periods of the system are common to the system before and after fixing the part under consideration, those periods will not be altered by the addition to the inertia.

77.   Ex. 1.   If the force function be increased by a positive quantity

$$\mu\, (\theta + b\phi + \&c.)^2$$

prove that all the roots of Lagrange's determinant are decreased but continue to be separated by the roots of the minor I. The periodic times of such of the oscillations as are real are therefore all increased.

Ex. 2.   Suppose all the periods of oscillation of a system to be known and let them be indicated as usual by the values of p. Let these be p_1, p_2, &c. Suppose all the periods to be also known when some particular mode of motion is prevented and let the corresponding values of p be q_1, q_2, &c. When the constraint is partly loosened, i.e. when the system is allowed to move in the particular manner formerly restricted but with more inertia than when free, show that the periods are given by the equation $(p^2 - p_1{}^2)\,(p^2 - p_2{}^2)$ &c. $+ Mp^2\,(p^2 - q_1{}^2)\,(p^2 - q_2{}^2)$ &c. $= 0$, where M is a quantity proportional to the mass added on to increase the inertia.

Ex. 3.   Let the system be referred to any co-ordinates θ, ϕ, &c., and let the inertia be increased by the addition of $\mu\,(a\theta' + b\phi' + ...)^2$. Let Δ be the discriminant of T before the addition to the inertia, and Δ' the same discriminant when bordered in the usual symmetrical manner by a, b, &c. with zero in the corner. Prove that the quantity M in Ex. (2) is given by $M = -\mu\Delta'/\Delta$.

78.   **Effect of introducing a constraint.** Supposing a system to be oscillating about a position of equilibrium with any number of independent co-ordinates θ, ϕ, &c.,

---

* Lord Rayleigh shows in his *Theory of Sound*, Vol. I., Art. 88, that any indefinitely small increment of mass is attended by a prolongation of all the natural periods or at any rate that no period is diminished. Thence by integration a similar theorem is true for any finite increment.

it is required to find the effect on the periods of introducing a geometrical relation between the co-ordinates.

Let this geometrical relation be $f(\theta, \phi,...)=0$, then since the system is in equilibrium for displacements represented by any values of θ, ϕ, &c., the coefficients of the first powers of θ, ϕ, &c. in the expansion of U will be zero. We may therefore (Art. 51) write this equation in the form $f(\theta, \phi...)=a\theta + b\phi + ... =0$.

We now use the method of indeterminate multipliers as already explained in Art. 48. We write down the equations of oscillation as if there were no geometrical constraint and then add to their right-hand sides $\lambda df/d\theta$ and $\lambda df/d\phi$, &c. In our case these additions are simply λa and λb, &c. The new determinant found by eliminating θ, ϕ, &c. and the additional unknown quantity λ will be the same as Lagrange's determinant bordered by a, b, &c. We thus have

$$\begin{vmatrix} A_{11}p^2+C_{11}, & A_{12}p^2+C_{12}......a \\ \&c. & \&c. & b \\ a & b & 0 \end{vmatrix}=0.$$

This equation will give the periods after introducing the geometrical relation between the formerly independent co-ordinates of the system.

The properties of this determinant have been discussed in Art. 64. We see that the system will have one principal oscillation fewer than it had before, and the periods of these principal oscillations will lie between or separate the periods of its former oscillations.

79. **Ex. 1.** Two independent systems whose principal co-ordinates (Art. 56) are respectively (θ, ϕ) and (ξ, η) vibrate in different periods. If they are connected by introducing a geometrical relation which may be represented by

$$a\theta + b\phi + a\xi + \beta\eta = 0,$$

show that the periods of the connected system are given by

$$\frac{a^2}{p^2-p_1{}^2} + \frac{b^2}{p^2-p_2{}^2} + \frac{a^2}{p^2-\pi_1{}^2} + \frac{\beta^2}{p^2-\pi_2{}^2} = 0,$$

where (p_1, p_2) (π_1, π_2) are the values of p for the two disconnected systems.

Ex. 2. Two independent systems referred to any co-ordinates (θ, ϕ) (ξ, η) are connected together so that the co-ordinates ϕ and ξ are made equal. If the letters have the meaning given in Art. 48 unaccented letters referring to the first and accented letters to the second, show that the periods are given by

$$(A_{11}p^2+C_{11})\begin{vmatrix} A'_{11}p^2+C'_{11}, & A'_{12}p^2+C'_{12} \\ A'_{12}p^2+C'_{12}, & A'_{22}p^2+C'_{22} \end{vmatrix} + (A'_{11}p^2+C'_{11})\begin{vmatrix} A_{11}p^2+C_{11}, & A_{12}p^2+C_{12} \\ A_{12}p^2+C_{12}, & A_{22}p^2+C_{22} \end{vmatrix}=0.$$

Composition and Analysis of Oscillations.

80. The position of a system being defined by several co-ordinates x, y, &c. the oscillations of that system will be generally given by equations of the form

$$x = N_1 \sin(p_1 t + \nu_1) + N_2 \sin(p_2 t + \nu_2) + \&c.$$

with similar expressions for y, z, &c.

In order to obtain a clear insight into the changes of the motion indicated by these series it will sometimes be necessary to combine these separate oscillations or to find some simple geometrical methods of representing these terms which may enable us to realize the nature of the motion.

To obtain a geometrical representation we use a representative point whose co-ordinates whether Cartesian or polar are made to depend in some convenient manner on the co-ordinates x, y, z, &c. The motion of this representative point will then exhibit to the eye the motion of the system.

81. **Commensurable Periods.** Suppose for example we wish to trace a motion represented by $x = N \sin pt + N \sin 2pt$, the coefficients being equal in magnitude. Choosing Cartesian co-ordinates we may let the abscissa of a point P represent on any scale the time elapsed since some epoch, and let the ordinate represent the value of x. There will be no difficulty in tracing the two curves $x_1 = N \sin pt$ and $x_2 = N \sin 2pt$. Let these be the two dotted lines. We obtain the required curve by adding the ordinates corresponding to each abscissa. Let this be the continuous line.

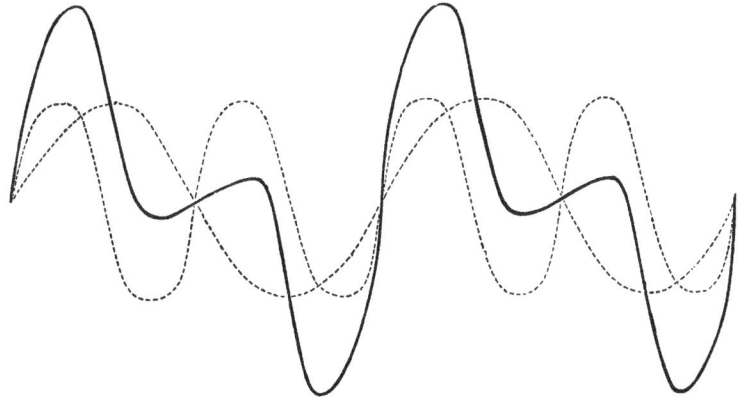

In the figure the axis of the abscissæ is not drawn. It clearly joins the two extreme points on the right and left-hand sides.

We see from a simple inspection of the figure that the motion consists of a violent oscillation to each side of the mean position followed by a very slight one and so on alternately. This figure resembles that used in Astronomy to trace the changes in the magnitude of the equation of time throughout the year.

82. Ex. 1. Show that the motion represented by $x = N \sin pt + N \sin 3pt$ consists of two large oscillations to one side of the mean position followed by two equally large ones to the other side, and so on continually.

Ex. 2. Trace the motion represented by $x = N \sin 2pt + N \sin 3pt$, and point out the difference between the two parts of the large oscillation.

83. When we combine together an *infinite number of commensurable oscillations* we obtain some interesting results by the use of Fourier's theorem. Thus, if we

examine the motion indicated by the series $y = N \sin pt - \frac{1}{2}N \sin 2pt + \frac{1}{3}N \sin 3pt - \&c.$
we shall prove that the representative point has an oscillatory motion whose period
is the same as that of the first term. This series is shown in treatises on the Integral
Calculus to be the expansion according to Fourier's theorem of $\frac{1}{2}Npt$ between the
limits $pt = -\pi$ to $pt = \pi$. Returning to the motion indicated by the series, we see
that y increases uniformly from $-\frac{1}{2}\pi N$ to $\frac{1}{2}\pi N$ during the time $2\pi/p$, and then suddenly or rapidly changes to $-\frac{1}{2}\pi N$, to repeat again its gradual increase during the
next oscillation.

As the series is convergent it will usually be sufficient to consider the motion as
represented by a limited number of terms. The expression for y is thus rendered
perfectly continuous.

84.   Ex.   Examine the motion represented by the series

$$y = N \sin pt + \frac{1}{3}N \sin 3pt + \frac{1}{5}N \sin 5pt + \&c.,$$

show that the representative point rapidly changes from one side of its mean position
to the other, remaining stationary for half the period of the first term in each of
these extreme positions.

85. **Analysis of Oscillations.** When the position of a
system is indicated by the sum of a number of oscillatory terms
whose periods are commensurable it is clear that the motion continually repeats itself at a constant interval. This interval is the
least common multiple of the periods of the several oscillatory
terms. Thus this compound oscillation resembles a principal
oscillation at least in one important feature. See Art. 53. Such
a compound oscillation might even be used as a new kind of
simple or principal oscillation by the help of which more complicated oscillations of the system might be analysed.

We are thus led to perceive that the single trigonometrical
oscillation is not the only one by which we may analyze a
complicated motion. We may sometimes find it advantageous
to combine many of these oscillations into larger units to obtain a
clear idea of the motion. This may even prove to be a necessity
when the number of coexistent oscillations is infinite.

86. **Analysis by Waves.** When the surface of still water
is disturbed by throwing a stone into it, or when a piano string
or a drum head is struck at some one point, the parts of the system
remote from the impact do not begin to move at once, but appear
to wait until the effect of the impulse has reached them. In
other words, the motion appears to diverge from the centre of
disturbance in the form of waves. These waves may be taken as
new simple oscillations. The convenience of this new elementary
motion is evident, for if several disturbances are given to different
parts of the medium each will produce a wave and the actual
motion at any point is the resultant of all these waves.

87.   The following illustration will put this theory in a clearer light. Let AOB
be a tight string, such as a piano string, whose extremities A and B are fixed and
whose length $AB = 2\pi l$, and let this string be vibrating transversely about its mean

position AB. Since the deviation of each particle from its position of equilibrium will require a separate co-ordinate to express its value, it is clear that the string has an infinite number of co-ordinates. Hence, by Lagrange's rule, the deviation of each particle will be expressed by an infinite number of trigonometrical terms. Let y represent the deviation from the straight line AB of the particle whose distance from the middle point O is x. Let the part of the string, viz. EOF, bounded by $x = -\epsilon$ and $x = +\epsilon$ be plucked aside and arranged so as to form the curve $y = f(x)$, the rest of the string being undisturbed, and let the whole string start from rest. By Fourier's theorem we may represent this initial state of the string by an equation which may sometimes be written in the form

$$y = 2 \left\{ N_1 \sin \frac{x}{l} + N_2 \sin 2\frac{x}{l} + N_3 \sin 3\frac{x}{l} + \&\mathrm{c.} \right\} \dots \dots \dots (1).$$

It will be shown in another chapter that the motion of the string at the time t is given by

$$y = 2 \left\{ N_1 \sin \frac{x}{l} \cos pt + N_2 \sin 2\frac{x}{l} \cos 2pt + \&\mathrm{c.} \right\} \dots \dots \dots (2),$$

where p is a constant which depends on the nature of the string.

Since the particles of the string are oscillating about their positions of equilibrium, their motions may be resolved into Lagrangian oscillations which of course are represented by the several terms of this series. Taking any one periodical term by itself (say the one containing $\cos \kappa pt$) we see that all the characteristics of a principal oscillation are satisfied. Thus the displacement of any one particle (defined by $x = x_1$) bears a ratio to the displacement of any other (defined by $x = x_2$) which is equal to $\sin \frac{\kappa x_1}{l} \Big/ \sin \frac{\kappa x_2}{l}$, and is therefore constant throughout the motion, Art. 53. In other words the phases of the oscillations of all the particles are the same.

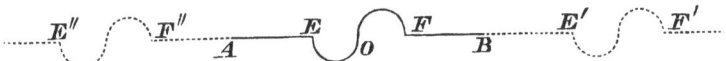

But if we recur to the expression (2) and examine how the string appears to move, we find something very different. If we trace the curve

$$y = N_1 \sin \frac{x}{l} + N_2 \sin \frac{2x}{l} + \&\mathrm{c.} \dots \dots \dots (3),$$

we find it represented in the accompanying figure. We have $y = 0$ for all values of x except those which lie between $x = 2il\pi \pm \epsilon$ where i is any integer; between these limits we have $y = \frac{1}{2}f(x)$. Since $2\pi l$ is the length of the string, x is practically limited to lie between $OA = -\pi l$ and $OB = \pi l$. This portion is represented by the thick line, while the dotted line exhibits the form of the curve for all values of x and should of course be continued to infinity on both the right and left-hand sides.

Comparing equations (1) and (3) we see that the form of the string at the time $t = 0$ is represented by the portion of this curve between A and B, the ordinates being doubled. To discover the motion at the time t, we write the equation (2) in the form

$$y = \Sigma N_\kappa \sin \kappa \left(\frac{x}{l} + pt \right) + \Sigma N_\kappa \sin \kappa \left(\frac{x}{l} - pt \right).$$

The first of these series may be derived from (3) by writing $x + lpt$ for x. This may be represented by moving the curve towards the left a distance equal to lpt, the origin O being fixed. Thus the disturbance EF travels towards the end A of the string and passes off, a new disturbance $E'F'$ entering the string at B. The second series may be represented by moving an equal and similar curve to the right of O

through a distance equal to lpt. The sum of the ordinates of these two curves represents the displacement at the time t of that particle of the string whose position in equilibrium is the foot of the ordinate.

Thus the original single disturbance has separated into two disturbances, one of which travels to the right and the other to the left. Each travels without change of form and with uniform velocity. This wave-like motion may be treated as a simple motion, by means of which we may construct other more complicated wave-motions. In this new simple oscillation all the particles have the same period, but they are not all in the same phase. One particle is at the crest of the wave at the same instant that another is in the hollow.

The case in which the particles of the string have any initial velocities may be treated in the same way. If the elements bounded by $x = -\epsilon$ and $x = \epsilon$ have an initial velocity represented by $f(t)$, the rest of the string being undisturbed, we obtain y by simply writing dy/dt for y in equation (1) and integrating the result. If the elements be both displaced from their initial position and have initial velocities, we merely add the two separate values of y.

88. Composition of oscillations of nearly equal periods.

Trace the motion represented by $x = N_1 \sin (pt + \nu_1) + N_2 \sin (qt + \nu_2)$, *where* N_1 *and* N_2 *are both positive and* p *and* q *are nearly equal.*

In the first place, consider any time at which $pt + \nu_1$ and $qt + \nu_2$ differ from each other by an even multiple of π. At this instant the two trigonometrical terms have the same sign, and, since p and q are nearly equal, they will increase and decrease together for several oscillations, how many will depend on the nearness of p and q to each other. The value of x will therefore vary between the limits $\pm (N_1 + N_2)$. Next consider any time at which $pt + \nu_1$ and $qt + \nu_2$ differ by an odd multiple of π. The two trigonometrical terms have opposite signs and will continue to have opposite signs for several oscillations. The value of x will therefore vary between the limits $\pm (N_1 - N_2)$. We see that the motion of that part of the dynamical system which depends on the co-ordinate x undergoes a periodic change of character. At one time, this part of the system is oscillating with an arc $N_1 + N_2$, after an interval equal to $\pi/(p - q)$, the arc of oscillation is $N_1 - N_2$. If N_1 and N_2 are nearly equal, this last may be so small, that the motion is invisible to the eye. Thus there will be alternate periods of comparative activity and rest. This alternation is sometimes called *beats*.

89. Transference of Oscillations.

When a system has two degrees of freedom, two co-ordinates x and y will be necessary to determine its position in space. Suppose the oscillation of x to be given by exactly the same expression as before, while that of y is the same with the opposite sign given to N_2. Let us also suppose that N_1 and N_2 are nearly equal. Each of these co-ordinates will have alternate periods of comparative rest and comparative activity. But the period of rest in one will synchronise with the period of activity in the other co-ordinate. If now the visible motion of one part of the system depend on x

and the visible motion of another on y, these parts will be in alternate rest and oscillation. Thus *there will appear to be a transference of energy from one part of the system to another and back again.*

90. This peculiarity of the resultant of two oscillations of nearly equal periods renders it important to determine when two roots of Lagrange's determinant are nearly equal. This point however has been practically discussed in Art. 62. It is there shown that when two roots are equal every first minor must be zero. If two roots are nearly equal, it follows from the principle of continuity that every minor is nearly equal to zero. By equating to zero some minor whose roots may be found as in Art. 62, we obtain some quantities which must be nearly equal to the roots sought, if any such exist. To settle this last point we substitute these quantities in turn in Lagrange's determinant and in the other minors. If all these nearly vanish for any one of these substitutions there will be nearly equal roots in Lagrange's determinant and these will be nearly equal to the quantity substituted.

91. **Composition of Oscillations of very unequal periods.** *Trace the motion represented by* $x = N_1 \sin(pt + \nu_1) + N_2 \sin(qt + \nu_2)$ *where* N_1 *and* N_2 *are both positive and* p *is small compared with* q.

In this case $qt + \nu_2$ increases by 2π, while $pt + \nu_1$ alters only by $2\pi p/q$, so that the second trigonometrical term goes through all its changes while the first is only very slightly altered. The system will therefore appear to oscillate about a mean position determined by the instantaneous value of the first trigonometrical term. Thus *the oscillations will appear to be simply harmonic with a period* $2\pi/q$ *and an extent of oscillation equal to* N_2. *At the same time the apparent mean position will travel slowly, first to one side and then to the other of the real mean, in the comparatively long period* $2\pi/p$.

92. **Resultant Oscillation.** We may compound any number of oscillations represented by the terms of the series

$$x = N_1 \sin(p_1 t + \nu_1) + N_2 \sin(p_2 t + \nu_2) + \&c. \quad\quad\quad (1)$$

in the following manner.

Let n be a quantity to be chosen at our convenience, and let $p_1 = n + q_1, p_2 = n + q_2, \&c.$ Suppose the resultant oscillation to be represented by

$$x = R \sin(nt + \rho) \quad\quad\quad\quad\quad (2),$$

then we have

$$\left. \begin{array}{l} R \cos \rho = \Sigma N \cos(qt + \nu) \\ R \sin \rho = \Sigma N \sin(qt + \nu) \end{array} \right\} \quad\quad\quad (3),$$

whence R and ρ may be found without difficulty.

This method of compounding oscillations is of great advantage when *their periods are equal*. In this case all the p's are equal, and by choosing $n = p$ we have all the q's equal to zero. We thus replace the series (1) by the simple harmonic form (2) in which R and ρ are absolute constants.

If the *periods are nearly equal*, we can choose n so that all the q's are small. The values of the elements R and ρ will now vary, but only slowly. The resultant oscillation is therefore very nearly a harmonic one. The elements of the resultant oscillation, being found at any one moment, will be nearly constant for a considerable time, and their small changes all follow known laws. These laws are determined by equation (3). We may thus still obtain a clearer insight into the changes of the values of x by examining the single term (2) than the series (1).

93. Geometrical Construction. We may represent any oscillation such as $x = N \sin (pt + \nu)$ by a simple geometrical construction which is sometimes useful. From any origin O draw a straight line OA whose length shall represent N on any scale we please, and let ν be the inclination of OA to a straight line OL fixed in space. We may call OL the axis of reference. With centre O and radius equal to OA describe a circle. If a particle P, starting from A, describe this circle with a uniform angular velocity equal to p it is clear that the distance of P from the axis of reference is equal to $N \sin (pt + \nu)$. Thus, by the help of this circle, when the straight line OA is given, the whole oscillation is determined. We may therefore *by a straight line* OA *represent any harmonic oscillation*.

In this manner we may replace the oscillations to be compounded by a series of straight lines OA_1, OA_2, &c. The circles on OA_1, OA_2, &c. are to be described by points P_1, P_2, &c., and the sum of their distances from the axis of reference is the quantity to be represented by the resultant oscillation. Let us also for the sake of simplicity, suppose that the periods are all equal, so that the q's in equations (3) are all zero.

Let OB represent the resultant of OA_1, OA_2, &c. found by the "parallelogram law," i.e. found as if OA_1, OA_2, &c. were forces to be compounded as in statics. Then by interpretation of equations (3) we see that OB will represent the resultant oscillation.

We may therefore find the resultant of any number of oscillations in the same co-ordinate, if of equal periods, by a geometrical construction. Representing each oscillation by a straight line, the resultant is found by compounding these straight lines according to the "parallelogram law."

94. Examples on Transference of Oscillations. Ex. 1. A uniform rod AB is suspended from a fixed point O by a short rod OC which is attached to it at right angles at its middle point. Equal weights are suspended from A and B by strings of equal lengths, the whole system forming a somewhat sluggish balance. If one weight be drawn slightly aside from the vertical and allowed to oscillate, the system starting from rest, find the subsequent motion *.

---

* D. Bernoulli in the *Nova Commen. Petrop.* Vol. XIX. p. 281 describes an experiment which he made on the motion of pendulums. Happening to pull aside one scale of a rather sluggish balance he noticed that it immediately began to swing to and fro, but that the opposite scale was not disturbed. Shortly however the latter scale began to move and to make sensibly greater and greater oscillations while the first scale gradually lost its oscillatory motion. At length the two appeared to have interchanged their motions, the scale first disturbed being almost at rest when the other attained its greatest extent of oscillation. The same movements were then repeated in the opposite order until the first scale had resumed its original motion and the second was again at rest.

Euler contributes two papers to the same volume of the Petersburgh memoirs with the object of explaining theoretically the cause of the motions observed by Bernoulli. In his first paper he assumes that the point of support of the balance lies in the straight line joining the points of attachment of the strings and finds that the motions observed by Bernoulli do not occur. He thus fails to find the explanation. In his second paper he rejects this limitation and has better success.

Let $CA = CB = a$, let l be the length of either of the strings AP, BQ. Let G be the centre of gravity of the beam, $OG = c$. Let Mk^2 be the moment of inertia of the beam about O, m the mass of either scale treated as a particle. Let $OC = b$.

Let ϕ, η, θ be the inclinations to the vertical of OC and the strings AP, BQ respectively; mP and mQ the tensions of the strings. Let Ox, Oy be horizontal and vertical axes, (x, y), (x_1, y_1) the coordinates of P and Q. The equations of motion are then

$$\left. \begin{aligned} x &= b\phi + a + l\eta \\ y &= b - a\phi + l \end{aligned} \right\} \dots\dots(1), \qquad \left. \begin{aligned} x_1 &= b\phi - a + l\theta \\ y_1 &= b + a\phi + l \end{aligned} \right\} \dots\dots(2),$$

$$\left. \begin{aligned} b\phi'' + l\eta'' &= -g\eta \\ b\phi'' + l\theta'' &= -g\theta \end{aligned} \right\} \dots(3), \qquad \left. \begin{aligned} -a\phi'' &= -P+g \\ a\phi'' &= -Q+g \end{aligned} \right\} \dots(4),$$

$$Mk^2\phi'' = -Mcg\phi - mP(b\phi + a) + mP\eta b + mQ(-b\phi + a) + mQ\theta b,$$

where accents as usual denote differential coefficients with regard to the time.

Eliminating P and Q, the last equation becomes

$$(Mk^2 + 2ma^2)\phi'' + (Mc + 2mb)g\phi = mbg(\eta + \theta)\dots\dots\dots\dots\dots(5).$$

To shorten the solution, we write

$$\frac{b}{l} = h, \qquad \frac{g}{l} = n^2, \qquad \frac{(Mc + 2mb)g}{Mk^2 + 2ma^2} = p^2, \qquad \frac{mbg}{Mk^2 + 2ma^2} = q.$$

The equations (3) and (5) then become

$$\left. \begin{aligned} h\frac{d^2\phi}{dt} + \left(\frac{d^2}{dt^2} + n^2\right)\eta &= 0 \\ h\frac{d^2\phi}{dt^2} + \left(\frac{d^2}{dt^2} + n^2\right)\theta &= 0 \\ \left(\frac{d^2}{dt^2} + p^2\right)\phi - q(\eta + \theta) &= 0 \end{aligned} \right\} \dots\dots\dots\dots\dots\dots\dots\dots(6).$$

Eliminating η and θ we obtain an equation to determine ϕ. To solve this we put $\phi = C\cos\mu t$ and thence find that μ must satisfy the quadratic

$$(\mu^2 - n^2)(\mu^2 - p^2) - 2hq\mu^2 = 0 \dots\dots\dots\dots\dots\dots(7).$$

In the *Cambridge Mathematical Journal*, Vol. II. p. 120 there is a paper signed D. G. S. on the sympathy of pendulums with special reference to Bernoulli's problem. Owing to numerical errors in all these investigations, the results obtained do not properly illustrate Bernoulli's problem. For instance Euler substitutes for the tensions of the strings, in the large as well as in the small terms, the weights of the scales and this substitution is also made by the writers in the *Cambridge Journal*.

If $\mu_1{}^2$, $\mu_2{}^2$ are the roots of this quadratic, the values of ϕ, η, θ are easily seen to be

$$\phi = C_1 \cos \mu_1 t + C_2 \cos \mu_2 t$$

$$\theta = \frac{C_1 h \mu_1{}^2}{n^2 - \mu_1{}^2} \cos \mu_1 t + \frac{C_2 h \mu_2{}^2}{n^2 - \mu_2{}^2} \cos \mu_2 t - C \cos nt$$

$$\eta = \text{same} \qquad + \text{same} \qquad + C \cos nt.$$

To find the values of the constants C, C_1, C_2 we have the initial conditions

$$\phi = 0, \quad \theta = 0, \quad \eta = \epsilon; \quad \phi' = 0, \quad \theta' = 0, \quad \eta' = 0.$$

We therefore find

$$\phi = \frac{\epsilon}{2h} \frac{(n^2 - \mu_1{}^2)(n^2 - \mu_2{}^2)}{n^2 (\mu_1{}^2 - \mu_2{}^2)} (\cos \mu_1 t - \cos \mu_2 t)$$

$$\theta = \frac{\epsilon}{2} \frac{(n^2 - \mu_2{}^2) \mu_1{}^2 \cos \mu_1 t - (n^2 - \mu_1{}^2) \mu_2{}^2 \cos \mu_2 t}{n^2 (\mu_1{}^2 - \mu_2{}^2)} - \tfrac{1}{2}\epsilon \cos nt$$

$$\eta = \text{same} + \tfrac{1}{2}\epsilon \cos nt.$$

In a pair of ordinary scales, m/M and therefore q will sometimes be small and $h = b/l$ will generally be small. Without assuming either of these to be small, we shall suppose that the product hq is small. We then find from (7)

$$\mu_1{}^2 - n^2 = \frac{2hqn^2}{n^2 - p^2}, \qquad \mu_2{}^2 - p^2 = -\frac{2hqp^2}{n^2 - p^2} \dots\dots\dots\dots\dots\dots(8).$$

Substituting and keeping only the principal terms, the values of ϕ, θ, η become

$$\left. \begin{aligned} 2\phi &= -\frac{2q\epsilon}{n^2 - p^2} (\cos \mu_1 t - \cos \mu_2 t) \\ 2\theta &= \epsilon \cos \mu_1 t - \epsilon \cos nt \\ 2\eta &= \epsilon \cos \mu_1 t + \epsilon \cos nt \end{aligned} \right\} \dots\dots\dots\dots\dots (9),$$

provided n^2 and p^2 are not so nearly equal that their difference is of the order hq.

Looking at the expressions for θ and η we see by the reasoning in Arts. 88 and 89 that the transference of oscillations from one scale to the other will take place in the manner described by Bernoulli.

We also notice that the beam will remain stationary if q, i.e. m/M be small whatever h, i.e. b/l may be. On the other hand the beam will oscillate if h is small but m/M not small.

Before leaving the discussion of the equation (7) we may remark that it gives the condition of stability of an ordinary balance, when a balance is disturbed it should return readily to its horizontal position. The beam oscillates about its position of equilibrium and the quicker the oscillation the more readily can it be determined by the eye whether the mean position of the beam is or is not horizontal. The balance should therefore be so constructed that the two times of oscillation are as short as possible. These times of oscillation are obviously $2\pi/\mu_1$ and $2\pi/\mu_2$ and hence μ_1 and μ_2 must be as large as possible. This requires that both n^2 and p^2 should be large, i.e. (1) the time of oscillation of either particle suspended from a fixed point by its string should be short; (2) the time of oscillation about the fulcrum O of the rigid body formed by attaching the particles to the extremities of the rods and removing the strings should be short.

Ex. 2. Supposing one scale of the balance described in the last example to be acted on by a small periodic force equal to $fl \cos \lambda t$ in a direction parallel to the arm AB, prove (1) that if λ is nearly equal to n, a large oscillation will be produced in the scales while the arm will not be much disturbed, (2) that if λ is nearly equal to μ_1 or μ_2 there will be large oscillations in all the parts of the system.

Ex. 3. A rod AB, length $2a$, can turn freely round a vertical axis through its centre of gravity C which bisects AB. At A and B are suspended two equal particles, each of mass m, by unequal strings of lengths l and l'. One of these strings is now slightly displaced through an angle ϵ in a plane perpendicular to the vertical plane through the rod. Find the motion.

If z, $z+x$, $z+y$ be the horizontal displacements at the time t of the extremity A and of the two particles respectively and if $g/l = n^2$, $g/l' = n'^2$, $ma^2/Mk^2 = p^2$, prove that

$$x = \frac{\lambda^2 C}{n^2 - \lambda^2} \cos \lambda t + \frac{\lambda'^2 C'}{n^2 - \lambda'^2} \cos \lambda' t \qquad y = \frac{\lambda^2 C}{n'^2 - \lambda^2} \cos \lambda t + \frac{\lambda'^2 C'}{n'^2 - \lambda'^2} \cos \lambda' t$$

$$z = C \cos \lambda t + C' \cos \lambda' t + C'',$$

where λ^2, λ'^2 are the roots of the quadratic

$$(\lambda^2 - n^2)(\lambda'^2 - n'^2) - (\lambda^2 - n^2) p^2 n'^2 - (\lambda^2 - n'^2) p^2 n^2 = 0.$$

The conditions that there may be a complete transference of oscillation from one string to the other are (1) λ and λ' must be nearly equal, (2) the coefficients of $\cos \lambda t$ and $\cos \lambda' t$ in the expression for x and y must be nearly equal, Arts. 88 and 89. Show that these conditions require that n and n' should be nearly equal and p small.

If the lengths of the strings are equal prove that

$$2x = \epsilon (\cos nt + \cos \lambda' t), \quad 2y = \epsilon (\cos nt - \cos \lambda' t) \quad 2z (1 + 2p^2) = - 2p^2 (\cos \lambda' t - 1),$$

where $\lambda'^2 = n^2 (1 + 2p^2)$. Thence show that the conditions for complete transference are satisfied if p is small.

Ex. 4. The middle points of two equal rods AB, $A'B'$, are fixed at C, C', about which they are capable of turning freely in one plane, the rods being without mass and the length of either rod small compared with CC'. Four particles of equal mass are placed at A, B, A', B'; and A and B', A' and B mutually attract each other, A and A', B and B' mutually repel each other according to the law of the inverse square. Prove that the rods will be in stable equilibrium when they lie in the same straight line, two mutually attracting particles being between C and C', and that if they be slightly disturbed the system will have a double oscillation whose periods are $2\pi (4c^3/\mu)^{\frac{1}{2}}$ and $2\pi (4c^3/3\mu)^{\frac{1}{2}}$ respectively ; μ being the absolute force of any particle and $CC' = 2c$. [Coll. Exam.

Show also that there will be no complete transference of motion from one bar to the other.

Ex. 5. Determine the small motions (in the magnetic meridian) of two permanent bar-magnets of equal mass suspended each by its extremities by parallel strings, all four of equal length, from points in a horizontal line, the mutual action of the magnets being slight compared with the other forces. The magnets being at rest, one only is set in motion, show that its whole energy will in time be communicated to the other. [Math. Tripos, 1875.

CHAPTER III.

The Energy Test of Stability.

95. It has been proved in Vol. I. that, when we know one first integral of the equations of motion of a system disturbed from a position of equilibrium, such as the equation of energy, we may sometimes from that one integral determine whether the position of equilibrium is stable or not. Thus, when the potential energy is a minimum in the position of equilibrium, it immediately follows from the equation of vis viva that the position of equilibrium is stable. But when the potential energy is not a minimum, the equation of vis viva alone is not sufficient to determine whether the equilibrium is stable or unstable. But by taking into consideration the other equations of motion this position of equilibrium is proved to be unstable.

We may apply an "energy test" of stability to a given state of motion as well as to a given position of equilibrium, but with a similar limitation. When a certain function derived from such of the first integrals as we may happen to know is an absolute minimum or maximum we may be able to prove that the system cannot depart far from the given state of motion. But when that function is neither a maximum nor a minimum we only infer that there is apparently nothing in these equations to restrict the deviations of the system. To determine this point we must examine more minutely the equations we already have or we must discover the remaining equations of motion. This latter part of the question will therefore be postponed until we discuss the oscillations about a state of motion. Meantime we shall consider the "energy test" with a view to determine how far it can be made to decide the question of stability.

96. **Stability of a State of Motion.** *Let a dynamical system be in motion in any manner under a conservative system of forces, and let* E *be its energy. Then* E *is a known function of the co-ordinates* θ, ϕ, &c. *and their first differential coefficients*

θ', ϕ', &c.: *this is constant and equal to* h *for the given motion. Suppose that either some or all of the other first integrals of the equations of motion are also known, let these be*

$$F_1(\theta,\ \theta',\ \&c.) = C_1, \quad F_2(\theta,\ \theta',\ \&c.) = C_2, \quad \&c. = \&c.$$

For the purposes of this proposition, let us regard θ *and* θ', ϕ *and* ϕ', &c. *as independent variables, except so far as they are connected by the equations just written down. Then, if* E *be an absolute maximum, or an absolute minimum, for all variations of* θ, θ', &c. *(those corresponding to the given motion making* E *constant), the motion is stable for all disturbances which do not alter the constants* C_1, C_2, &c.

Let as many of the letters as is possible be found from the first integrals in terms of the rest, and substituted in the expression for E. Let ψ, ψ', &c. be these remaining letters, then we have

$$E = f\,(\psi,\ \psi',\ \&c.,\ C_1,\ C_2,\ \&c.) = h.$$

Let the system be started in some manner slightly different from that given, then the constant h is altered into $h + \delta h$. First let E be a minimum along the given motion, then any change whatever of the letters ψ, ψ', &c. increases E, and it follows that the disturbed motion cannot deviate so far from the given motion that the change in E becomes greater than δh. Similarly, if E be an absolute maximum, the same result follows.

The same argument will apply to any first integral of the equations of motion, besides the energy integral. If any one of the functions F_1, F_2, &c., which contains all the letters, be an absolute maximum or minimum, then the motion is stable for all displacements which do not alter the constants of the other integrals used.

97.   When the system is disturbed from a position of equilibrium which is defined, as in Vol. I., by the vanishing of the co-ordinates θ, ϕ, &c., we have

$$E = \tfrac{1}{2}A_{11}\theta'^{2} + A_{12}\theta'\phi' + \&c. - U$$

where A_{11}, A_{12}, &c. are all constants, and U is independent of θ', ϕ', &c.   Here the terms which constitute the kinetic energy, being necessarily positive and vanishing with θ', ϕ', &c., are evidently a minimum for all variations of θ', ϕ', &c.   We see, without the use of any other integrals, that if $-U$ be a minimum for all variations of θ, ϕ, &c., E is an absolute minimum, and that therefore the equilibrium is stable.

In what follows a similar result will be obtained when the system is disturbed from a state of steady motion. It will be shown that, when a function represented by $F - U$ is a minimum under certain conditions, this state of steady motion is stable under the same conditions. The function F of course reduces to zero when the state of motion reduces to a state of rest.

98. To find a steady motion. It often happens that the motion whose stability is in question is a state of steady motion. This generally occurs when some of the co-ordinates are absent from the Lagrangian function, though present in the form of velocities. Let us represent by x, y, &c. the co-ordinates which are absent from the Lagrangian function, and let ξ, η, &c. be the remaining co-ordinates. Thus the Lagrangian function L will be a function of ξ, ξ', η, η', &c., x', y', &c., but not of x, y, &c. The Lagrangian equations will therefore take the forms

$$\frac{d}{dt}\frac{dL}{d\xi'} = \frac{dL}{d\xi} \text{ \&c.,} \quad \frac{dL}{dx'} = u, \quad \frac{dL}{dy'} = v, \text{ \&c.}$$

where u, v, &c. are constants introduced by integration. These equations will contain ξ, ξ', ξ'', η, η', η'', &c., x', x'', y', y'', &c., and do not contain t explicitly. They may therefore be satisfied by putting $x' = a$, $y' = b$, &c., $\xi = \alpha$, $\eta = \beta$, &c., where a, b, &c., α, β, &c. are constants to be determined by substituting in the equations. If θ stand for any one of the co-ordinates, it is evident that $dT/d\theta$ and $dT/d\theta'$ will both be constants after the substitution is made. Omitting the equations which contain u, v, &c., as they do not assist in finding the constants a, b, &c., α, β, &c.

we have the equations $$\frac{dL}{d\xi} = 0, \quad \frac{dL}{d\eta} = 0, \quad \text{\&c.} = 0 \dots\dots\dots\dots\dots\dots\dots\dots(1),$$

where $L = T + U$. Thus we have as many equations as there are co-ordinates ξ, η, &c. directly present (i.e. not merely present as velocities) in the expressions for T and U. The quantities a, b, &c. are therefore undetermined except by the initial conditions, while α, β, &c. may be found in terms of a, b, &c. by these equations. These equations may be conveniently remembered by the following rule.

In the Lagrangian function, which is the difference between the kinetic and potential energies, write for all the differential coefficients their assumed constant values in the steady motion, viz. $\mathrm{x}' = a$, $\mathrm{y}' = b$, *&c.,* $\xi' = 0$, $\eta' = 0$, *&c. The Lagrangian function is now a function of the co-ordinates* ξ, η, *&c. only. Differentiating this result partially with regard to each of these co-ordinates and equating the results to zero, we obtain the equations of steady motion.*

99. Stability of a steady motion. To determine if this motion is stable we use the method indicated in Art. 96. The equation of energy may be written in the form $$E = T - U = h.$$

Since T is not a function of the co-ordinates x, y, &c. the Lagrangian equations for these co-ordinates lead as before to the integrals $dT/dx' = u$, $dT/dy' = v$, &c., where u, v, &c. are constants. By the help of these integrals we shall eliminate x', y', &c., and thus obtain E as a function of the other co-ordinates. If E be an absolute maximum or minimum, this motion is stable for all disturbances which do not alter the constants u, v, &c. There can be no difficulty in effecting the elimination in any particular case, but we may perform the process once for all. The process is a repetition of that called *Modification* in Vol. I.

To effect the elimination, let

$$T = \tfrac{1}{2}(xx)\,x'^2 + (x\xi)\,x'\xi' + \text{\&c.} \quad \dots\dots\dots\dots\dots\dots(2),$$

where the coefficients of the accented letters, viz. the quantities in brackets, are all known functions of ξ, η, &c., but not of x, y, &c. The integrals may then be written in the form

$$\begin{aligned}
(xx)\,x' + (xy)\,y' + \dots &= u - (x\xi)\,\xi' - (x\eta)\,\eta' - \text{\&c.} \\
(xy)\,x' + (yy)\,y' + \dots &= v - (y\xi)\,\xi' - (y\eta)\,\eta' - \text{\&c.} \\
\text{\&c.} &= \text{\&c.}
\end{aligned} \right\} \quad \dots\dots\dots\dots(3).$$

For the sake of brevity, let us call the right-hand sides of these equations $u - X$,

$v - Y$, &c.   Since T is a quadratic function of the accented letters, we may write it in the form

$$T = \tfrac{1}{2} (\xi\xi) \, \xi'^2 + (\xi\eta) \, \xi'\eta' + \&c. + \tfrac{1}{2} x' \, (u + X) + \tfrac{1}{2} y' \, (v + Y) + \&c.$$

If we substitute in the terms after the first &c. the values of x', y' given by (3), we obtain the result

$$T = \tfrac{1}{2} (\xi\xi) \, \xi'^2 + (\xi\eta) \, \xi'\eta' + \&c. - \frac{1}{2\Delta} \begin{vmatrix} 0, & u+X, & v+Y, & \&c. \\ u-X, & (xx), & (xy), & \&c. \\ v-Y, & (xy), & (yy), & \&c. \\ \&c. & & & \end{vmatrix}$$

where Δ is the discriminant of T, when ξ', η', &c. have been put zero. If we change the signs of X, Y, &c., this determinant is unaltered, hence when expanded such terms as uX, vX, &c. cannot occur. If therefore, we put

$$F = - \frac{1}{2\Delta} \begin{vmatrix} 0 & u & v & \dots \\ u & (xx) & (xy) & \dots \\ \dots & \dots & \dots & \dots \end{vmatrix} \qquad \dots\dots\dots\dots\dots\dots\dots\dots(4),$$

and expand the first determinant, we have as the result of the elimination

$$T = F + \tfrac{1}{2} B_{11} \xi'^2 + B_{12} \xi'\eta' + \dots \qquad \dots\dots\dots\dots\dots\dots(5),$$

where the terms after F express some homogeneous quadratic function of ξ', η', &c.

Now T is essentially positive for all values of x', y', &c. and therefore for such as make u, v, &c. all zero. Hence the quadratic expression $B_{11} \xi'^2 + \&c.$ is a minimum when ξ', η', &c. are zero. *If then the function* F $-$ U *is a minimum for all variations of ξ, η, &c., the steady motion given by (1) is stable for all disturbances which do not alter the momenta* u, v, *&c.*

100.   When ξ', η', &c. are put zero, the process indicated by the successive equations (2), (3), (4), (5) is exactly that described in Vol. I. as the Hamiltonian method of forming the reciprocal function of T for the co-ordinates x, y, &c. We may therefore enunciate the rule in the following manner.

Suppose a steady motion to be given by $\xi' = 0$, $\eta' = 0$, &c., x' = a, y' = b, &c., so that the momenta u, v, *&c. with regard to* x, y, *&c. are constants. Form the reciprocal function of* T *with regard to* x', y', *&c., putting zero for each of the letters ξ', η', &c. Let* F *be this reciprocal function, and* $-$ U *or* V *be the potential energy. Then if* F $-$ U *or* F $+$ V *is a minimum for all variations of ξ, η, &c. this steady motion is stable for all disturbances which do not alter the momenta* u, v, *&c.*

When the reciprocal function F has been found, we may put the equations (1) which determine the steady motion into another form. The function F is the reciprocal of T with regard to x', y', &c., and ξ, η, &c. are merely other letters present during the process of transformation, hence, as explained in Vol. I., we have $\dfrac{dT}{d\xi} = - \dfrac{dF}{d\xi}$ with similar equations for η, &c. *The equations of steady motion* (1) *therefore become*

$$\frac{d (F - U)}{d\xi} = 0 \qquad \frac{d (F - U)}{d\eta} = 0$$

$$x' = \frac{d (F - U)}{du} \qquad y' = \frac{d (F - U)}{dv} \Bigg\} \qquad \dots\dots\dots\dots\dots\dots(6),$$

where F $-$ U *or* F $+$ V *is the energy expressed as a function of the momenta* u, v, *&c. instead of* x', y', *&c., the other accented letters ξ', η', &c. being put equal to zero either before or after the differentiation.*

101.   **Special case of Motion.**   *If the energy be a function of one only of the co-ordinates, though it is a function of the differential coefficients of all of them, we*

may show conversely that the steady motion will not be stable unless F − U *is a minimum.*

Let ξ be this single co-ordinate, then, following the same notation as before, we have by vis viva
$$\tfrac{1}{2} B_{11} \xi'^2 + F - U = h.$$

Differentiating with regard to t, and treating B_{11} as constant because we shall neglect the square of ξ', we obtain
$$B_{11} \xi'' + \frac{d}{d\xi}(F - U) = 0.$$

To find the oscillation, let $\xi = a + p$, then by (6) we have
$$B_{11} \frac{d^2 p}{dt^2} + \left[\frac{d^2 (F - U)}{d\xi^2}\right] p = 0,$$

where a is to be written for ξ after differentiation in the quantity in square brackets. The motion is clearly stable or unstable according as the coefficient of p is positive or negative, i.e. according as $F - U$ is a minimum or maximum.

Further information on this subject will be found in the author's *Essay on the Stability of Steady Motion*, 1877.

102. Examples of stability of motion. Ex. 1. Let us consider the simple case of a particle describing a circular orbit about a centre of attraction whose acceleration at a distance r is μr^n. If θ be the angle the radius vector r makes with the axis of x, we have here a steady motion in which $r' = 0$ and θ' is constant. Also
$$E = \tfrac{1}{2}(r'^2 + r^2 \theta'^2) + \frac{\mu r^{n+1}}{n+1}.$$

We notice that θ is absent from this expression, hence by the rule we eliminate θ' also by the integral $r^2 \theta' = h$, where h is the constant called u in Art. 99. We have then
$$E = \tfrac{1}{2} r'^2 + \tfrac{1}{2} \frac{h^2}{r^2} + \frac{\mu r^{n+1}}{n+1}.$$

Putting the remaining accented letters equal to zero according to the rule, we have in steady motion
$$\frac{dE}{dr} = -\frac{h^2}{r^3} + \mu r^n = 0,$$

and, since
$$\frac{d^2 E}{dr^2} = \frac{3h^2}{r^4} + \mu n r^{n-1} = \mu (n+3) r^{n-1},$$

this steady motion is stable or unstable according as $n+3$ is positive or negative for all disturbances which do not alter the angular momentum of the particle.

Ex. 2. A top, two of whose principal moments at the vertex O are equal, turns about its vertex under the action of gravity. If OC be the axis of unequal moment, and θ, ϕ, ψ the Eulerian angular co-ordinates of the body referred to a vertical axis measured upwards, we have (as in the chapter on vis viva, Vol. I.)
$$2T = A\,(\theta'^2 + \sin^2\theta \psi'^2) + C\,(\phi' + \psi' \cos \theta)^2$$
$$U = -Mgh \cos \theta + \text{constant},$$

where h is the distance of the centre of gravity from O, and M is the mass of the top. We have therefore the two integrals $\phi' + \psi' \cos \theta = n$ and $Cn \cos \theta + A \sin^2\theta \psi' = m$, where n and m are two constants, the former representing the angular velocity of the top about its axis and the latter the angular momentum about the vertical. By eliminating ϕ' and ψ' and making the energy E a minimum, show (1) that a state of steady motion, with real values of the constants m and n, is given by $\theta = a$ provided $C^2 n^2 - 4MghA \cos a$ is positive. Show (2), by examining the sign of $d^2 E / d\theta^2$, that this motion is stable. Thus the axis of the top will describe a right cone of semi-angle α round the vertical through the point of support with an angular velocity given by the value of ψ'.

Ex. 3. A solid of revolution moves in steady motion on a smooth horizontal plane, so that the inclination θ of its axis to the vertical is constant. Prove that the angular velocity μ of the axis about the vertical is given by

$$\mu^2 - \frac{Cn}{A\cos\theta}\mu - \frac{Mg}{A\sin\theta\cos\theta}\frac{dz}{d\theta} = 0,$$

where z is the altitude of the centre of gravity above the horizontal plane, n the angular velocity of the body about the axis, C, A and A the principal moments of inertia at the centre of gravity, and M the mass. Find the least value of n which makes μ real, and determine if the steady motion is stable.

Examples of Oscillations about Steady Motion.

103. The oscillations of a system about a state of steady motion may be found by methods analogous to those used in the oscillations about a position of equilibrium. Let the general equations of motion of the bodies be formed by any of the methods already described. If any reactions enter into these equations it will be generally found advantageous to eliminate them. Let the co-ordinates used in these equations to fix the positions of the bodies be called θ, ϕ, &c. Suppose the motion, about which the oscillation is required, to be determined by $\theta = f(t)$, $\phi = F(t)$, &c. We then substitute $\theta = f(t) + x$, $\phi = F(t) + y$, &c., in the equations of motion. The squares of x, y, &c. being neglected, we have certain linear equations to find x, y, &c. These equations can, however, seldom be solved unless we can make t disappear explicitly from them. When this can be done the linear equations can be solved by the usual known methods, and the required oscillations are then found.

In what follows we shall first illustrate the method just described by forming the equations in a few interesting cases from the beginning. We shall then generalize the process and obtain a determinantal equation analogous to that given by Lagrange for oscillations about a position of equilibrium. This equation will be adapted to all cases which lead to differential equations with constant coefficients.

104. **Theory of Watt's governor.** *To find the motion of the balls in Watt's governor of the steam engine.*

The mode in which this works to moderate the fluctuations of the engine is well known. A somewhat similar apparatus has been used to regulate the motion of clocks, and in other cases where uniformity of motion is required. If there be any increase in the driving power of the engine, or any diminution of the load, so that the engine begins to move too fast, the balls, by their increased centrifugal force, open outwards, and by means of a lever either cut off the driving power or increase the load by a quantity proportional to the angle opened out. If on the other hand the engine goes too slowly, the balls fall inward, and more driving power is called into action. In the case of the steam engine the lever is attached to the throttle-valve, and thus regulates the supply of steam. It is clear that a complete adaptation of the driving power to the load cannot take place instantaneously, but the machine will make a series of small oscillations about a mean state of steady motion. The problem to be considered may therefore be stated thus :—

Two equal rods OA, OA', each of length l, are connected with a vertical spindle by means of a hinge at O which permits free motion in the vertical plane AOA'. At A and A' are attached two balls, each of mass m. To represent the inertia of the other parts of the engine we shall suppose a horizontal fly-wheel attached to the spindle, whose moment of inertia about the spindle is I. When the machine is in uniform motion, the rods are inclined at some angle a to the vertical, and turn round it with uniform angular velocity n. If, owing to any disturbance of the motion, the rods have opened out to an angle θ with the vertical, a force is called into play whose moment about the spindle is some function of $(\theta - a)$. We may expand this function in powers of $(\theta - a)$ and, as it will be sufficient to retain only the first power, we shall represent it by $-\beta(\theta - a)$. It is required to find the oscillations about the state of steady motion.

Let ϕ be the angle which the plane AOA' makes with some vertical plane fixed in space. The equation of angular momentum about the spindle is

$$\frac{d}{dt}\left\{(I + 2mk^2\sin^2\theta)\frac{d\phi}{dt}\right\} = -\beta(\theta - a) \dots\dots\dots\dots\dots\dots(1),$$

where mk^2 is the moment of inertia of a rod and ball about a perpendicular to the rod through O, the balls being regarded as indefinitely small heavy particles. The semi vis viva of the system is

$$T = \tfrac{1}{2}I\left(\frac{d\phi}{dt}\right)^2 + mk^2\left\{\left(\frac{d\theta}{dt}\right)^2 + \sin^2\theta\left(\frac{d\phi}{dt}\right)^2\right\},$$

and the moment of the impressed forces on either rod and ball about a horizontal through O perpendicular to the plane AOA' is $\tfrac{1}{2}dU/d\theta = -mgh\sin\theta$, where h is the distance of the centre of gravity of a rod and ball from O. Hence, by Lagrange's equation $\dfrac{d}{dt}\dfrac{dT}{d\theta'} - \dfrac{dT}{d\theta} = \dfrac{dU}{d\theta}$, we have

$$\frac{d^2\theta}{dt^2} - \sin\theta\cos\theta\left(\frac{d\phi}{dt}\right)^2 = -\frac{g}{a}\sin\theta \dots\dots\dots\dots\dots\dots(2),$$

where a has been written for k^2/h. This equation might also have been obtained by taking the acceleration of either ball, treated as a particle, in a direction perpendicular to the rod in the plane in which θ is measured.

To find the steady motion we put $\theta = a$, $d\phi/dt = n$, the second equation then gives $n^2\cos a = g/a$. To find the oscillations, we put $\theta = a + x$, $d\phi/dt = n + y$. The two equations then become

$$(I + 2mk^2\sin^2 a)\frac{dy}{dt} + 2mk^2 n\sin 2a\,\frac{dx}{dt} = -\beta x \left.\vphantom{\frac{d^2x}{dt^2}}\right\}$$
$$\frac{d^2x}{dt^2} - n\sin 2a\,y = \left(n^2\cos 2a - \frac{g}{a}\cos a\right)x \quad\left.\vphantom{\frac{d^2x}{dt^2}}\right\}$$

To solve these equations, we must write them in the form

$$\left(\sin 2a\,\delta + \frac{\beta}{2mk^2 n}\right)nx + \left(\frac{I}{2mk^2} + \sin^2 a\right)\delta y = 0 \left.\vphantom{\frac{I}{2mk^2}}\right\},$$
$$(\delta^2 + n^2\sin^2 a)x - n\sin 2a\,y = 0 \left.\vphantom{\frac{I}{2}}\right)$$

where the symbol δ stands for the operation d/dt. Eliminating y by cross multiplication we have

$$\left[\left(\frac{I}{2mk^2} + \sin^2 a\right)\delta^3 + n^2\sin^2 a\left(1 + 3\cos^2 a + \frac{I}{2mk^2}\right)\delta + \frac{\beta}{2mk^2}n\sin 2a\right]x = 0.$$

The real root of this cubic equation is necessarily negative, because the last term is positive. The other two roots are imaginary because the term δ^2 has disappeared between two terms of like signs. Also, the sum of the three roots being zero, the real parts of the two imaginary roots must be positive. Let these roots therefore be $-2p$ and $p \pm q \sqrt{-1}$. Then

$$x = He^{-2pt} + Ke^{pt} \sin (qt + L),$$

where H, K, L are three undetermined constants depending on the nature of the initial disturbance. Thus it appears that the oscillation is unstable. The balls will alternately approach and recede from the vertical spindle with increasing violence.

105. The defect of a governor is therefore that it acts too quickly, and thus produces considerable oscillation of speed in the engine. If the engine is working too violently, the governor cuts off the steam, but owing to the inertia of the parts of the machinery, the engine does not immediately take up the proper speed. The consequence is that the balls continue to separate after they have reduced the supply of steam to the proper amount, and thus too much steam is cut off. Similar remarks apply when the balls are approaching each other, and a considerable oscillation is thereby produced. This of course is but an incomplete explanation, but that the oscillation thus produced is of considerable magnitude has been strictly proved in Art. 104. It will be presently shown that this fault may be very much modified by applying some resistance to the motion of the governor.

In the same way when the motion of clock-work is regulated by centrifugal balls, it is found as a matter of observation that there is a strong tendency to irregularity. If the balls once receive in the slightest degree an elliptic motion, the resistance $\beta (\theta - \alpha)$ by which the motion of the balls is regulated may tend to render the ellipse more and more elliptical. To correct this some other resistance must be called into play. This resistance should be of such a character that it does not affect the circular motion and is only produced by the ellipticity of the movement.

One method of effecting this has been suggested by Sir G. B. Airy. The elliptic motion of the balls may be made to cause a slider on the vertical spindle to rise and fall. If this be connected with a horizontal circular plate in a vertical cylinder of slightly greater radius, and filled with water, the slider may be made to move the plate up and down by its oscillations. Thus the slider may be subjected to a very great resistance, tending to diminish its oscillations, while its place of rest, as depending on statical, or slowly altering forces, is totally unaffected. *Memoirs of the Astronomical Society of London*, Vol. xx., 1851.

The general effect of the water will be to produce a resistance varying as the velocity, and may therefore be represented by a term $- \gamma d\theta/dt$ on the right hand of equation (2). The solution being continued as before, the cubic will now take the form

$$\left[\left(\frac{I}{2mk^2} + \sin^2 \alpha \right) (\delta^3 + \gamma \delta^2) + n \sin^2 \alpha \left(1 + 3 \cos^2 \alpha + \frac{I}{2mk^2} \right) \delta + \frac{\beta}{2mk^2} n \sin 2\alpha \right] x = 0.$$

If the roots of this cubic are real, they are all negative, and the value of x takes the form
$$x = Ae^{-\lambda t} + Be^{-\mu t} + Ce^{-\nu t},$$
where $-\lambda$, $-\mu$, $-\nu$ are the roots, and A, B, C are three undetermined constants. If one root only is real, that root is negative, and if the other two be $p \pm q \sqrt{-1}$ the value of x takes the form

$$x = He^{-rt} + Ke^{pt} \sin (qt + L),$$

where H, K, L as before are undetermined constants.

In order that the motion may be stable it is necessary that p should be negative. The analytical condition* for this is

$$\gamma \left(1 + 3\cos^2 a + \frac{I}{2mk^2} \right) > \frac{\beta}{2mk^2} 2 \cot a.$$

If γ be sufficiently great this condition may be satisfied. The uniformity of motion of the rods round the vertical will then be disturbed by an oscillation whose magnitude is continually decreasing and whose period is $2\pi/q$. By properly choosing the magnitude of I when constructing the instrument, the period may sometimes be so arranged as to produce the least possible ill effect. If the period be made very long the instrument will work smoothly. If it can be made very short there will be less deviation from circular motion.

In this investigation no notice has been taken of the frictions at the hinge and at the mechanical appliances of the governor, which may not be inconsiderable. These in many cases tend to reduce the oscillation and keep it within bounds.

106. In the case of Watt's governor if any permanent change be made in the relation between the driving power and the load, the state of uniform motion which the engine will finally assume is different from that which it had before the change. Thus, when the engine is driving a given number of looms, let the rods OA, OA' of the governor be inclined to each other at an angle $2a$ and be revolving about the vertical with an angular velocity n. If some large number of the looms is suddenly disconnected from the engine, the balls will separate from each other, and the rods will become inclined at some other angle $2a'$. In this case, if n' be the angular velocity about the vertical, $n'^2 \cos a' = n^2 \cos a$. The rate of the engine is therefore altered, it works quicker with a less load than with a greater. This is a great defect of Watt's governor. For this reason it has been suggested that the term *governor* is inappropriate, the instrument being in fact only a *moderator* of the fluctuations of the engine.

This defect may be considerably decreased by the use of Huyghens' parabolic pendulum. In this instrument the centres of gravity A, A' of the balls are made to move along the arc of a parabola whose axis is the axis of revolution. Let AN be an ordinate of the parabola, AG the normal, then NG is constant and equal to L, where $2L$ is the latus rectum. Regarding the balls as particles, and neglecting the inertia of the rods which connect them with the throttle valve, we see by the triangle of forces that the balls will rest in any positions on the parabola, if $n^2 L = g$, where n is the angular velocity of the balls about the vertical through O. It is also clear that when the angular velocity is not that given by this formula, the balls (unless placed at the vertex) must slide along the arc. Let us now consider how this modification of the governor affects the working of the engine. When the load is diminished the engine begins to quicken; the balls separate and the steam is cut off. It is clear that equilibrium will not be established until the quantity of steam admitted is just such as to cause the engine to move at exactly the same rate as before.

Ex. Show that when the inertia of the rod and balls are taken account of, the centre of gravity of either ball and rod must be constrained to describe a

---

* If the roots of the cubic $ax^3 + bx^2 + cx + d = 0$ be $x = a \pm \beta \sqrt{(-1)}$ and γ, we have $-b/a = 2a + \gamma$, $c/a = 2\gamma a + a^2 + \beta^2$, $-d/a = (a^2 + \beta^2)\gamma$, whence we easily deduce $(bc - ad)/a^2 = -2a\{(a + \gamma)^2 + \beta^2\}$; hence $bc - ad$ and a have always opposite signs.

parabola whose latus rectum is independent of the radius of the ball, if the governor is to cause the engine always to move at a given rate.

It should be mentioned that several other methods of avoiding this defect have been invented besides the parabolic pendulum. But any further description of these would be here out of place.

107. A speed governor, similar to that invented by Sir G. Airy, but with a spring instead of a pendulum, was described by Prof. J. A. Ewing in *Nature*, Vol. xxiii. 1881. He applied it to a clock driving a recording seismograph whose motion was required to be continuous and fairly uniform.

Another governor was invented by the brothers Siemens which is remarkable because it does not require the use of Watt's governor. A short description of it is given in the *Life of Sir William Siemens*, by William Pole, 1888; see also a paper by Mr Wood, *Institution of Civil Engineers*, March 10, 1846.

The reader who is interested in the subject of governors may refer to an article by Sir G. B. Airy, Vol. xi of the *Memoirs of the Astronomical Society*, 1840, where four different constructions are considered. He may also consult an article by Mr Siemens in the *Phil. Trans.* for 1866, and a brief sketch of several kinds of governors by *Prof. Maxwell* in the *Phil. Mag. for* 1868. An account of some experiments by *Mr Ellery*, on Huyghens' parabolic pendulum, may be found in the *Astronomical Notices for December*, 1875.

108. LAPLACE'S THREE PARTICLES. *It has been shown in Vol. I. Chap. VI., that if three particles be placed at the corners of an equilateral triangle and properly projected, they will move under their mutual attractions so as always to remain at the angular points of an equilateral triangle. These we may call Laplace's three particles. It is our present object to determine if this motion is stable or unstable.*

We shall begin by assuming that the three particles remain always very nearly at the corners of an equilateral triangle. We shall then have to determine whether their oscillations about these corners are real or imaginary. To effect this we might choose their common centre of gravity as a fixed origin of co-ordinates. But the triangles formed by joining the particles to their common centre of gravity are not marked by any simplicity of form. Instead of referring the motion to the centre of gravity it will be more convenient to reduce one of the particles to rest, and to consider the relative motion of the other two. We have thus only one triangle to examine, and that one nearly equilateral.

Let the mass M of the particle to be reduced to rest be taken as unity, and let m, m' be the masses of the other two. Let r, r', R be the distances between the particles Mm, Mm', mm'; and let ϕ', ϕ, ψ be the angles opposite to these distances. If θ, θ' be the angles which r, r' make with a straight line fixed in space, and if the law of attraction be the inverse κth power of the distance, the equations of motion are

$$\left. \begin{array}{l} \dfrac{d^2r}{dt^2} - r\left(\dfrac{d\theta}{dt}\right)^2 + \dfrac{1+m}{r^\kappa} + \dfrac{m'\cos\psi}{r'^\kappa} + \dfrac{m'\cos\phi}{R^\kappa} = 0 \\[2mm] \dfrac{1}{r}\dfrac{d}{dt}\left(r^2\dfrac{d\theta}{dt}\right) + \dfrac{m'\sin\psi}{r'^\kappa} - \dfrac{m'\sin\phi}{R^\kappa} = 0 \end{array} \right\},$$

with two similar equations for the motion of m'.

Let us now put $r = a + x$, $r' = a + x + X$, and let the angle between these radii vectores be $\frac{1}{3}\pi + Y$, also let $\theta = nt + y$, where x, y, X and Y, are all small quantities whose squares are to be neglected. It should be noticed that a variation of x, y alone, X and Y being zero, will represent a variation of steady motion in which the particles always keep at the corners of an equilateral triangle, while a variation of X, Y will represent a change from the equilateral form. The former of these by hypothesis is a possible motion, hence the equations can be satisfied by some values of x, y joined to $X=0$, $Y=0$. By this choice of variables we may hope to discover some roots of the fundamental determinant previous to expansion, and thus save a great amount of numerical labour. If δ stand for d/dt, and $b=a^{\kappa+1}$, the four equations will now become

$$\{b\delta^2 - (\kappa+1)(1+m+m')\}x - 2abn\delta y - \tfrac{3}{4}m'(\kappa+1)X - \tfrac{1}{4}\sqrt{3}\,m'(\kappa+1)\,aY = 0,$$

$$2bn\delta x + ab\delta^2 y - \tfrac{1}{4}\sqrt{3}m'(\kappa+1)X + \tfrac{3}{4}m'(\kappa+1)\,aY = 0,$$

$$\{b\delta^2 - (\kappa+1)(1+m+m')\}x - 2abn\delta y + \{b\delta^2 - (\kappa+1)(1+\tfrac{1}{4}m+m')\}X - \{2abn\delta + \tfrac{1}{4}\sqrt{3}m(\kappa+1)a\}Y = 0,$$

$$2bn\delta x + ab\delta^2 y + \{2bn\delta - \tfrac{1}{4}\sqrt{3}(\kappa+1)\,m\}X + \{ab\delta^2 - \tfrac{3}{4}m(\kappa+1)\,a\}Y = 0.$$

109. To solve these we put $x=Ae^{\lambda t}$, $y=Be^{\lambda t}$, $X=Ge^{\lambda t}$, $Y=He^{\lambda t}$. Substituting and eliminating the ratios of A, B, G and H we obtain a determinantal equation whose constituents are the coefficients of x, y, X and Y with λ written for δ. This equation will give eight values of λ. We see at once that one factor is λ. This might have been expected, because we know that a variation of y, (with x, X and Y all zero,) is a possible motion. Again, some variation of x and y, (with X and Y both zero,) is also a possible motion, hence some factor of the determinant can be found by examining the first two columns. By subtracting from the first $2n$ times the second column we find that this factor is $b\lambda^2 - (\kappa - 3)(1+m+m') = 0$.

To find the other factors we divide the determinant by the factors already found. Then, subtracting the first row from the third and the second from the fourth, we have three zeros in the first column and two in the second. The expansion is then easy. We then see that there is another factor λ, and also that

$$b^2\lambda^4 + b\lambda^2(3-\kappa)(1+m+m') + \tfrac{3}{4}(1+\kappa)^2(m+m'+mm') = 0.$$

The two zero roots give $x=A_1+A_2t$ with similar expressions y, X and Y. But by substitution in the equations of motion we see that $x=A_1$, $y=B_1-\tfrac{1}{2}(\kappa+1)A_1nt/a$, $X=0$ and $Y=0$. These roots therefore indicate merely a permanent change in the size of the triangle. On examining the other values of λ^2, we find (1) The motion cannot be stable unless κ is less than 3. (2) The motion is stable whatever the masses may be, if the law of force be expressed by any positive power of the distance or any negative power less than unity. For other powers the stability depends on the relation between the masses. (3) The motion is stable to a first

approximation if $\qquad \dfrac{(M+m+m')^2}{Mm+Mm'+mm'} > 3\left(\dfrac{1+\kappa}{3-\kappa}\right)^2,$

where M, m, m' are the masses*. To express the co-ordinates in terms of the time, we must return to the differential equations of the second order. The results are rather long, and it may be sufficient to state that when, as in the solar system, two of the masses are much smaller than the third, the inequalities in their angular

---

* In a brief note in Jullien's Problems, Vol. II. p. 29 it is mentioned that this question has been discussed by M. Gascheau in a Thèse de Mécanique, the particles being supposed to attract each other according to the law of nature. The result arrived at is that the motion is stable when the square of the sum of the masses is greater than 27 times the sum of the products of the masses taken two and two. No reference is given to where M. Gascheau's work can be found, and the author is therefore unable to give a description of the process employed.

distances, as seen from the large body, have much greater coefficients than the inequalities in their linear distances from the same body.

The reader will find a more complete discussion of this problem in a paper by the author published in the sixth volume of the *Proceedings of the London Mathematical Society*, 1875. The co-ordinates x, y, X, Y are expressed in terms of the time and the possibility of any small term rising into importance is shortly treated.

Theory of oscillations about steady motion.

110. Having illustrated by two important examples the methods of practically finding the oscillations about a state of motion, we pass on to the general theory of the subject.

111. **The Determinantal Equation of steady motion.** *To form the general equations of oscillation of a dynamical system about a state of steady motion.*

Let the system be referred to any co-ordinates θ, ϕ, ψ, &c. If the geometrical equations do not contain the time explicitly the vis viva $2T$ may be represented by the expression

$$2T = P_{11}\theta'^2 + 2P_{12}\theta'\phi' + P_{22}\phi'^2 + \&c.$$

where P_{11}, P_{12}, &c. are known functions of the co-ordinates θ, ϕ, &c. Let the force function be U. Let the state of motion about which the system is oscillating be determined by $\theta = f(t)$, $\phi = F(t)$, &c. To determine these oscillations we put $\theta = f(t) + x$, $\phi = F(t) + y$, &c. Let the Lagrangian function $L = T + U$ be expanded in powers of x, y, &c. as follows:

$$L = L_0 + A_1x' + A_2y' + \&c. + C_1x + C_2y + \&c.$$
$$+ \tfrac{1}{2}(A_{11}x'^2 + 2A_{12}x'y' + \&c.) + \tfrac{1}{2}(C_{11}x^2 + 2C_{12}xy + \&c.)$$
$$+ G_{11}xx' + G_{12}xy' + G_{21}yx' + \&c.$$

It will afterwards be found convenient to write $E_{12} = G_{12} - G_{21}$, $E_{13} = G_{13} - G_{31}$, and so on.

We shall now define a *steady motion* to be one in which all the coefficients in this expansion are independent of the time. The physical characteristic of such a motion is that when referred to proper co-ordinates the same oscillations follow from the same disturbance of the same co-ordinate at whatever instant it may be applied to the motion. If the coefficients are not constant for the co-ordinates chosen it may be possible to make them constant by a change of co-ordinates. There are obviously many systems of co-ordinates which may be chosen, and a set may generally be found by a simple examination of the steady motion. If there are any quantities which are constant during the steady motion, such as those called ξ, η, &c. in Art. 98, these may serve for some of the co-ordinates, others may be found by considering what quantities appear only as differential coefficients or velocities, for example those called x, y, &c. in the same article. If none of these are obvious, we may sometimes obtain them by combining the existing co-ordinates. Practically these will be the most convenient methods of discovering the proper co-ordinates.

To obtain the equations of motion we must now substitute the value of L in the Lagrangian equations

$$\frac{d}{dt}\frac{dL}{dx'} - \frac{dL}{dx} = 0, \ \&c. = 0,$$

and reject the squares of small quantities. The steady motion being given by x, y, &c. all zero, each of these must be satisfied when we omit the terms containing $x, \cdot y$, &c. We thus obtain the equations of steady motion, viz.

$$C_1 = 0, \quad C_2 = 0, \quad \&c. = 0,$$

which by Taylor's theorem are the same as the equations (1) of steady motion given in Art. 98.

Omitting these terms and retaining the first powers of all the small quantities we obtain the equations of small oscillations. Representing differentiations with regard to t by the letter δ, we have

$$(A_{11}\delta^2 - C_{11})x + (A_{12}\delta^2 - E_{12}\delta - C_{12})y + (A_{13}\delta^2 - E_{13}\delta - C_{13})z + \&c. = 0,$$
$$(A_{12}\delta^2 + E_{12}\delta - C_{12})x + (A_{22}\delta^2 - C_{22})y + (A_{23}\delta^2 - E_{23}\delta - C_{23})z + \&c. = 0,$$
$$\&c. + \qquad\qquad \&c. + \qquad\qquad \&c. = 0.$$

112. To solve these we write $x = Le^{\lambda t}$, $y = Me^{\lambda t}$, &c. Substituting and eliminating the ratios L, M, &c. we obtain the following determinantal equation

$$\begin{vmatrix} A_{11}\lambda^2 - C_{11}, & A_{12}\lambda^2 - E_{12}\lambda - C_{12}, & A_{13}\lambda^2 - E_{13}\lambda - C_{13}, \&c. \\ A_{12}\lambda^2 + E_{12}\lambda - C_{12}, & A_{22}\lambda^2 - C_{22}, & A_{23}\lambda^2 - E_{23}\lambda - C_{23}, \&c. \\ A_{13}\lambda^2 + E_{13}\lambda - C_{13}, & A_{23}\lambda^2 + E_{23}\lambda - C_{23}, & A_{33}\lambda^2 - C_{33}, \qquad \&c. \\ \&c. & \&c. & \&c. \qquad\qquad \&c. \end{vmatrix} = 0.$$

If in this equation we write $-\lambda$ for λ the rows of the new determinant are the same as the columns of the old, so that the determinant is unaltered. We therefore infer that the *determinantal equation when expanded contains only even powers of* λ.

We notice that if we remove from this determinant the terms which contain the letter E, the remaining determinant is the same as that which gives the oscillation about a position of equilibrium, Art. 58. We may therefore say that the terms which depend on E are due to the centrifugal forces of the steady motion.

113. **Conditions of Stability.** Regarding this as an equation to find λ^2, we notice that if the roots are all real and negative, each of the co-ordinates x, y, &c. can be expressed in a series of trigonometrical terms having different periods; the motion will therefore be stable. If any one of the roots is imaginary or if

any one is real and positive, there will be both positive and negative real exponentials entering into the expressions for x, y, &c. and therefore the motion will be unstable. *The condition of dynamical stability is therefore that the roots of this equation must all be of the form* $\lambda = \pm \mu \sqrt{-1}$, *where μ is some real quantity.*

114. Number of Oscillations. It follows also that when a system, under the action of forces which have a potential, oscillates about a stable state of steady motion, the oscillations of the co-ordinates are represented by trigonometrical terms of the form $A \sin (\lambda t + \alpha)$ which are not accompanied by any real exponential factors such as those which occurred in the problem of the Governor.

We see further that there will in general be as many finite values of λ^2 and therefore as many trigonometrical terms of different periods as there are co-ordinates. It often happens, as explained in Art. 111, that some of the co-ordinates are absent from the expression for L, appearing only as differential coefficients. Suppose for example θ to be absent; then C_{11}, C_{12}, &c. are all zero, and we may divide λ both out of the first line and the first column of the fundamental determinant. We therefore have two zero values of λ, while at the same time the number of finite values of λ^2 is diminished by unity. *Hence the number of trigonometrical terms of different periods cannot exceed the number of co-ordinates which explicitly enter into the Lagrangian function.* Thus, in Ex. 2 of Art. 102, the function $T + U$ has only the co-ordinate θ explicitly expressed, the others ϕ' and ψ' appearing only as differential coefficients. It follows that if a top is disturbed from a state of steady motion, there will be but one period in the oscillation.

115. The relations between the coefficients L, M, &c. in the exponential values of x, y, &c. may be obtained without difficulty if we remember that the several lines of the fundamental determinant are really the equations of motion. Taking any one line; multiply the first constituent by L, the second by M, &c. and equate the sum to zero. If n be the number of co-ordinates, we thus obtain $n - 1$ independent equations to find the ratios of $L : M :$ &c.; so that we have one undetermined constant for each value of λ. On the whole therefore we have, exactly as in Lagrange's equations, Chap. II., twice as many arbitrary constants as there are co-ordinates, all the other constants being determined by the equations just found. The arbitrary constants are determined by the initial values of the co-ordinates and their differential coefficients.

But, unlike Lagrange's equations, the quantity λ occurs in the first power in each of these equations, so that the ratios of $L, M,$ &c. thus found may be imaginary. If $-p_1^2$, $-p_2^2$, &c. be the values of λ^2, the expressions for the co-ordinates when rationalized may therefore take the form

$$x = A_1 \sin (p_1 t + \alpha_1) + A_2 \sin (p_2 t + \alpha_2) + \dots$$
$$y = B_1 \sin (p_1 t + \beta_1) + B_2 \sin (p_2 t + \beta_2) + \dots$$
$$z = \&\text{c.}$$

where α_1 is not necessarily equal to β_1, nor α_2 to β_2, &c., though they are connected together.

116. **Principal Oscillations.** When the initial conditions are such that every co-ordinate is expressed by a trigonometrical term of one and the same period, the system is said to be performing a *principal* or *harmonic oscillation*. Thus each trigonometrical term corresponds to a principal oscillation, and any oscillation of the system is therefore said to be *compounded* of its principal oscillations. *The physical characteristic of a principal oscillation is that the motion of every part of the system is repeated at a constant interval.* If the type of the principal oscillation be $\lambda^2 = -p_1^2$, we see that throughout the motion we shall have $x'' = -p_1^2 x$, $y'' = -p_1^2 y$, &c.

117. **Ex.** A homogeneous sphere of unit mass and radius a is suspended from a fixed point by a string of length b and is set in rotation about the vertical diameter. When the sphere is slightly disturbed from this state of steady motion, let bx, by and b be the co-ordinates of the point on the surface to which the string is attached; $bx + a\xi$, $by + a\eta$ and $b + a$ the co-ordinates of the centre, the fixed point being the origin and the axis of z vertical and downwards. Also let $\chi = \phi + \psi$ where ϕ and ψ have the meanings usually given to them in Euler's geometrical equations, see Vol. I. Chap. v. Thus before disturbance $\chi' = n$. Prove that the Lagrangian function is

$$L = \frac{a^2}{5} \left\{ \left(\chi' - \frac{\xi \eta'}{2} + \frac{\xi' \eta}{2} \right)^2 + \xi'^2 + \eta'^2 \right\} + \tfrac{1}{2}(a\xi' + bx')^2 + \tfrac{1}{2}(a\eta' + by')^2 - g \left\{ b\frac{x^2 + y^2}{2} + a\frac{\xi^2 + \eta^2}{2} \right\}.$$

If the motion of the centre of gravity be represented by a series of terms of the form $M \cos (pt + \alpha)$, prove that the values of p are given by

$$\left(p^2 - \frac{g}{b} \right) \left(p^2 - np - \frac{5g}{2a} \right) = \frac{5g}{2b} p^2.$$

Show that, whatever sign n may have, this equation has two positive and two negative roots which are separated by the roots of either of the factors on the left-hand side.

118. **Impulsive Forces.** If we regard an impulse as the limit of a force acting for a very short time, we may deduce from Art. 111 the equations of motion of a system moving in steady motion and suddenly disturbed by an impulse. Integrating the equations of motion given in Art. 111 with regard to the time during the limits of the impulse, the integrals of all the terms except those of the form $A\delta^2 x$ will be zero. This follows from the definition of an impulse given in Chapter II. of Vol. I. or from the argument given in adjusting Lagrange's equations to impulses in Chapter VIII. of Vol. I.

The equations of motion for impulses are therefore

$$A_{11}(\delta x_1 - \delta x_0) + A_{12}(\delta y_1 - \delta y_0) + \ldots\ldots = X,$$
$$A_{12}(\delta x_1 - \delta x_0) + A_{22}(\delta y_1 - \delta y_0) + \ldots\ldots = Y,$$
$$\&c. = \&c.$$

Here $\delta x_1 - \delta x_0$, &c. are the changes in the velocities of the co-ordinates produced by the jerks. The quantities X, Y, &c. are the integrals of the disturbing forces and therefore measure the jerks. If U be the force function of the impulses as explained in Vol. I. Chap. VIII. we have $X = dU/dx$, $Y = dU/dy$, &c.

119. **Analysis of the roots of the determinantal equation.** If the determinantal equation of Art. 112 is not very complicated we may expand it in powers of λ. We thus have an equation with only even powers of λ. The important point to settle is the number of real negative values of λ^2 which satisfy the equation. To determine this, we may use Sturm's theorem. Since the equation has only alternate powers of λ, we may use the short rule which will be given in the chapter on the Conditions of Stability to find the successive remainders.

But if it be inconvenient to follow this process, we may use some of the following theorems.

120. We shall first show that the *quadratic expression*

$$2A = A_{11}x'^2 + 2A_{12}x'y' + A_{22}y'^2 + \&c.$$

is a one-signed positive function. To prove this we notice that the coefficients A_{11}, &c. are what the coefficients P_{11}, &c. of the vis viva become when we write for the co-ordinates θ, ϕ, &c. their values in the steady motion. If then, by any linear relation between the variables, we could make A equal to zero, we could by introducing a constraint into the motion represented by a similar relation between θ', ϕ', &c. cause the vis viva to be zero. But since the vis viva is essentially positive, this is impossible.

When a given quadratic function is a one-signed positive function, it is known (Art. 60) that its discriminant is positive. It follows immediately that every discriminant formed after putting any of the variables x', y', &c. equal to zero must also be positive.

121. *Theorem* I. It frequently happens that there are but two independent co-ordinates, so that the determinant is reduced to two rows. If we write

$$D = A_{11}A_{22} - A_{12}^2, \qquad D' = C_{11}C_{22} - C_{12}^2, \qquad \Theta = A_{11}C_{22} + A_{22}C_{11} - 2A_{12}C_{12},$$

the determinantal equation when expanded reduces to $D\lambda^4 + (-\Theta + E_{12}^2)\lambda^2 + D' = 0$. The conditions of stability are therefore (1) D' is positive, (2) $E_{12}^2 - \Theta$ is positive and greater than $2\sqrt{DD'}$. See Art. 113.

122. *Theorem* II. Whatever be the number of co-ordinates the steady motion cannot be stable unless all the values of λ^2 given by the determinantal equation are real and negative. The coefficient of the highest power of λ^2 (Art. 120) is positive, hence the term independent of λ^2 must also be positive. We therefore infer *that the steady motion cannot be stable unless the discriminant of the quadratic expression*

$$2C = -C_{11}x^2 - 2C_{12}xy - C_{22}y^2 + \ldots\ldots$$

is positive.

123. *Theorem* III. Let there be n co-ordinates and let Δ be the determinant given in Art. 112. Beginning with this determinant we may form a series of determinants each being obtained from the preceding by erasing the first line and the first column. Let us represent these by Δ_1, Δ_2, &c. The determinant Δ is not

altered if we border it with a column of zeros on the right-hand side and a row of zeros at the bottom, provided we put unity in the corner. We may therefore consider $\Delta_n = 1$. Thus we have a series of determinantal functions of λ^2 analogous to those used in connection with Lagrange's determinant. See Art. 58.

Let us substitute in this series of determinants any *negative* value of λ^2 and count the number of variations of sign. If as λ^2 passes from $\lambda^2 = -\alpha$ to $\lambda^2 = -\beta$, κ variations of sign are lost, then *the number of real roots between* $-\alpha$ *and* $-\beta$ *is either exactly equal to* κ *or exceeds* κ *by an even number.*

To prove this, we let I_{11}, I_{12}, &c. be the minors of the several constituents of the determinant Δ. We notice that I_{12} is changed into I_{21} by changing the sign of λ.

Hence if $I_{12} = \phi(\lambda^2) + \lambda\psi(\lambda^2)$,

then $I_{21} = \phi(\lambda^2) - \lambda\psi(\lambda^2)$.

Thus the product $I_{12}I_{21}$ is necessarily *positive* for all *negative* values of λ^2. It also follows that if I_{12} vanishes for any negative value of λ^2, then I_{21} vanishes for the same value of λ^2.

Starting with the equation $\Delta\Delta_2 = I_{11}I_{22} - I_{12}I_{21}$ the rest of the proof is so nearly the same as that for the corresponding theorem in Lagrange's determinant (Art. 58) that it seems unnecessary to reproduce it here. Passing over therefore this proof we notice the following applications.

124. *Theorem* IV. The coefficients of the highest powers of λ^2 in the series of determinants Δ, Δ_1, &c. are the discriminants of the quadric A (Art. 120), and are therefore necessarily positive. The signs of the series of determinants when $\lambda^2 = -\infty$ are therefore alternatively positive and negative. If the discriminants of the quadric $2C = -C_{11}x^2 - 2C_{12}xy - C_{22}y^2 - \&c.$
be also all positive, the signs of the series of determinants when $\lambda^2 = 0$ are all positive. Thus the full number, viz. n, of variations of signs have been lost in the passage from $\lambda^2 = -\infty$ to $\lambda^2 = 0$. It immediately follows from the theorem just stated that *when the quadric* C *is a one-signed positive function all the roots of the determinantal equation are real and negative.*

We may also express this by saying that when the quadric function C is a *minimum for all displacements from the steady motion, that steady motion is stable.*

125. When this occurs *the roots of each of the series of determinants* Δ, Δ_1, Δ_2, &c. *are all real and negative, and the roots of each separate or lie between the roots of the determinant next above it.*

This follows from the mode of proof adopted in discussing Lagrange's determinant.

126. *Theorem* V. **Equal roots.** The existence of equal roots usually indicates that there are terms in the solution with t as a factor, but it will be shown in another chapter that this is not the case when the minors of the determinant Δ are also zero.

Suppose, as in the last proposition, that the full number of variations of sign have been lost in the passage from $\lambda^2 = -\infty$ to $\lambda^2 = 0$. Then it may be shown, as in the corresponding proposition in Lagrange's determinant, that *if the fundamental determinant have* r *equal roots, then every first minor has* r − 1 *roots equal to each of these, and every second minor has* r − 2 *roots equal to each of these, and so on.*

We therefore infer that the existence of equal roots merely indicates a corresponding indeterminateness in the coefficients of the principal oscillation which is derived from these equal roots.

Thus in Art. 115 we have $n - 1$ independent equations to find the ratios of the coefficients L, M, &c. of any exponential. But when there are r equal roots we have only $n - r$ independent equations leaving r of the coefficients independent.

127. *Theorem VI.* If we remove the terms which contain the centrifugal forces the remaining determinant has the same form as Lagrange's determinant. Thus we have two determinantal equations each of which, for its own use, may be regarded as an equation to find λ^2. From each of these we may derive a series of determinants formed by the rule given in Art. 58. If we count the number of variations of sign when $\lambda^2 = -\infty$ and when $\lambda^2 = 0$, it is evident that each of the two series exhibit the same loss. It therefore follows that the equation with the centrifugal forces has at least as many negative roots as the corresponding Lagrange's equation, and if it have more, the excess is an even number. If therefore all the roots of the corresponding Lagrange's determinants are negative, then all the roots of the equation with the centrifugal forces are also real and negative. Thus the general effect of these centrifugal forces is to increase the stability.

The substance of this section may be found partly in a paper by the author published by the *London Mathematical Society*, 1875, and partly in the author's *Essay on the Stability of Motion*, 1877.

128. **The Representative Point.** When a dynamical system has not more than three co-ordinates, we may obtain a geometrical representation of the oscillation. Let these independent co-ordinates be x, y, z. If we regard these as the Cartesian co-ordinates of some point P, it is clear that the positions of P as it moves about will exhibit to the eye the motion of the system. We may call this point the *representative point*. The importance of this point has been already shown by the use made of it in Art. 80.

129. **Oscillation about equilibrium.** Let us first suppose the system to be oscillating about a position of equilibrium, and let it be performing any principal oscillation. Then throughout the motion the co-ordinates x, y, z bear a constant ratio to each other (Art. 53). We therefore infer that the path of the representative particle is a straight line passing through the origin. If the oscillation be defined by the type $\sin(pt + a)$ we have also (by Art. 55) $x'' = -p^2 x$, $y'' = -p^2 y$, &c. Hence the *representative point oscillates in a straight line with an acceleration tending to the origin and varying as the distance therefrom.*

130. To find the position of this straight line let the vis viva $2T$ and the force function U be represented by

$$\left.\begin{array}{l} 2T = A_{11} x'^2 + 2A_{12} x'y' + \&c. \\ 2(U - U_0) = C_{11} x^2 + 2C_{12} xy + \&c. \end{array}\right\} \ \ldots\ldots\ldots\ldots\ldots\ldots (1).$$

Then by Lagrange's equations, since $x'' = -p^2 x$, &c., we have

$$\left.\begin{array}{l} -p^2(A_{11}x + A_{12}y + \&c.) = C_{11}x + C_{12}y + \&c. \\ -p^2(A_{12}x + A_{22}y + \&c.) = C_{12}x + C_{22}y + \&c. \\ \&c. = \&c. \end{array}\right\} \ \ldots\ldots\ldots\ldots (2).$$

Omitting the accents in T and the constant term U_0, let us put

$$\left.\begin{array}{l} 2A = A_{11}x^2 + 2A_{12}xy + \&c. \\ -2C = C_{11}x^2 + 2C_{12}xy + \&c. \end{array}\right\} \ \ldots\ldots\ldots\ldots\ldots\ldots (3).$$

We also construct the two quadrics $A = a$, $C = \gamma$ where a and γ are any constants. These quadrics have their centre at the origin and have a common set of conjugate diameters which may be found by the following process. Let x, y, z be the Cartesian co-ordinates of any point on one of the three conjugates. Then, since the diametral planes of this point in the two quadrics are parallel, we have

$$\mu \frac{dA}{dx} = \frac{dC}{dx}, \quad \mu \frac{dA}{dy} = \frac{dC}{dy}, \quad \mu \frac{dA}{dz} = \frac{dC}{dz}.$$

Comparing these with the equations (2) we see that when the system is performing a principal oscillation the *representative point* P *oscillates in one of the common conjugate diameters of the quadrics.*

131. By Euler's theorem on homogeneous functions we have $\mu A = C$. Applying the same reasoning to equations (2) we have $p^2 A = C$. Hence $\mu = p^2$. Let the diameter described by the representative point cut the quadrics $A = a$ and $C = \gamma$ in the points D and D' and let O be the origin. Then putting P at D we have $A = a$, and since C is a homogeneous function we have $C = (OD/OD')^2 \gamma$.

Hence $p^2 = (OD/OD')^2 \gamma/a$. *The value of $2\pi/p$ gives the period of oscillation corresponding to any common conjugate diameter* ODD'.

132. The quadric $C = \gamma$ possesses the property that if x, y, z be the co-ordinates referred to any axes of a point P on its surface the work done by such a displacement from the position of equilibrium is constant and equal to $-\gamma$.

133. As an example of this geometrical analogy let us consider the following problem. *A rigid body, free to move about a fixed point* O, *is under the action of any forces and makes small oscillations about a position of equilibrium; find the principal oscillations.*

Let OA, OB, OC be the positions of the principal axes in the position of equilibrium, OA', OB', OC' their positions at the time t. The position of the body may be defined by the angles between (1) the planes AOC, AOC', (2) the planes BOC, BOC', (3) the planes COA, COA'. Let these be called θ, ϕ, ψ respectively. Then θ, ϕ, ψ are angular displacements of the body about OA, OB, OC. Taking these as the axes of co-ordinates in the geometrical analogy; a small displacement of P from the origin to a point $x = \theta$, $y = \phi$, $z = \psi$ represents a rotation of the body about the straight line described by P and whose magnitude is measured by the distance traversed by P.

If I_1, I_2, I_3 be the principal moments of inertia at O, the vis viva of the body is clearly $\qquad\qquad 2T = I_1 \theta'^2 + I_2 \phi'^2 + I_3 \psi'^2$.

Writing x, y, z for θ', ϕ', ψ' as before, the quadric $T = a$ or $A = a$ is evidently the momental ellipsoid at the fixed point.

Let the work of the forces as the co-ordinates change from zero to θ, ϕ, ψ, or x, y, z be given by $\qquad\qquad 2U = C_{11}x^2 + 2C_{12}xy + \&c.$

Then, following the analogy, as P moves along a radius vector OD' of the quadric $U = -\gamma$ or $C = \gamma$, the work is $-(OP/OD')^2 \gamma$. Hence this quadric possesses the property that the work done by the forces when the body is twisted through a given angle round any radius vector varies inversely as the square of that radius vector. If the equilibrium is stable, the work due to a rotation about every diameter must be negative, the quadric must therefore be an ellipsoid.

It now follows from the general theorem that *the body will perform a principal oscillation if it is set in rotation about any one of the three conjugate diameters of the momental ellipsoid and the ellipsoid* $U = -\gamma$, *and will therefore continue to oscillate as if that diameter were fixed in space.*

The quadric U has been called the *ellipsoid of the potential*. This name was given to it by Prof. Ball, who arrived at the theorem just proved by a different course of reasoning. See his *Theory of Screws*, Art. 126.

134. Oscillation about steady motion. To determine the motion of the representative point we must have recourse to the equations of motion written down in Art. 111. Since we must follow the same line of argument as in Art. 131, it is unnecessary to do more than state the result. The symbols A and C having the same meaning as before the path of the representative point is given by the equations $Ap^2 - C = \beta.$

$$[(A_{11}E_{23} - A_{12}E_{13} + A_{13}E_{12})\,x + \&c.]\,p^2 + [(C_{11}E_{23} - C_{12}E_{13} + C_{13}E_{12})\,x + \&c.] = 0.$$

The path of the representative point is therefore a plane section of a quadric. We infer that *when a system is performing a principal oscillation about a state of steady motion the representative point describes an ellipse. The ellipse is described with an acceleration tending to the centre and varying as the distance therefrom. The periodic time in the ellipse is by definition the same as that in which the system performs its principal oscillation.*

135. Ex. Show that the three planes of these harmonic ellipses are diametral planes of the *same straight line* with regard to the three quadrics represented by $Ap^2 - C = \beta$, where p^2 has any one of the three values given by the determinant of motion. The direction cosines of this straight line are proportional to E_{23}, $-E_{13}$, E_{12} and it may be called the axis of the centrifugal forces.

136. The introduction of the representative point to exhibit the motion of the system may appear somewhat artificial. But there is a closer connection than has yet been mentioned. Suppose for example that a system is oscillating about a position of equilibrium. Let us transform the co-ordinates x, y, z into others ξ, η, ζ so that

$$A_{11}x'^2 + 2A_{12}x'y' + \&c. = \xi'^2 + \eta'^2 + \zeta'^2.$$

The equations of motion take a simplified form and become those of a single particle of unit mass acted on by forces with a known force function U. Thus when the co-ordinates are properly chosen some kinds of motion may be completely found by replacing the system by its representative particle. In other kinds of motion constraints have to be placed on the particle that it may represent the motion. The single particle used by Fresnel in his theory of double refraction is practically the representative particle, and it is necessary to impose imaginary constraints that its motion may represent that of the medium.

A more complete account of the theory of the representative point is given in the essay on *Stability of motion* already referred to.

CHAPTER IV.

Solution of Euler's Equations.

137. *To determine the motion of a body about a fixed point, in the case in which there are no impressed forces.*

Euler's equations of motion are

$$
\left.
\begin{aligned}
A \frac{d\omega_1}{dt} - (B - C)\,\omega_2\omega_3 &= 0 \\
B \frac{d\omega_2}{dt} - (C - A)\,\omega_3\omega_1 &= 0 \\
C \frac{d\omega_3}{dt} - (A - B)\,\omega_1\omega_2 &= 0
\end{aligned}
\right\} ;
$$

multiplying these respectively by ω_1, ω_2, ω_3; adding and integrating, we get

$$A\omega_1{}^2 + B\omega_2{}^2 + C\omega_3{}^2 = T \dots\dots\dots\dots\dots(1),$$

where T is an arbitrary constant.

Again, multiplying the equations respectively by $A\omega_1$, $B\omega_2$, $C\omega_3$, we get, similarly,

$$A^2\omega_1{}^2 + B^2\omega_2{}^2 + C^2\omega_3{}^2 = G^2 \dots\dots\dots\dots\dots(2),$$

where G is an arbitrary constant.

To find a third integral, let

$$\omega_1{}^2 + \omega_2{}^2 + \omega_3{}^2 = \omega^2 \dots\dots\dots\dots\dots(3);$$

$$\therefore \omega_1 \frac{d\omega_1}{dt} + \omega_2 \frac{d\omega_2}{dt} + \omega_3 \frac{d\omega_3}{dt} = \omega \frac{d\omega}{dt};$$

then multiplying the original equations respectively by ω_1/A, ω_2/B, ω_3/C, and adding, we get

$$\omega \frac{d\omega}{dt} = \left(\frac{B-C}{A} + \frac{C-A}{B} + \frac{A-B}{C} \right) \omega_1\omega_2\omega_3 \dots\dots(4)$$

$$= - \frac{(B-C)(C-A)(A-B)}{ABC}\, \omega_1\omega_2\omega_3.$$

But solving the equations (1), (2), (3), we get

$$\left.\begin{aligned}
\omega_1{}^2 &= \frac{BC}{(A-C)(A-B)} \cdot (-\lambda_1 + \omega^2) \\
\omega_2{}^2 &= \frac{CA}{(B-A)(B-C)} \cdot (-\lambda_2 + \omega^2) \\
\omega_3{}^2 &= \frac{AB}{(C-B)(C-A)} \cdot (-\lambda_3 + \omega^2)
\end{aligned}\right\} \dots\dots\dots\dots (5),$$

where $\lambda_1 = \dfrac{T(B+C) - G^2}{BC}$, with similar expressions for λ_2 and λ_3.
Substituting in equation (4), we have

$$\omega \frac{d\omega}{dt} = \sqrt{(\lambda_1 - \omega_2)(\lambda_2 - \omega^2)(\lambda_3 - \omega^2)} \dots\dots\dots (6).$$

The integration of equation (6)* can be reduced without difficulty to depend on an elliptic integral. The integration can be
effected in finite terms in two cases; when $A = B$, and when
$G^2 = TB$, where B is neither the greatest nor the least of the three
quantities A, B, C. Both these cases will be discussed further on.

138. It will generally be supposed that A, B, C are in order of magnitude, so
that A is greater than B, and B than C. The axis of B will be called the axis of
mean moment. If we eliminate ω_1 from the equations (1) and (2), we have

$$AT - G^2 = B(A-B)\omega_2{}^2 + C(A-C)\omega_3{}^2,$$

which is essentially positive. In the same way we can show that $CT - G^2$ is
negative. Thus the quantity G^2/T may have any value lying between the greatest
and least moments of inertia.

The three quantities λ_1, λ_2, λ_3 in Art. 137 are all positive quantities; for since
$B + C - A$ is positive, and $G^2/T < A$, it follows that λ_1 is positive. The numerators
of λ_2 and λ_3 are each greater than that of λ_1, and are therefore positive, the
denominators are also positive; hence λ_2 and λ_3 are both positive. Also we have
$ABC(\lambda_1 - \lambda_2) = (TC - G^2)(A - B)$, with similar expressions for $\lambda_2 - \lambda_3$ and $\lambda_3 - \lambda_1$.
It easily follows that λ_2 is the greatest of the three, and λ_1 or λ_3 is the least according
as G^2/T is greater or less than B.

It follows from equations (5) that throughout the motion ω^2 must lie between λ_2
and the greater of the quantities λ_1 and λ_3.

139. **Kirchhoff's solution.** The solution in terms of elliptic integrals has
been effected in the following manner by Kirchhoff. If we put

$$\Delta(\phi) = \sqrt{1 - k^2 \sin^2 \phi}, \qquad F(\phi) = \int_0^\phi \frac{d\phi}{\sqrt{1 - k^2 \sin^2 \phi}},$$

then k is called the modulus of F, and must be less than unity if F is to be real for
all values of ϕ. The upper limit ϕ is called the amplitude of the elliptic integral F

---

* Euler's solution of these equations is given in the *ninth volume of the Quarterly
Journal*, p. 361, by Prof. Cayley. Kirchhoff's and Jacobi's integrations by elliptic
functions are given in an improved form by Prof. Greenhill in the fourteenth
volume, pages 182 and 265. 1876.

and is usually written am F. In the same way $\sin\phi$, $\cos\phi$, and $\Delta(\phi)$ are written $\sin$ am F, $\cos$ am F, and Δ am F.

We have by differentiation

$$\left.\begin{aligned}
\frac{d\cos\phi}{dF} &= -\sin\phi\frac{d\phi}{dF} = -\sin\phi\Delta(\phi) \\[2mm]
\frac{d\sin\phi}{dF} &= \cos\phi\frac{d\phi}{dF} = \cos\phi\Delta(\phi) \\[2mm]
\frac{d\Delta(\phi)}{dF} &= -\frac{k^2\sin\phi\cos\phi}{\Delta(\phi)}\frac{d\phi}{dF} = -k^2\sin\phi\cos\phi
\end{aligned}\right\} \quad\ldots\ldots\ldots\ldots(1).$$

These equations may be made identical with Euler's equations if we put $F = \lambda(t-\tau)$ and

$$\left.\begin{aligned}
\omega_1 &= a\Delta\ \text{am}\ \lambda(t-\tau) \\
\omega_2 &= b\sin\ \text{am}\ \lambda(t-\tau) \\
\omega_3 &= c\cos\ \text{am}\ \lambda(t-\tau)
\end{aligned}\right\} \quad\ldots\ldots\ldots\ldots\ldots\ldots\ldots\ldots\ldots\ldots(2),$$

$$\frac{A-B}{C} = -\frac{c\lambda}{ab}, \quad \frac{A-C}{B} = -\frac{b\lambda}{ca}, \quad \frac{B-C}{A} = -k^2\frac{a\lambda}{bc} \quad\ldots\ldots\ldots\ (3).$$

We have introduced here six new constants, viz. a, b, c, λ, k and τ. With these we may satisfy the three last equations and also any initial values of ω_1, ω_2, ω_3. The solution if real will also be complete.

When $t = \tau$ we have from (2) $\omega_1 = a$, $\omega_2 = 0$, and $\omega_3 = c$. Hence by Art. 137

$$Aa^2 + Cc^2 = T, \quad A^2a^2 + C^2c^2 = G^2;$$

$$\therefore a^2 = \frac{G^2 - CT}{A(A-C)}, \quad c^2 = \frac{AT - G^2}{C(A-C)}.$$

Dividing the second of equations (3) by the first, we have

$$\frac{b^2}{c^2} = \frac{A-C}{A-B}\frac{C}{B}; \quad \therefore b^2 = \frac{AT-G^2}{B(A-B)}.$$

Multiplying the first and second of equations (3), we obtain

$$\lambda^2 = \frac{(A-B)(G^2 - CT)}{ABC}.$$

The ratios of the right-hand sides of (3) are as $c^2 : b^2 : k^2 a^2$, and these have just been found. Hence if the signs of a, b, c, λ be chosen to satisfy any one of the three equalities, the signs of all will be satisfied.

Dividing the last of equations (3) by either of the other two, we find

$$k^2 = \frac{B-C}{A-B}\frac{AT-G^2}{G^2 - CT}; \quad \therefore 1 - k^2 = \frac{A-C}{A-B}\frac{G^2 - BT}{G^2 - CT}.$$

If $G^2 > BT$ and A, B, C are in descending order of magnitude, the values of a^2, b^2, c^2 and λ^2 are all positive. Also k^2 is positive and less than unity. The solution is therefore real and complete.

If $G^2 < BT$ we must suppose A, B, C to be in ascending order of magnitude to obtain a real solution. If we may anticipate a phrase used by Poinsot, and which will be explained a little further on, we may say that the expression for ω_1 in this solution is to be taken for the angular velocity about that principal axis which is enclosed by the polhode.

If $G^2 = BT$ we have $k^2 = 1$ and

$$F = \int_0^\phi \frac{d\phi}{\cos\phi} = \tfrac{1}{2}\log\frac{1+\sin\phi}{1-\sin\phi}; \quad \therefore \sin\ \text{am}\ F = \frac{e^F - e^{-F}}{e^F + e^{-F}}.$$

Substituting in equations (2) the elliptic functions become exponential.

If $B = C$ we have $k^2 = 0$ and in this case $F = \phi$, so that am $F = F$. If we again substitute in equations (2) the elliptic functions become trigonometrical.

The geometrical meaning of this solution will be given a little further on.

Poinsot's and MacCullagh's constructions for the motion.

140. The fundamental equations of motion of a body about a fixed point are

$$A^2\omega_1^2 + B^2\omega_2^2 + C^2\omega_3^2 = G^2 \dots\dots\dots\dots\dots(1),$$

$$A\omega_1^2 + B\omega_2^2 + C\omega_3^2 = T \dots\dots\dots\dots\dots(2).$$

These have been already obtained by integrating Euler's equations, but they also follow very easily from the principles of Angular Momentum, and Vis Viva.

Let the body be set in motion by an impulsive couple whose moment is G. Then we know by Vol. 1. Chap. vi., that throughout the whole of the subsequent motion, the moment of the momentum about every straight line which is fixed in space, and passes through the fixed point O, is constant, and is equal to the moment of the couple G about that line. Now by Art. 10, the moments of the momentum about the principal axes at any instant are $A\omega_1$, $B\omega_2$, $C\omega_3$. Let α, β, γ be the direction angles of the normal to the plane of the couple G referred to these principal axes as co-ordinate axes. Then we have

$$\left.\begin{array}{l} A\omega_1 = G\cos\alpha \\ B\omega_2 = G\cos\beta \\ C\omega_3 = G\cos\gamma \end{array}\right\} \dots\dots\dots\dots\dots(3),$$

adding the squares of these we get equation (1).

Throughout the subsequent motion the whole momentum of the body is equivalent to the couple G. It is therefore clear that if at any instant the body were acted on by an impulsive couple equal and opposite to the couple G, the body would be reduced to rest.

141. It follows from the definition given in Vol. I. Chap. vi. that the plane of this couple is the Invariable plane and the normal to it the Invariable line. This line is absolutely fixed in space, and the equations (3) give the direction cosines of this line referred to axes moving in the body.

It appears from these equations, that if the body be set in rotation about an axis whose direction cosines are (l, m, n) when referred to the principal axes at the fixed point, then the direction cosines of the invariable line are proportional to Al, Bm, Cn. If the axes of reference are not the principal axes of the body at the fixed point, the direction cosines of the invariable line will, by Art. 10, be proportional to $Al - Fm - En$, $Bm - Dn - Fl$, and

$Cn - El - Dm$, where A, F &c. are the moments and products of inertia*.

142. Since the body moves under the action of no impressed forces, we know that the Vis Viva will be constant throughout the motion. We have therefore

$$A\omega_1^2 + B\omega_2^2 + C\omega_3^2 = T,$$

where T† is a constant to be determined from the initial values of ω_1, ω_2, ω_3.

The equations (1), (2), (3) will suffice to determine the path in space described by every particle of the body, but not the position at any given time.

143. **Poinsot's construction.** *To explain Poinsot's representation of the motion by means of the momental ellipsoid.*

Let the momental ellipsoid at the fixed point be constructed, and let its equation be

$$Ax^2 + By^2 + Cz^2 = M\epsilon^4.$$

Let r be the radius vector of this ellipsoid coinciding with the instantaneous axis, and p the perpendicular from the centre on the tangent plane at the extremity of r. Also let ω be the angular velocity about the instantaneous axis.

The equations to the instantaneous axis are $\dfrac{x}{\omega_1} = \dfrac{y}{\omega_2} = \dfrac{z}{\omega_3}$, and if (x, y, z) be the co-ordinates of the extremity of the length r, each of these fractions is equal to r/ω. Substituting in the equation to the ellipsoid, we have

$$(A\omega_1^2 + B\omega_2^2 + C\omega_3^2)\frac{r^2}{\omega^2} = M\epsilon^4 ; \therefore \omega = \sqrt{\frac{T}{M\epsilon^2}}\frac{r}{\epsilon}.$$

The equation to the tangent plane at the point (x, y, z) is

$$Ax\xi + By\eta + Cz\zeta = M\epsilon^4,$$

substituting again for (x, y, z) we see that the equations to the perpendicular from the origin are $\dfrac{\xi}{A\omega_1} = \dfrac{\eta}{B\omega_2} = \dfrac{\zeta}{C\omega_3}$;

* That the straight line whose equations referred to the moving principal axes are $x/A\omega_1 = y/B\omega_2 = z/C\omega_3$ is absolutely fixed in space may be also proved thus, if we assume the truth of equation (1) in the text. Let x, y, z be the co-ordinates of any point P in the straight line at a given distance r from the origin, then each of the equalities in the equation to the straight line is equal to r/G and is therefore constant. The actual velocity of P in space resolved parallel to the instantaneous position of the axis of x is $=\dfrac{dx}{dt} - y\omega_3 + z\omega_2 = \dfrac{r}{G}\left\{A\dfrac{d\omega_1}{dt} - (B-C)\omega_2\omega_3\right\}$. But this is zero, by Euler's equation. Similarly the velocities parallel to the other axes are zero.

† It should be observed that in this Chapter T represents the whole vis viva of the body. In treating of Lagrange's equations in Chapter II. it was convenient to let T represent *half* the vis viva of the system.

but these are the equations to the invariable line. Hence this perpendicular is fixed in space.

The expression for the length of the perpendicular on the tangent plane at (x, y, z) is known to be $\dfrac{1}{p^2} = \dfrac{A^2x^2 + B^2y^2 + C^2z^2}{M^2\epsilon^8}$, substituting as before, we get

$$\frac{1}{p^2} = \frac{A^2\omega_1^2 + B^2\omega_2^2 + C^2\omega_3^2}{M^2\epsilon^8} \cdot \frac{r^2}{\omega^2} = \frac{G^2}{M^2\epsilon^8} \cdot \frac{M\epsilon^4}{T};$$

$$\therefore p = \frac{\sqrt{MT}}{G} \cdot \epsilon^2.$$

From these equations we infer

(1) *The angular velocity about the radius vector round which the body is turning varies as that radius vector.*

(2) *The resolved part of the angular velocity about the perpendicular on the tangent plane at the extremity of the instantaneous axis is constant.* This theorem is due to Lagrange.

For the cosine of the angle between the perpendicular and the radius vector $= p/r$. Hence the resolved angular velocity is $= \omega\, p/r = T/G$, which is constant.

(3) *The perpendicular on the tangent plane at the extremity of the instantaneous axis is fixed in direction, viz. normal to the invariable plane, and constant in length.*

The motion of the momental ellipsoid is therefore such that, its centre being fixed, it always touches a fixed plane, and the point of contact, being in the instantaneous axis, has no velocity. *Hence the motion may be represented by supposing the momental ellipsoid to roll on the fixed plane with its centre fixed.*

144. **Ex. 1.** If the body while in motion be acted on by any impulsive couple whose plane is perpendicular to the invariable line, show that the momental ellipsoid will continue to roll on the same plane as before, but the rate of motion will be altered.

Ex. 2. If a plane be drawn through the fixed point parallel to the invariable plane, prove that the area of the section of the momental ellipsoid cut off by this plane is constant throughout the motion.

Ex. 3. The sum of the squares of the distances of the extremities of the principal diameters of the momental ellipsoid from the invariable line is constant throughout the motion. This result is due to Poinsot.

Ex. 4. A body moves about a fixed point O under the action of no forces. Show that if the surface $Ax^2 + By^2 + Cz^2 = M(x^2 + y^2 + z^2)^2$ be traced in the body, the principal axes at O being the axes of co-ordinates, this surface throughout the motion will roll on a fixed sphere.

145. These theorems have been proved on the supposition that the quantities T and G are constant, but when the body is acted on by forces and both T and G vary, the theorems do not altogether lose their significance. It is still true that at

each instant during the motion the axis of the resultant couple of angular momentum, i.e. the invariable line, is coincident in direction with the perpendicular on the tangent plane to the momental ellipsoid at its point of intersection with the instantaneous axis; also the angular velocity about the invariable line is always equal to T/G though this ratio may not be constant. At any instant the values of the vis viva T and the couple G are given by the equations

$$T = K \left(\frac{\omega}{r} \right)^2, \qquad \frac{G^2}{T} = K \cdot \frac{1}{p^2}, \qquad G = K \frac{\omega}{pr},$$

where K has been written for $M\epsilon^4$.

Conversely, we may enquire what conditions must hold amongst the impressed forces that any one of Poinsot's theorems may hold throughout the motion. Let us suppose the body to be acted on by a couple Q whose components about the axes are L, M, N.

146. (1) If we examine the proof in Art. 137 by which T is proved constant when no forces act on the body, we see that

$$\frac{1}{2} \frac{dT}{dt} = L\omega_1 + M\omega_2 + N\omega_3 = Q\omega \cos QOI,$$

where QOI is the angle between the axis OQ of the couple Q and the instantaneous axis OI. It immediately follows that T is constant when the moment of the impressed forces about the instantaneous axis is always zero. When this is the case ω is proportional to r throughout the motion.

147. (2) Referring again to Art. 137 we see in the same way that

$$\frac{dG}{dt} = \frac{1}{G}(LA\omega_1 + MB\omega_2 + NC\omega_3) = Q \cos QOL,$$

where QOL is the angle between the axis OQ of the couple and the invariable line OL. It follows that G is constant when the impressed forces have no moment about the invariable line. When this happens, ω varies as the product pr throughout the motion.

148. (3) The plane containing the invariable line OL and the instantaneous axis OI may be called the *plane of the centrifugal forces* for the reasons given in Vol. I. Chap. v. Art. 260.

We see that both T and G are constant when the plane of the impressed couple coincides with the plane of the centrifugal forces. When this is the case, ω varies as r, and p is constant throughout the motion.

Ex. 1. Show that $\qquad \dfrac{1}{p}\dfrac{dp}{dt} = \dfrac{r^2}{K} \cdot \dfrac{Q}{\omega} \sin IOL . \sin QOL . \cos ILQ,$

where ILQ is the angle between the planes IOL and QOL. It immediately follows that p and therefore G^2/T is constant when the projection of the axis of the impressed couple on the plane of the centrifugal couple is the invariable line.

Ex. 2. Show also that if the instantaneous axis is near a principal axis, the angular displacement of p is not made small by the presence of the small factor IOL. It is also necessary that one of the other factors should be small.

149. **The Polhode.** To assist our conception of the motion of the body, let us suppose it so placed, that the plane of the couple G, which would set it in motion, is horizontal. Let a tangent plane to the momental ellipsoid be drawn parallel to the

plane of the couple G, and let this plane be fixed in space. Let the ellipsoid roll on this fixed plane, its centre remaining fixed, with an angular velocity which varies as the radius vector to the point of contact, and let it carry the given body with it. We shall then have constructed the motion which the body would have assumed if it had been left to itself after the initial action of the impulsive couple $G*$. See Fig. (1).

The point of contact of the ellipsoid with the plane on which it rolls traces out two curves, one on the surface of the ellipsoid, and one on the plane. The first of these is fixed in the body and is called the *polhode*, the second is fixed in space and is called the *herpolhode*. The equations to any polhode referred to the principal axes of the body may be found from the consideration that the length of the perpendicular on the tangent plane to the ellipsoid at any point of the polhode is constant. Taking the expressions for this perpendicular given in Art. 143 we see that the equations of the polhode are

$$\left. \begin{aligned} A^2x^2 + B^2y^2 + C^2z^2 &= \frac{MG^2\epsilon^4}{T} \\ Ax^2 + By^2 + Cz^2 &= M\epsilon^4 \end{aligned} \right\}.$$

Eliminating y, we have

$$A\,(A - B)\,x^2 + C\,(C - B)\,z^2 = \left(\frac{G^2}{T} - B\right)M\epsilon^4.$$

Hence if B be the axis of greatest or least moment of inertia, the signs of the coefficients of x^2 and z^2 will be the same, and the projection of the polhode will be an ellipse. But if B be the axis of mean moment of inertia, the projection is a hyperbola.

A polhode is therefore a closed curve drawn round the axis of greatest or least moment, and the concavity is turned towards the axis of greatest or least moment according as G^2/T is greater or less than the mean moment of inertia. The boundary line which separates the two sets of polhodes is that polhode whose projection on the plane perpendicular to the axis of mean moment is a hyperbola whose concavity is turned neither to the axis of greatest, nor to the axis of least moment. In this case $G^2 = BT$, and the projection consists of two straight lines whose equation is

$$A\,(A - B)\,x^2 - C\,(B - C)\,z^2 = 0.$$

---

* Prof. Sylvester has pointed out a *dynamical* relation between the free rotating body and the ellipsoidal top, as he calls Poinsot's central ellipsoid. If a *material* ellipsoidal top be constructed of uniform density, similar to Poinsot's central ellipsoid, and if with its centre fixed it be set rolling on a perfectly rough horizontal plane, it will represent the motion of the free rotating body, not in space only, but also in time: the body and the top may be conceived as continually moving round the same axis, and at the same rate, at each moment of time. The reader is referred to the memoir in the *Philosophical Transactions* for 1866.

This polhode consists of two ellipses passing through the axis of mean moment, and corresponds to the case in which the perpendicular on the tangent plane is equal to the mean axis of the ellipsoid. This polhode is called the *separating polhode.*

Since the projection of the polhode on one of the principal planes is always an ellipse, the polhode must be a re-entering curve.

Supposing the principal moments A, B, C to be in descending order and the axis of C placed in a vertical position, figure (2) is a rough sketch of that half of the polhodes which is viewed by an eye placed in the positive octant not far from the axis of B. The arcs ABA', CBC', ACA' represent the principal sections, B being the positive end of the mean axis. The remaining arcs represent the two sets of polhodes separated from each other by the separating polhodes SS', TT'.

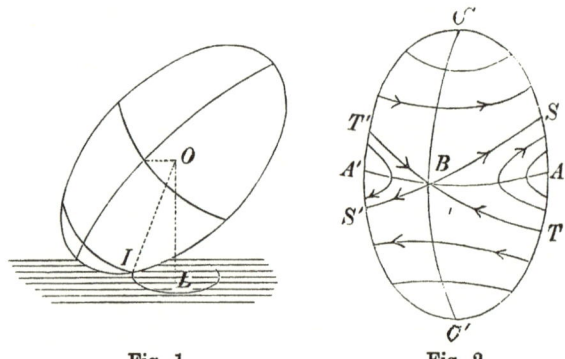

Fig. 1.   Fig. 2.

The terms polhode and herpolhode are due to Poinsot, *Théorie nouvelle de la rotation des corps* 1834 and 1852.

150. To find the motion of the extremity of the instantaneous axis along the polhode which it describes we have merely to substitute from the equations

$$\frac{\omega_1}{x} = \frac{\omega_2}{y} = \frac{\omega_3}{z} = \frac{\omega}{r} = \sqrt{\frac{T}{M}} \frac{1}{\epsilon^2},$$

in any of the equations of Art. 137. For example we thus obtain

$$\frac{dx}{dt} = \sqrt{\frac{T}{M}} \frac{B-C}{A} \frac{yz}{\epsilon^2}, \ \&c., \ \&c., \ x^2 = \frac{BC}{(A-C)(A-B)}(-\lambda_1' + r^2), \ \&c., \ \&c.$$

Since $dx/dt, dy/dt, dz/dt$ cannot vanish simultaneously it is evident from these equations that the instantaneous axis moves continuously along its polhode without any halting or change in the direction of its motion. This is, of course, also obvious from Fig. (1) for as the angular velocity about the instantaneous axis OI cannot change sign without vanishing and therefore contradicting the equation of vis viva (Art. 137, (2)), the point I must continuously describe both its polhode and herpolhode.

Again since the sign of dz/dt for every polhode is positive or negative according as the product xy is positive or negative we see that for that portion of the polhodes represented in the figure the extremity of the instantaneous axis moves upwards or downwards according as it is on the right or the left hand side of the arc CC'. These directions are indicated by the arrows. Thus the positive extremity of the instantaneous axis moves in the positive direction round the axis of greatest moment and in the negative direction round the axis of least moment of inertia.

Ex. 1. A point P moves along a polhode traced on an ellipsoid, show that the length of the normal between P and any one of the principal planes at the centre is constant. Show also that the normal traces out on a principal plane a conic similar to the focal conic in that plane. Also the measure of curvature of an ellipsoid along any polhode is constant.

Ex. 2. Show that the straight line OJ whose direction cosines are proportional to $d\omega_1/dt$, $d\omega_2/dt$, $d\omega_3/dt$ lies in the diametral plane of the invariable line and is at right angles to the invariable line. Show also that the sum of the squares of these quantities is

$$\Omega'^4 = -\omega^4 + (2Tp_2 - G^2 p_1)\,\omega^2/p_3 - \{p_2{}^2 T^2 - (p_1 p_2 + p_3)\,G^2 T + p_2 G^4\}/p_3{}^2,$$

where p_1, p_2, p_3 are the sum of the products of the quantities A, B, C taken respectively one, two and three together.

Ex. 3. Show that the resolved pressures P, Q, R on the fixed point O in the directions of the principal axes at O are given by

$$P = -\omega_1 \omega_2 y\,(A-B)/C + \omega_1 \omega_3 z\,(C-A)/B + \omega_1\,(\omega_2 y + \omega_3 z) - (\omega_2{}^2 + \omega_3{}^2)\,x$$

with similar expressions for Q and R, where x, y, z are the co-ordinates of the centre of gravity G, and A, B, C are the principal moments of inertia at O.

Thence show that the pressure on O is equivalent to two forces (1) a force $\Omega'^2 . GK$ which acts perpendicular to the plane OGK, where GK is the perpendicular drawn from G on the straight line OJ described in the last example, (2) a force $\omega^2 . GH$ acting parallel to GH where GH is a perpendicular from G on the instantaneous axis.

151. **The Herpolhode.** Since the herpolhode is traced out by the points of contact of an ellipsoid rolling about its centre on a fixed plane, it is clear that the herpolhode must always lie between two circles which it alternately touches. The common centre of

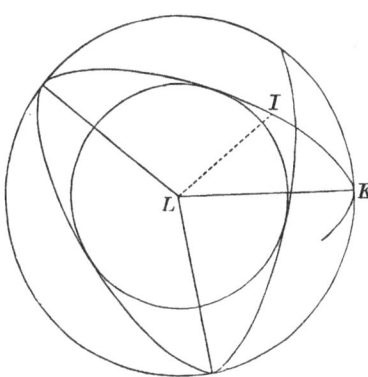

these circles will be the foot of the perpendicular from the fixed centre O on the fixed plane. To find the radii let OL be this perpendicular, and I be the point of contact. Let $LI = \rho$. Then we have by Art. 143, $\rho^2 = r^2 - p^2 = \dfrac{M\epsilon^4}{T}\left(\omega^2 - \dfrac{T'^2}{G^2}\right)$.

The radii will therefore be found by substituting for ω^2 its greatest and least values. But by Art. 138, these limits are λ_2, and the greater of the two quantities λ_1, λ_3.

The herpolhode is not in general a re-entering curve; but if the angular distance of the two points in which it successively touches the same circle be commensurable with 2π, it will be re-entering, i.e. the same path will be traced out repeatedly on the fixed plane by the point of contact.

152. MacCullagh's Construction. *To explain MacCullagh's representation of the motion by means of the ellipsoid of gyration.*

This ellipsoid is the reciprocal of the momental ellipsoid with regard to a sphere of radius ϵ, and the motion of the one ellipsoid may be deduced from that of the other by reciprocating the properties proved in the preceding Articles. We find,

(1) The equation to the ellipsoid referred to its principal axes is
$$\frac{x^2}{A} + \frac{y^2}{B} + \frac{z^2}{C} = \frac{1}{M}.$$

(2) *This ellipsoid moves so that its superficies always passes through a point fixed in space.* The point lies in the invariable line at a distance $G/\sqrt{MT}$ from the fixed point. By Art. 138 we know that this distance is less than the greatest, and greater than the least semi-diameter of the ellipsoid.

(3) *The perpendicular on the tangent plane at the fixed point is the instantaneous axis of rotation, and the angular velocity of the body varies inversely as the length of this perpendicular.* If p be the length of this perpendicular, then $\omega = \dfrac{1}{p}\sqrt{\dfrac{T}{M}}$.

(4) *The angular velocity about the invariable line is constant and* $= T/G$.

The corresponding curve to a polhode is the path described on the moving surface of the ellipsoid by the point fixed in space. This curve is clearly a sphero-conic. The equations to the sphero-conic described under any given initial conditions are easily seen to be
$$x^2 + y^2 + z^2 = \frac{G^2}{MT}, \quad \frac{x^2}{A} + \frac{y^2}{B} + \frac{z^2}{C} = \frac{1}{M}.$$

These sphero-conics may be shown to be closed curves round the axes of greatest and least moment. But in one case, viz. when $G^2/T = B$, where B is neither the greatest nor the least moment of inertia, the sphero-conic becomes the two central circular sections of the ellipsoid of gyration.

The motion of the body may thus be constructed by means of

either of these ellipsoids.   The momental ellipsoid resembles the general shape of the body more nearly than the ellipsoid of gyration.   It is protuberant where the body is protuberant, and compressed where the body is compressed.   The exact reverse of this is the case in the ellipsoid of gyration.   See Vol. I. Art. 27.

153. **MacCullagh's geometrical interpretation.**   MacCullagh has used the ellipsoid of gyration to obtain a geometrical interpretation of the solution of Euler's equations in terms of elliptic integrals.

The ellipsoid of gyration moves so as always to touch a point L fixed in space. Let us now project the point L on a plane passing through the axis of mean moment and making an angle a with the axis of greatest moment.   This projection may be effected by drawing a straight line parallel to either the axis of greatest moment or least moment.   We thus obtain two projections which we will call P and Q.   These points will be in a plane PQL which is always perpendicular to the axis of mean moment.   As the body moves about O the point L describes on the surface of the ellipsoid of gyration a sphero-conic KK', and the points P, Q describe two curves pp', qq' on the plane of projection OBD.   If the sphero-conic, as in the figure, enclose the extremity A of the axis of greatest moment, the curve inside the ellipsoid is formed by the projection parallel to the axis of greatest moment, but if the sphero-conic enclose the axis of least moment, the inner curve is formed by the projection parallel to that axis.   The point P which describes the inner curve will obviously travel round its projection, while the point Q which describes the outer curve will oscillate between two limits obtained by drawing tangents to the inner projection at the points where it cuts the axis of mean moment.

Since the direction-cosines of OL are proportional to $A\omega_1$, $B\omega_2$, $C\omega_3$ it is easy to see that, if x, y, z are the co-ordinates of L,

$$\frac{x}{A\omega_1} = \frac{y}{B\omega_2} = \frac{z}{C\omega_3} = \frac{r}{G} = \frac{1}{\sqrt{MT}} \dots\dots\dots\dots\dots(1).$$

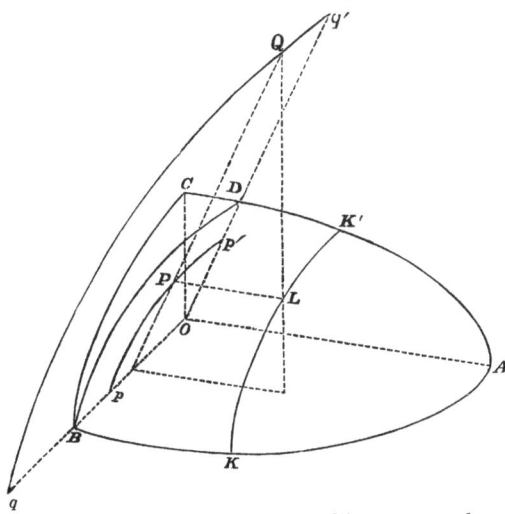

Let $OP = \rho$, $OQ = \rho'$, and let the angles these radii vectores make with the plane containing the axes of greatest and least moment be ϕ and ϕ' measured in the direction BD so that $DOP = -\phi$, $DOQ = -\phi'$: we then have

$$\left. \begin{aligned} -\rho \sin \phi &= y = B\omega_2 \, (MT)^{-\frac{1}{2}} \\ \rho \cos \phi \sin \alpha &= z = C\omega_3 \, (MT)^{-\frac{1}{2}} \end{aligned} \right\} \quad \dots\dots\dots\dots\dots\dots (2),$$

$$\left. \begin{aligned} \rho' \cos \phi' \cos \alpha &= x = A\omega_1 \, (MT)^{-\frac{1}{2}} \\ -\rho' \sin \phi' \quad &= y = B\omega_2 \, (MT)^{-\frac{1}{2}} \end{aligned} \right\} \quad \dots\dots\dots\dots\dots\dots (3).$$

It is proved in treatises on solid geometry that, if the plane on which the projection is made is one of the circular sections of the ellipsoid, the projections will be circles. This result may be verified by finding ρ or ρ' from these equations. Remembering that ρ and ρ' are constants, let us substitute in Euler's equation

$$B \frac{d\omega_2}{dt} - (C - A) \omega_3 \omega_1 = 0$$

from (2) and the first of equations (3). We have

$$\rho \frac{d\phi}{dt} = \frac{A - C}{AC} \sqrt{MT} \, \rho\rho' \sin \alpha \cos \alpha \cos \phi'.$$

Since $\rho' \cos \phi'$ is the ordinate of Q, we see that *the velocity of* P *varies as the ordinate of* Q, *and in the same way the velocity of* Q *varies as the ordinate of* P.

To find the constants ρ, ρ' we notice that ρ is the value of y obtained from the equations to the sphero-conic when $z = 0$. We thus have

$$\rho^2 = \frac{(AT - G^2) B}{MT (A - B)}, \qquad \rho'^2 = \frac{(G^2 - CT) B}{MT (B - C)},$$

the latter being obtained from the former by interchanging the letters A and C.

Hence

$$\left(\begin{matrix} \text{velocity} \\ \text{of P} \end{matrix} \right) = \frac{\sqrt{B - C}}{\sqrt{ABC}} \sqrt{AT - G^2} \left(\begin{matrix} \text{ordinate} \\ \text{of Q} \end{matrix} \right),$$

$$\left(\begin{matrix} \text{velocity} \\ \text{of Q} \end{matrix} \right) = \frac{\sqrt{A - B}}{\sqrt{ABC}} \sqrt{G^2 - CT} \left(\begin{matrix} \text{ordinate} \\ \text{of P} \end{matrix} \right).$$

154.   Since $\rho' \sin \phi' = \rho \sin \phi$, we have by substitution

$$\frac{d\phi}{dt} = \lambda \sqrt{1 - \frac{\rho^2}{\rho'^2} \sin^2 \phi},$$

where λ^2 has the same value as in Art. 139. Let us suppose ϕ expressed in terms of t by the elliptic integral

$$\lambda (t - \tau) = \int_0^\phi \frac{d\phi}{\sqrt{1 - \frac{\rho^2}{\rho'^2} \sin^2 \phi}},$$

so that $\phi = \operatorname{am} \lambda (t - \tau)$. Substituting this value of ϕ in equations (2) or (3), we obtain the values of ω_1, ω_2, ω_3 expressed in terms of the time.

155.   **Stability of Rotation.** If a body be set in rotation about any principal axis at a fixed point, it will continue to rotate about that axis as a permanent axis. But the three principal axes at the fixed point do not possess equal degrees of stability. If any small disturbing cause act on the body, the axis of rotation will be moved into a neighbouring polhode. If this polhode be a small nearly circular curve enclosing the original axis of rotation, the instantaneous axis will never deviate far in the body from the principal axis which was its original position. The herpolhode also

will be a curve of small dimensions, so that the principal axis will never deviate far from a straight line fixed in space. In this case the rotation is said to be *stable*. But if the neighbouring polhode be not nearly circular, the instantaneous axis will deviate far from its original position in the body. In this case a very small disturbance may produce a very great change in the subsequent motion, and the rotation is said to be *unstable*.

If the initial axis of rotation be the axis OB of mean moment, all the neighbouring polhodes have their convexities turned towards B. Unless, therefore, the cause of disturbance be such that the axis of rotation is displaced along the separating polhode, the rotation must be unstable. If the displacement be along the separating polhode, the axis may have a tendency to return to its original position. This case will be considered a little further on, and for this particular displacement the rotation may be said to be stable.

If the initial axis of rotation be the axis of greatest or least moment, the neighbouring polhodes are ellipses of greater or less eccentricity. If they be nearly circular, the rotation will certainly be stable; if very elliptical, the axis will recede far from its initial position, and the rotation may be called unstable. If OC be the axis of initial rotation, the ratio of the squares of the axes of the neighbouring polhode is ultimately $\dfrac{A(A-C)}{B(B-C)}$. It is therefore necessary for the stability of the rotation that this ratio should not differ much from unity.

156. It is well known that the steadiness or stability of a moving body is much increased by a rapid rotation about a principal axis.

The reason of this is evident from what precedes. If the body be set rotating about an axis very near the principal axis of greatest or least moment, both the polhode and herpolhode will generally be very small curves, and the direction of that principal axis of the body will be very nearly fixed in space. If now a small impulse f act on the body, the effect will be to alter slightly the position of the instantaneous axis. It will be moved from one polhode to another very near the former, and thus the angular position of the axis in space will not be much affected. Let Ω be the angular velocity of the body, ω that generated by the impulse, then, by the parallelogram of angular velocities, the change in the position of the instantaneous axis cannot be greater than $\sin^{-1}(\omega/\Omega)$. If therefore Ω be great, ω must also be great, to produce any considerable change in the axis of rotation. But if the body have no initial rotation Ω, the impulse may generate an angular velocity ω about an axis not nearly coincident with a principal axis. Both the polhode and the herpolhode may then be large curves, and the instantaneous axis of rotation will move about both in the body and in space. The motion will then appear very unsteady. In this manner, for example, we may explain why in the game of cup and ball, spinning the ball about a vertical axis makes it more easy to catch on the spike. Any motion caused by a wrong pull of the string or by gravity will not produce so great a change of motion as it would have done if the ball had been initially at rest. The fixed direction of the earth's axis in space is also due to its rotation about its axis of figure. In rifles, a rapid rotation is communicated to the bullet about an axis in the direction in which the bullet is moving. It follows, from what precedes, that the axis of rotation will be nearly unchanged throughout the motion. One consequence is that the resistance of the air acts in a known manner on the bullet, the amount of which may therefore be calculated and allowed for.

On the Cones described by the Invariable and Instantaneous Axes treated by Spherical Trigonometry.

157.   There are various ways in which we may study the motion of a body about a fixed point. We may have recourse to the properties of an ellipsoid as Poinsot and MacCullagh have done. But we may also use a sphere whose centre is at the fixed point and which is either fixed in the body or fixed in space at our pleasure. This method is particularly useful when we wish to find the angular motion of any line in space or in the body. By referring these angles to arcs drawn on the surface of the sphere we are enabled to shorten our processes by using such formulæ of spherical trigonometry as may suit our purpose.

The cones described by the invariable line and the instantaneous axis intersect this sphere in sphero-conics. The properties of such cones are not usually given with sufficient fulness in our treatises on solid geometry. For this reason we have added a list of several properties likely to be useful. In order not to interrupt the general line of the argument this list has been placed at the end of the chapter.

158.   It is clear from what precedes that there are two important straight lines whose motions we should consider. These are the invariable line and the instantaneous axis. The first of these is fixed in space, but as the body moves the invariable line describes a cone in the body, which by Art. 152 intersects the ellipsoid of gyration in a sphero-conic. This cone is usually called the *Invariable Cone*. The instantaneous axis describes both a cone in the body and a cone in space. By Art. 143, the cone described in the body intersects the momental ellipsoid in a polhode, and the cone described in space intersects the fixed plane on which the momental ellipsoid rolls in a herpolhode. These two

cones may be called respectively the *instantaneous cone* and the *cone of the herpolhode*.

159. **The Cones**. Let the principal axes at the fixed point be taken as the axes of co-ordinates. The axes of reference are therefore fixed in the body but moving in space. By Art. 140, the direction-cosines of the invariable line are $A\omega_1/G$, $B\omega_2/G$, $C\omega_3/G$; and the direction-cosines of the instantaneous axis are ω_1/ω, ω_2/ω, ω_3/ω. From the equations (1) and (2) of Art. 140, we easily find

$$(A\omega_1^2 + B\omega_2^2 + C\omega_3^2)\, G^2 = (A^2\omega_1^2 + B^2\omega_2^2 + C^2\omega_3^2)\, T.$$

If we take the co-ordinates x, y, z to be proportional to the direction-cosines of either of these straight lines and eliminate ω_1, ω_2, ω_3 by the help of this equation, we obtain the equation to the corresponding cone described by that straight line. In this way we find that the cones described in the body by the invariable line and the instantaneous axis are respectively

$$\frac{AT - G^2}{A}\, x^2 + \frac{BT - G^2}{B}\, y^2 + \frac{CT - G^2}{C}\, z^2 = 0,$$

$$A\,(AT - G^2)\, x^2 + B\,(BT - G^2)\, y^2 + C\,(CT - G^2)\, z^2 = 0.$$

These cones become two planes when the initial conditions are such that $G^2 = BT$.

Ex. 1. Show that the circular sections of the invariable cone are parallel to those of the ellipsoid of gyration and perpendicular to the asymptotes of the focal conic of the momental ellipsoid.

160. There is a third straight line whose motion it is sometimes convenient to consider, though it is not nearly so important as either the invariable line or the instantaneous axis. If x, y, z be the co-ordinates of the extremity of a radius vector of an ellipsoid referred to its principal diameters as axes and if a, b, c be the semi-axes, the straight line whose direction-cosines are x/a, y/b, z/c is called the *eccentric line* of that radius vector. Taking this definition, it is easy to see that the direction-cosines of the eccentric line of the instantaneous axis with regard to the momental ellipsoid are $\omega_1 \sqrt{A/T}$, $\omega_2 \sqrt{B/T}$, $\omega_3 \sqrt{C/T}$. These are also the direction-cosines of the eccentric line of the invariable line with regard to the ellipsoid of gyration. This straight line may therefore be called simply the *eccentric line* and the cone described by it in the body may be called the *eccentric cone*.

Ex. 1. The equation to the eccentric cone referred to the principal axes at the fixed point is $(AT - G^2)\, x^2 + (BT - G^2)\, y^2 + (CT - G^2)\, z^2 = 0$. This cone has the same circular sections as the momental ellipsoid and cuts that ellipsoid in a sphero-conic.

Ex. 2. The polar plane of the instantaneous axis with regard to the eccentric cone touches the invariable cone along the corresponding position of the invariable line. Thus the invariable and instantaneous cones are reciprocals of each other with regard to the eccentric cone.

161. **The sphero-conics**. Let a sphere of radius unity be described with its centre at the fixed point O about which the

body is free to turn. Let this sphere be fixed in the body, and therefore move with it in space. Let the invariable line, the instantaneous axis, and the eccentric line cut this sphere in the points L, I, and E respectively. Also let the principal axes cut the sphere in A, B, C. It is clear that the intersections of the invariable, instantaneous, and eccentric cones with this sphere will be three sphero-conics which are represented in the figure by the

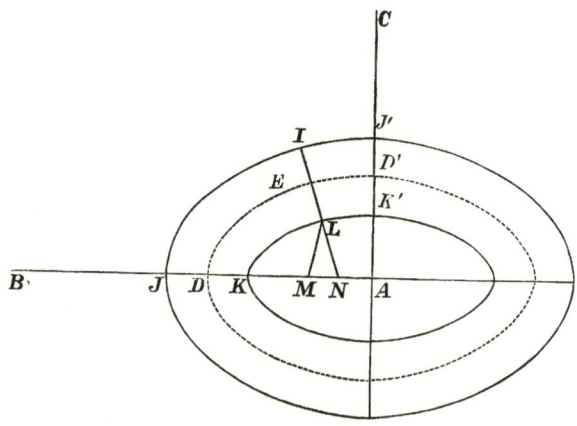

lines KK', JJ', DD', respectively. The eye is supposed to be situated on the axis OA, viewing the sphere from a considerable distance. All great circles on the sphere are represented by straight lines. Since the cones are co-axial with the momental ellipsoid, these sphero-conics are symmetrical about the principal planes of the body. The intersections of these principal planes with the sphere will be three arcs of great circles, and the portions of these arcs cut off by any sphero-conic are called axes of that sphero-conic. If we put $z = 0$ in the equations to any one of the three cones, the value of y/x is the tangent of that semi-axis of the sphero-conic which lies in the plane of xy. Similarly, putting $y = 0$, we find the axis in the plane of xz. If (a, b), (a', b'), (a, β) be the semi-axes of the invariable, instantaneous, and eccentric sphero-conics respectively, we thus find

$$\frac{\tan a}{B} = \frac{\tan a'}{A} = \frac{\tan \alpha}{\sqrt{AB}} = \frac{\sqrt{AT - G^2}}{\sqrt{G^2 - BT}} \frac{1}{\sqrt{AB}},$$

$$\frac{\tan b}{C} = \frac{\tan b'}{A} = \frac{\tan \beta}{\sqrt{AC}} = \frac{\sqrt{AT - G^2}}{\sqrt{G^2 - CT}} \frac{1}{\sqrt{AC}}.$$

The first of these two sets gives the axes in the plane AOB, the second those in the plane AOC. The former will be imaginary if $G^2 < BT$. In this case the sphero-conics do not cut the plane AOB. The sphero-conics will therefore have their concavities turned towards the extremities of the axes OA or OC, i.e. towards the extremities of the axes of greatest or least moment according as G^2 is $>$ or $< BT$. Since $\tan b/\tan b' = C/A$ it is clear that the invariable cone and the axis of greatest moment of inertia always lie on the same side of the instantaneous cone.

162.　**Ex. 1.**　If we put $1 - e^2 = \sin^2 b/\sin^2 a$ we may define e to be the eccentricity of the sphero-conic whose semi-axes are a and b. If e and e' be the eccentricities of the invariable and eccentric sphero-conics respectively, prove that

$$e^2 = A(B - C)/B(A - C) \text{ and } e'^2 = (B - C)/(A - C)$$

so that both these eccentricities are independent of the initial conditions.

Ex. 2.　If the radius of the sphere had been taken equal to $(G^2/MT)^{\frac{1}{2}}$ instead of unity, show that it would have intersected the ellipsoid of gyration along the invariable cone, and if the radius had been $(MT\epsilon^4/G^2)^{\frac{1}{2}}$, it would have intersected the momental ellipsoid along the eccentric cone.

Ex. 3.　A body is set rotating with an initial angular velocity n about an axis which very nearly coincides with a principal axis OC at a fixed point O. The motion of the instantaneous axis in the body may be found by the following formulæ. Let a sphere be described whose centre is O, and let I be the extremity of the radius vector which is the instantaneous axis at the time t. If (x, y) be the co-ordinates of the projection of I on the plane AOB referred to the principal axes OA, OB, then
$$x = \sqrt{B(B - C)}\, L \sin(pnt + M),$$
$$y = \sqrt{A(A - C)}\, L \cos(pnt + M),$$
where $p^2 = (B - C)(A - C)/AB$, and L, M are two arbitrary constants depending on the initial values of x, y.

Ex. 4.　If in the last question L be the point in which the sphere cuts the invariable line, if (ρ, θ) be the spherical polar co-ordinates of C with regard to L as origin, and a the radius of the sphere, then

$$\rho^2 = n^2 \frac{AB}{2G^2} L^2 \{2AB - C(A + B) + (A - B)C \cos 2(pnt + M)\}, \quad \theta = \frac{T}{G}t + \frac{CT - G^2}{CG}\int \frac{a^2 dt}{\rho^2}.$$

163.　*To find the motion of the invariable line and of the instantaneous axis in the body.*

Since the invariable line OL is fixed in space and the body is turning about OI as instantaneous axis, it is evident that the direction of motion of OL in the body is perpendicular to the plane IOL. Hence on a sphere whose centre is at O the arc IL *is normal to the sphero-conic described by the invariable line.* This simple relation will serve to connect the motions of the invariable line and the instantaneous axis along their respective spheroconics.

Supposing ω_1, ω_2, ω_3 to be all positive the axis OI lies in the positive octant, and the body is turning round OI in the direction ABC (Fig. Art. 161). Since OL is fixed in space, it appears to move in the body in the direction opposite to rotation.

If then L and A lie on the same side of the sphero-conic JJ' (as is the case when A, B, C are in descending order of magnitude), L moves in the body along its sphero-conic in the direction KK'. On the other hand, if L and A lie on opposite sides of the sphero-conic JJ', L moves in the opposite direction. See also Art. 150.

164. Let v be the velocity of the invariable line along its sphero-conic, then since the body is turning about OI with angular velocity ω, and OL is unity, we have $v = \omega \sin LOI$. But by Art. 143 $T/G = \omega \cos LOI$. Eliminating ω we have

$$v = (T/G) \tan LOI.$$

165. Produce the arc IL to cut the axis AK in N, so that LN is a normal to the sphero-conic described by the invariable line. Taking the principal axes at the fixed point O as axes of reference, the direction-cosines of OL and OI are respectively proportional to $A\omega_1$, $B\omega_2$, $C\omega_3$, and ω_1, ω_2, ω_3. The equation to the plane LOI is

$$(B - C)\, \omega_2\omega_3 x + (C - A)\, \omega_3\omega_1 y + (A - B)\, \omega_1\omega_2 z = 0.$$

This plane intersects the plane of xy in the straight line ON, hence putting $z = 0$, we find the direction-cosines of ON to be proportional to $(A - C)\, \omega_1$, $(B - C)\, \omega_2$, and 0. Hence

$$\cos LON = \frac{A\,(A - C)\,\omega_1^2 + B\,(B - C)\,\omega_2^2}{G\sqrt{(A - C)^2\,\omega_1^2 + (B - C)^2\,\omega_2^2}}.$$

The numerator of this expression is easily seen to be $G^2 - CT$. Expanding the quantity under the root we have

$$A^2\omega_1^2 + B^2\omega_2^2 - 2C\,(A\omega_1^2 + B\omega_2^2) + C^2\,(\omega_1^2 + \omega_2^2),$$

which is clearly the same as

$$G^2 - C^2\omega_3^2 - 2C\,(T - C\omega_3^2) + C^2\,(\omega^2 - \omega_3^2).$$

Substituting we find

$$\cos LON = \frac{G^2 - CT}{G\sqrt{G^2 - 2CT + C^2\omega^2}};$$

$$\therefore\ \tan LON = \frac{C\sqrt{G^2\omega^2 - T^2}}{G^2 - CT}.$$

But $T/G = \omega \cos LOI$, $\therefore\ T \tan LOI = \sqrt{G^2\omega^2 - T^2}$. *Hence the ratio* $\dfrac{\tan LOI}{\tan LON} = \dfrac{G^2 - CT}{CT}$, *and is therefore constant throughout the motion.*

Combining this result with that given in the last Article, we see that the

$$\left.\begin{array}{r}\text{velocity of } L\\ \text{along its conic}\end{array}\right\} = \frac{G^2 - CT}{CG}\tan n,$$

where n is the angle LON. If we adopt the conventions of spherical trigonometry, n is also the length of the arc normal to the sphero-conic intercepted between the curve and the principal plane AB of the body.

166. **Ex. 1.** If the focal lines of the invariable cone cut the sphere in S and S', these points are called the foci of the sphero-conic. Prove that *the velocity*

of L *resolved perpendicular to the arc* SL *is constant throughout the motion* and equal to $\{(G^2 - BT)(AT - G^2)/ABG^2\}^{\frac{1}{2}}$. If LM be an arc of a great circle perpendicular to the axis containing the foci, and ρ be the arc SL, prove also that

$$\frac{d\rho}{dt} = -\frac{G}{C}\left\{\frac{(A-C)(B-C)}{AB}\right\}^{\frac{1}{2}}\sin LM.$$

Ex. 2.   Prove that the velocity of L resolved perpendicularly to the central radius vector AL is $\dfrac{AT - G^2}{AG}\cot AL$.

Ex. 3.   If r, r', r'' be the lengths of the arcs joining the extremity A of a principal axis to the extremities L, I, E of the invariable line, instantaneous axis, and eccentric line respectively; θ, θ', θ'' the angles these arcs make with any principal plane AOB, prove that

$$\frac{\cos r}{AT} = \frac{\cos r'}{G^2\cos\zeta} = \frac{\cos r''}{G\sqrt{AT}}, \qquad \frac{\tan\theta}{C} = \frac{\tan\theta'}{B} = \frac{\tan\theta''}{\sqrt{BC}}.$$

where $\zeta = $ arc LI. This theorem will enable us to discover in what manner the motions of the three points L, I, E are related to each other.

Ex. 4.   Show that the velocity of the instantaneous axis along its sphero-conic is $\dfrac{G}{T}\dfrac{G^2 - CT}{AB}\tan n'\cos\zeta$, where n' is the length of the normal to the instantaneous sphero-conic intercepted between the curve and the arc AB, and $\zeta = $ arc LI.

Comparing this result with the corresponding formula for the motion of L given in Art. 165, we see that for every theorem relating to the motion of L in its sphero-conic there is a corresponding theorem for the motion of I. For example, if S' be a focus of the instantaneous sphero-conic, we see by Ex. 1 that the velocity of I resolved perpendicular to the focal radius vector $S'I$ bears a constant ratio to $\cos LI$. This constant ratio is equal to that given in Ex. 1 multiplied by G^2C/TAB.

Ex. 5.   Show that the velocity of the eccentric line along its sphero-conic is $\{(G^2 - CT)/\sqrt{ABCT}\}\tan n''$, where n'' is the length of the arc normal to the sphero-conic intercepted between the curve and the principal arc AB.

Ex. 6.   Prove that (velocity of E)^2 − (velocity of L)^2 = constant. Show also that this constant $= (AT - G^2)(BT - G^2)(CT - G^2)/ABCG^2T$.

Ex. 7.   The motion of L along its sphero-conic is the same as that of a particle acted on by two forces whose directions are the tangents at L to the arcs LS, LS' joining L to the foci of the sphero-conic and whose magnitudes are respectively proportional to $\sin LS\cos LS'$ and $\sin LS'\cos LS$.

Solutions of these examples and proofs of other theorems in this section may be found in a paper contributed by the author to the *Proceedings of the Royal Society*, 1873.

167.   The instantaneous axis describes a cone in space, which has been called the cone of the herpolhode. The equation of this cone cannot generally be found, but when it can be determined we have another geometrical representation of the motion. For suppose the two cones described by the instantaneous axis in space and in the body to be constructed. Since each of these cones will contain two consecutive positions of their common generator, they will touch each other along the instantaneous axis. Then, the points of contact having no velocity, the motion

will be represented by making the cone fixed in the body roll on the cone fixed in space.

168. **Poinsot's theorem.** *To find the motion of the instantaneous axis in space.*

Since the invariable line OL is fixed in space, it will be convenient to refer the motion to OL as one axis of co-ordinates. Let the angle the instantaneous axis OI makes with OL be called ζ, and let ϕ be the angle the plane IOL makes with any plane passing through OL and fixed in space.

During the motion the cone described by OI in the body rolls on the cone described by OI in space. It is therefore clear that the angular velocity of the instantaneous axis in space is the same as its angular velocity in the body. Describe a sphere whose centre is at O and radius unity, and let this sphere be fixed in the body. Let L, I be the intersections of the invariable line and instantaneous axis with the sphere at the time t, L', I' their intersections at the time $t+dt$. Then $IL, I'L'$ are consecutive normals to the sphero-conic KK' traced out by the invariable line and therefore intersect each other in some point P

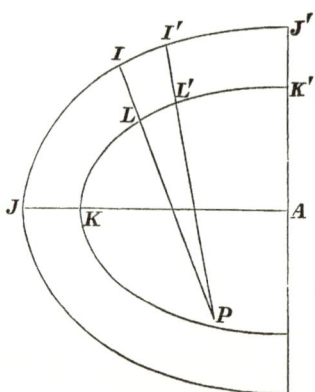

which may be regarded as a centre of curvature of the spheroconic. Let $\rho = PL$. Then clearly

$$\begin{matrix}\text{velocity of } I \text{ resolved} \\ \text{perpendicularly to } IL\end{matrix} \Big\} = \begin{pmatrix}\text{velocity} \\ \text{of } L\end{pmatrix} \cdot \frac{\sin(\rho+\zeta)}{\sin\rho}.$$

Therefore by Art. 164 we have

$$\sin\zeta\frac{d\phi}{dt} = \frac{T}{G}\tan\zeta(\cos\zeta + \cot\rho\sin\zeta);$$

$$\therefore \frac{d\phi}{dt} = \frac{T}{G}\left(1 + \frac{\tan\zeta}{\tan\rho}\right).$$

But in any sphero-conic $\tan\rho = \tan^3 n/\tan^2 l$, where n is the length of the normal intercepted between the curve and that axis which contains the foci, and $2l$ is the length of the ordinate through either focus, and is usually called the latus rectum. Substituting for $\tan \rho$, and remembering that

$$\frac{\tan \zeta}{\tan n} = \frac{G^2 - CT}{CT}, \text{ by Art. 165, and } \tan l = \frac{\tan^2 b}{\tan a}, \text{ we get}$$

$$\frac{d\phi}{dt} = \frac{T}{G} + \frac{T}{G}\left(\frac{G^2 - CT}{CT}\right)^3 \cdot \left(\frac{\tan^2 b}{\tan a}\right)^2 \cot^2 \zeta.$$

If we substitute for $\tan a$ and $\tan b$ their values, we get

$$\frac{d\phi}{dt} = \frac{T}{G} + \frac{(AT - G^2)(BT - G^2)(CT - G^2)}{ABCGT^2}\cot^2 \zeta.$$

169.   A simple geometrical construction for this result has been given by Dr Ferrers, Master of Caius College, in a Smith's Prize paper (1882). If OH be the projection of the instantaneous axis OI on the invariable plane drawn through the fixed point O, and if OH intersect the momental ellipsoid in H, then

$$\frac{d\phi}{dt} = \frac{G^3 M \epsilon^4}{TABC} \frac{1}{OH^2}.$$

170.   Since the resolved angular velocity about the invariable line is constant, we easily find $\omega = \sec \zeta \, T/G$. Substituting this value of ω in equation (6) of Art. 137, we find a relation between ζ and $d\zeta/dt$, which however is too complicated to be of much use.

The values of $d\phi/dt$ and $d\zeta/dt$ in terms of ζ have now both been found; from these the motion of the instantaneous axis in space can be deduced.

171.   Ex. 1.   Show that the angular velocity v' of the instantaneous axis in space or in the body is given by $\omega^2 v'^2 = \frac{T^2}{ABC}\left(A + B + C - 2\frac{G^2}{T}\right) - \frac{\lambda_1 \lambda_2 \lambda_3}{\omega^2}$, where ω is the resultant angular velocity of the body and λ_1, λ_2, λ_3 have the meanings given to them in Art. 137.   This result is due to Poinsot.

Ex. 2.   The length of the spiral between two of its successive apsides, described in absolute space, on the surface of a fixed concentric sphere, by the instantaneous axis of rotation, is equal to a quadrant of the spherical ellipse described by the same axis on an equal sphere moving with the body.   This is Booth's Theorem.

Ex. 3.   If the eccentric line intersect in the point E the unit sphere which is fixed in the body and has its centre at the fixed point, prove that

$$\left(\begin{matrix}\text{velocity} \\ \text{of } E\end{matrix}\right)^2 = \frac{T}{G}\frac{d\phi}{dt}\tan^2 \zeta,$$

where the letters have the meanings given to them in Art. 168.

172.   **The Rolling and Sliding Cone.**   Let O be the fixed point, OI the instantaneous axis.   Let the angular velocity ω about OI be resolved into two, viz. a uniform angular velocity T/G about the invariable line OL, and an angular velocity $\omega \sin IOL$

about a line OH lying in a plane fixed in space perpendicular to
the invariable line, and passing through the fixed point O. Let
this fixed plane be called the invariable plane at O. As the body
moves, OH will describe a cone in the body which will always touch
this fixed plane. The velocity of any point of the body lying for a
moment in OH is unaffected by the rotation about OH, and the
point has therefore only the motion due to the uniform angular
velocity about OL. We have thus a new representation of the
motion of the body. Let the cone described by OH in the body
be constructed, and let it roll on the invariable plane at O with the
proper angular velocity, while at the same time this plane turns
round the invariable line with a uniform angular velocity T/G.
The cone described by OH in the body has been called by Poinsot
the *Rolling and Sliding Cone*.

To find a construction for the sliding cone. Its generator
OH is at right angles to OL, and lies in the plane IOL. Now
OL is fixed in space; let OL' be the line in the body which, after
an interval of time dt, will come into the position OL. Since the
body is turning about OI, the plane LOL' is perpendicular to the
plane LOI, and hence OH is perpendicular to both OL and OL'.
That is, OH is perpendicular to the tangent plane to the cone
described by OL in the body. The cone described by OH in the
body is therefore the *reciprocal cone* of that described by OL.
The equation to the cone described by OL has been found in Art.
159. Turning therefore its coefficients upside down, we see that
the equation to the cone described by OH is

$$\frac{A}{AT - G^2} x^2 + \frac{B}{BT - G^2} y^2 + \frac{C}{CT - G^2} z^2 = 0.$$

The focal lines of the cone described by OH are perpendicular
to the circular sections of the reciprocal cone, that is the cone
described by OL. And these circular sections are the same as
the circular sections of the ellipsoid of gyration. Hence the focal
lines lie in the plane containing the axes of greatest and least
moment, and are independent of the initial conditions.

This cone becomes a straight line in the case in which the
cone described by OL becomes a plane, viz. when the initial
conditions are such that $G^2 = BT$.

173. *To find the motion of* OH *in space and in the body.*

Since OL, OH and OI are always in the same plane the
motion of OH in space round the fixed straight line OL is the
same as that of OI, and is given by the expression for $d\phi/dt$ in
Art. 168.

To find the motion of OH in the body it will be convenient
to refer to the figure of Art. 168. Produce the arcs PL, PL'
to H and H' so that LH and $L'H'$ are each quadrants. Then

H and H' are the points in which the axis OH intersects the unit sphere at the times t and $t + dt$. We have therefore

$$\begin{pmatrix}\text{velocity} \\ \text{of } H\end{pmatrix} = \begin{pmatrix}\text{velocity} \\ \text{of } L\end{pmatrix} \cdot \frac{\sin (\rho + \frac{1}{2}\pi)}{\sin \rho} = \frac{T}{G} \tan \zeta \cot \rho.$$

Substituting for $\tan \rho$ as before we may express the result in terms of ζ or ω at our pleasure.

Since the cone described by OH in the body rolls on a plane which also turns round a normal to itself at O, it is clear that the angular velocity of OH in the body is less than the angular velocity of OH in space by the angular velocity of the plane, i.e.

$$\begin{pmatrix}\text{velocity} \\ \text{of } H\end{pmatrix} = \frac{d\phi}{dt} - \frac{T}{G}.$$

Ex. If l, m, n be the direction-cosines of OH referred to the principal axes of the body, prove $\dfrac{l}{(AT - G^2)\,\omega_1} = \dfrac{m}{(BT - G^2)\,\omega_2} = \dfrac{n}{(CT - G^2)\,\omega_3} = \dfrac{1}{G\sqrt{G^2\omega^2 - T^2}}.$

The conjugate Ellipsoid and the conjugate line.

174. Let the momental ellipsoid at the fixed point be

$$Ax^2 + By^2 + Cz^2 = K \dots\dots\dots\dots\dots(1),$$

where $K = M\epsilon^4$. We also have

$$\left.\begin{array}{l} A\,\omega_1^2 + B\,\omega_2^2 + C\,\omega_3^2 = T \\ A^2\omega_1^2 + B^2\omega_2^2 + C^2\omega_3^2 = G^2 \end{array}\right\}\dots\dots\dots\dots(2).$$

These give

$$\left.\begin{array}{l} (\lambda A - A^2)\,\omega_1^2 + (\lambda B - B^2)\,\omega_2^2 + (\lambda C - C^2)\,\omega_3^2 = \lambda T - G^2 \\ (\mu A - A^2)\,\omega_1^2 + (\mu B - B^2)\,\omega_2^2 + (\mu C - C^2)\,\omega_3^2 = \mu T - G^2 \end{array}\right\}\cdot\cdot\,(3).$$

If we now choose three quantities A', B', C', such that

$$\left.\begin{array}{ll} A' = (\lambda A - A^2)\,i, & A'^2 = (\mu A - A^2)j, \\ B' = (\lambda B - B^2)\,i, & B'^2 = (\mu B - B^2)j, \\ C' = (\lambda C - C^2)\,i, & C'^2 = (\mu C - C^2)j, \end{array}\right\}\dots\dots(4),$$

we may construct in the body another conicoid, viz.

$$A'x^2 + B'y^2 + C'z^2 = K' \dots\dots\dots\dots(5),$$

which will afterwards be shewn to be an ellipsoid. We shall also have

$$\left.\begin{array}{l} A'\,\omega_1^2 + B'\,\omega_2^2 + C'\,\omega_3^2 = T' \\ A'^2\omega_1^2 + B'^2\omega_2^2 + C'^2\omega_3^2 = G'^2 \end{array}\right\}\dots\dots\dots\dots(6),$$

where T' and G' are two new constants.

This second ellipsoid will possess some properties analogous to those of the momental ellipsoid. Thus :

(1) The angular velocity about the radius vector round which the body is turning varies as that radius vector.

(2)   The length of the perpendicular on the tangent plane at the extremity of the instantaneous axis is constant.

(3)   The resolved angular velocity of the body about this perpendicular on the tangent plane is constant, and $= T'/G'$.

It is not generally true that the position in space of this perpendicular is fixed.

3.   To determine if this transformation is possible we must examine the constants λ and μ.   Solving (4) we find

$$\lambda = \tfrac{1}{2} (A + B + C),$$

$$\frac{4ABC}{\mu} = 2AB + 2BC + 2CA - A^2 - B^2 - C^2 = \frac{4j}{i^2}.$$

We have therefore the following results:

$$A' = \tfrac{1}{2} iA (B + C - A), \qquad B' = \tfrac{1}{2} i B (C + A - B), \qquad C' = \tfrac{1}{2} iC (A + B - C),$$

$$T' = i (\lambda T - G^2), \qquad G'^2 = i^2 \frac{ABC}{\mu} (\mu T - G^2).$$

Since A, B, C are moments of inertia, they are all positive, and the sum of any two is greater than the third.   We infer (1) that A', B', C' are also all positive, (2) that λ and μ are positive and greater than the greatest of the three A, B, C, (3) that T' and G'^2 are real and positive.

175.   Since this analysis gives only *one* value each to λ and μ, it follows that if we perform the same operations on the second ellipsoid we shall obtain the first ellipsoid and no other.   *Hence the two ellipsoids are conjugate to each other.*   Thus we have

$$A = \tfrac{1}{2} i' A' (B' + C' - A'), \quad \&c., \quad \&c.,$$

and by substitution $i'/i = ABC/A'B'C'$.

Either of the two bodies whose moments of inertia are A, B, C and A', B', C' may be called the *conjugate* of the other.   When we consider only the motion of one body, we suppose that body to carry with it the two ellipsoids as if rigidly connected to it.   The perpendicular on the tangent plane to the momental ellipsoid of the body at its intersection with the instantaneous axis is the invariable line, while the corresponding perpendicular on the tangent plane to the conjugate ellipsoid at its intersection with the instantaneous axis is called the *conjugate line*.   The direction cosines of the conjugate line are therefore $A'\omega_1/G'$, $B'\omega_2/G'$, $C'\omega_3/G'$. See a paper by the author in the *Quarterly Journal*, 1888.

Ex. 1.   Show that $\dfrac{B' - C'}{A'} = - \dfrac{B - C}{A}$, $\dfrac{A' (A'T' - G'^2)}{A (AT - G'^2)} = - \dfrac{A'B'C'}{ABC}$
with similar equations for the other letters.

Show also that if A, B, C are in descending order of magnitude, A', B', C' are in ascending order.

Ex. 2.   Show that the motion in space of any point situated in the conjugate line is in the same direction as if that point fixed (for the moment) on the body, but its velocity is twice as great. See Art. 5, (1), and Art. 140, note.

Ex. 3. Many of the theorems which govern the motion of the conjugate line OL' are similar to those which govern the motion of the invariable line OL.

The following are examples :—

(1) The straight lines OL, OI, OL' describe quadric cones in the body in the same direction, the cone described by the instantaneous axis OI being between the cones described by the invariable line OL and the conjugate line OL'.

(2) The normal planes to the cones described by OL, OL' intersect each other along the instantaneous axis OI.

(3) The velocity of OL' along its cone varies as the tangent of the inclination to OI, and the ratio is equal to T'/G'. It also varies as the tangent of the angle OL' makes with the intersection of the plane $L'OI$ with any principal section of the conjugate ellipsoid. See Art. 165.

(4) The cosines of the angles IOL, IOL' are always in a constant ratio.

Ex. 4. If θ, ψ be the angular co-ordinates of the conjugate line OL' referred to the invariable line OL as the axis of z, show that

$$\sin^2 \theta \frac{d\psi}{dt} = 2 \left(\frac{T}{G} - \frac{T'}{G'} \cos \theta \right)$$

$$\sin^2 \theta \left(\frac{d\psi}{dt} \right)^2 + \left(\frac{d\theta}{dt} \right)^2 = H - 4 \left(\frac{T'}{G'} \right)^2 - \frac{4}{ABC} \frac{GG'}{i} \cos \theta$$

where
$$\tfrac{1}{4} ABCH = T \left(BC + CA + AB \right) - \tfrac{1}{2} G^2 \left(A + B + C \right)$$

$$\sin^2 \theta \left(\frac{d\theta}{dt} \right)^2 = \frac{4}{ABC} \frac{GG'}{i} (\cos \theta - \alpha)(\cos \theta - \beta)(\cos \theta - \gamma)$$

where $\alpha G G'/i = TBC + G^2 (A - \lambda)$, &c., &c.

It should be noticed that α, β, γ are real.

Ex. 5. Two bodies each turning about a fixed point have angular velocities ω_1, ω_2, ω_3 and ω_1', ω_2', ω_3' about their principal axes and their principal moments are A, B, C and A', B', C'. If these bodies move so that $\omega_1 = \omega_1'$, $\omega_2 = \omega_2'$, $\omega_3 = \omega_3'$ prove from Euler's equations that $A'/A = B'/B = C'/C$. If they move so that $\omega_1 = -\omega_1'$, $\omega_2 = -\omega_2'$, $\omega_3 = -\omega_3'$, prove that the bodies are conjugate.

Prove also that if the relations given between the angular velocities hold initially it will always hold and that the cones described by the instantaneous axes are equal and similar.

Motion of the Principal Axes.

176. *To find the angular motions in space of the principal axes.*

Since the invariable line OL is fixed in space it will be convenient to refer the motion to this straight line as axis of z. Let OA, OB, OC be the principal axes at the fixed point O, and let, as before, α, β, γ be their inclinations to the axis OL or OZ. Let λ, μ, ν be the angles the planes LOA, LOB, LOC make with some fixed plane LOX passing through OL. Our object is to find $d\alpha/dt$ and $d\lambda/dt$ with similar expressions for the other axes. We might here refer to Euler's geometrical equations given in Vol. I. chap. 5 and by writing α, λ for θ, ψ respectively obtain the required expressions, but it will be found advantageous to make a slight variation in the argument.

Describe a sphere whose centre is at the fixed point, and whose radius is unity. Let the invariable line, the instantaneous axis and the principal axes cut this sphere in the points L, I, A, B, C respectively. The velocity of A resolved perpendicular to LA will then be $\sin \alpha \, d\lambda/dt$. But since the body is turning round OI as instantaneous axis, the point A is moving perpendicularly to the arc IA, and its velocity is $\omega \sin IA$. Resolving this perpendicularly to the arc LA, we have

$$\sin \alpha \frac{d\lambda}{dt} = \omega \sin AI \cos LAI = \omega \frac{\cos LI - \cos LA \cos IA}{\sin LA},$$

by a fundamental formula in spherical trigonometry. But $\omega \cos LI$ is the resolved part of the angular velocity about OL, which is equal to T/G, and $\omega \cos IA$ is the resolved part of the angular

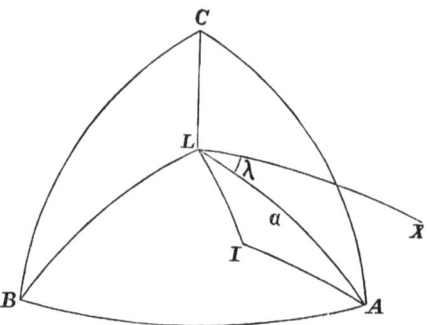

velocity about OA, which is ω_1. We have therefore

$$\sin^2 \alpha \frac{d\lambda}{dt} = \frac{T}{G} - \omega_1 \cos \alpha,$$

a result which follows immediately from Art. 19. Since $G \cos \alpha = A\omega_1$,

we have
$$\sin^2 \alpha \frac{d\lambda}{dt} = \frac{T}{G} - \frac{G \cos^2 \alpha}{A} \dots\dots\dots\dots\dots(1).$$

This result may also be written in the form

$$\frac{d\lambda}{dt} = \frac{T}{G} + \frac{AT - G^2}{AG} \cot^2 \alpha \dots\dots\dots\dots(2).$$

177. To find $\frac{d\alpha}{dt}$ we may proceed in the following manner. By Art. 140, we have $\cos \alpha = A\omega_1/G$, $\cos \beta = B\omega_2/G$, $\cos \gamma = C\omega_3/G$. Substituting in Euler's equation

$$A \frac{d\omega_1}{dt} - (B - C) \omega_2 \omega_3 = 0,$$

we have $$\sin\alpha\,\frac{d\alpha}{dt} = \left(\frac{1}{B}-\frac{1}{C}\right)G\cos\beta\cos\gamma\ldots\ldots\ldots(3).$$

But by Art. 137 $\cos\alpha$, $\cos\beta$, $\cos\gamma$ are connected by the equations

$$\left.\begin{array}{c}\dfrac{\cos^2\alpha}{A}+\dfrac{\cos^2\beta}{B}+\dfrac{\cos^2\gamma}{C}=\dfrac{T}{G^2}\\[2mm]\cos^2\alpha+\cos^2\beta+\cos^2\gamma=1\end{array}\right\}\ldots\ldots\ldots(4).$$

If we solve these equations so as to express $\cos\beta$, $\cos\gamma$ in terms of $\cos\alpha$, we easily find

$$\sin^2\alpha\left(\frac{d\alpha}{dt}\right)^2 = -\frac{G^2}{BC}\left(\frac{G^2-CT}{G^2}-\frac{A-C}{A}\cos^2\alpha\right)\left(\frac{G^2-BT}{G^2}-\frac{A-B}{A}\cos^2\alpha\right)\ldots(5).$$

178. Since the left-hand side of equation (5) is necessarily real, we see that the values of $\cos^2\alpha$ are restricted to lie between certain limits. If the axis whose motion we are considering is the axis of greatest or least moment let B be the axis of mean moment. In this case $\cos^2\alpha$ must lie *between* the two limits $\dfrac{G^2-CT}{G^2}\dfrac{A}{A-C}$ and $\dfrac{G^2-BT}{G^2}\dfrac{A}{A-B}$ if both are positive. By Art. 138 the former of these two is positive and less than unity; this is easily shown by dividing the numerator and the denominator by ACG^2. If the latter is positive the spiral described by the principal axis on the surface of a sphere whose centre is at the fixed point lies between two concentric circles which it alternately touches. If the latter limit is negative $\cos\alpha$ has no inferior limit. In this case the spiral always lies between two small circles on the sphere, one of which is exactly opposite the other.

If the axis considered is the axis of mean moment, $\cos^2\alpha$ must lie *outside* the same two limits as before. Both these are positive, but one is greater and the other less than unity. The spiral therefore lies between two small circles opposite each other.

In order that $d\lambda/dt$ may vanish we must have $G^2\cos^2\alpha=AT$, but this by substitution makes $d\alpha/dt$ imaginary. Thus $d\lambda/dt$ always keeps one sign. It is easy to see that if the initial conditions are such that G^2/T is less than the moment of inertia about the axis which describes the spiral we are considering, the angular velocity will be greatest when the axis is nearest the invariable line and least when the axis is furthest. The reverse is the case if G^2/T is greater than the moment of inertia.

179. Ex. 1. Let OM be any straight line fixed in the body and passing through O and let it cut the ellipsoid of gyration at M. Let OM' be the perpendicular from O on the tangent plane at M. If $OM=r$, $OM'=p$, and if i, i' be the angles OM, OM' make with the invariable line OL, prove that

$$\sin^2 i\,\frac{dj}{dt} = \frac{T}{G}-\frac{G}{mpr}\cos i\cos i',$$

where j is the angle the plane LOM makes with some plane fixed in space passing through OL and m is the mass of the body. This follows from Art. 19.

Ex. 2. If KLK' be the conic traced out by the invariable line in the manner described in Art. 161, show that $\lambda=(T/G)\,t+(\text{angle }LAK)-(\text{vectorial area }LAK)$, where λ is the angle described by the plane containing the invariable line and the principal axis OA.

Ex. 3. If we draw three straight lines OA, OB, OC along the principal axes at the fixed point O of equal lengths, the sum of the areas conserved by these lines on the invariable plane is proportional to the time. [Poinsot.]

Ex. 4. If the lengths OA, OB, OC be proportional to the radii of gyration about the axes respectively, the sum of the areas conserved by these lines on the invariable plane will also be proportional to the time. [Poinsot.]

Motion of the body when two principal axes are equal.

180. Let the body be rotating with an angular velocity ω about an instantaneous axis OI. Let OL be the perpendicular on the invariable plane. The momental ellipsoid is in this case a spheroid, the axis of which is the axis of unequal moment in the body. Let the equal moments of inertia be A and B. From the symmetry of the figure it is evident that as the spheroid rolls on the invariable plane, the angles LOC, LOI are *constant*, and the three axes OI, OL, OC are always in one plane. Let the angles $LOC = \gamma$, $IOC = i$.

Following the same notation as in Art. 137, we have

$$\omega_3 = \omega \cos i, \quad \omega_1^2 + \omega_2^2 = \omega^2 \sin^2 i,$$

$$G^2 = (A^2 \sin^2 i + C^2 \cos^2 i)\, \omega^2,$$

$$T = (A \sin^2 i + C \cos^2 i)\omega^2.$$

We therefore have

$$\cos \gamma = \frac{C\omega_3}{G} = \frac{C \cos i}{\sqrt{A^2 \sin^2 i + C^2 \cos^2 i}}.$$

This result may also be obtained as follows. In any conic if i and γ be the angles a central radius vector and the perpendicular on the tangent at its extremity make with the minor axis, and if a, b be the semi-axes, then $\tan \gamma = \tan i \,.\, b^2/a^2$. Applying this to the momental spheroid, we have

$$\tan \gamma = \frac{A}{C} \tan i.$$

The angle i being known from the initial conditions, the angle γ can be found from either of these expressions. The peculiarities of the motion will then be as follows.

The invariable line describes a right cone in the body whose axis is the axis of unequal moment, and whose semi-angle is γ.

The instantaneous axis describes a right cone in the body whose axis is the axis of unequal moment, and whose semi-angle is i.

The instantaneous axis describes a right cone in space, whose axis is the invariable line, and whose semi-angle is $i \sim \gamma$.

The axis of unequal moment describes a right cone in space whose axis is the invariable line, and whose semi-angle is γ.

The angular velocity of the body about the instantaneous

axis varies as the radius vector of the spheroid, and is therefore constant.

181. *To find the common angular velocity in space of the instantaneous axis and the axis of unequal moment round the invariable line.*

Let C be the extremity of the axis of figure of the momental ellipsoid, and let Ω be the rate at which the plane LOC is turning round OL. Let CM, CN be perpendiculars on OL and OI. Then since the body is turning round OI, the velocity of C is $CN \cdot \omega$. But this is also $CM \cdot \Omega$. Since $CM = OC \sin \gamma$, $CN = OC \sin i$, we have at once $\Omega \sin \gamma = \omega \sin i$, whence Ω can be found.

182. *To find the common angular velocity in the body of the invariable line and the instantaneous axis round the axis of unequal moment.*

Let Ω' be the rate at which the plane LOC is turning round OC in the body. Let LM, LN be perpendiculars from any point L in the invariable line on OC and OI. Then since OL is fixed in space and the body is turning round OI, the velocity of L in the body is $LN \cdot \omega$. But this is also $LM \cdot \Omega'$. Since $LM = OL \sin \gamma$, $LN = OL \sin (i - \gamma)$, we have at once $\Omega' \sin \gamma = \omega \sin (i - \gamma)$, whence Ω' can be found.

183. Ex. 1. If a right circular cone, whose altitude a is double the radius of its base, turn about its centre of gravity as a fixed point, and be originally set in motion about an axis inclined at an angle a to the axis of figure, the vertex of the cone will describe a circle whose radius is $\frac{3}{4} a \sin a$. [Coll. Exam.]

Ex. 2. A circular plate revolves about its centre of gravity as a fixed point. If an angular velocity ω were originally impressed on it about an axis making an angle a with its plane, a normal to the plane of the disc will make a revolution in space in a time τ given by $2\pi/\tau = \omega \sqrt{1 + 3 \sin^2 a}$. [Coll. Exam.]

Ex. 3. A body which can turn freely about a fixed point at which two of the principal moments are equal and less than the third, is set in rotation about any axis. Owing to the resistance of the air and other causes, it is continually acted on by a retarding couple whose axis is the instantaneous axis of rotation and whose magnitude is proportional to the angular velocity. Show that the axis of rotation will continually tend to become coincident with the axis of unequal moment. In the case of the earth therefore, a near coincidence of the axis of rotation and axis of figure is not a proof that such coincidence has always held. [Astronomical Notices, March 8, 1867.]

Ex. 4. When $A = B$, show that the conjugate ellipsoid is a spheroid the axis of which is the axis OC of unequal moment in the body.

Show also that the conjugate line OL' lies in the plane which contains OC, OI and OL; and if γ' be the angle COL', $\tan \gamma' = A \tan i/(2A - C)$ so that

$$\cot \gamma + \cot \gamma' = 2 \cot i.$$

Motion when $G^2 = BT$.

184. The peculiarities of this case have been already alluded to in Art. 137. When the initial conditions are such that this relation holds between the Vis Viva and the Momentum of the body the whole discussion of the motion becomes more simple*.

The fundamental equations of motion are

$$A\omega_1^2 + B\omega_2^2 + C\omega_3^2 = T \atop A^2\omega_1^2 + B^2\omega_2^2 + C^2\omega_3^2 = G^2 = BT \Big\} \dots\dots\dots(1).$$

Solving these, we have

$$\omega_1^2 = \frac{B-C}{A-C} \cdot \frac{G^2 - B^2\omega_2^2}{AB} \atop \omega_3^2 = \frac{A-B}{A-C} \cdot \frac{G^2 - B^2\omega_2^2}{BC} \Bigg\} \dots\dots\dots(2).$$

But

$$\frac{d\omega_2}{dt} = \frac{C-A}{B} \omega_1\omega_3;$$

$$\therefore \frac{d\omega_2}{dt} = \mp \sqrt{\frac{(A-B)(B-C)}{AC}} \cdot \frac{G^2 - B^2\omega_2^2}{B^2}.$$

When the initial values of ω_1 and ω_3 have like signs, $(C-A)\omega_1\omega_3$ is negative and therefore $d\omega_2/dt$ must be negative, hence in this expression the upper or lower sign is to be used according as the initial values of ω_1, ω_3 have like or unlike signs.

$$\therefore \frac{B^2}{G^2 - B^2\omega_2^2} \frac{d\omega_2}{dt} = \mp \sqrt{\frac{(A-B)(B-C)}{AC}}$$

If we put $\mp n$ for the right-hand side and integrate we have

$$\frac{G + B\omega_2}{G - B\omega_2} = E \cdot e^{\mp \frac{2G}{B}nt}, \quad \therefore \frac{B\omega_2}{G} = \frac{E \cdot e^{\mp \frac{2G}{B}nt} - 1}{E \cdot e^{\mp \frac{2G}{B}nt} + 1},$$

where E is some undetermined constant. As t increases indefinitely, ω_2 approaches $\mp G/B$ as its limit and therefore by (2) ω_1 and ω_3 approach zero.

The conclusion is that the instantaneous axis ultimately approaches to coincidence with the mean axis of principal moment, but never actually coincides with it. It approaches the positive or negative end of the mean axis according as the initial value of $(C-A)\omega_1\omega_3$ is positive or negative.

185. *To find what the cones traced out in the body by the invariable line and instantaneous axis become when* $G^2 = BT$.

* This case appears to have been considered by nearly every writer on this subject. As examples of different methods of treatment the reader may consult *Legendre, Traité des Fonctions Elliptiques*, 1825, Vol. I. page 382, and *Poinsot, Théorie Nouvelle de la Rotation des corps*, 1852, page 104.

Eliminating ω_2 from the fundamental equations of the last Article we have $A(A-B)\omega_1^2 = C(B-C)\omega_3^2$.

Taking the principal axes at the fixed point as axes of reference, the equations of the invariable line are $x/A\omega_1 = y/B\omega_2 = z/C\omega_3$. Eliminating ω_1 and ω_3 the locus of the invariable line is one of the two planes
$$\sqrt{\frac{A-B}{A}}\,x = \pm\sqrt{\frac{B-C}{C}}\,z.$$

The equations of the instantaneous axes are $x/\omega_1 = y/\omega_2 = z/\omega_3$. Eliminating ω_1 and ω_3 the locus of the instantaneous axis is one of the two planes.
$$\sqrt{A(A-B)}\,x = \pm\sqrt{(B-C)}\,z.$$

In these equations since z/x follows the sign of ω_3/ω_1 the upper or lower sign is to be taken according as the initial values of ω_1, ω_3 have like or unlike signs. These planes pass through the mean axis, and are independent of the initial conditions except so far that $G^2 = BT$.

The rolling and sliding cone is the reciprocal of that described by the invariable plane Art. 172, and is therefore the straight line perpendicular to that plane which is traced out by the invariable line.

Ex. 1. Show that the planes described by the invariable line coincide with the central circular sections of the ellipsoid of gyration and are perpendicular to the asymptotes of that focal conic of the momental ellipsoid which lies in the plane of the greatest and least moments.

Ex. 2. The planes described by the instantaneous axis are perpendicular to the umbilical diameters of the ellipsoid of gyration and are the diametral planes of the asymptotes of the focal conic in the momental ellipsoid.

186. The relations to each other of the several planes fixed in the body may be exhibited by the following figure. Let A, B, C be the points in which the principal axes of the body cut a sphere whose centre is O, and radius unity. Let BLK', BIJ' be the planes traced out by the invariable line and the instantaneous axis respectively. Then by the last Article
$$\tan CK' = \sqrt{\frac{A}{C}\cdot\frac{B-C}{A-B}},\quad \tan CJ' = \sqrt{\frac{C}{A}\cdot\frac{B-C}{A-B}}.$$
Hence we find
$$\tan K'J' = \tan LBI = \sqrt{\frac{(B-C)(A-B)}{AC}}.$$

This is the quantity which has been called n in Art. 184.

Exactly as in Art. 163 the direction of motion of L is perpendicular to IL and hence the angle ILB is a right angle. Thus the spherical triangle ILB has one angle right, and another constant and independent of all initial conditions.

Exactly as in Art. 163, the velocity of L along LB is equal to

$\omega \sin IL$ which, by Art. 143, is equal to $\tan IL \cdot T/G$. But from the spherical triangle ILB we have $n \sin BL = \tan IL$. If then we put, as before, $\beta = BL$, we have

$$\frac{d\beta}{dt} = \pm \frac{T}{G} n \sin \beta.$$

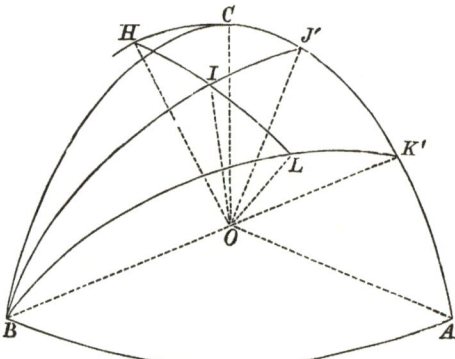

If the initial values of ω_1, ω_3 have the same sign, the body is turning round I from K' to B. Hence, since L is fixed in space, BL is increasing and therefore the upper sign must be used in this figure. See also Art. 184.

We may also find an expression for β in terms of the time. Since $\cos \beta = B\omega_2/G$ we have, by Art 184,

$$\frac{1 + \cos \beta}{1 - \cos \beta} = Ee^{\mp \frac{2G}{B}nt}, \quad \therefore \cot \frac{\beta}{2} = \sqrt{Ee^{\mp \frac{G}{B}nt}}.$$

Ex. 1. When the body moves so that $G^2 = BT$, prove that the conjugate body (Art. 174) also moves so that $G'^2 = B'T'$. Thence shew that the conjugate line OL' describes a great circle BQ' passing through B such that BQ' and BK' make equal angles on opposite sides with BJ'.

Show also that the spherical triangle $IL'B$ has one angle (viz. $IL'B$) right and another (viz. IBL') constant and equal to $\tan^{-1} n$, where n has the meaning above given.

Ex. 2. Show that the eccentric line describes a great circle passing through B and cutting AC in some point D' where $\tan^2 CD' = \tan CJ' \tan CK'$. If E be the intersection of the eccentric line with the sphere, show that the arcs BE and BL are always equal.

187. *To find the motion of the body in space.*

We have already seen that the motion is such that a plane fixed in the body, viz. the plane BK', contains a straight line fixed in space, viz. the invariable line OL. Since the body is brought from any position into the next by an angular velocity $\omega \cos IOL = T/G$ about OL, and an angular velocity $\omega \sin IOL$ about a perpendicular to OL, viz. OH, it follows that the plane fixed in the body turns round the line fixed in space with a uniform angular velocity T/G or G/B. At the same time the plane moves so that the line fixed in space appears to describe the plane with a variable velocity $\omega \sin IOL$. If β be the angle BL, this has been proved in the last Article to be $n \sin \beta T/G$.

188.   The cone described by OH in the body is the reciprocal cone of that described by OL, and from it we may deduce reciprocal theorems. The motion is therefore such that a straight line fixed in the body, viz. OH, describes a plane fixed in space, viz. the plane perpendicular to OL. The straight line moves along this plane with a uniform angular velocity equal to T/G or G/B, while the angular velocity of the body about this straight line is $\pm\, n \sin \beta G/B$.

189.   The motion of the principal axes may be deduced from the general results given in Art. 176. But we may also proceed thus. Since the body is turning about OI, the point B on the sphere is moving perpendicularly to the arc IB. Hence the tangent to the path of B makes with LB an angle which is the complement of the constant angle IBL. The path traced out by the axis of mean moment on a sphere whose centre is at O is a rhumb line which cuts all the great circles through L at an angle whose cotangent is $\pm\, n$.

190.   *To find the motion of the instantaneous axis in space.*

This problem is the same as that considered in Art. 168. We may however deduce the result at once from Art. 187. The angle ILB is always a right angle, it therefore follows that the angular velocity of I round L is the same as that of the arc BL round L. But the angular velocity of the latter is constant and equal to T/G. If then ϕ be the angle the plane LOI, containing the instantaneous axis and the invariable line, makes with some fixed plane passing through the invariable line, we have $\dfrac{d\phi}{dt} = \dfrac{T}{G}$.

191.   To find the equation of the cone described by the instantaneous axis in space, we require a relation between ζ and ϕ, where ζ is the arc IL on the sphere. From the right-angled triangle ILB we have $n \sin \beta = \tan \zeta$, and by Art. 186,

$$\cot \frac{\beta}{2} = \sqrt{E} e^{\mp \frac{G}{B} nt}.$$

Eliminating β, we shall have an expression for ζ in terms of t.

We find $\dfrac{2n}{\tan \zeta} = \cot \dfrac{\beta}{2} + \tan \dfrac{\beta}{2} = \sqrt{E} e^{\mp \frac{G}{B} nt} + \dfrac{1}{\sqrt{E}} e^{\pm \frac{G}{B} nt}$.

By the last Article $\phi = (T/G)\, t + F$, where F is some constant.

Let us substitute for t in terms of ϕ, and let us choose the plane from which ϕ is measured so that $\sqrt{E}e^{\mp nF} = 1$.

The equation to the cone traced out in space by the instantaneous axis is

$$2n \cot \zeta = e^{n\phi} + e^{-n\phi}.$$

When $\phi = 0$, we have $\tan \zeta = n$. Therefore the plane fixed in space from which ϕ is measured is the plane containing the axes of greatest and least moment at the instant when that plane contains the invariable line.

On tracing this cone, we see that it cuts a sphere whose centre is at the fixed point in a spiral curve. The branches determined by positive and negative values of ϕ are perfectly equal. As ϕ increases positively the radial arc ζ continually decreases, the spiral therefore makes an infinite number of turns round the point L, the last turn being infinitely small.

Ex. In the herpolhode $\dfrac{2mb}{r} = e^{m\theta} + e^{-m\theta}$, if the locus of the extremity of the polar subtangent of this curve be found and another curve be similarly generated from this locus, the curve thus obtained will be similar to the herpolhode. [Math. Tripos, 1863.]

On Correlated and Contrarelated Bodies.

192. *To compare the motions of different bodies acted on by initial couples whose planes are parallel.*

Let α, β, γ be the angles the principal axes OA, OB, OC of a body at the fixed point O make with the invariable line OL. Then by Art. 140, Euler's equations may be put into the form

$$\frac{d\cos\alpha}{dt} + G\left(\frac{1}{B} - \frac{1}{C}\right)\cos\beta\cos\gamma = 0 \ldots\ldots\ldots(1),$$

with two similar equations. Let λ, μ, ν be the angles the planes LOA, LOB, LOC make with any plane fixed in space, and passing through OL. Then

$$\sin^2\alpha\frac{d\lambda}{dt} = \frac{T}{G} - \frac{G\cos^2\alpha}{A} \ldots\ldots\ldots\ldots(2),$$

with similar equations for μ and ν.

If accented letters denote similar quantities for some other body, the corresponding equations will be

$$\frac{d\cos\alpha'}{dt} + G'\left(\frac{1}{B'} - \frac{1}{C'}\right)\cos\beta'\cos\gamma' = 0 \ldots\ldots(3),$$

$$\sin^2\alpha'\frac{d\lambda'}{dt} = \frac{T'}{G'} - \frac{G'\cos^2\alpha'}{A'} \ldots\ldots\ldots(4).$$

If then the bodies are such that

$$G\left(\frac{1}{B} - \frac{1}{C}\right) = G'\left(\frac{1}{B'} - \frac{1}{C'}\right), \&c. = \&c. \ldots\ldots\ldots(5),$$

the equations (1) to find α, β, γ are the same as the equations (3) to find α', β', γ'. Therefore if these two bodies be initially placed with their principal axes parallel and be set in motion by impulsive couples whose magnitudes are G and G', and whose planes are parallel, then after the lapse of any time t the principal axes of the two bodies will still be equally inclined to the common axis of the couples.

The equations (5) may be put into the form

$$\frac{G}{A} - \frac{G'}{A'} = \frac{G}{B} - \frac{G'}{B'} = \frac{G}{C} - \frac{G'}{C'} \dots\dots\dots\dots(6).$$

Since by Art. 142 the vis viva is given by

$$\frac{T}{G^2} = \frac{\cos^2\alpha}{A} + \frac{\cos^2\beta}{B} + \frac{\cos^2\gamma}{C} \dots\dots\dots\dots(7),$$

we see that each of the expressions in (6) is equal to $T/G - T'/G'$.

It immediately follows by subtracting equations (2) and (4) and dividing by $\sin^2\alpha$ that

$$\frac{d\lambda}{dt} - \frac{d\lambda'}{dt} = \frac{T}{G} - \frac{T'}{G'} \dots\dots\dots\dots\dots(8),$$

with similar equations for μ and ν. Thus the two bodies being started as before with their principal axes parallel each to each, the parallelism of the principal axes may be restored by turning the body whose principal axes are A', B', C' about the common axis of the impulsive couples through an angle $(T/G - T'/G') t$ in the direction in which positive impulsive couples act.

193. When the couples G and G' are equal the condition (6) becomes

$$\frac{1}{A} - \frac{1}{A'} = \frac{1}{B} - \frac{1}{B'} = \frac{1}{C} - \frac{1}{C'} = \frac{T - T''}{G^2} \dots\dots\dots(9),$$

the bodies are then said to be *correlated*. If momental ellipsoids of the two bodies be taken so that the moment of inertia in each bears the same ratio to the square of the reciprocal of the radius vector these ellipsoids are clearly confocal.

When the couples G and G' are equal and opposite, the equation (6) becomes

$$\frac{1}{A} + \frac{1}{A'} = \frac{1}{B} + \frac{1}{B'} = \frac{1}{C} + \frac{1}{C'} = \frac{T + T'}{G^2} \dots\dots\dots (10),$$

and the bodies are said to be *contrarelated*.

194. *To compare the angular velocities of the two bodies at any instant.*

Let ω be the angular velocity of one body at any instant, then following the usual notation we have

$$\omega^2 = \omega_1^2 + \omega_2^2 + \omega_3^2 = G^2\left(\frac{\cos^2\alpha}{A^2} + \frac{\cos^2\beta}{B^2} + \frac{\cos^2\gamma}{C^2}\right).$$

If the same letters accented denote similar quantities for the other body $\qquad \omega'^2 = G'^2 \left(\dfrac{\cos^2 \alpha}{A'^2} + \dfrac{\cos^2 \beta}{B'^2} + \dfrac{\cos^2 \gamma}{C'^2} \right).$

But remembering the condition (6) these give

$$\omega^2 - \omega'^2 = \left(\frac{T}{G} - \frac{T'}{G'} \right) \left[\cos^2 \alpha \left(\frac{G}{A} + \frac{G'}{A'} \right) + \cos^2 \beta \left(\frac{G}{B} + \frac{G'}{B'} \right) + \cos^2 \gamma \left(\frac{G}{C} + \frac{G'}{C'} \right) \right].$$

By referring to (7) the quantity in square brackets is easily seen to be $T/G + T'/G'$,

$$\therefore \ \omega^2 - \omega'^2 = \frac{T^2}{G^2} - \frac{T'^2}{G'^2} \quad \dots\dots\dots\dots (11).$$

195. Ex. 1. If two bodies be so related that their ellipsoids of gyration are confocal, and be initially so placed that the angles (α, β, γ) $(\alpha', \beta', \gamma')$ their principal axes make with the invariable line of each are connected by the equations

$$\frac{\cos \alpha}{\sqrt{A}} = \frac{\cos \alpha'}{\sqrt{A'}}, \ \frac{\cos \beta}{\sqrt{B}} = \frac{\cos \beta'}{\sqrt{B'}}, \ \frac{\cos \gamma}{\sqrt{C}} = \frac{\cos \gamma'}{\sqrt{C'}},$$

and if these bodies be set in motion by two impulsive couples G, G' respectively proportional to $\sqrt{ABC}$ and $\sqrt{A'B'C'}$, then the above relations will always hold between the angles (α, β, γ) $(\alpha', \beta', \gamma')$. If p and p' be the reciprocals of $d\lambda/dt$ and $d\lambda'/dt$, then $Gp - G'p'$ will be constant throughout the motion, where λ, λ', &c., are the angles the planes LOA, $L'O'A'$ make at the time t with their positions at the time $t = 0$.

Ex. 2. In order that the angles which the principal axes make with the axis of the couple may be the same in each body, it is necessary that the invariable cones and therefore also their reciprocals, i.e. Poinsot's rolling and sliding cones, should be the same in each body. Thus in the two bodies the rolling motions of these cones are equal, but the sliding motions may be different. Thence deduce equations (8) and (11). This mode of proof is partly due to Cayley.

196. Sylvester's measure of the time. When a body turns about a fixed point its motion in space is represented by making its momental ellipsoid roll on a fixed plane. This gives no representation of the *time* occupied by the body in passing from any position to any other. The preceding Articles will enable us to supply this defect.

To give distinctness to our ideas let us suppose the momental ellipsoid to be rolling on a horizontal plane underneath the fixed point O, and that the instantaneous axis OI is describing a polhode about the axis of A. Let us now remove that half of the ellipsoid which is bounded by the plane of BC, and which does not touch the fixed plane. Let us replace this half by the half of another smaller ellipsoid which is confocal with the first. Let a plane be drawn parallel to the invariable plane to touch this ellipsoid in I' and suppose this plane also to be fixed in space. These two semi-ellipsoids may be considered as the momental ellipsoids of two correlated bodies. If they were not attached to each other and were free to move without interference, each would roll, the one on the fixed plane which touches at I, and the other on that which touches at I'. By Arts. 192 and 193 the upper ellipsoid (being the smallest) may be brought into parallelism with the lower by a rotation $Gt(1/A - 1/A')$ about the invariable line. If then the upper plane on which the upper ellipsoid rolls be made to turn round the invariable line as a fixed axis with an angular velocity $G(1/A - 1/A')$, the two ellipsoids will always be in a state of parallelism, and may be supposed to be rigidly attached to each other.

Suppose then the upper tangent plane to be perfectly rough and capable of turning in a horizontal plane about a vertical axis which passes through the fixed point. As the nucleus is made to roll with the under part of its surface on the fixed plane below, the friction between the upper surface and the plane will cause the latter* to rotate about its axis. Then the time elapsed will be in a constant ratio to this motion of rotation, which may be measured off on an absolutely fixed dial face immediately over the rotating plane.

197. The preceding theory, so far as it relates to correlated and contrarelated bodies, is taken from a memoir by Prof. Sylvester in the *Philosophical Transactions* for 1866. He proceeds to investigate in what cases the upper ellipsoid may be reduced to a disc. It appears that there are always two such discs and no more, except in the case of two of the principal moments being equal, when the solution becomes unique. Of these two discs one is correlated and the other contrarelated to the given body, and they will be respectively perpendicular to the axes of greatest and least moments of inertia.

198. **Poinsot's measure of the time.** Poinsot has shown that the motion of the body may be constructed by a cone fixed in the body rolling on a plane which turns uniformly round the invariable line. If, as in the preceding theory, we suppose the plane rough, and to be turned by the cone as it rolls on the plane, the angle turned through by the plane will measure the time elapsed.

The Sphero-Conic or Spherical Ellipse.

199. The following properties of a sphero-conic will be found useful in connexion with the theorems of Art. 157. They appear to be new. The curve is

---

* As the ellipsoid rolls on the lower plane, a certain geometrical condition must be satisfied that the nucleus may not quit the upper plane or tend to force it upwards. This condition is that the plane containing OI, OI', must contain the invariable line, for then and then only the rotation about OI can be resolved into a component about OI' and a component about the invariable line. That this condition must be satisfied is clear from the reasoning in the text. But it is also clear from the known properties of confocal ellipsoids.

represented by the line $DED'E'$. As before, the eye is supposed to be situated in the radius through A, viewing the sphere from a considerable distance. The three principal planes of the cone intersect the sphere in the three quadrants AB, BC, CA, and any one of the three points A, B, C might be called the centre. The arcs AD and AE are represented by a and b.

The letters are not always the same as those used in the dynamical applications of the curve, but have been chosen to agree as far as possible with those usually employed in plane conics. In this way the analogy between the plane and the spherical ellipse will be made more apparent.

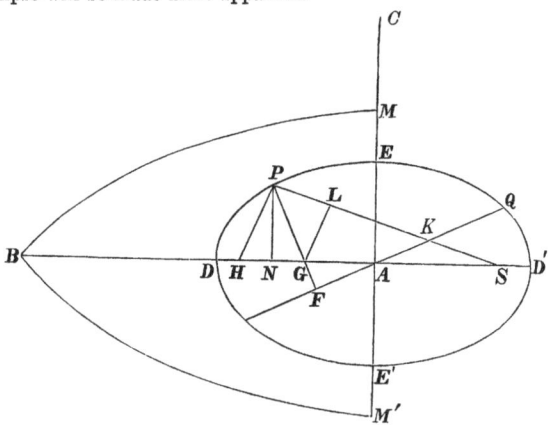

1. Equation to the conic. Draw the arc PN perpendicular to AD and let $PN=y$, $AN=x$. Let NP produced cut the small circle described on DD' as diameter in P', let NP' be called the eccentric ordinate and be represented by y'. We then have

$$\frac{\tan y}{\tan y'} = \text{constant} = \frac{\tan b}{\tan a}, \qquad \cos a = \cos y' \cos x.$$

2. The projection of the normal PG on the focal radius vector SP, i.e. PL, is constant and equal to half the latus rectum. Also $\dfrac{\tan GL}{\sin PN} = \text{constant}$.

If $2l$ be the latus rectum, then $\tan l = \dfrac{\tan^2 b}{\tan a}$.

3. If QAF be an arc cutting PG at right angles, QA may be called the semi-conjugate of AP. Then $\tan PG \cdot \tan PF = \tan^2 b$.

4. The length PK cut off the focal radius vector by the conjugate diameter is constant and equal to a. This follows from (2) and (3).

5. If $1 - e^2 = \dfrac{\sin^2 b}{\sin^2 a}$, e may be called the eccentricity of the sphero-conic. Then
$$\tan AG = e^2 \tan AN.$$

6. Also S being a focus $SE = HE = a$, and $\tan SA = e \tan a$
$$\tan (SP - a) = e \tan AN.$$

7. Polar equations to the conic
$$\frac{\tan l}{\tan SP} = 1 - \frac{e}{\cos^2 b} \cos P\hat{S}A. \qquad \frac{\sin^2 b}{\sin^2 AP} = 1 - e^2 \cos^2 PAD.$$

8. If ρ be the radius of curvature at P, then $\tan \rho = \dfrac{\tan^3 n}{\tan^2 l}$.

9. Regarding AP, AQ as conjugate semi-diameters, defined as above,

$$\begin{aligned}\sin^2 AP + \sin^2 AQ &= \sin^2 a + \sin^2 b \\ \sin AQ \cdot \sin PF &= \sin a \cdot \sin b\end{aligned}\Big\} \quad . \qquad \tan PAD \cdot \tan QAD = -\frac{\sin^2 b}{\sin^2 a}.$$

10. If p be the perpendicular from the centre A on the tangent at P,

$$\frac{\tan^2 a \tan^2 b}{\tan^2 p} = \tan^2 a + \tan^2 b - \tan^2 AP.$$

11. Also $\tan^2 PG - \tan^2 l = \dfrac{e^2}{\cos^4 b} \sin^2 PN.$ $\dfrac{\tan^2 PG}{\sin SP \cdot \sin HP} = \dfrac{\tan^2 b}{\sin^2 a}.$

12. $\left.\begin{aligned}\sin^2 a - \sin^2 AP \\ = \sin^2 AQ - \sin^2 b\end{aligned}\right\} = \dfrac{e^2}{1 - e^2} \sin^2 PN.$

Cor. $\tan^2 PG = \dfrac{\tan^2 b}{\cos^2 b \sin^2 a}(\cos^2 AP - \cos^2 a \cos^2 b).$

If $\sin AM = \sin AM' = \dfrac{\sin b}{\sin a}$, the planes of the arcs BM and BM' are parallel to the circular sections of the cone. Some of the properties of these arcs resemble those of asymptotes when B is regarded as the centre of the conic. The properties which connect the sphero-conic with the arcs BM and BM' will be found in Dr Salmon's *Solid Geometry*.

Many other properties of sphero-conics will also be found in Dr Frost's *Solid Geometry*.

EXAMPLES*.

1. A right cone the base of which is an ellipse is supported at G the centre of gravity, and has a motion communicated to it about an axis through G perpendicular to the line joining G, and the extremity B of the axis minor of the base, and in the plane through B and the axis of the cone. Determine the position of the invariable plane.

Result. The normal to the invariable plane lies in the plane passing through the axis of the cone and the axis of instantaneous rotation, and makes an angle whose tangent is $h (h^2 + 4a^2)/16b (a^2 + b^2)$.

2. A spheroid has a particle of mass m fastened at each extremity of the axis of revolution, and the centre of gravity is fixed. If the body be set rotating about any axis, show that the spheroid will roll on a fixed plane during the motion provided $m/M = \frac{1}{10}(1 - a^2/c^2)$, where M is the mass of the spheroid, a and c are the axes of the generating ellipse, c being the axis of figure.

3. A lamina of any form rotating with an angular velocity a about an axis through its centre of gravity perpendicular to its plane has an angular velocity $a (B + C)^{\frac{1}{2}}/(B - C)^{\frac{1}{2}}$ impressed upon it about its principal axis of least moment, A, B, C being arranged in descending order of magnitude: show that at any time t the angular velocities about the principal axes are respectively

$$\frac{2a}{e^{at} + e^{-at}}, \quad -\sqrt{\frac{B+C}{B-C}}\, a\, \frac{e^{at} - e^{-at}}{e^{at} + e^{-at}} \text{ and } \sqrt{\frac{B+C}{B-C}}\frac{2a}{e^{at} + e^{-at}},$$

and that it will ultimately revolve about the axis of mean moment.

4. A rigid body, not acted on by any forces, is in motion about its centre of gravity: prove that if the instantaneous axis be at any moment situated in the plane of contact of either of the right circular cylinders described about the central ellipsoid, it will be so throughout the motion.

* These examples are taken from the Examination Papers which have been set in the University and in the Colleges.

If a, b, c be the semi-axes of the central ellipsoid, arranged in descending order of magnitude, e_1, e_2, e_3 the eccentricities of its principal sections, Ω_1, Ω_2, Ω_3 the initial component angular velocities of the body about its principal axes, prove that the condition that the instantaneous axis should be situated in the plane above described is $\Omega_1/e_1 = (ab/c^2) (\Omega_3/e_3)$.

5. A rigid lamina not acted on by any forces has one point fixed about which it can turn freely. It is started about a line in the plane of the lamina the moment of inertia about which is Q. Show that the ratio of the greatest to the least angular velocity is $\sqrt{A+B} : \sqrt{B+Q}$, where A, B are the principal moments of inertia about axes in the plane of the lamina.

6. If the earth were a rigid body acted on by no forces rotating about a diameter which is not a principal axis, show that the latitudes of places would vary and that the values would recur whenever $\sqrt{A-B}\sqrt{A-C}\int\omega_1 dt$ is a multiple of $2\pi\sqrt{BC}$. If a man were to lie down when his latitude is a minimum and to rise when it becomes a maximum, show that he would increase the vis viva, and so cause the pole of the earth to travel from the axis of greatest moment of inertia towards that of least moment of inertia.

7. If $d\theta$ be the angle between two consecutive positions of the instantaneous axis, prove that $\omega^2 \left(\dfrac{d\theta}{dt}\right)^2 = \left(\dfrac{d\omega_1}{dt}\right)^2 + \left(\dfrac{d\omega_2}{dt}\right)^2 + \left(\dfrac{d\omega_3}{dt}\right)^2 - \left(\dfrac{d\omega}{dt}\right)^2$.

8. If n be the angular velocity of the plane through the invariable line and the instantaneous axis about the invariable line and λ the component angular velocity of the body about the invariable line, prove that

$$\left(\frac{1}{2}\frac{dn}{dt}\right)^2 + (n-\lambda)\left(n-\frac{G}{A}\right)\left(n-\frac{G}{B}\right)\left(n-\frac{G}{C}\right) = 0.$$

9. If a body move in any manner, and all the forces pass through the centre of gravity, prove that $\dfrac{d(\omega^2)}{dt} + 2\dfrac{d}{dt}(\log\omega_1)\dfrac{d}{dt}(\log\omega_2)\dfrac{d}{dt}(\log\omega_3) = 0$, where ω_1, ω_2, ω_3 are the angular velocities about the principal axes at the centre of gravity, and ω is the resultant angular velocity.

CHAPTER V.

200. In this Chapter it is proposed to discuss some cases of the motion of a rigid body in three dimensions as examples of the processes explained in Chapter I. The reader will find it an instructive exercise to attempt their solution by other methods; for example, the equations of Lagrange might be applied with advantage in some cases.

In each section of the Chapter the general method of proceeding will first be explained and a number of examples will then be considered. These have been chosen as being apparently the most interesting cases of the motion of a body which occur. But of course all the results obtained are not equally valuable. Besides this, some of the processes are only slight variations of those which have been already explained. Accordingly it has not been thought necessary in every case to give the whole of the algebraical work. The plan of the solution is sketched more or less fully and the results are stated. It is believed that the reader will be able to supply the omitted steps for himself. The student will find his interest in the subject greatly increased if, after reading the first few articles in each section, he will attack the problems which follow in his own way. He may then profitably compare his results with the solutions here sketched out.

Motion of a Top.

201. *A body two of whose principal moments at the centre of gravity are equal moves about some fixed point O in the axis of unequal moment under the action of gravity. Determine the motion*[*].

To give distinctness to our ideas we may consider the body to be a top spinning on a perfectly rough horizontal plane.

---

[*] A partial solution of this problem by Lagrange's equations is given in Vol. I., Chap. VIII.

9

Let the axis OZ be vertical. Let the axis of unequal moment at the centre of gravity be the axis OC and let this be called the axis of the body. Let h be the distance of the centre of gravity G of the body from the fixed point O and *let the mass of the body be taken as unity.* Let OA be that principal axis at O which lies in the plane ZOC, OB the principal axis perpendicular to this plane.

If we take moments about the axis OC we have by Euler's equations (Vol. I. Chap. V.),

$$C\frac{d\omega_3}{dt} - (A - B)\,\omega_1\omega_2 = N.$$

But in our case $A = B$, and since the centre of gravity lies in the axis OC, we have $N = 0$. Hence ω_3 is constant and equal to its initial value. Let this be called n.

Let us measure along the axis OC in the direction OG a length $OP = A/h$. Then, by Vol. I. Chap. III., P is the centre* of oscillation of the body. This length we shall call l. Let θ be the inclination of the axis OC to the vertical, ψ the angle the plane ZOC makes with some plane fixed in space passing through OZ. Then by the same reasoning as that used in Euler's geometrical equations (Vol. I. Chap. V.) we find that the velocities of P resolved

$$\left.\begin{array}{l}\text{perpendicular to plane } ZOC = -\,l\omega_1 = l\sin\theta\,d\psi/dt\\ \text{parallel to plane } ZOC = \quad l\omega_2 = l\,d\theta/dt\end{array}\right\}\dots(1).$$

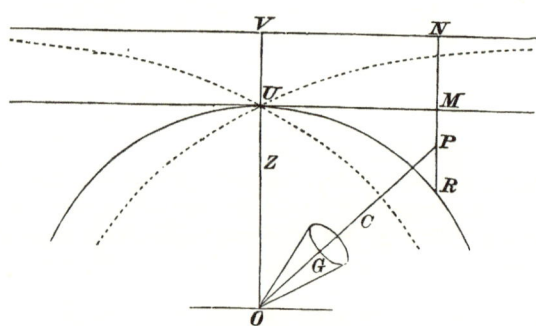

It is clear that the moment of the momentum about OZ will be constant throughout the motion. Since the direction-cosines of OZ referred to OA, OB, OC are $-\sin\theta$, 0 and $\cos\theta$, this principle gives

$$-A\omega_1\sin\theta + Cn\cos\theta = E\dots\dots\dots\dots\dots(2),$$

where E is some constant depending on the initial conditions,

* To avoid confusion in the figure, the body, which is represented by a top, is drawn smaller than it should be.

and whose value may be found from this equation by substituting the initial values of ω_1, and θ.

The equation of Vis Viva gives

$$A\,(\omega_1{}^2 + \omega_2{}^2) + Cn^2 = F - 2gh\cos\theta\ldots\ldots\ldots\ldots(3),$$

where F is some constant, whose value may be found by substituting in this equation the initial values of ω_1, ω_2, and θ*.

202. **Motion of the centre of Oscillation.** Let us measure along the vertical OZ, in the direction opposite to gravity as the positive direction, two lengths $OU = El/Cn$, $OV = l\,(F - Cn^2)/2gh$. These lengths we shall write briefly $OU = a$, and $OV = b$. Draw through U and V two horizontal planes, and let the vertical through P intersect these planes in M and N. Then the equations (2) and (3) give by (1),   transverse velocity of $P = (Cn/h)\tan PUM$ ....(4).

$$(\text{velocity of }P)^2 = 2gPN\ldots\ldots\ldots\ldots\ldots\ldots\ldots\ldots\ldots\ldots\ldots\ldots(5).$$

Thus the resultant velocity of P *is that due to the depth of* P *below the horizontal plane through* V, *and the velocity of* P *resolved perpendicular to the plane* ZOP *is proportional to the tangent of the angle* PU *makes with a horizontal plane.*

It appears from this last result that when P is below the horizontal plane through U, the plane POV turns round the vertical in the same direction as the body turns round its axis, i.e. according to the usual rule, OV and OP are the positive directions of the axes of rotation. When P passes above the horizontal plane through U, the plane POV turns round the vertical in the opposite direction. If P be below both the horizontal planes through O and U these results are still true, but if a top is viewed from above, the axis will appear to turn round the vertical in the direction opposite to the rotation of the top. In all the cases in which P is below the plane UM the lowest point of the rim of the top moves round the vertical in the same direction as the axis of the top.

If we substitute for ω_1, ω_2, E and F in (2) and (3) their values, we easily obtain

$$\left.\begin{array}{l} hl\sin^2\theta\,\dfrac{d\psi}{dt} + Cn\cos\theta = Cn\,\dfrac{a}{l} \\[2mm] l^2\left\{\left(\dfrac{d\theta}{dt}\right)^2 + \sin^2\theta\left(\dfrac{d\psi}{dt}\right)^2\right\} = 2g\,(b - l\cos\theta) \end{array}\right\}\quad\ldots\ldots\ldots\ldots\ldots(6).$$

These equations give in a convenient analytical form the whole motion. We see from the last equation, what is indeed obvious otherwise, that $b - l\cos\theta$ is always positive. The horizontal plane through V is therefore above the initial position of P and remains above P throughout the whole motion.

---

* If we eliminate ω_1, ω_2 from equations (1), (2), (3) we have two equations from which θ and ψ may be found by quadratures. These were first obtained by Lagrange in his *Mécanique Analytique*, and were afterwards given by Poisson in his *Traité de Mécanique*. The former passes them over with but slight notice, and proceeds to discuss the small oscillations of a body of any form suspended under the action of gravity from a fixed point. The latter limits the equations to the case in which the body has an initial angular velocity only about its axis, and applies them to determine directly the small oscillations of a top (1) when its axis is nearly vertical, and (2) when its axis makes a nearly constant angle with the vertical. His results are necessarily more limited than those given in this treatise.

Ex. 1. If ω be the resultant angular velocity of the body and v the velocity of P show that $\omega^2 = n^2 + (v/l)^2$.

Ex. 2. Show that the cosine of the inclination of the instantaneous axis to the vertical is $\{E + (A - C) n \cos \theta\}/A\omega$.

203. **Rise and Fall of a Top.** As the axis of the body goes round the vertical its inclination to the vertical is continually changing. These changes may be found by eliminating $d\psi/dt$ between the equations (6). We thus obtain

$$\left(l \frac{d\theta}{dt}\right)^2 = 2g\,(b - l\cos\theta) - \frac{C^2 n^2}{h^2}\left(\frac{a - l\cos\theta}{l\sin\theta}\right)^2 \quad\ldots\ldots\ldots\ldots(7).$$

It appears from this equation that θ can never vanish unless $a = l$, for in any other case the right-hand side of this equation would become infinite. This may be proved otherwise. Since a/l is equal to the ratio of the angular momentum about the vertical to that about the axis of the body, it is clear that the axis could not become vertical unless the ratio is unity.

Suppose the body to be set in motion in any way with its axis at an inclination i to the vertical. The axis will begin to approach or to fall away from the vertical according as the initial value of $d\theta/dt$ or ω_2 is negative or positive. The axis will then oscillate between two limiting angles given by the equation

$$0 = 2gh^2 l^2\,(b - l\cos\theta)\,(1 - \cos^2\theta) - C^2 n^2\,(a - l\cos\theta)^2 \ldots\ldots(8).$$

This is a cubic equation to determine $\cos\theta$. It will be necessary to examine its roots. When $\cos\theta = -1$ the right-hand side is *negative*; when $\cos\theta = \cos i$, since the initial value of $(d\theta/dt)^2$ is essentially positive, the right-hand side is either zero or *positive*; hence the equation has one real root between $\cos\theta = -1$ and $\cos\theta = \cos i$. Again, the right-hand side is *negative* when $\cos\theta = +1$ and *positive* when $\cos\theta = \infty$. Hence there is another real root between $\cos\theta = \cos i$, and $\cos\theta = 1$, and a third root greater than unity. This last root is inadmissible.

204. These limits may be conveniently expressed geometrically. The equation (7) may evidently be written in the form

$$\left(l\frac{d\theta}{dt}\right)^2 = 2g \cdot PN - \frac{C^2 n^2}{h^2}\left(\frac{PM}{UM}\right)^2 \ldots\ldots\ldots\ldots\ldots\ldots(9).$$

Describe a parabola with its vertex at U, its axis vertically downwards and its latus rectum equal to $C^2 n^2/2gh^2$. Let the vertical PMN cut this parabola in R, we then have

$$\frac{2g}{(l\,d\theta/dt)^2 - 2gMN} = \frac{1}{PM} + \frac{1}{PR} \ldots\ldots\ldots\ldots\ldots\ldots(10).$$

The point P oscillates between the two positions in which the harmonic mean of PM and PR is equal to $-2 \cdot MN$. In the figure V is drawn above U, and in this case one of the limits of P is above UM, and the other below the parabola. If we take U as origin and UO as the axis of x, we have $PM = x$, $UM = y$. Let $2pl$ be the latus rectum of the parabola, and $UV = c$, then the axis of the body oscillates between the two positions in which P lies on the cubic curve

$$y^2 (x + c) = 2plx^2 \quad \dots\dots\dots\dots\dots\dots\dots\dots(11).$$

When c is positive, i.e. when V is above U, the form of the curve is indicated in the figure by the dotted line. The tangents at U cut each other at a finite angle and the tangent of the angle either makes with the vertical is $(2pl/c)^{\frac{1}{2}}$. When c is negative the curve has two branches, one on each side of the vertical, with a conjugate point at the origin. It is clear from what precedes that the upper branch will lie above, and the lower branch below, the initial position of P, and that P must always lie between the two branches.

205. . In the case of a top, the initial motion is generally given by a rotation n about the axis. We have initially $\omega_1 = 0$, $\omega_2 = 0$, and therefore by (2) and (3) $E = Cn \cos i$, and $F - Cn^2 = 2gh \cos i$. This gives $a = b = l \cos i$. Putting $C^2n^2/2gh^2 = 2pl$, as before, the roots of equation (8) are $\cos \theta = \cos i$, and $\cos \theta = p - \sqrt{1 - 2p \cos i + p^2}$. The value $\cos \theta = p + \sqrt{1 - 2p \cos i + p^2}$ is always greater than unity, for it is clearly decreased by putting unity for $\cos i$, and its value is then not less than unity. The axis of the body will therefore oscillate between the values of θ just found.

Since $a = b$, the horizontal planes through U and V coincide, and $c = 0$. The cubic curve which determines the limits of oscillation, becomes the parabola UR and the straight line UM. The axis of the body will then oscillate between the two positions in which P lies on the horizontal through U and on the parabola, beginning at the former.

Generally the angular velocity n about the axis of figure is very great. In this case p is very great, and if we reject the squares of $1/p$ we see that $\cos \theta$ will vary between the limits $\cos i$ and $\cos i - \sin^2 i \cdot /2p$.

If the initial value of i is zero, we see that the two limits of $\cos i$ are the same. The axis of the body will therefore remain vertical.

EXAMPLES. Ex. 1. When the limiting angles between which θ varies are equal to each other, so that θ is constant throughout the motion and equal to a, show that $\tan^2 \phi - \tan \phi \tan a + \tan^2 a \cos a / 4p = 0$, where ϕ is the angle PUM.

Ex. 2. A top is set in motion on a smooth horizontal plane with an initial resultant angular velocity about its axis of figure. Show that the path traced out by the apex on the horizontal plane lies beween two circles, one of which it touches and the other it cuts at right angles. [*M. Finck, Nouvelles Annales de Mathématiques*, Tom. IX. 1850.]

Ex. 3. Show that the vertical pressure of a top on the ground is greater than its weight by $\frac{1}{2} h \dfrac{d}{d \cos \theta} \left(\sin \theta \dfrac{d\theta}{dt} \right)^2$. Hence by equation (7) of Art. 203 show that R is a quadratic function of $\cos \theta$ with constant coefficients.

206. If we compare the equations (6) of Art. 202 with those giving the motion of the conjugate line in Art. 175, Ex. 4. we see that they are analogous. It follows that *the motion of the axis of*

a top can be represented by the motion of the conjugate line of a body moving about a fixed point under no forces, with the proper initial conditions. It may be shown that the comparison leads to real values of the constants of the body moving under no forces.

See a paper by the author in the *Quarterly Journal*, 1888. *On a theorem of Jacobi in dynamics.*

207. **Precession and Nutation of a top.** *A body, two of whose principal moments at the centre of gravity* G *are equal, turns about a fixed point* O *in the axis of unequal moment under the action of gravity. The axis* OG *being inclined to the vertical at an angle* a, *and revolving about it with a uniform angular velocity, find the condition that the motion may be steady, and the time of a small oscillation.*

The equations (2) and (3) of Art. 201 contain the solution of this problem. But if we use the equation of Vis Viva in the form (3) we shall have to take into account the squares of small quantities. It will be found more convenient to replace it by one of the equations of the second order from which it has been derived. The simplest method of obtaining this equation is to use Lagrange's Rule as given in Vol. I. Chap. VIII. We thus obtain

$$A\theta'' - A\cos\theta\sin\theta\psi'^2 + Cn\sin\theta\psi' = gh\sin\theta \quad\ldots\ldots\ldots\ldots(12),$$

where accents denote differentiations with regard to the time.

This equation might also have been obtained by differentiating both (2) and (3) and eliminating $d^2\psi/dt^2$.

When the motion is steady both θ and $d\psi/dt$ are constants. Let $\theta = a$, $d\psi/dt = \mu$, then the equation (2) only determines the constant E and (12) becomes

$$\sin a\,(-A\cos a\mu^2 + Cn\mu - gh) = 0 \quad\ldots\ldots\ldots\ldots\ldots\ldots(13).$$

This indicates two states of steady motion, one in which $a = 0$ or π, and the other in which

$$\mu = \frac{Cn \pm \sqrt{C^2n^2 - 4ghA\cos a}}{2A\cos a} \quad\ldots\ldots\ldots\ldots\ldots\ldots(14),$$

a relation which does not necessarily hold when $a = 0$ or π.

In the former of these two motions the axis of the body will oscillate about the vertical and $d\psi/dt$ will not be small or nearly constant. It will therefore be more convenient to discuss the oscillations about this state of steady motion with other co-ordinates than θ and ψ.

In the latter of these two motions, if the centre of gravity of the body be above the horizontal plane through the fixed point O, $h\cos a$ will be positive. In this case the angular velocity n of the top round its axis of figure must be sufficiently great to make the quantity under the radical positive. We must therefore have n^2 not less than $4ghA\cos a/C^2$.

When a and n are given we can make the body move with either of these two values of μ by giving the proper initial angular velocities to the body. By equations (1) we see that the conditions of steady motion are $\omega_1 = -\mu\sin a$, $\omega_2 = 0$. When a top is set in motion by unwinding a string from the axis, the value of n is very great while the initial values of ω_1 and ω_2 are zero. The steady motion about which the top makes small oscillations will therefore have μ small. Hence the radical in (14) will have the negative sign. We have therefore very nearly $\mu = gh/Cn$.

208. *To find the small oscillation.* Let $\theta = a + x$, and $d\psi/dt = \mu + dy/dt$, where x and dy/dt are small quantities whose squares are to be neglected. Let a and μ be

such that they contain the whole of the constant parts of θ and $d\psi/dt$, so that x and dy/dt contain only trigonometrical terms. Then when we substitute these values in equations (2) and (12), the constant parts must vanish of themselves. The equations thus obtained determine E and μ, and shew that their values are the same as those determined when the motion is steady. The variable parts of the two equations become, after writing for Cn its value obtained from (13),

$$A\mu \sin a\, y' - (gh - A\mu^2 \cos a)\, x = 0$$

$$A\mu x'' + \sin a\, (gh - A\mu^2 \cos a)\, y' + \mu^3 A \sin^2 a x = 0.$$

To solve these we put $x = F \sin (pt + f)$, and $y = G \cos (pt + f)$. Substituting, we have

$$\left.\begin{array}{l} - A\mu \sin a \,.\, pG = (gh - A\mu^2 \cos a)\, F \\ (A\mu p^2 - \mu^3 A \sin^2 a)\, F = - (gh - A\mu^2 \cos a) \sin a \,.\, Gp \end{array}\right\}.$$

Multiplying these equations together, we have

$$p^2 = \frac{A^2\mu^4 - 2ghA \cos a\mu^2 + g^2h^2}{A^2\mu^2},$$

and the required time is $2\pi/p$. It is evident that p^2 is always positive, and therefore both the values of μ given by (14) correspond to stable motions. This expression was given by Dr Ferrers, now Master of Gonville and Caius College, as the result of a problem proposed by him for solution in the Mathematical Tripos, 1859.

We notice that in these results the precession of the axis in the steady motion is less the greater is the angular velocity of the top and is nearly given by $\mu = gh/Cn$ when n is very large. On the same supposition we have $p = gh/A\mu$ nearly, or which is the same thing $p = Cn/A$. It follows that the nutation or oscillation of the axis about the steady motion is very rapid, its period $2\pi/p$ being very short.

It is to be observed that this investigation does not apply if a be very small, for in that case some of the terms rejected are of the same order of magnitude as those retained. A different mode of investigation is therefore required, this case will be considered in Art. 212.

Ex. 1. The angular velocity n of a top is communicated to it by unwinding rapidly a string from the axis when at an inclination i to the vertical. Prove that the inclination of the axis at any time t is given by

$$\theta = i + r \sin i\, (1 - \cos pt) \quad \text{and} \quad \psi = \mu t - r \sin pt,$$

where $rgh = A\mu^2$. Thence show that the axis describes very nearly a right cone round its position in the steady motion, in the same direction as the top rotates. Find also the friction and pressure at the apex.

Ex. 2. A top two of whose principal moments at O are equal is set in rotation about its axis of figure, viz. OC, with an angular velocity n, the point O being fixed. If OC be *horizontal*, and if the proper initial angular velocity be communicated to the top about the vertical through O, prove that the top will not fall down, but that the axis of figure will revolve round the vertical, in steady motion, with an angular velocity $\mu = gh/Cn$, where h is the distance of the centre of gravity of the top from O, and C is the moment of inertia about the axis of figure. Show also that if the top be initially placed with OC nearly horizontal and if a very great angular velocity be communicated to it about OC without any initial angular velocity about OA or OB, then OC will revolve round the vertical, remaining very nearly in a horizontal plane, with an angular velocity μ given by the same formula as before, and the time of the vertical oscillations of OC about its mean position will be $2\pi A/Cn$.

Ex. 3. The gyroscope in one form consists of a hemispherical shell with an external axis through the vertex upon which a weight may be moved up and down so as to raise or lower the centre of gravity. The weight being in a certain position and the gyroscope being supported with the vertex on a pivot, a rapid rotation is imparted to it by unwinding a string from the axis, and the motion of the axis about the vertical is found to be precessional. Examine whether the weight must be moved up or down to reverse the direction of the motion. Is the motion of the axis of a top precessional or the reverse? [Math. Tripos.

Ex. 4. A gyrostat symmetrical about its axis of rotation is suspended from a fixed point by a string whose length is a. The string being fastened to a point on the axis of rotation, prove that when the gyrostat is moving steadily with its axis of rotation horizontal the circular measure of the angle which the string makes with the vertical is given by the equation $C^2 n^2 \tan a = (h + a \sin a) g h^2$ where n is the angular velocity of the gyrostat, h the distance from the point of attachment of the string to the centre of gravity of the gyrostat, and MC its moment of inertia about its axis of rotation, M being the mass of the gyrostat. [Math Tripos, 1888, Part II.

Ex. 5. A symmetrical top is set in motion on a rough horizontal plane with an angular velocity n about its axis of figure, the axis itself being inclined at an angle a to the vertical. Prove that between the greatest approach to and recess from the vertical, the centre of gravity describes an arc $h\beta$, where $(p - \cos a) \tan \beta = \sin a$, and $p = C^2 n^2 / 4gh \, AM$. [Math. Tripos, 1880.

209. **General considerations on the motion of a top.** We see from the example of the top in Art. 203 how greatly the effect of forces acting on a body is modified by an existing rotation in the body. If the top were initially at rest with its apex O fixed, gravity would cause it to turn round OB and fall downwards. When the top is in rapid rotation about its axis OC the effect of gravity is, not to alter sensibly the inclination of the axis to the vertical, but to make that axis describe a right cone round the vertical. In order the better to understand the cause of this difference, it will be useful to consider the motion from a different point of view. Assuming, then, Poinsot's construction for the motion of a body under no forces we shall endeavour to trace how that construction is modified by the action of gravity.

Let a body be in rotation about an axis OI nearly coincident with the axis of figure, then the invariable line OL is also nearly coincident with the axis of figure and would describe a small polhode round it, if the top were left to itself. We know by Art. 148, Ex. 1, that the polhode is only slightly altered by an impressed couple Q if either the angular velocity of the top is very great or the projection of the axis of the couple on the plane LOI is close to OL. When either of these conditions is satisfied the invariable line OL, the instantaneous axis OI and the axis of figure OC closely accompany each other in their motion through space.

Let us next consider how the invariable line is moved in space by the action of the impressed couple Q. The existing angular momentum of the top is equivalent to some couple G whose axis is the invariable line. The angular momentum generated about the axis of the impressed couple in the time dt is Qdt. Compounding these couples, we see that *the positive extremity of the invariable line is always moving towards the positive extremity of the axis of the impressed couple.*

Ex. 1. *To determine the steady motion of a swiftly rotating top with its apex O fixed.*

Let the figure represent the upper half of the momental spheroid at O. Then when the motion is steady the straight lines OL, OI, OC lie in a vertical plane which revolves round OZ with a uniform angular velocity μ. The force of gravity is continually generating an angular momentum about the horizontal diameter OB, so that OL, closely accompanied by OC, moves towards OB. This again causes OB to revolve round the vertical OZ. If these two motions are properly adjusted to each other the axis of the top will steadily revolve round the vertical in the same direction that the top rotates about its axis of figure.

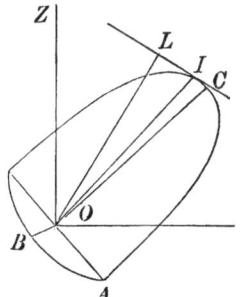

The angular displacement of OC in the time dt is $\mu \sin a\, dt$ where a is the angle ZOC, but since the body is turning round OI with an angular velocity ω, the same displacement is also $\omega \sin IOC$. Equating these we have, as in Art. 181,

$$\omega \sin IOC = \mu \sin a \dots\dots\dots\dots\dots\dots\dots\dots\dots\dots\dots (1).$$

In the time dt, gravity generates an angular momentum equal to $gh \sin a\, dt$ about the axis OB; the existing angular momentum being G, the displacement of the invariable line OL towards OB is $gh \sin a\, dt/G$. But since OL moves round OZ with an angular velocity μ, this is also equal to $\mu \sin ZOL\, dt$. We therefore have

$$\mu \cdot G \sin ZOL = gh \sin a \dots\dots\dots\dots\dots\dots\dots\dots\dots \dots\dots(2).$$

Now $G \sin ZOL$ is the angular momentum of the top about a horizontal line in the plane ZOC. Let n be the resolved part of ω about OC, then since the angular momenta about OC and OA are respectively Cn and $-A\omega \sin IOC$, we have by a simple resolution,

$$G \sin ZOL = Cn \sin a - A\omega \sin IOC \cdot \cos a \dots\dots\dots\dots\dots\dots(3).$$

Substituting from (1) and (3) in (2) we have, after division by $\sin a$,

$$gh/\mu = Cn - A\mu \cos a,$$

which is the same expression as in Art. 207.

It will be noticed that in this general explanation we have only shown that a steady motion is possible, that this steady motion is also *stable* is proved by the analysis in Art. 208.

Ex. 2. Let the resistance of the air on the top be represented by a retarding couple whose axis is the instantaneous axis. Show that the instantaneous axis will approach to or recede from the axis of figure OC according as the moment of inertia C is greater or less than A. See Art. 183, Ex. 3.

Ex. 3. A homogeneous sphere of radius a is loaded at a point of its surface by a particle whose mass is $1/p$th of its own. If it move steadily on a smooth horizontal plane, the diameter through the particle making a constant angle a with the vertical, and the sphere rotating about it with uniform angular velocity n, prove that $n^2ap\,(p+1)$ must not be less than $5\,(2p+7)\,g\cos a$ and show that the particle will revolve round the vertical in one or other of two periods whose sum is $4\pi apn/5g$.   [Math. Tripos.

210. Ex. **The boomerang.** As another illustration of how the apparent effect of a force is modified by a rapid rotation of the body we may consider the flight of a boomerang. This is a stick cut flat and bent in that plane; it is usually bulged out on one side, flat on the other, with a sharp edge along the convexity. The missile is so projected by a jerk of the hand that it has a rapid rotation about an axis perpendicular to its plane. Since this is a principal axis the body after projection will so move that the direction of the axis is sensibly fixed in space, Art. 156. Let GC be this axis, GA a perpendicular horizontal axis and let GB be perpendicular to both.

The resistance of the air at the edge is very small, but the flat side of the instrument being downwards, the pressure on the lower part tends to support the body in its flight. To make a rather vague comparison, the body moves as if projected upwards on a fixed inclined plane along the line of greatest slope, the pressure of the plane representing the supporting power of the air. The body advances upwards until the translational motion is destroyed by gravity. If this occurs before the rotation is much modified by the action of the air, the missile begins to descend in the same plane towards the point of projection. The explanation requires (1) that the rotation should be so great that the direction of the axis is sensibly fixed in space and in the body; (2) that the resistance of the air should prevent any great motion perpendicular to the plane of the bent stick.

According to some experiments of Prof. S. P. Langley on the motion of a heavy disc placed with its plane horizontal the resistance of the air to vertical descent is much increased by a horizontal motion of the disc, so much so that the time of falling through a given space may be indefinitely prolonged by lateral motion. This perhaps is due to the inertia of the undisturbed air over which the disc passes. *Paris, Academy of Sciences*, translated in *Nature*, July 23, 1891.

In many specimens also of the boomerang the fore-part is slightly hollowed or the curve has a slight lateral twist by means of which the instrument is caused to rise or screw itself up in the air by virtue of its rotation.

Ex. It is stated by Col. Lane Fox in his lecture on *Primitive Warfare* that the plane of rotation instead of continuing perfectly parallel to its original position is slightly raised as the projectile advances; (*Journal of the United Service Institute*, vol. XII., 1868). A diagram is given, which is reproduced by Sir Richard Burton in his book on *The Sword* (1884), and shows that the boomerang should therefore be projected towards a point under the object intended to be hit. Show that this may be explained on the principles of Art. 209, if we suppose that the pressure of the air is greatest on that part of the under side which is moving in the same direction as the centre of gravity.

In the lecture already referred to, Col. Lane Fox (now Major-Gen. A. Pitt Rivers) remarks that the Australians cannot be said to have *invented* the boomerang. By giving a series of diagrams of the intermediate forms between it and the club, he shows that the savage may have been led to the adoption of the instrument "purely through the laws of accidental variation guided by the natural grain of the material in which he worked."

211. **Unsymmetrical tops.** We now pass on to the important and general problem of finding the oscillations of a heavy,

not necessarily uniaxal body, about a fixed point. We begin with the general equations of motion and proceed to apply them to two cases; (1) when the vertical through the fixed point is a principal axis at that point and the body has an initial rotation given to it about the vertical. (2) When the vertical is not necessarily a principal axis and the body has no initial rotation. Finally we shall examine what cases of steady motion are possible.

A body whose principal moments of inertia are not necessarily equal has a point O fixed in space and moves about O under the action of gravity. It is required to form the general equations of motion.

Let OA, OB, OC be the principal axes at the fixed point O, and let these be taken as axes of reference. Let h, k, l be the co-ordinates of the centre of gravity G, and let the mass of the body be taken as unity. Let OV be drawn vertically *upwards* and let p, q, r be the direction-cosines of OV referred to OA, OB, OC. Then we have by Euler's equations

$$\left.\begin{aligned}
A\omega_1' - (B - C)\,\omega_2\omega_3 &= -g\,(kr - lq) \\
B\omega_2' - (C - A)\,\omega_3\omega_1 &= -g\,(lp - hr) \\
C\omega_3' - (A - B)\,\omega_1\omega_2 &= -g\,(hq - kp)
\end{aligned}\right\} \dots\dots\dots\dots(1),$$

where accents denote differentiations with regard to the time.

Also p, q, r may be regarded as the co-ordinates of a point in OV, distant unity from O. This point is fixed in space, and therefore its velocities as given by Art. 17 are zero. We have

$$p' = \omega_3 q - \omega_2 r, \quad q' = \omega_1 r - \omega_3 p, \quad r' = \omega_2 p - \omega_1 q \dots\dots(2).$$

It is obvious that two integrals of these equations are supplied by the principles of Angular Momentum and Vis Viva. These give

$$A\omega_1 p + B\omega_2 q + C\omega_3 r = E,$$

$$A\omega_1^2 + B\omega_2^2 + C\omega_3^2 = F - 2g\,(ph + qk + rl),$$

where E and F are two arbitrary constants. The first of these might also have been obtained by multiplying the equations (1) by p, q, r respectively, and (2) by $A\omega_1$, $B\omega_2$, $C\omega_3$, and adding all six results. The second might have been obtained by multiplying the equations (1) by ω_1, ω_2, ω_3 respectively, adding and simplifying the right-hand side by (2).

212. *A body whose principal moments of inertia at the centre of gravity* G *are not necessarily equal, has a point* O *in one of the principal axes at* G *fixed in space and can move about* O *under the action of gravity. It is set in rotation about* OG *which is supposed to be vertical. Find the small oscillations.*

Referring to the general equations of Art. 211, we see that in this case $h=0$, $k=0$. Since OC remains always nearly vertical, ω_1 and ω_2 are small quantities, we may therefore reject the product $\omega_1\omega_2$ in the last of equations (1). This gives ω_3 constant. Let this constant value be called n. For the same reason $r=1$ nearly and p, q are both small quantities. Substituting we get the following linear equations,

$$Aω_1' - (B - C) nω_2 = lgq \atop Bω_2' - (C - A) nω_1 = - lgp \Big\} \quad\dots\dots(3), \qquad p' = qn - ω_2 \atop q' = - pn + ω_1 \Big\} \quad\dots\dots\dots(4).$$

If we substitute for $ω_1$, $ω_2$ in equations (3) their values given by (4) we should obtain two linear equations to find p, q which might have been at once deduced from the general equations of small oscillations given in Art. 15. But we may also solve these by assuming

$$ω_1 = F \sin (λt + f) \atop ω_2 = G \cos (λt + f) \Big\}, \qquad p = P \sin (λt + f) \atop q = Q \cos (λt + f) \Big\}.$$

Substituting, we get

$$AλF - (B - C) nG = glQ \atop BλG - (A - C) nF = glP \Big\} \quad\dots\dots\dots(5), \qquad λP = Qn - G \atop λQ = Pn - F \Big\} \quad\dots\dots\dots(6).$$

Eliminating the ratios $F : G : P : Q$ we have

$$λ^2 n^2 (A + B - C)^2 = \{gl + Aλ^2 + (B - C) n^2\} \{gl + Bλ^2 + (A - C) n^2\}.$$

If the values of $λ$ thus found should be real, the body will make small oscillations about the position in which OG is vertical. If C be the greatest moment, and n^2 sufficiently great to make both $gl - (C - A) n^2$ and $gl - (C - B) n^2$ negative, then all the values of $λ$ are real and the body will continue to spin with OG vertical. If G be beneath O, l is negative and it will be sufficient that OC should be the axis of greatest moment.

In order that the values of $λ^2$ may be real, we must have

$$\{gl (A + B) + n^2 (AC + BC - 2AB - C^2)\}^2 > 4 \{(B - C) n^2 + gl\} \{(A - C) n^2 + gl\} AB,$$

and in order that the two values of $λ^2$ may have the same sign we must have the last term of the quadratic positive; $\therefore \{(B - C) n^2 + gl\} \{(A - C) n^2 + gl\}$ is positive, and in order that the values of $λ^2$ may be both positive, we must have the coefficient of $λ^2$ in the quadratic negative; $\therefore gl (A + B) < n^2 (B - C) (A - C)$.

In the particular case in which $A = B$, each side of the quadratic becomes a perfect square and we have

$$Aλ^2 ± (2A - C) nλ + (A - C) n^2 + gl = 0 ;$$

$$\therefore λ = \mp \frac{2A - C}{2A} n ± \frac{\sqrt{C^2 n^2 - 4Agl}}{2A}.$$

In this case the conditions of stability reduce to $n > 2 \sqrt{Agl}/C$. By referring to equations (5) and (6) it will be seen that when $A = B$ we have $F = G$ and $P = Q$. If $λ_1$, $λ_2$ be the two values of $λ$ found above, we have

$$p = P_1 \sin (λ_1 t + f_1) + P_2 \sin (λ_2 t + f_2) \atop q = P_1 \cos (λ_1 t + f_1) + P_2 \cos (λ_2 t + f_2) \Big\}.$$

Following the notation used in Euler's geometrical equations Vol. I. Chap. v., let $θ$ be the angle OC makes with the vertical taken as axis of z, then $r^2 = \cos^2 θ = 1 - θ^2$, and hence $\qquad θ^2 = p^2 + q^2 = P_1^2 + P_2^2 + 2P_1 P_2 \cos \{(λ_1 - λ_2) t + f_1 - f_2\}.$

Let $φ$ be the angle the plane containing OA, OC makes with the plane containing OC and the vertical OV, we have $p = - \sin θ \cos φ$, and $q = \sin θ \sin φ$, and hence

$$- \tan φ = \frac{P_1 \cos (λ_1 t + f_1) + P_2 \cos (λ_2 t + f_2)}{P_1 \sin (λ_1 t + f_1) + P_2 \sin (λ_2 t + f_2)}.$$

Since $θ$ is very small we have, still following the same notation, $ψ = nt + a - φ$, where a is some constant, depending on the position of the arbitrary plane from which $ψ$ is measured.

When the axis of the top is inclined at an angle a to the vertical, the period of oscillation about the steady motion is found in Art. 208 to be $2\pi/p$. But this period is different from either of the periods found in Art. 212 when the axis is supposed to be nearly vertical. We easily see by eliminating μ from the expression for p that $p = \lambda_1 - \lambda_2$, so that the period of oscillation of θ when the axis is inclined is the same as the period of oscillation of θ^2 when the axis is vertical.

The periods of oscillation found by this method do not seem to agree with those found by a different process in Vol. I. Art. 268. But the difference is only apparent. In Vol. I., the axis OC is referred to axes fixed in space, we have

$$\xi = \theta \cos \psi = P_1 \cos (\mu_1 t + f_1) + P_2 \cos (\mu_2 t + f_2)$$
$$\eta = \theta \sin \psi = P_1 \sin (\mu_1 t + f_1) + P_2 \sin (\mu_2 t + f_2).$$

In the method of this volume the vertical is referred to axes fixed in the body by the direction cosines $p = -\theta \cos \phi$, $q = \theta \sin \phi$, and 1. Now by Euler's third geometrical equation $\omega_3 = \cos \theta d\psi/dt + d\phi/dt$ hence $\phi = nt - \psi +$ terms with P^2, see Vol. I. We have therefore

$$-p = \theta \cos (nt - \psi) = \cos nt (P_1 \cos \mu_1 t + P_2 \cos \mu_2 t) + \sin nt (P_1 \sin \mu_1 t + P_2 \sin \mu_2 t)$$
$$= P_1 \cos (n - \mu_1) t + P_2 \cos (n - \mu_2) t$$
$$q = \theta \sin (nt - \psi) = P_1 \sin (n - \mu_1) t + P_2 \sin (n - \mu_2) t.$$

Thus, since $n - \mu = \pm \lambda$, the expressions found here for p, q follow easily from those for ξ, η found in Vol. I.

213. *A body whose principal moments at the centre of gravity are not necessarily equal is free to turn about a fixed point* O, *and is in equilibrium under the action of gravity. A small disturbance being given, find the oscillations.*

Referring to the general equations in Art. 211 we see that in this case ω_1, ω_2, ω_3, are small, hence in equations (1) we may omit the terms containing the products $\omega_1\omega_2$, $\omega_2\omega_3$, $\omega_3\omega_1$. Also since in equilibrium OG is vertical, p, q, r are always nearly in the ratio $h : k : l$; hence if $OG = a$, we may write h/a, k/a, l/a for p, q, r on the right-hand sides of equations (2). The six equations are now all linear. To solve these we put $\omega_1 = H \sin (\lambda t + \mu)$ and $p = h/a + P \cos (\lambda t + \mu)$ ....(3), ω_2, ω_3, q and r being represented by similar expressions with K and L written for H; Q, k and R, l written for P and h. Substituting these in the equations we get six linear equations. Eliminating P, Q, R we have

$$\left(\frac{a}{g} A\lambda^2 + k^2 + l^2\right) H - hkK - lhL = 0$$
$$-hkH + \left(\frac{a}{g} B\lambda^2 + l^2 + h^2\right) K - lkL = 0$$ ....(4).
$$-lhH - lkK + \left(\frac{a}{g} C\lambda^2 + h^2 + k^2\right) L = 0$$

Eliminating the ratios of H, K, L we have an equation to find λ^2. One root is $\lambda^2 = 0$, the others are given by the quadratic

$$\lambda^4 + \left(\frac{k^2 + l^2}{A} + \frac{l^2 + h^2}{B} + \frac{h^2 + k^2}{C}\right) \frac{g}{a} \lambda^2 + g^2 \frac{Ah^2 + Bk^2 + Cl^2}{ABC} = 0 \, ... (5).$$

To ascertain if the roots are real we must apply the usual criterion for a quadratic. This requires that

$$\{A (B - C) h^2 + B (C - A) k^2 - C (A - B) l^2\}^2 + 4AB (B - C) (A - C) h^2k^2......(6)$$

should be positive. Since A, B, C can be chosen to be in descending order, we see that the condition is satisfied. See also Art. 58.

If G is above O, a is positive and the values of λ^2 are both negative. The equilibrium is therefore unstable. If G is below O, a is negative and the values of λ^2 are both positive. If the roots are equal, the two positive terms in (6) must be separately zero, this gives $k=0$ and $A(B-C)h^2=C(A-B)l^2$, i.e. the centre of gravity lies in the asymptote to the focal hyperbola of the momental ellipsoid. In this case we find $\lambda^2 = ag/B$. The case in which $k=0$, $l=0$, $B=C$ has been considered in Art. 212.

If the values of λ^2 are written 0, $\lambda_1{}^2$, $\lambda_2{}^2$ we have

$$\omega_1 = H_0 + H_0't + H_1 \sin(\lambda_1 t + \mu_1) + H_2 \sin(\lambda_2 t + \mu_2),$$

with similar expressions for ω_2, ω_3. Equations (2) then give p, q, r. But substituting in (1) we find that all the non-periodic terms which contain t are zero. Remembering that $p^2 + q^2 + r^2 = 1$ we have finally

$$\omega_1 = \Omega h/a + H_1 \sin(\lambda_1 t + \mu_1) + H_2 \sin(\lambda_2 t + \mu_2),$$

ω_2 and ω_3 being represented by similar expressions with k, K and l, L written for h, H. The values of K_1, L_1 and K_2, L_2 are determined by equations (4) in terms of H_1 and H_2 respectively. We also have

$$p = \frac{h}{a} + \frac{lK_1 - kL_1}{a\lambda_1} \cos(\lambda_1 t + \mu_1) + \frac{lK_2 - kL_2}{a\lambda_2} \cos(\lambda_2 t + \mu_2),$$

with similar expressions for q and r. There remain five constants, viz. Ω, H_1, H_2, μ_1, μ_2 to be determined by the initial values of ω_1, ω_2, ω_3, r and q.

When the roots are equal the equations depending on p, r, ω_2 separate from those depending on q, ω_1, ω_3, forming two sets; we find

$$\left.\begin{aligned}
\omega_1 &= \Omega\frac{h}{a} + H \sin(\lambda t + \mu_1) \\
\omega_3 &= \Omega\frac{l}{a} + H\,\frac{Aa\lambda^2 + gl^2}{ghl} \sin(\lambda t + \mu_1) \\
q &= \qquad H\frac{A\lambda}{gl} \cos(\lambda t + \mu_1)
\end{aligned}\right\}, \qquad
\left.\begin{aligned}
\omega_2 &= \qquad K \sin(\lambda t + \mu_2) \\
p &= \frac{h}{a} + K\,\frac{l}{a\lambda} \cos(\lambda t + \mu_2) \\
r &= \frac{l}{a} - K\,\frac{h}{a\lambda} \cos(\lambda t + \mu_2)
\end{aligned}\right\}.$$

A solution of this problem conducted in a totally different manner has been given by Lagrange in his *Mécanique Analytique*. His results do not altogether agree with those given here.

If we substitute the values of ω_1, ω_2, ω_3, p, q, r in the equation of angular momentum of Art. 211 and neglect the squares of small quantities, we evidently obtain 　　$(Ah^2 + Bk^2 + Cl^2)\,\Omega = Ea^2$, 　　$AHh + BKk + CLl = 0$.

The first of these equations shows that Ω vanishes when the initial conditions are such that the angular momentum about the vertical is zero. In this case the problem reduces to that considered in Art. 134.

214. *A body whose principal moments of inertia are not necessarily equal has a point O fixed in space and moves about O under the action of gravity. It is required to find what cases of steady motion are possible in which one principal axis OC at O describes a right cone round the vertical while the angular velocity of the body about OC is constant; and to find the small oscillations.*

Referring to the general equations of Art. 211, we see that it is given that r and ω_3 are constants. In this case the first two equations of (1) and (2) form a set of *linear* equations from which we have to find the four quantities p, q, ω_1, ω_2. The solution of these equations is therefore of the form

$$\left.\begin{aligned}
\omega_1 &= F_0 + F_1 \sin(\lambda t + f) \\
\omega_2 &= G_0 + G_1 \cos(\lambda t + f)
\end{aligned}\right\}, \qquad
\left.\begin{aligned}
p &= P_0 + P_1 \sin(\lambda t + f) \\
q &= Q_0 + Q_1 \cos(\lambda t + f)
\end{aligned}\right\}.$$

But these must also satisfy the last of the equations (1). Substituting we see that there will be a term on the left side of the form

$$- \tfrac{1}{2} (A - B) F_1 G_1 \sin 2 (\lambda t + f).$$

But there will be no such term on the right side. Hence we must have either $A = B$, $F_1 = 0$ or $G_1 = 0$. The motion in the case in which $A = B$ has already been considered in Art. 207. Again, substituting in the last of equations (2) and equating to zero the coefficient of $\sin 2 (\lambda t + f)$ we find

$$P_1 G_1 - F_1 Q_1 = 0.$$

Substituting in the first two of equations (1) and equating to zero the coefficients of $\cos (\lambda t + f)$ and $\sin (\lambda t + f)$, we find

$$A \lambda F_1 - (B - C) n G_1 = g l Q_1 \qquad - B \lambda G_1 - (C - A) n F_1 = - g l P_1;$$

from these equations we have F_1, G_1, P_1, Q_1 all equal to zero and therefore ω_1, ω_2, p, q are all constant as well as the given constants ω_3 and r.

In this case the equations (2) give $\omega_1/p = \omega_2/q = \omega_3/r$, so that the axis of revolution must be vertical. Let ω be the angular velocity about the vertical. Then $\omega_1 = p\omega$, $\omega_2 = q\omega$, $\omega_3 = r\omega$. Substituting in equations (1) we get

$$\frac{h}{p} - \frac{A\omega^2}{g} = \frac{k}{q} - \frac{B\omega^2}{g} = \frac{l}{r} - \frac{C\omega^2}{g} \quad \dots\dots\dots\dots\dots\dots (3).$$

Unless, therefore, two of the principal moments are equal, it is necessary for steady motion that the axis of rotation should be vertical and the centre of gravity (h, k, l) *must lie in the vertical straight line whose equations are* (3).

This straight line may be constructed geometrically in the following manner. Measure along the vertical a length $OV = g/\omega^2$ and draw a plane through V perpendicular to OV to touch an ellipsoid confocal with the ellipsoid of gyration. The centre of gravity must lie on the normal at the point of contact.

To find the small oscillations about the steady motion, i.e. to determine whether this motion be stable or not, we must put

$$p = \cos a + P_0 \sin \lambda t + P_1 \cos \lambda t,$$

with similar expressions for q, r, ω_1, ω_2, ω_3. Substituting we shall get twelve linear equations to determine eleven ratios. Eliminating these we have an equation to find λ. It is sufficient for stability that all the roots of this equation should be real.

Motion of a Sphere.

215. General equations of Motion. *To determine the motion of a sphere on any perfectly rough surface under the action of any forces whose resultant passes through the centre of the sphere.*

Let G be the centre of gravity of the body and let the moving axes GC, GA, GB be respectively a normal to the surface and some two lines at right angles to be afterwards chosen at our convenience. Let the motions of these axes be determined by the angular velocities θ_1, θ_2, θ_3 about their instantaneous positions in the manner explained in Art. 3. Let u, v, w be the velocities of G resolved parallel to the axes, and ω_1, ω_2, ω_3 the angular velocities of the body about these axes; then $w = 0$. Let F, F' be the resolved parts of the friction of the perfectly rough surface on the sphere parallel to the axes, GA, GB, and let R be the normal reaction. Let X, Y, Z be the resolved parts of the impressed forces on the centre of gravity. Let k be the radius of gyration of the sphere about a diameter, a its radius, and let its mass be unity. We shall suppose that in the standard case the sphere rolls on the convex side of the fixed surface and that the positive direction of the axis Z is drawn outwards from the surface. The equations of motion of the sphere are by Arts. 14 and 5,

$$\left.\begin{aligned}\frac{d\omega_1}{dt} - \theta_3\omega_2 + \theta_2\omega_3 &= \frac{F'a}{k^2}\\ \frac{d\omega_2}{dt} - \theta_1\omega_3 + \theta_3\omega_1 &= -\frac{Fa}{k^2}\\ \frac{d\omega_3}{dt} - \theta_2\omega_1 + \theta_1\omega_2 &= 0\end{aligned}\right\} \dots\dots \dots\dots (1).$$

$$\left.\begin{aligned}\frac{du}{dt} - \theta_3 v &= X + F\\ \frac{dv}{dt} + \theta_3 u &= Y + F'\\ -\theta_2 u + \theta_1 v &= Z + R\end{aligned}\right\} \dots\dots\dots (2),$$

and since the point of contact of the sphere and surface is at rest, we have $\qquad u - a\omega_2 = 0, \quad v + a\omega_1 = 0 \dots\dots\dots\dots (3).$

Eliminating F, F', ω_1, ω_2 from these equations, we get

$$\left.\begin{aligned}\frac{du}{dt} - \theta_3 v &= \frac{a^2}{a^2 + k^2} X + \frac{k^2}{a^2 + k^2} \theta_1 a\omega_3\\ \frac{dv}{dt} + \theta_3 u &= \frac{a^2}{a^2 + k^2} Y + \frac{k^2}{a^2 + k^2} \theta_2 a\omega_3\end{aligned}\right\} \dots\dots (4).$$

216. The meaning of these equations may be found as follows. They are the two equations of motion of the centre of gravity of the sphere, which we should have obtained if the given surface had been smooth and the centre of gravity had been acted on by accelerating forces $\frac{k^2}{a^2 + k^2} \theta_1 a\omega_3$ and $\frac{k^2}{a^2 + k^2} \theta_2 a\omega_3$ along the axes GA, GB, and by the same impressed forces as before reduced in the ratio $\frac{a^2}{a^2 + k^2}$. The motion therefore of the centre of gravity in these two cases with the same initial conditions will be the same. More convenient expressions for these two additional forces may be found thus. The centre of gravity moves along a surface formed by producing all the normals to the given surface a constant length equal to the radius of the sphere. Let us take the axes GA, GB to be tangents to the lines of curvature of this surface and let ρ_1, ρ_2 be the radii of curvature of the normal sections through these tangents respectively. Then

$$\theta_1 = -\frac{v}{\rho_2}, \quad \theta_2 = \frac{u}{\rho_1} \dots\dots\dots\dots (5).$$

If G be the position of the centre of gravity at the time t, the quantity $\theta_3 dt$ is the angle between the projections of two successive positions of GA on the tangent plane at G. Let χ_1, χ_2 be the angles the radii of the curvature of the lines of curvature at G make with the normal. The centre of the sphere may be brought from G to any neighbouring position G' by moving it first from G to H along one line of curvature and then from H to G' along the other. As the sphere moves from G to H, the angle turned round by GA is the product of the arc GH into the resolved curvature of GH in the tangent plane. By Meunier's theorem, the curvature is $\dfrac{1}{\rho_1 \cos \chi_1}$, multiplying this by $\sin \chi_1$ to resolve it into the tangent plane we find that the part of θ_3 due to the motion along GH is $\dfrac{u}{\rho_1} \tan \chi_1$. Treating the arc HG' in the same way, we have

$$\theta_3 = \frac{u}{\rho_1} \tan \chi_1 + \frac{v}{\rho_2} \tan \chi_2 \ldots\ldots\ldots\ldots(6).$$

This result follows also from that given in Art. 21, Ex. 2.

We have also an expression for ω_3 given by equations (1). Substituting for ω_1, ω_2 from the geometrical equations (3) we get

$$a \frac{d\omega_3}{dt} = uv \left(\frac{1}{\rho_2} - \frac{1}{\rho_1} \right) \ldots\ldots\ldots\ldots (7).$$

Many of the results in this section are deduced from equations (4) and (7) and in all these cases an apparently independent solution may be obtained by forming over again the equations (1), (2), (3), &c. (from which (4) and (7) have been derived), with such simplifications as suit the problem under consideration. An example of this process is given in Art. 221.

217. The solution of the equations may be conducted as follows. Let (x, y, z) be the co-ordinates of the centre of the sphere. Then u, v may be found from the equation to the surface in terms of dx/dt, dy/dt, dz/dt by resolving parallel to the axes of reference. If we eliminate $u, v, \theta_1, \theta_2, \theta_3$ by means of (4), (5), and (6), we shall get three equations containing x, y, z, ω_3, and their differential coefficients with respect to t. These, together with the equation to the surface, will be sufficient to determine the motion at any time. One integral can always be found by the principle of Vis Viva. Since the sphere is turning about the point of contact as an instantaneously fixed point we have

$$(a^2 + k^2)(\omega_1^2 + \omega_2^2) + k^2 \omega_3^2 = 2\phi,$$

where ϕ is the force function of the impressed forces. This is the same as

$$u^2 + v^2 + \frac{k^2 a^2}{a^2 + k^2} \omega_3^2 = 2 \frac{a^2}{a^2 + k^2} \phi \ldots\ldots\ldots\ldots(8),$$

and the right-hand side of this equation is twice the force function of the altered impressed forces.

218. It will sometimes be more convenient to take the axis GA to be a tangent to the path. Then $v = 0$ and therefore $\omega_1 = 0$. If U be the resultant velocity of the centre of the sphere we have $u = U$. Also if R be the radius of torsion of a geodesic touching the path at G and ρ the radius of curvature of the normal section at G through a tangent to the path, we have $\theta_1 = U/R$ and $\theta_2 = U/\rho$. In these expressions, as elsewhere, R is estimated positive when the torsion round GA is from the positive direction of GB to the positive direction of GC. If χ be the angle the radius of curvature of the path makes with the normal, we have as before $\theta_3 = \tan \chi \, U/\rho$. The equations (4) become

$$\left. \begin{aligned} \frac{dU}{dt} &= \frac{a^2}{a^2 + k^2} X + \frac{k^2}{a^2 + k^2} \frac{U}{R} a\omega_3 \\ \frac{U^2}{\rho} \tan \chi &= \frac{a^2}{a^2 + k^2} Y + \frac{k^2}{a^2 + k^2} \frac{U}{\rho} a\omega_3 \end{aligned} \right\} \quad \dots\dots\dots\dots\dots (\text{iv}).$$

The expression for ω_3 given by equations (1) now takes the form

$$a \frac{d\omega_3}{dt} = -\frac{U^2}{R} \quad \dots\dots\dots\dots\dots\dots\dots (\text{vii}).$$

It may be shown by geometrical considerations that this form is identical with that given in (7).

219. *To find the pressure on the surface* we use the last of equations (2). This may be written in either of the forms

$$\frac{U^2}{\rho} = \frac{u^2}{\rho_1} + \frac{v^2}{\rho_2} = -Z - R \quad \dots\dots\dots\dots\dots (9).$$

The sphere will leave the surface when R changes sign. This will generally occur when the velocity of the centre of the sphere is that due to one half of the projection of the radius of curvature of the normal section on the direction of the resultant force.

220. **Ex. 1.** Show that the angular velocity of the sphere about a normal to the surface, viz. ω_3, is constant when the direction of motion of the centre of gravity is a tangent to a line of curvature, and only then.

Ex. 2. A sphere is projected without initial angular velocity about the radius normal to the surface, so that its centre begins to move along a line of curvature. Show that it will continue to describe that line of curvature if the force transverse to the line of curvature and tangential to the surface is equal to seven-fifths of the centrifugal force of the whole mass collected into the centre, resolved in the tangent plane to the surface.

Ex. 3. If the sphere be not acted on by any forces, show that

$$U^2 \left(\tan^2 \chi + \frac{2}{7} \right) = \text{constant}, \quad a\omega_3 = \frac{7}{2} U \tan \chi, \quad \frac{d}{ds} \log \left(\tan^2 \chi + \frac{2}{7} \right) = -\frac{2}{R} \tan \chi.$$

Show also that the path will not be a geodesic unless the path is a plane curve.

221. **Motion on a rough plane.** *If the given surface on which the sphere rolls be a plane, we have ρ_1 and ρ_2 both infinite, hence θ_1, θ_2 are both zero. If therefore a homogeneous sphere roll on a perfectly rough plane under the action of any forces whatever the resultant of which passes through the centre of the sphere, the motion of the centre of gravity, with the same initial conditions, is the same as if the plane were smooth, and all the forces were reduced to five-sevenths of their former value. And it is also clear that the*

plane is the only surface which possesses this property for all initial conditions.

We may easily obtain the first part of this theorem from first principles. Taking the directions of the axes of x and y to be fixed in space and parallel to the rough plane we have (Arts. 14 and 236)

$$k^2\omega_1' = F'a \brace k^2\omega_2' = -Fa \qquad u' = X + F \brace v' = Y + F' \qquad u - a\omega_2 = 0 \brace v + a\omega_1 = 0 .$$

Eliminating F, F', ω_1, ω_2 we find $\dfrac{du}{dt} = \dfrac{a^2}{a^2+k^2}X$, $\dfrac{dv}{dt} = \dfrac{a^2}{a^2+k^2}Y$,

which is the analytical statement of the theorem. The six equations of motion from which this result is derived are obviously only simplified forms of equations (1), (2), (3) of Art. 215. See Vol. I. Art. 269.

222. **Ex. 1.** If the plane is imperfectly rough, prove that the sphere can roll only if two-sevenths of the resultant impressed force parallel to the plane is less than the greatest friction which can be called into play. Prove also that the direction of the friction is opposite to that of the resultant impressed force parallel to the plane.

Ex. 2. If the rough plane on which the sphere rolls rotate about a normal through any point O with a uniform angular velocity Ω, prove that the motion in space of the centre of gravity is the same as if the plane were smooth and the sphere acted on by the impressed forces reduced to five-sevenths of their former values, together with an accelerating force acting perpendicular to the tangent to the path and equal to $\frac{4}{7}\Omega U$, where U is the velocity of the centre of gravity. If the positive direction of rotation of Ω is the same as that of the hands of a watch, this additional force acts on the right-hand side of the tangent when an observer at the centre of gravity looks in the direction of motion.

223. **Motion on a rough spherical surface.** *If the given surface on which the sphere rolls is another sphere* of radius $b - a$, we have $\rho_1 = \rho_2 = b$. Hence ω_3 is constant; let this constant value be called n, and let U be the velocity of the centre of gravity. Since every normal section is a principal section, let us take GA a tangent to the path. *Hence the motion of the centre of gravity is the same as if the whole mass, collected at that point, were acted on by an accelerating force equal to* $\dfrac{k^2}{a^2+k^2}\dfrac{an\,U}{b}$ *in a direction perpendicular to the path, and all the impressed forces were reduced in the ratio* $a^2/(a^2+k^2)$. According to the usual convention as to the relative positions of the axes GA, GB, GC it is clear that if the positive direction of GA be in the direction of motion, the angular velocity n should be estimated positive when the part of the sphere in front is moving to the right of GA and the additional force when positive will also act toward the right-hand side of the tangent. Since this additional force acts perpendicularly to the path, it will not appear in the equation of Vis Viva. Hence the velocity of the centre of gravity in any position is the same as if it had arrived there simply under the action of the reduced forces. Let O be the centre of the fixed sphere, θ the angle OG makes with

the vertical OZ, and ψ the angle the plane ZOG makes with any fixed plane passing through OZ. Then by Vis Viva we have

$$\left(\frac{d\theta}{dt}\right)^2 + \sin^2\theta \left(\frac{d\psi}{dt}\right)^2 = F - \frac{2g}{b}\frac{a^2}{a^2+k^2}\cos\theta \dots\dots(\text{I}),$$

where F is some constant to be determined from the initial conditions. This also follows from equation (8),

Also taking moments about OZ, we have

$$\frac{b}{\sin\theta}\frac{d}{dt}\left(\sin^2\theta \frac{d\psi}{dt}\right) = \frac{k^2}{a^2+k^2}\,an\,\frac{d\theta}{dt} \dots\dots (\text{II}),$$

an equation which will be found to be a transformation of the second of equations (4). Integrating this equation we have

$$\sin^2\theta \frac{d\psi}{dt} = E - \frac{k^2}{a^2+k^2}\frac{an}{b}\cos\theta\dots\dots\dots(\text{III}),$$

where E is some constant. These two equations will suffice to determine $d\theta/dt$ and $d\psi/dt$ under any given initial conditions.

The pressure on either sphere is given by

$$\frac{R}{m} = g\cos\theta\,\frac{3a^2+k^2}{a^2+k^2} - bF\dots\dots\dots\dots(\text{IV}),$$

where m is the mass of the sphere. The spheres separate when R vanishes and changes sign.

If the sphere have no initial angular velocity about the normal to the surface it is clear that $n = 0$ and the additional impressed force is zero. *If therefore a homogeneous sphere roll on a perfectly rough fixed spherical surface, and if the sphere either start from rest, or have its initial angular velocity about the common normal equal to zero, the motion of the centre of the sphere is the same as if the fixed spherical surface were smooth and the forces on the rolling sphere were reduced to five-sevenths of their former value.*

It will be noticed that the equations (I) and (III) which determine the motion when gravity is the acting force are the same as those marked (6) in Art. 202 which give the motion of a top. The results obtained in Art. 203 therefore also apply to the motion of the sphere. If the sphere does not roll off it will roll round the fixed sphere oscillating between an upper and lower horizontal circle. In order that the sphere may not roll off it is necessary that the value of $\cos\theta$ found by equating the pressure R to zero should not lie between the limiting circles of motion. These results are given in greater detail in the examples immediately following.

Ex. 1. A homogeneous sphere rolls under the action of gravity in any manner on a perfectly rough fixed sphere whose centre is O. Prove that throughout the motion (1) the velocity of the centre G of the moving sphere is that due to five-sevenths of its depth below a fixed horizontal plane; (2) the moving sphere will leave the fixed sphere when the altitude of its centre above O is ten-seventeenths of the altitude of the fixed plane above the same point; (3) the transverse velocity of G is proportional to the tangent of the angle GU makes with the horizon, where U is a fixed point on a vertical through O.

Ex. 2. As in the corresponding problem for a top, Art. 205, let the initial motion of the sphere be simply a rotation n about the common normal. If i be the inclination of this normal to the vertical prove that the angular radii of the circles between which the point of contact oscillates are i and that value of θ between i and π which is given by the quadratic $\cos^2\theta - 1 + 2p\,(\cos i - \cos\theta) = 0$ and $4bg\,(a^2 + k^2)\,p = k^4 n^2$. The spheres separate when $\cos\theta = \cos i\,2a^2/(3a^2 + k^2)$, and it is supposed that the initial angular velocity n is so great that this value of θ does not lie between the angular radii of the limiting circles.

224. Sphere rolling on a moveable sphere. If the guiding sphere, hitherto fixed, is either constrained to rotate with a uniform angular velocity about a fixed diameter or is free to move about its centre as a fixed point the theorems given above are but slightly altered. The chief change is that the quantity n must be replaced by another constant which we shall represent by n'.

As the proofs are so nearly the same as when the guiding sphere is fixed, minute details are unnecessary. It is sufficient to enunciate the results in the following examples, the demonstrations of which are left to the reader.

If the guiding sphere is constrained to turn about an axis OZ with angular velocity Ω the equations (1) and (2) of Art. 215 are still true, but the geometrical equations (3) become

$$u - a\omega_2 = 0, \quad v + a\omega_1 = c\Omega \sin\theta,$$

where c is the radius of the sphere whose diameter OZ is fixed, θ is the inclination of the common normal OG of the two spheres to OZ and the axis GA lies in the plane ZOG.

If the guiding sphere is free to move about its centre O, its equations of motion are the same as (1) except that we write Ω_1, Ω_2, Ω_3 for ω_1, ω_2, ω_3; c for a; and MK^2 for mk^2. The geometrical equations (3) become

$$u - a\omega_2 = c\Omega_2, \quad v + a\omega_1 = -c\Omega_1.$$

Ex. 1. A sphere, radius a, rolls on a guiding sphere, radius c, which is constrained to turn about a fixed diameter, taken as the axis of reference, with a constant angular velocity Ω. If θ, ψ are the angular co-ordinates of the common normal OG, prove (1) that $a\omega_3 + c\Omega\cos\theta = an'$ where n' is a constant. The value of n' is therefore known from the initial conditions.

If U be the velocity in space of the centre G of the rolling sphere prove (2) that the velocity of the centre G is the same as if the whole mass, collected at that point, were acted on by the impressed forces, reduced in the ratio $a^2/(a^2 + k^2)$, together with an accelerating force equal to $\dfrac{k^2}{a^2 + k^2}\,\dfrac{an'U}{b}$, where $b = a + c$, acting in a direction perpendicular to the path and tending to the right-hand side of the tangent.

Prove (3) that the pressure R on the rolling sphere is given by $-R = Z + U^2/b$.

It follows from these results that the equations (I.) (III.) and (IV.) of Art. 223 hold also when the guiding sphere rotates uniformly about a vertical diameter and gravity is the only force acting.

Ex. 2. A sphere of radius a and mass m rolls on a guiding sphere of radius c and mass M which is free to turn about its centre O as a fixed point. Let Ω_1, Ω_2, Ω_3 be the angular velocities of the guiding sphere about axes meeting at O parallel to those about which ω_1, ω_2, ω_3 are the angular velocities of the rolling sphere, Art. 215. Prove (1) that $a\omega_3 + c\Omega_3 = an'$ where n' is a constant. The value of n' is therefore known from the initial conditions and is zero when both spheres start from rest.

Prove (2) that the motion of G is the same as if the mass of the sphere, collected at that point, were acted on by the impressed forces, reduced in the ratio $e/(1+e)$ where $e = \dfrac{a^2}{k^2} + \dfrac{m}{M}\dfrac{c^2}{K^2}$, together with a transverse accelerating force equal to $\dfrac{1}{1+e}\dfrac{an'U}{b}$ in a direction perpendicular to the tangent and tending to the right-hand side, where U is the velocity of G and $b = a + c$.

Ex. 3. A perfectly rough sphere of radius c is made to rotate about a vertical diameter which is fixed, with a constant angular velocity n. A uniform sphere of radius a is placed on it at a point distant ca from the highest point : investigate the motion and determine in any position the angular velocity of the sphere. Show that the sphere will leave the rotating sphere when the point of contact is at an angular distance θ from the vertex, where $\cos\theta = \dfrac{10}{17}\cos a + \dfrac{4}{119}\dfrac{c^2 n^2 \sin^2 a}{(a+c)g}$. [Notice the initial impact.]   [Math. Tripos, 1889.

225. Motion on a rough cylinder. *If the surface on which the sphere rolls is a cylinder* the lines of curvature are the generators and the transverse sections. Let the axis GA be directed parallel to the generators, then ρ_1 is infinite and $\rho_2 - a$ is the radius of curvature of the transverse section. We have $\theta_1 = -v/\rho_2$, $\theta_2 = 0$, and since $\chi_2 = 0$, $\theta_3 = 0$. The equations (4) and (7) therefore become

$$\left. \begin{aligned} \frac{du}{dt} &= \frac{a^2}{a^2+k^2}X - \frac{k^2}{a^2+k^2}\frac{v}{\rho_2}a\omega_3 \\ \frac{dv}{dt} &= \frac{a^2}{a^2+k^2}Y \\ \frac{d(a\omega_3)}{dt} &= \frac{uv}{\rho_2} \end{aligned} \right\}.$$

From these equations the motion may be found.

The second of these gives the motion transverse to the generators of the cylinder, and if Y be the same for all positions of the sphere on the same generator, this equation may be solved independently of the other two. *The transverse motion of the centre of the sphere is therefore the same, under the same initial circumstances, as that of a smooth sphere constrained to slide, in a plane perpendicular to the generators, on the transverse section of the cylinder and acted on by the same impressed forces but reduced in the ratio* $a^2/(a^2 + k^2)$.

Having found v we may proceed thus; let ϕ be the angle the normal plane to the cylinder through a generator and through the centre of the sphere makes with some fixed plane passing through a generator, then $v = \rho_2 d\phi/dt$. If $d\phi/dt$ is not zero, the first and third equations then become

$$\frac{du}{d\phi} + \frac{k^2}{a^2+k^2}a\omega_3 = \frac{a^2}{a^2+k^2}\frac{\rho_2}{v}X \qquad u = \frac{d(a\omega_3)}{d\phi}.$$

If X is the same for all positions of the sphere on the same generator these equations can be solved without difficulty. For v and ρ_2 being known in terms of ϕ, we have in this case two linear equations to find u and $a\omega_3$. If X is zero, and $k^2 = \frac{2}{5}a^2$, we find $a\omega_3 = A \sin\left(\sqrt{\frac{2}{7}}\,\phi + B\right)$, $u = A\sqrt{\frac{2}{7}} \cos\left(\sqrt{\frac{2}{7}}\,\phi + B\right)$,
where A and B are two arbitrary constants to be determined by the initial values of u and ω_3.

If X is not the same for all positions of the sphere on the same generator, let ξ be the space traversed by the sphere measured along a generator. Then

$$u = d\xi/dt = (d\xi/d\phi)\,(v/\rho_2).$$

Substituting this value of u, we have two equations to find ξ and $a\omega_3$ in terms of ϕ. One integral of these is equation (8) of Art. 217 which was obtained by the principle of Vis Viva.

226.   **Ex. 1.**   A sphere rolls under the action of gravity on a perfectly rough cylindrical surface with its axis inclined at an angle a to the horizon. The section of the cylinder is such that when the sphere rolls on it, the centre describes a cycloid with its cusps on the same horizontal line. If the sphere start from rest with its centre at a cusp, find the motion.

Let the position of the sphere be defined by ξ, the space described along a generator, and s, the arc of the cycloid measured from the vertex. If $4b$ is the radius of curvature of the cycloid at its vertex, we have $s = 4b \cos\sqrt{\dfrac{5g\cos a}{28b}}\,t.$

Since $v = ds/dt$ and $\rho_2{}^2 + s^2 = 16b^2$ we find that v/ρ_2 is constant. This gives without difficulty

$$\omega_3 = -\frac{\sin a}{a} \sqrt{\frac{35bg}{\cos a}} \left\{1 - \cos\frac{1}{7}\sqrt{\frac{5g\cos a}{2b}}\,t\right\},$$

$$u = \sin a \sqrt{\frac{10bg}{\cos a}} \sin\frac{1}{7}\sqrt{\frac{5g\cos a}{2b}}\,t.$$

Ex. 2.   If a rough inelastic sphere of radius c be dropped on to the lowest generating line on the interior of a circular cylinder radius a, which is revolving freely (with angular velocity Ω) about its axis which is fixed at an angle a to the horizon, prove that the plane through the axis of the cylinder and the centre of the sphere will move like a simple circular pendulum of length l where

$$l\,.\,(2m + 5M)\cos a = (a - c)\,(2m + 7M),$$

M and m being the masses of the cylinder and sphere respectively.

[May Exam. 1877.

227.   The relation, $v/\rho_2 =$ constant, holds whenever (1) the forces acting at the centre of the sphere, and the form of the section of the cylinder, are so related that the tangential component bears a constant ratio to $\rho_2 d\rho_2/ds$, and (2) the sphere starts from rest at a point where ρ_2 is zero. In such a case, the normal plane to the section through the centre of the sphere has a constant angular velocity in space and the resolved motion of the sphere perpendicular to the generators is independent of that along the generators.

Ex.   A sphere rolls on a perfectly rough right circular cylinder whose radius is c under the action of no forces, show that the path traced out by the point of contact becomes the curve $x = A \sin (2y/7c)^{\frac{1}{2}}$ when the cylinder is developed on a plane.

This result shows that the sphere cannot be made to travel continually in one direction along the length of the cylinder except when the point of contact describes a generator.

228.   **Motion on a rough cone.**   *If the surface on which the sphere rolls is a*

cone, the lines of curvature are the generators and their orthogonal trajectories. Let the axis GA be directed parallel to the generator, then ρ_1 is infinite and $\rho_2 - a$ is the radius of curvature of a normal section perpendicular to the generators. Also $\theta_1 = -v/\rho_2$, $\theta_2 = 0$. Let the position of the sphere be defined by the distance r of its centre from the vertex O of the cone on which the centre always lies and by an angle ϕ such that $d\phi$ is the angle between two consecutive positions of the distance r, $d\phi$ being taken as positive when the centre moves in the positive direction of GB. If the cone were developed on a plane it is clear that r and ϕ would be the ordinary polar co-ordinates of a point G. We have

$$\theta_3 = \frac{d\phi}{dt}, \quad u = \frac{dr}{dt}, \quad v = r\frac{d\phi}{dt}.$$

The equations (4) and (7) become therefore

$$\left.\begin{aligned}
\frac{d^2r}{dt^2} - r\left(\frac{d\phi}{dt}\right)^2 &= \frac{a^2}{a^2+k^2}X - \frac{k^2}{a^2+k^2}\frac{r}{\rho_2}a\omega_3\frac{d\phi}{dt} \\
\frac{1}{r}\frac{d}{dt}\left(r^2\frac{d\phi}{dt}\right) &= \frac{a^2}{a^2+k^2}Y \\
\frac{d(a\omega_3)}{dt} &= \frac{r}{\rho_2}\frac{d\phi}{dt}\frac{dr}{dt}
\end{aligned}\right\}.$$

If the impressed forces have no component perpendicular to the normal plane through a generator, $Y = 0$, and we have $r^2 d\phi/dt = h$, where h is some constant depending on the initial values of r and v.

If also the component X of the forces along a generator is a function of r only, another integral can be found by the principle of Vis Viva, viz.

$$\left(\frac{dr}{dt}\right)^2 + r^2\left(\frac{d\phi}{dt}\right)^2 + \frac{k^2}{a^2+k^2}a^2\omega_3^2 = \frac{2a^2}{a^2+k^2}\int X dr + h',$$

where h' is another constant depending on the initial values of u, v and r.

If, further, the cone be a right cone, $\rho_2 = r\tan a$ where a is the semi-angle, and we have

$$a\omega_3 = -\frac{h\cot a}{r} + h'',$$

where h'' is a third constant depending on the initial values of ω_3 and r. The equations of the motion of the centre of the sphere resemble those of a particle in central forces. Hence r and ϕ will be found as functions of the time if we regard them as the co-ordinates of a free particle moving in a plane under the action of a central force represented by $\dfrac{a^2}{a^2+k^2}\left\{X - k^2\omega_3\dfrac{d\omega_3}{dr}\right\}$, where ω_3 has the value just found.

229. **Ex.** A sphere rolls on a perfectly rough cone such that the equation to the cone on which the centre G always lies is $r = \rho_2 F(\phi)$. If the centre is acted on by a force tending to the vertex, find the law of force that any given path may be described. If the equation to the path be $1/r = f(\phi)$, prove that the force X is

$$X = k^2\omega_3\frac{d\omega_3}{dr} + \frac{a^2+k^2}{a^2}h^2 f^2\left(f + \frac{d^2f}{d\phi^2}\right), \text{ where } \omega_3 \text{ is given by } \frac{d\omega_3}{d\phi} = -\frac{h}{a}F\frac{df}{d\phi}.$$

230. **Motion on a surface of revolution.** *Let the given rough surface be any surface of revolution placed with its axis of figure vertical and vertex upwards, and let gravity be the only impressed force.* In this case the meridians and parallels are the lines of curvature. Let the axis of figure be the axis of Z. Let θ be the angle the axis GC makes with the axis of Z, ψ the angle the plane containing Z and GC makes with any fixed vertical plane.

Then
$$\theta_1 = -\sin\theta\frac{d\psi}{dt}, \qquad \theta_2 = \frac{d\theta}{dt}, \qquad \theta_3 = \cos\theta\frac{d\psi}{dt}.$$

Hence the equations (4) become

$$\frac{du}{dt} - \cos\theta \frac{d\psi}{dt} v = \frac{a^2}{a^2+k^2} g\sin\theta - \frac{k^2}{a^2+k^2} a\omega_3 \sin\theta \frac{d\psi}{dt} \quad\text{................(i)},$$

$$\frac{dv}{dt} + \cos\theta \frac{d\psi}{dt} u = \frac{k^2}{k^2+k^2} a\omega_3 \frac{d\theta}{dt} \quad\text{...............................(ii)},$$

and equation (8) becomes

$$u^2 + v^2 + \frac{k^2}{a^2+k^2} a^2\omega_3{}^2 = E + 2g\frac{a^2}{a^2+k^2} \int\rho\sin\theta\, d\theta \quad\text{.................(iii)},$$

where E is some constant, and ρ is the radius of curvature of the meridian. Also we have by (7)

$$\frac{d\omega_3}{dt} = -\frac{uv}{a}\left(\frac{1}{\rho} - \frac{\sin\theta}{r}\right) \quad\text{..............................(iv)},$$

where r is the distance of the centre of the sphere from the axis of z. The geometrical equations (5) become

$$u = \rho\frac{d\theta}{dt}, \qquad\qquad v = \rho\frac{d\psi}{dt} \quad\text{...........................(v)}.$$

To solve these, we put (ii) into the form

$$\frac{dv}{d\theta} + \cos\theta \frac{d\psi}{d\theta} u = \frac{k^2}{a^2+k^2} a\omega_3,$$

which by (v) becomes

$$\frac{dv}{d\theta} + \frac{\rho\cos\theta}{r} v = \frac{k^2}{a^2+k^2} a\omega_3 ;$$

differentiating this, we have by (iv),

$$\frac{d^2v}{d\theta^2} + \frac{\rho\cos\theta}{r}\frac{dv}{d\theta} + Pv = 0 \quad\text{................................(vi)},$$

where

$$P = \frac{d}{d\theta}\left(\frac{\rho\cos\theta}{r}\right) + \frac{k^2}{k^2+a^2}\left(1 - \frac{\rho\sin\theta}{r}\right).$$

Now ρ and r may be found from the equation to the meridian curve as functions of θ. Hence P is a known function of θ. Solving this linear equation we have v found as a function of θ. Then by (iv) we have

$$\frac{d\omega_3}{d\theta} = -\frac{v}{a}\left(1 - \frac{\rho\sin\theta}{r}\right),$$

and thence having found ω_3 we have u by equation (iii). Knowing u and v; θ and ψ may be found by equations (v).

231. Oscillations on the summit of a rough fixed surface. *A heavy sphere rotating about a vertical axis is placed in equilibrium on the highest point of a surface of any form and being slightly disturbed makes small oscillations, find the motion.*

Let O be the highest point of the surface on which the centre of gravity G always lies. Let the tangents to the lines of curvature at O be taken as the axes of x and y, and let (x, y, z) be the co-ordinates of G. We shall assume that O is not a singular point on the surface. In order to simplify the general equations of motion (4) of Art. 215 we shall take as the axes GA and GB the tangents to the lines of curvature at G. But since G always remains very near O, the tangents to the lines of curvature at G will be nearly parallel to those at O. So that to the first order of small quantities we have

$$\theta_1 = -\frac{1}{\rho_2}\frac{dy}{dt}, \qquad \theta_2 = \frac{1}{\rho_1}\frac{dx}{dt}, \qquad u = \frac{dx}{dt}, \qquad v = \frac{dy}{dt},$$

and θ_3 will be a small quantity of at least the first order. Also since the sphere is supposed not to deviate far from the highest point of the surface, we have ω_3 constant, let this constant be called n.

The equation to the surface on which G moves, in the neighbourhood of the highest point, is $z = -\frac{1}{2}\left(\dfrac{x^2}{\rho_1} + \dfrac{y^2}{\rho_2}\right)$. The direction cosines of the normal at x, y, z are $x/\rho_1, y/\rho_2, 1$. Hence the resolved parts parallel to the axes of the normal pressure R on the sphere are Rx/ρ_1, Ry/ρ_2 and R. The equations of motion (4) therefore become

$$\left.\begin{array}{l} \dfrac{d^2x}{dt^2} = \dfrac{a^2}{a^2+k^2}R\dfrac{x}{\rho_1} - \dfrac{k^2}{a^2+k^2}\dfrac{dy}{dt}\dfrac{an}{\rho_2} \\[2mm] \dfrac{d^2y}{dt^2} = \dfrac{a^2}{a^2+k^2}R\dfrac{y}{\rho_2} + \dfrac{k^2}{a^2+k^2}\dfrac{dx}{dt}\dfrac{an}{\rho_1} \\[2mm] \dfrac{d^2z}{dt^2} = R - g \end{array}\right\} \quad \dots\dots\dots\dots\dots\dots\text{(iv)}.$$

But z is a small quantity of the second order, hence the last equation gives $R = g$. To solve these equations, we put $x = F\cos(\lambda t + f)$, $y = G\sin(\lambda t + f)$.

These give
$$\left.\begin{array}{l} \left(\lambda^2 + \dfrac{a^2}{a^2+k^2}\dfrac{g}{\rho_1}\right)F = \dfrac{k^2}{a^2+k^2}\dfrac{a\lambda n}{\rho_2}G \\[3mm] \left(\lambda^2 + \dfrac{a^2}{a^2+k^2}\dfrac{g}{\rho_2}\right)G = \dfrac{k^2}{a^2+k^2}\dfrac{a\lambda n}{\rho_1}F \end{array}\right\}.$$

The equation to find λ is therefore
$$\left(\lambda^2 + \dfrac{a^2}{a^2+k^2}\dfrac{g}{\rho_1}\right)\left(\lambda^2 + \dfrac{a^2}{a^2+k^2}\dfrac{g}{\rho_2}\right) = \dfrac{k^4}{(a^2+k^2)^2}\dfrac{a^2\lambda^2 n^2}{\rho_1\rho_2}.$$

This is a quadratic equation to determine λ^2. In order that the motion may be oscillatory it is necessary and sufficient that the roots should be both positive. If ρ_1, ρ_2 are both negative, so that the sphere is placed like a ball inside a cup, the roots of the quadratic are positive for all values of n. If ρ_1, ρ_2 have opposite signs the roots cannot be both positive. If ρ_1, ρ_2 are both positive the two conditions of stability will be found to reduce to $n^2 > \dfrac{a^2+k^2}{k^4}g\,(\sqrt{\rho_1} + \sqrt{\rho_2})^2$.

If ρ_1 is infinite, it is necessary that ρ_2 should be negative, and in that case the two values of λ^2 are $-\dfrac{a^2}{a^2+k^2}\dfrac{g}{\rho_2}$ and zero, which are both independent of n. If $\rho_1 = \rho_2$, we have $F = G$. In this case if θ is the inclination of the normal to the vertical, we have $\theta^2 = (x^2+y^2)/\rho^2$ and, as in Art. 212, we find
$$\theta^2 = F_1{}^2 + F_2{}^2 + 2F_1F_2\cos\{(\lambda_1 - \lambda_2)t + f_1 - f_2\},$$
where λ_1, λ_2 are the roots of the quadratic
$$\lambda^2 \pm \dfrac{k^2}{a^2+k^2}\dfrac{an}{\rho}\lambda + \dfrac{a^2}{a^2+k^2}\dfrac{g}{\rho} = 0.$$

232. This problem may also be solved by Lagrange's method although the geometrical equations contain differential coefficients with regard to the time. To effect this we have recourse to the method of indeterminate multipliers as explained in Vol. I. Chap. VIII. Let the axes of reference Ox, Oy, Oz be the same as before. Let GC be that diameter which is vertical when the sphere is in equilibrium on the summit. Let GA, GB be two other diameters forming with GC a system of rectangular axes fixed in the sphere. Let the position of these with reference to the axes fixed in space be defined by the angular co-ordinates θ, ϕ, ψ in Euler's manner. The vis viva of the sphere will then be
$$2T = x'^2 + y'^2 + z'^2 + k^2(\phi' + \psi'\cos\theta)^2 + k^2(\theta'^2 + \sin^2\theta\,\psi'^2).$$
If we put $\sin\theta\cos\psi = \xi$, $\sin\theta\sin\psi = \eta$, $\phi + \psi = \chi$, and reject all small quantities above the second order, we find that the Lagrangian function is
$$L = \tfrac{1}{2}(x'^2 + y'^2) + \tfrac{1}{2}k^2\{\chi'^2 - \chi'(\xi\eta' - \xi'\eta) + \xi'^2 + \eta'^2\} + \tfrac{1}{2}g\left(\dfrac{x^2}{\rho_1} + \dfrac{y^2}{\rho_2}\right).$$

It is easy to see, by reference to the figure for Euler's geometrical equations given in Vol. I. Chap. v., that ξ and η are the cosines of the angles the diameter GC makes with the axes Ox, Oy. See also Vol. II., Art. 15.

If ω_x, ω_y, ω_z are the angular velocities of the sphere about parallels to the axes fixed in space, the geometrical equations are

$$x' - a\left(\omega_y - \omega_z \frac{y}{\rho_2}\right) = 0, \qquad y' + a\left(\omega_x - \omega_z \frac{x}{\rho_1}\right) = 0.$$

These are found by making the resolved velocities of the point of contact in the directions of the axes of x and y equal to zero. See the expressions in Vol. I. Art. 238 for *the velocity of any point*. The angular velocities ω_x, ω_y, ω_z may be expressed in terms of θ, ϕ, ψ by formulæ analogous to those of Euler. See Vol. I. Art. 257. Thus

$$\left.\begin{aligned} \omega_x &= -\theta' \sin\psi + \phi' \sin\theta \cos\psi \\ \omega_y &= \theta' \cos\psi + \phi' \sin\theta \sin\psi \\ \omega_z &= \phi' \cos\theta + \psi' \end{aligned}\right\}.$$

Substituting and expressing the result in terms of the new co-ordinates ξ, η, χ, the geometrical equations become

$$L_1 = -\frac{x'}{a} + \chi'\eta + \xi' - \chi'\frac{y}{\rho_2} = 0, \qquad L_2 = \frac{y'}{a} + \chi'\xi - \eta' - \chi'\frac{x}{\rho_1} = 0.$$

Lagrange's equations of motion modified by the indeterminate multipliers λ and μ are represented by the typical form

$$\frac{d}{dt}\frac{dL}{dq'} - \frac{dL}{dq} = \lambda\frac{dL_1}{dq'} + \mu\frac{dL_2}{dq'},$$

where q stands for any one of the five co-ordinates x, y, ξ, η, χ. The steady motion is given by x, y, ξ, η all zero and $\chi' = n$. Taking $q = x$ and $q = y$ and giving the several co-ordinates their values in the steady motion, we find that λ and μ are both zero in the steady motion.

To find the oscillations, we write for q in turn x, y, χ, ξ and η, and retain the first powers of the small quantities. Remembering that λ and μ are small quantities (Art. 51), we find

$$x'' - g\frac{x}{\rho_1} + \frac{\lambda}{a} = 0, \qquad y'' - g\frac{y}{\rho_2} - \frac{\mu}{a} = 0, \qquad k^2\chi'' = 0,$$

$$k^2(\xi'' + \chi'\eta') - \lambda = 0, \qquad\qquad k^2(\eta'' - \chi'\xi') + \mu = 0.$$

These and the two geometrical equations L_1 and L_2 are all linear, and may be solved in the usual manner. If we put $\chi' = n$ and eliminate first λ and μ and then ξ and η we get two equations to find x and y, which are the same as those marked (iv) in the solution of Art. 231.

233. **Ex.** A perfectly rough sphere is placed on a perfectly rough fixed sphere near the highest point. The upper sphere has an angular velocity n about the diameter through the point of contact; prove that its equilibrium will be stable if $n^2 > 35g(a+b)/a^2$, where b is the radius of the fixed sphere, and a is the radius of the moving sphere.

234. **Oscillations about steady motion.** *A perfectly rough surface of revolution is placed with its axis vertical. Determine the circumstances of motion that a heavy sphere may roll on it so that its centre describes a horizontal circle. And this state of steady motion being disturbed, find the small oscillations.*

In this case we must recur to the equations of Art. 230. We shall adopt the notation of that article, except that to shorten the expressions we shall put for k^2 its value $\frac{2}{5}a^2$.

To find the steady motion. We must put u, v, ω_3, θ, $d\psi/dt$ all constant. Let a, μ and n be the constant values of θ, $d\psi/dt$ and ω_3. Then we have $u=0$, $v=b\mu$, where b is the constant value of r. The equation (1) becomes

$$-b\cos a\mu^2 = \tfrac{5}{7} g \sin a - \tfrac{2}{7} an \sin a\mu.$$

The other dynamical equations are satisfied without giving any relation between the constants. If the motion be steady, we have therefore

$$n = \frac{5}{2}\frac{g}{a\mu} + \frac{7}{2}\frac{b}{a}\mu \cot a \dots\dots\dots \dots\dots\dots\dots\dots \text{(1)},$$

thus for the same value of n we have two values of μ, which correspond to different initial values of v.

Elementary determination of the steady motion.

As the steady motion of a sphere on a rough solid of revolution is often required, it will be useful to give a separate investigation of this result. The centre of gravity G describes with uniform velocity v a horizontal circle whose radius GN is the perpendicular on the axis of the solid. The friction perpendicular to the meridian plane is therefore zero and we have

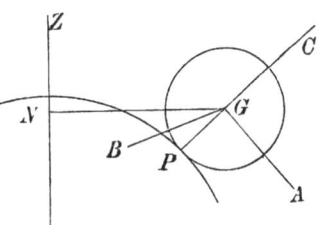

$$\left.\begin{aligned}-\mu^2 b &= R \sin a + F \cos a\\g &= R \cos a - F \sin a\end{aligned}\right\} .$$

We also have $v=\mu b$, where $GN=b$. Since the point of contact of the sphere and solid is at rest, we have the geometrical equations $\omega_2=0$, $v+a\omega_1=0$.

Let the axes of reference for the rotations be the normal GC to the solid and GA, GB respectively in and perpendicular to the meridian plane through G. These axes move round G with angular velocities $\theta_1 = -\mu \sin a$, $\theta_2 = 0$, $\theta_3 = \mu \cos a$. The equation of moments about OB is obviously

$$k^2 (\omega_2' - \theta_1\omega_3 + \theta_3\omega_1) = -Fa.$$

See Art. 215. Substituting for θ_1, θ_3, ω_1 and F we obtain at once the required result.

We have the geometrical relation $a\omega_1 = -v$, so that ω_1 and n have opposite signs if b/a and a are positive. Hence the axis of rotation, which necessarily passes through the point of contact of the sphere with the rough surface, makes an angle with the vertical less than that made by the normal at the point of contact.

If the sphere roll on a surface of revolution so that the axis GC is turned from the axis of symmetry, the angle a must be positive. By inspecting the expression for n and making $dn/d\mu=0$ it will be seen that the least value of the angular velocity n of the sphere is given by $n^2 = 35 \cot a . bg/a^2$. In this case the precessional motion of the sphere is given by $\mu^2 = \tfrac{5}{7} \tan a . g/b$. If the sphere roll on the inner and upper side of such a surface as an anchor ring held with its axis vertical the angle a is negative, and there is no inferior limit to the value of n.

To find the small oscillation.

Put $\theta=a+x$, $d\psi/dt=\mu+dy/dt$ where a and μ are supposed to contain *all* the constants parts of θ and $d\psi/dt$, so that x and dy/dt only contain trigonometrical terms. Let $c-a$ be the radius of curvature of the surface of revolution at the point of contact of the sphere in steady motion, so that ρ differs from c only by small quantities, and may be put equal to c in the small terms. Also we have $r=b+c\cos a . x$.

Now by equations (iv) and (v) of Art. 230 we have

$$\frac{d\omega_3}{dt} = \frac{d\theta}{dt}\frac{d\psi}{dt}\frac{\rho \sin\theta - r}{a} = \frac{dx}{dt}\mu \frac{c\sin a - b}{a} \; ;$$

$$\therefore \; \omega_3 = \mu \frac{c\sin a - b}{a} x + n \dots\dots\dots\dots\dots\dots(2),$$

when n is the whole of the constant part of ω_3.

Again, from equation (ii), we have

$$-\frac{1}{a}\frac{d}{dt}\left(r\frac{d\psi}{dt}\right) - \frac{\rho}{a}\frac{d\theta}{dt}\cos\theta \frac{d\psi}{dt} + \frac{k^2}{a^2+k^2}\omega_3\frac{d\theta}{dt} = 0 \; ;$$

$$\therefore \; -\frac{\mu}{a}c\cos a \frac{dx}{dt} - \frac{b}{a}\frac{d^2y}{dt^2} - \frac{c\cos a\mu}{a}\frac{dx}{dt} + \frac{2}{7}n\frac{dx}{dt} = 0 \; ;$$

integrating we have $\left(\dfrac{2}{7}n - \dfrac{2\mu c\cos a}{a}\right)x = \dfrac{b}{a}\dfrac{dy}{dt}$ .... (3),

the constant being put zero because x and y only contain trigonometrical terms.

Thirdly, from equation (i), we have

$$\frac{1}{a}\frac{d}{dt}\left(\rho\frac{d\theta}{dt}\right) - \frac{r}{a}\left(\frac{d\psi}{dt}\right)^2\cos\theta + \frac{2}{7}\omega_3\sin\theta\frac{d\psi}{dt} = \frac{5}{7}\frac{g}{a}\sin\theta \; ;$$

$$\therefore \; \frac{c}{a}\frac{d^2x}{dt^2} - \frac{b+c\cos ax}{a}(\cos a - \sin ax)\left(\mu^2 + 2\mu\frac{dy}{dt}\right)$$

$$+ \frac{2}{7}(\sin a + \cos ax)\left(\mu + \frac{dy}{dt}\right)\left(n + \mu\frac{c\sin a - b}{a}x\right) = \frac{5}{7}\frac{g}{a}(\sin a + \cos ax).$$

This expression must be expanded and expressed in the form

$$\frac{d^2x}{dt^2} + Ax = B \dots\dots\dots\dots\dots\dots\dots (4).$$

In this case, since x contains only trigonometrical expressions, we must have $B = 0$. Putting $x = 0$ in the above expression, we find the same value for n as in steady motion. After expanding the preceding equation we find

$$A = \mu^2\left(-\cos^2 a + \frac{2}{7}\sin^2 a\right) + \mu^2\frac{b}{c\sin a}\left(2\cos^2 a + \frac{5}{7}\sin^2 a\right)$$

$$+ \frac{25}{49}\frac{g^2\sin a}{\mu^2 bc} - \frac{10}{7}\frac{g}{b}\sin a\cos a + \frac{10}{7}\frac{g}{c}\cos a \dots\dots\dots\dots (5).$$

In order that the steady motion may be stable, it is sufficient and necessary that this value of A should be positive. And the time of oscillation is then $2\pi/\sqrt{A}$.

It is to be observed that this investigation does not apply if a and therefore b be small, for some terms which have been rejected have b in their denominators, and may become important.

Ex. A heavy sphere rolls round the inside of a rough horizontal circular wire, the normal to the sphere at the point of contact being inclined at a constant angle a to the vertical; prove that the angular velocity μ of the point of contact of the sphere is given by $\mu^2 = \frac{5}{7}g\tan a/(h - a\sin a)$, h being the radius of the ring, and a that of the sphere.   [Math. Tripos, 1881.

In this problem the rough surface on which the sphere moves is an anchor ring in which the radius of the generating circle is zero. Supposing the sphere to roll, but not to spin about the normal at the point of contact, the result follows by writing $n = 0$ in equation (1).

235. **Motion on an Imperfectly rough surface.** The general equations of the motion of a sphere on an imperfectly

rough surface may be obtained on principles similar to those adopted in Vol. I. Chap. VI. to determine the motion of rough elastic bodies impinging on each other. The difference in the theory will be made clear by the following example, in which a method of proceeding is explained which is generally applicable, whenever the integrations can be effected.

236. *A homogeneous sphere moves on an imperfectly rough inclined plane with any initial conditions, find the direction of the motion and the velocity of its centre at any time.*

Let G be the centre of gravity of the sphere. Let the axes of reference GA, GB, GC have their directions fixed in space, the first being directed down the inclined plane, and the last normal to the plane. Let u, v, w be the velocities of G resolved parallel to these axes, and ω_1, ω_2, ω_3 the angular velocities of the body about these axes. Let F, F' be the resolved parts of the frictions of the plane on the sphere parallel to the axes GA, GB, but taken *negatively* in those directions. Let k be the radius of gyration of the sphere about the diameter, a its radius, and let the mass be unity. Let α be the inclination of the plane to the horizon.

Whether the sphere rolls or slides the equations of motion are

$$\left. \begin{array}{l} k^2 \omega_1' = - F'a \\ k^2 \omega_2' = \quad Fa \end{array} \right\} \dots (1), \qquad \left. \begin{array}{l} u' = -F + g \sin \alpha \\ v' = -F' \end{array} \right\} \dots (2).$$

Eliminating F and F' from these equations and integrating we have

$$u + \frac{k^2}{a^2} a\omega_2 = U_0 + g \sin \alpha t \qquad v - \frac{k^2}{a^2} a\omega_1 = V_0 \dots (3),$$

where U_0 and V_0 are two constants determined by the initial values of u, v, ω_1, ω_2.

The meaning of these equations may be found as follows. Let P be the point of contact of the sphere and plane, let Q be a point within the sphere on the normal at P so that $PQ = (a^2 + k^2)/a$. Then Q is the centre of oscillation of the sphere when suspended from P. It is clear that the left-hand sides of the equations (3) express the components of the velocity of Q parallel to the axes. The equations assert that the frictional impulses at P cannot affect the motion of Q, and this also readily follows from Vol. I. Chap. III., because Q is in the axis of spontaneous rotation for a blow at P.

237. The friction at the point of contact P always acts opposite to the direction of sliding and tends to reduce this point to rest. When sliding ceases the friction (see Vol. I. Chap. IV.) also ceases to be limiting friction and becomes only of sufficient magnitude to keep the point of contact at rest. If sliding ever does cease, we then have

$$u - a\omega_2 = 0, \qquad v + a\omega_1 = 0 \dots (4).$$

The equations (3) and (4) suffice to determine these final values of u, v, ω_1 and ω_2. *Thus the direction of the motion and the velocity of the centre of gravity after sliding has ceased have been found in terms of the time.* It appears that both these elements are independent of the friction.

If the equations (4) hold initially the sphere will begin to move without sliding provided the friction found from the equations (1), (2) and (4) is less than the limiting friction. To determine this point we must find the magnitude of the friction necessary to prevent sliding. If the sphere does not slide we may differentiate the equations (4); then substituting from (1) and (2) we find $F'=0$ and $F = g \sin \alpha . k^2/(a^2 + k^2)$. But, since the pressure on the plane is $g \cos \alpha$, this requires that the coefficient of friction $\mu > \tan \alpha \dfrac{k^2}{a^2 + k^2}$. Supposing this inequality to hold the friction called into play will be always less than, or not greater than, the limiting friction, and therefore equations (3) and (4) give the whole motion.

This method of finding the inferior limit to the value of μ is the same as that used in Vol. I. Chap. IV. in the corresponding problem where the sphere rolls down the inclined plane along the line of greatest slope.

238. If the equations (4) do not hold initially or if the inequality just mentioned is not satisfied, let S be the velocity of sliding and let θ be the angle the direction of sliding makes with GA. To fix the signs we shall take S to be positive while θ may have any value from $-\pi$ to π. Then

$$S \cos \theta = u - a\omega_2, \qquad S \sin \theta = v + a\omega_1 \ldots\ldots\ldots(5).$$

The friction is equal to $\mu g \cos \alpha$ and acts in the direction opposite to sliding, hence

$$F = \mu g \cos \alpha \cos \theta, \qquad F' = \mu g \cos \alpha \sin \theta.$$

The equations (1), (2) and (5) therefore give

$$\left. \begin{aligned} \frac{d(S \cos \theta)}{dt} &= -\left(1 + \frac{a^2}{k^2}\right) \mu g \cos \alpha \cos \theta + g \sin \alpha \\ \frac{d(S \sin \theta)}{dt} &= -\left(1 + \frac{a^2}{k^2}\right) \mu g \cos \alpha \sin \theta \end{aligned} \right\} \ldots\ldots(6).$$

Expanding we find

$$\left. \begin{aligned} \frac{dS}{dt} &= -\left(1 + \frac{a^2}{k^2}\right) \mu g \cos \alpha + g \sin \alpha \cos \theta \\ S \frac{d\theta}{dt} &= -g \sin \alpha \sin \theta \end{aligned} \right\} \ldots\ldots\ldots(7).$$

If θ is not constant, we may eliminate t and integrate with regard to θ, this gives $S \sin \theta = 2A \left(\tan \dfrac{\theta}{2}\right)^n \ldots\ldots\ldots\ldots(8),$

where $n = (1 + a^2/k^2) \mu \cot \alpha$, and A is the constant of integration. If S_0 and θ_0 are the initial values of S and θ determined by equations (5), we have $\quad 2A = S_0 \sin \theta_0 \left(\cot \dfrac{\theta_0}{2} \right)^n$ ....(9).

Substituting the value of S given by (8) in the second of equations (7) and integrating we find

$$\frac{\left(\tan \dfrac{\theta}{2} \right)^{n-1}}{n-1} + \frac{\left(\tan \dfrac{\theta}{2} \right)^{n+1}}{n+1} = \frac{\left(\tan \dfrac{\theta_0}{2} \right)^{n-1}}{n-1} + \frac{\left(\tan \dfrac{\theta_0}{2} \right)^{n+1}}{n+1} - \frac{g \sin \alpha}{A} t \ldots (10),$$

the constant of integration being determined from the condition that $\theta = \theta_0$ when $t = 0$. The equations (8), (9) and (10) give S and θ in terms of t. The equations (3) and (5) then give u, v, ω_1 and ω_2 in terms of t.

The second of equations (7) shows that $d\theta/dt$ has an opposite sign to θ, hence θ beginning at any initial value except $\pm \pi$ continually approaches zero. It follows that, unless α is zero, θ will be constant only when $\theta_0 = 0$ or $\pm \pi$, i.e. *the direction of sliding on the plane is not fixed in space but continually approaches the line of greatest slope.* On a horizontal plane $\alpha = 0$, and the direction of sliding is fixed.

If $n > 1$, i.e. $\mu > \tan \alpha \cdot k^2/(a^2 + k^2)$, we see from (8) that sliding will cease when θ vanishes. This, by (10) will occur when

$$t = \frac{S_0}{g \sin \alpha} \left(\frac{\cos^2 \tfrac{1}{2} \theta_0}{n-1} + \frac{\sin^2 \tfrac{1}{2} \theta_0}{n+1} \right).$$

The subsequent motion has already been found.

If $n < 1$ we see by (8) that S increases as θ decreases, so that sliding will never cease. It also follows from (10) that θ vanishes only at the end of an infinite time.

If $S_0 = 0$, sliding will never begin if $n > 1$, but will immediately begin and never cease if $n < 1$.

239. **Billiard Balls.** The theory of the motion of a sphere on an imperfectly rough *horizontal* plane is so much simpler than when the plane is inclined or than when the sphere rolls on any other surface, that it seems unnecessary to consider this case in detail. At the same time the game of billiards supplies many problems which it would be unsatisfactory to pass over in silence. The following examples have been arranged so as both to indicate the mode of proof to be adopted and to supply some results which may be submitted to experiment.

The result given in Ex. 1, was first obtained by J. A. Euler, the son of the celebrated Euler, and published in the *Mém. de l'Acad. de Berlin*, 1758. Most, possibly all, of the other results may be found in the *Jeu de Billard par G. Coriolis*, published at Paris in 1835.

Ex. 1. A billiard-ball is set in motion on an imperfectly rough horizontal plane, show that the direction and magnitude of the friction are constant throughout the motion. The path of the centre of gravity is therefore an arc of a parabola while sliding continues, and finally a straight line. The parabola is described with the given initial motion of the centre of gravity under an acceleration equal to μg tending in a direction opposite to the initial direction of sliding.

Ex. 2. If S_0 be the initial velocity of sliding prove that the parabolic path lasts for a time $\frac{2}{7} S_0/\mu g$. From some experiments of Coriolis it appears that $\mu = \frac{1}{5}$ nearly. If the initial velocity of sliding be one foot per second, the parabolic path lasts therefore less than a twentieth part of a second.

Ex. 3. If P be the point of contact in any position and Q the centre of oscillation with regard to P, prove that the velocity of Q is always the same in direction and magnitude. Thence show that the final rectilinear path of the centre of gravity is parallel to the initial direction of the motion of Q and the final velocity of the centre of gravity is five-sevenths of the initial velocity of Q. If PP' be the initial direction of motion and V the initial velocity of the centre of gravity and t the time given by Ex. 2, prove that the final rectilinear path of the centre of gravity intersects PP' in a point P' so that $PP' = \frac{1}{2} Vt$.

Ex. 4. A billiard-ball, at rest on an imperfectly rough horizontal table, is struck by a cue in a horizontal direction at any point whose altitude above the table is h, and the cue is withdrawn as soon as it has delivered its blow. Supposing the cue to be sufficiently rough to prevent sliding, show that the centre of the ball will move in the direction of the blow and that its velocity will become uniform and equal to $\dfrac{5}{7} \dfrac{h}{a} B$ after a time $\dfrac{5h \sim 7a}{7a} \dfrac{B}{\mu g}$ where B is the ratio of the blow to the mass of the sphere and a is the radius.

In order that there should be no sliding the distance of the cue from the centre of the ball must be less than $a \sin \epsilon$ where $\tan \epsilon$ is the coefficient of friction between the cue and ball.

Ex. 5. A billiard-ball, initially at rest and touching the table at a point P, is struck by a cue making an angle β with the horizon. Show that the final rectilinear motion of the centre of gravity is parallel to the straight line PS joining P to the point S where the direction of the blow meets the table, and the final velocity of the centre of gravity is $\frac{5}{7} B \sin \beta \cdot PS/a$ in the direction of the projection of the blow on the horizon. It should be noticed that these results are independent of the friction.

Ex. 6. Measure $ST = \frac{7}{5} a \cot \beta$ along the projection of the blow on the horizontal table, then TS measures the horizontal component of the blow referred to a unit of mass, on the same scale that PS measures the final velocity of the centre of gravity. Prove that, during the impact and the whole of the subsequent motion, the friction acts along PT and that the whole friction called into play is measured by PT on the scale just mentioned. Thence show that unless $\mu < \frac{5}{7} PT/a$ the parabolic arc of the path is suppressed. Show also that PT is the direction in which the lowest point of the ball would begin to move if the horizontal plane were smooth and the ball were acted on by the same blow as before.

Motion of a Solid Body on a plane.

240. **Historical Summary.** The motion of a heavy body of any form on a horizontal plane seems to have been studied first by Poisson. The body is supposed to be either bounded by a continuous surface which touches the plane in a single

point or to be terminated by an apex as in a top, while the plane is regarded as perfectly smooth. Poisson uses Euler's equations to find the rotations about the principal axes, and refers these axes to others fixed in space by means of the formulæ usually called Euler's geometrical equations. He finds one integral by the principle of vis viva and another by that of angular momentum about the vertical straight line through the centre of gravity. These equations are then applied to find how the motion of a vertical top is disturbed by a slow movement of the smooth plane on which it rests. See the *Traité de Mécanique*.

In three papers in the fifth and eighth volumes of *Crelle's Journal* (1830 and 1832) M. Cournot repeated Poisson's equations, and expressed the corresponding geometrical conditions when the body rests on more than one point or rolls on an edge such as the base of a cylinder. He also considers the two cases in which the plane is (1) perfectly rough, and (2) imperfectly rough. He proceeds on the same general plan as Poisson, having two sets of rectangular axes, one fixed in the body and the other in space connected together by the formulæ usually given for transformation of co-ordinates. As may be supposed, the equations obtained are extremely complicated. M. Cournot also forms the corresponding equations for impulsive forces. Those however which include the effects of friction do not agree with the equations given in this treatise.

In the thirteenth and seventeenth volumes of *Liouville's Journal* (1848 and 1852) there are two papers by M. Puiseux on this subject. In the first of these he repeats Poisson's equations and applies them to the case of a solid of revolution on a smooth plane. He shows that the inclination of the axis of the solid to the vertical remains very nearly constant provided a sufficiently great initial angular velocity is communicated to the body about that axis. An inferior limit to this angular velocity is found only in the case in which the axis is vertical. In the second memoir he applies Poisson's equations to determine the conditions of stability of a solid of any form placed on a *smooth* plane with a principal axis at its centre of gravity vertical, the body rotating about that axis. He also determines the small oscillations of a body resting on a smooth plane about a position of equilibrium.

In the fourth volume of the *Quarterly Journal of Mathematics*, 1861, Mr G. M. Slesser forms the equations of motion of a body on a perfectly rough horizontal plane and applies them to the problem considered at the end of Art. 251. He uses moving axes, and his analysis is almost exactly the same as that which the author independently adopted.

241. **Oscillations about steady motion.** *A solid of revolution rolls on a perfectly rough horizontal plane under the action of gravity. To find the steady motion and the small oscillations.*

Let G be the centre of gravity of the body, GC the axis of figure, P the point of contact. Let GA be that principal axis which lies in the plane PGC and GB the axis at right angles to GA, GC. Let GM be a perpendicular from G on the horizontal plane, and PN a perpendicular from P on GC. Let R be the normal reaction at P; F, F' the resolved parts of the frictions respectively in and perpendicular to the plane PGC. Let the mass of the body be unity.

Let θ be the angle GC makes with the vertical, ψ the angle MP makes with any fixed straight line in the horizontal plane. Then θ and ψ are two of the angles used in Euler's geometrical equations to refer the *moving axes* GA, GB, GC to an axis fixed in space, viz. the vertical (Vol. I. Chap. v.). The third Eulerian angle ϕ is here zero. The moving axes GA, GB, GC are therefore the same as those described in

Art. 13. Since GC is fixed in the body we have $\theta_1=\omega_1$, $\theta_2=\omega_2$. Since $\phi=0$ the third of Euler's geometrical equations gives $\theta_3=\cos\theta d\psi/dt$. Remembering that the

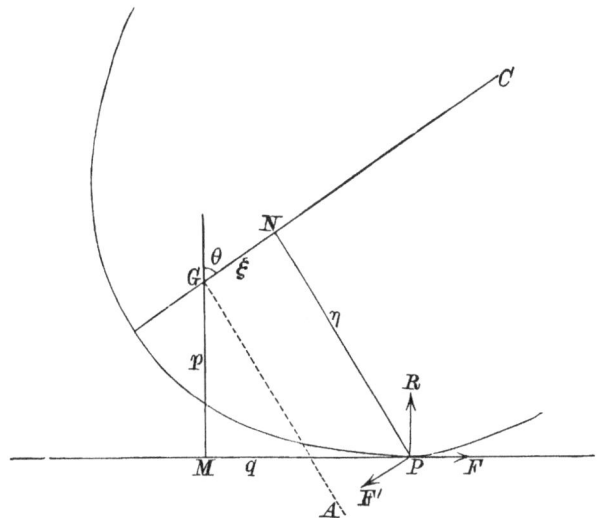

angular momenta about the axes are $h_1=A\omega_1$, $h_2=A\omega_2$, $h_3=C\omega_3$ as in Art. 12, the equations of moments of Art. 10 become

$$A\frac{d\omega_1}{dt}-A\omega_2\frac{d\psi}{dt}\cos\theta+C\omega_3\omega_2=-F'\,.\,GN\dots\dots\dots\dots\dots(1).$$

$$A\frac{d\omega_2}{dt}-C\omega_3\omega_1+A\omega_1\frac{d\psi}{dt}\cos\theta=-F\,.\,GM-R\,.\,MP\dots\dots(2).$$

$$C\frac{d\omega_3}{dt}=F'\,.\,PN\dots\dots\dots\dots\dots\dots\dots\dots\dots\dots(3).$$

The first two of Euler's geometrical equations give the relations between θ_1, θ_2 and the angles θ, ψ. Since $\theta_1=\omega_1$, $\theta_2=\omega_2$ and $\phi=0$, these become

$$\frac{d\theta}{dt}=\omega_2\dots\dots\dots\dots\dots(4). \qquad \sin\theta\frac{d\psi}{dt}=-\omega_1\dots\dots\dots\dots\dots(5).$$

The Eulerian geometrical equations which refer *the body* to the axes fixed in space are not required. We may also notice that the equations (4) and (5) are sufficiently obvious from the geometry of the figure to render any reference to Euler's equations unnecessary.

Let u and v be the velocities of the centre of gravity respectively along and perpendicular to MP, both being parallel to the horizontal plane. The accelerations of the centre of gravity along these moving axes will be

$$\frac{du}{dt}-v\frac{d\psi}{dt}=F\dots\dots\dots\dots\dots\dots(6),$$

$$\frac{dv}{dt}+u\frac{d\psi}{dt}=F'\dots\dots\dots\dots\dots\dots(7).$$

And if z be the altitude of G above the horizontal plane, i.e. $z=GM$, we have

$$\frac{d^2z}{dt^2}=-g+R\dots\dots\dots\dots\dots\dots\dots(8).$$

11—2

Also since the point P is at rest, we have

$$u - GM\omega_2 = 0 \quad \dotfill (9),$$

$$v + PN\omega_3 - GN\omega_1 = 0 \quad \dotfill (10),$$

$$z = - GN \cos\theta + PN \sin\theta \quad \dotfill (11).$$

These are the general equations of motion of a solid of revolution moving on a perfectly rough horizontal plane. If the plane is not perfectly rough the first eight equations will still hold, but the remaining three must be modified in the manner explained in the next proposition.

When the form of the solid of revolution is given these equations may admit of considerable simplification, and may generally be formed in any special case without much difficulty. Thus if the solid is a plane hoop or disc of radius a, we have $GN = 0$, $GM = z = a \sin\theta$, $MP = a \cos\theta$, and the radius of curvature $\rho = 0$.

242. *To find the steady motion.*

When the motion is steady, the surface of revolution rolls on the plane so that its axis makes a constant angle with the vertical. In this state of motion, let $\theta = a$, $d\psi/dt = \mu$, $\omega_3 = n$, $GM = p$, $MP = q$, $GN = \xi$, $NP = \eta$, and let ρ be the radius of curvature of the rolling body at P. The relations between these quantities may be found by substitution in the above equations.

Suppose it were required to find the conditions that the surface may roll with a given angular velocity n, its axis of figure making a given angle a with the vertical. Here n and a are given, and p, q, ξ, η, ρ may be found from the equations to the surface. We have to find μ, ω_1, ω_2, u, v and the radius of the circle described by G in space. Eliminating F and R, we have $F' = 0$, and

$$\mu^2 \sin a \, (A \cos a - p\xi) - n\mu \, (C \sin a + p\eta) - gq = 0 \quad \dotfill (12),$$

$$\omega_1 = - \mu \sin a, \qquad \omega_2 = 0,$$

$$u = 0, \qquad v = - n\eta - \xi\mu \sin a.$$

Let r be the radius of the circle described by G as the surface rolls on the plane. Since G describes its circle with angular velocity μ, we have $r\mu = v$, and hence

$$r = - \frac{\eta n}{\mu} - \xi \sin a.$$

Eliminating n we may also find r from the equation

$$\mu^2 \{ A\eta \sin a \cos a + C\xi \sin^2 a + r \, (C \sin a + p\eta) \} = gq\eta.$$

For every value of n and a there are two values of μ, which however correspond to different initial conditions. In order that a steady motion may be possible, it is necessary that the roots of the quadratic (12) should be real. This gives

$$(C \sin a + p\eta)^2 n^2 + 4gq \sin a \, (A \cos a - p\xi) = \text{a positive quantity.}$$

The instantaneous axis passes through P and intersects the axis of figure in some point E, not drawn. Let EH be a perpendicular on the horizontal plane, then as the body rolls in steady motion, the vertical EH is fixed and G revolves round it with an angular velocity μ. To find E, we notice that the velocity of N is $\mu . EN . \sin a$, but since EP is the instantaneous axis it is also $\omega . EN . \sin NEP$. Equating these, we see that E is determined by the equation $\mu \sin a = n \tan NEP$.

Elementary determination of the steady motion.

In many problems only the steady motion is required, and then the process just described becomes very simple. Since all the quantities, except ψ, are constant, we omit all the differential coefficients in the general equations of motion of Art. 10. Since G describes its circle uniformly under the influence of all the forces transferred to that point, $u = 0$ and $F' = 0$. Thus to find the steady motion we have merely to substitute in Art. 10 the values $h_1 = A\omega_1$, $h_2 = 0$, $h_3 = Cn$, $\omega_2 = 0$, $\theta_1 = \omega_1$, $\theta_2 = 0$, $\theta_3 = \mu \cos a$. In this way we find, writing g for R, a for θ,

$$- Cn\omega_1 + A\omega_1\mu \cos a = -F \cdot GM - g \cdot MP \dots\dots\dots\dots (2).$$

$$\mu \sin a = -\omega_1 \dots\dots(5). \qquad\qquad -v\mu = F \dots\dots(6).$$

Joining these to the geometrical equation (10) the results given above for the steady motion follow at once.

Why some tops rise.

If the body is set in motion by unwinding a string, as in a top, the initial values of ω_1, ω_2, u, v are small. If therefore the oscillations of the body are to be small, the steady motion, by (5), must be such that μ is small. Referring to (12) we see that this condition can be satisfied by making n large. We then have

$$n\mu (C \sin a + p\eta) = -gq.$$

Since a large value of ω_3 or n renders (10) impossible, we infer that *a steady motion cannot be established in this manner unless* $n \cdot PN$ *is small or* GN *is large*.

If the body is a top with its apex rounded off PN is small, but nevertheless the angular velocity n communicated by unwinding the string may still be too great to satisfy the equation (10). If the plane is perfectly rough an impulsive friction is called into play at P sufficiently great to reduce that point to rest. If the plane is imperfectly rough the point P slides on the plane and there is therefore a component of friction acting perpendicularly to the plane GPM in the direction opposite to that marked F' in the figure. This produces a new couple acting on the body in the plane GPF' besides those which act in the steady motion. Since in a top G is on the other side of N to that represented in the figure, the positive direction of the axis of the couple is to the left of GC. Comparing this couple with the couple Q in Art. 209, we see that its general effect is to make the invariable line at G, accompanied by GC, approach nearer to the vertical drawn through G. If the initial value of n is only a little greater than that required by (10) the top slightly rises to adjust itself to the equation. But if n is much too great the top rises until the axis is sensibly vertical.

243.   *To find the small oscillation.*

We put $\theta = a + x$, $d\psi/dt = \mu + dy/dt$, $\omega_3 = n + z$. Then we have by geometry,

$$z = GM = p + qx, \qquad\qquad PM = q + (\rho - p)\,x,$$

$$GN = \xi + \rho x \sin a, \qquad\qquad PN = \eta + \rho x \cos a,$$

and substituting in (5), (9), (10), (6), (7) respectively, we find

$$\omega_1 = -\mu \sin a - \mu \cos ax - \sin ay', \qquad\qquad u = px',$$

$$v = -\mu \sin a\xi - n\eta - (\mu \cos a\xi + \mu\rho \sin^2 a + n\rho \cos a)\,x - \sin a\xi y' - \eta z,$$

$$F = px'' + \mu^2 \sin a\xi + n\mu\eta + 2 \sin a\mu\xi y' + \eta ny' + \mu(\mu \cos a\xi + \mu\rho \sin^2 a + n\rho \cos a)\,x + \eta\mu z,$$

$$F' = -(\mu \cos a\xi - p\mu + \mu\rho \sin^2 a + n\rho \cos a)\,x' - \sin a\xi y'' - \eta z',$$

where accents, except in the case of F', denote differentiations with regard to t.

Substituting these in equation (3) and integrating, we have

$$(C + \eta^2)z = (p\mu - \mu\xi \cos a - \mu\rho \sin^2 a - n\rho \cos a)\,\eta x - \eta \sin a\xi y' \dots\dots\dots(A),$$

the constant being omitted because n, a and μ are supposed to contain all the constant parts of ω_3, θ, and $d\psi/dt$.

Again substituting in (1) and integrating, we have

$$\{Cn - 2A\mu \cos a + \xi\,(p\mu - \mu \cos a\xi - \mu \sin^2 a\rho - n\rho \cos a)\}x - (A + \xi^2) \sin ay' = \xi\eta z \dots(B).$$

Also substituting in (2), we have

$$
\left.
\begin{aligned}
&(A+p^2+q^2)\,x'' + x\{A\mu^2\,(\sin^2\alpha-\cos^2\alpha)+Cn\mu\cos\alpha+(\rho-p)\,g \\
&\quad +\mu^2\sin\alpha\xi q+n\mu\eta q+\mu^2\cos\alpha\xi p+n\mu\rho p\cos\alpha+\mu^2\sin^2\alpha\rho p\} \\
&\quad +y'\{-2A\mu\sin\alpha\cos\alpha+Cn\sin\alpha+2\xi p\mu\sin\alpha+np\eta\} \\
&\quad +z\{C\mu\sin\alpha+\mu p\eta\} \\
&\quad +\{-A\sin\alpha\cos\alpha\mu^2+Cn\mu\sin\alpha+gq+\sin\alpha\mu^2p\xi+n\mu p\eta\}
\end{aligned}
\right\}=0\ldots\text{(C)}.
$$

The last term of this equation must vanish since x, y', z contain only periodic terms. It is the equation thus formed which determines the steady motion and gives us the value of μ.

To solve these equations we may put

$$x=L\sin(\lambda t+f),\qquad y'=M\sin(\lambda t+f),\qquad z=N(\lambda t+f).$$

If we substitute these in (A), (B), (C) we shall get three equations to eliminate the ratios $L:M:N$. Before substitution it will be found convenient to simplify the equations, firstly by multiplying (A) by ξ, (B) by η and subtracting the latter result from the former, secondly by multiplying (A) by $\mu\rho/\eta$ and adding the result to (C). We then obtain the following determinant,

$$
\begin{vmatrix}
\begin{aligned}&-(A+p^2+q^2)\,\lambda^2+(\rho-p)\,g\\&+\mu^2\,(p^2-A\cos2\alpha-qr)\\&+n\mu C\cos\alpha\end{aligned} & \begin{aligned}&A\mu\sin\alpha\cos\alpha\\&\quad+\dfrac{qg}{\mu}\end{aligned} & C\mu\,(\eta\sin\alpha-p) \\[2ex]
Cn-2A\mu\cos\alpha & A\sin\alpha & C\xi \\[1ex]
\begin{aligned}&(p-\xi\cos\alpha-\rho\sin^2\alpha)\,\mu\\&\quad-\rho n\cos\alpha\end{aligned} & \xi\sin\alpha & -(C+\eta^2)
\end{vmatrix}=0.
$$

244. Examples. Ex. 1. To find the least angular velocity which will make a hoop roll in a straight line.

In this case r is infinite and therefore μ must be zero. It follows from the equation of steady motion that $q=0$, or the hoop must be upright. We have $p=a$, $q=0$, $\xi=0$, $\eta=a$, $\mu=0$, and $C=2A$. The determinant becomes

$$(A+a^2)\,\lambda^2=2n^2\,(2A+a^2)-ag,$$

so that the least angular velocity which will make λ a real quantity is given by

$$2\,(C+a^2)\,n^2=ag.$$

Let the hoop be an arc, we have $C=a^2$, and if V be the least velocity of the centre of gravity, this equation gives $V^2>\tfrac14 ag$. Let the hoop be a disc, then $C=\tfrac12 a^2$, and we have $V^2>\tfrac13 ag$.

Though these results have been deduced from the determinantal equation, it is nevertheless intended that the student should repeat the process given at length in Arts. 242, 243 with the simplifications which result when the rolling solid becomes a hoop. An easy independent solution may also be obtained by using the second set of equations given in Art. 15.

Ex. 2. A circular disc is placed with its rim resting on a perfectly rough horizontal table and is spun with an angular velocity Ω about the diameter through the point of contact. Prove that in steady motion the centre is at rest at an altitude $k^2\Omega^2/g$ above the horizontal plane, where k is the radius of gyration about a diameter; and, if α be the inclination of the plane to the horizon, the point of contact has made a complete circuit in the time $2\pi\sin\alpha/\Omega$. If the disc be slightly

disturbed from this state of steady motion, show that the time of a small oscillation
is $2\pi \left\{ \dfrac{k^2}{ga} \dfrac{(k^2+a^2)\sin\alpha}{3k^2\cos^2\alpha+a^2\sin^2\alpha} \right\}^{\frac{1}{2}}$.

Ex. 3. An infinitely thin circular disc moves on a perfectly rough horizontal
plane in such a manner as to preserve a constant inclination α to the horizon.
Find the condition that the motion may be steady and the time of a small oscillation.

Let the radius of the disc be a, and the radius of gyration about a diameter k.
Let ω_3 be the angular velocity about the axis, μ the angular velocity of the centre
of gravity about the centre of the circle described by it, r the radius of this circle,
then in steady motion

$$(2k^2+a^2)\,\omega_3 = k^2\mu\cos\alpha - \frac{ga}{\mu}\cot\alpha, \quad (2k^2+a^2)\,r = -k^2a\cos\alpha + \frac{ga^2}{\mu^2}\cot\alpha.$$

If T be the time of a small oscillation

$$\left(\frac{2\pi}{T}\right)^2 (k^2+a^2) = \mu^2\{k^2(1+2\cos^2\alpha)+a^2\sin^2\alpha\} - n\mu\cos\alpha\,(6k^2+a^2)+2n^2(2k^2+a^2) - ga\sin\alpha.$$

Ex. 4. A homogeneous right circular cylinder, whose altitude is twice the radius
of the base, rolls on a rough horizontal plane with its axis inclined at an angle $45°$
to the vertical. If n be the angular velocity about its axis, prove that in steady
motion the vertical plane through its axis turns round a fixed vertical line with an
angular velocity $\mu = n \cdot 30\sqrt{2}/31$. Show that the instantaneous axis divides the axis
of the top in the ratio $31:29$. Prove also that the period $2\pi/\lambda$ of the small oscil-
lations about the steady motion is given by $\lambda^2 + \dfrac{12\sqrt{2}}{31}\dfrac{g}{h} = \dfrac{1800}{(31)^2}n^2$ where h is the
radius of the base.

The motion of a cylinder rolling on its edge may be deduced from that of the solid
of revolution by putting the radius of curvature $\rho = 0$. The general results for the
cylinder are rather long, but when $\alpha = 45$, $\xi = -h$, $\eta = h$ we have $p = h\sqrt{2}$ and $q = 0$;
putting also $C = \frac{1}{2}h^2$, $A = \frac{7}{12}h^2$ the results are considerably simplified.

Ex. 5. A heavy body is attached to the plane face of a hemisphere so as to form
a solid of revolution, the radius of the hemisphere being a and the distance of the
centre of gravity of the whole body from the centre of the hemisphere being h. The
body is placed with its spherical surface resting on a horizontal plane, and is set
in motion in any manner. Show that one integral of the equations of motion is
$A\sin^2\theta\,\dfrac{d\psi}{dt} + C\omega_3\left(\cos\theta + \dfrac{h}{a}\right) = $ constant whether the plane be smooth, imperfectly
rough, or perfectly rough.

It is clear that the first two terms on the left-hand side of this equation is the
angular momentum about the vertical through G. Let this be called I. Since we
may take moments about any axis through G as if G were fixed in space, we have
$dI/dt = F' \cdot PM$. But $PM = -PN \cdot h/a$, hence eliminating F' by equation (3) and integrating, we get the required result.

This integral is given by Prof. Jellett in his *Theory of friction*, chap. VIII. 1872:
though he obtains it by a very different process. He also assumes that the rotation
about the axis of figure is so rapid that the friction acts perpendicularly to the
vertical plane containing the axis of figure. He proceeds to apply the integral to
show that when a top is placed on an imperfectly rough horizontal plane "the axis
will soon become vertical, assuming that the other motions are slow compared with
the rotation round the axis."

Ex. 6. A surface of revolution rolls on another perfectly rough surface of
revolution with its axis vertical. The centre of gravity of the rolling surface lies

in its axis.   Find the cases of steady motion in which it is possible for the axes of both the surfaces to lie in a vertical plane throughout the motion.

Let θ be the inclination of the axes of the two surfaces, P the point of contact, GM a perpendicular on the tangent plane at P, PN a perpendicular on the axis GC of the rolling body; F the friction, R the reaction at P; n the angular velocity of the rolling body about its axis GC, μ the angular rate at which G describes its circular path in space, r the radius of this circle.   Then in steady motion
$$M\mu \sin\theta\,(Cn - A\mu\cos\theta) = -F\,.\,GM - R\,.\,MP,$$
$$R = -Mr\mu^2 \sin a + Mg \cos a,$$
$$F = -Mr\mu^2 \cos a - Mg \sin a,$$
$$n\,.\,PN + \mu \sin\theta\,.\,GN = -r\mu,$$

where M is the mass of the body.   These results were set by the author in an examination paper in the University of London, 1860.

245.   General equations of motion. *A surface of any form rolls on a fixed horizontal plane under the action of gravity.   To form the equations of motion.*

Let GA, GB, GC, the principal axes at the centre of gravity, be the axes of

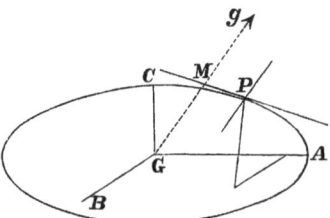

reference and let the mass be unity.   Let $\phi\,(\xi,\ \eta,\ \zeta) = 0$ be the equation to the bounding surface, $(\xi,\ \eta,\ \zeta)$ the co-ordinates of the point P of contact.   Let $(p,\ q,\ r)$ be the direction-cosines of the *outward* direction of the normal to the surface at the point $\xi,\ \eta,\ \zeta$, then $\qquad p\left|\dfrac{d\phi}{d\xi} = q\right/\dfrac{d\phi}{d\eta} = r\left|\dfrac{d\phi}{d\zeta}\right.$.

Firstly, let the plane be perfectly rough.   Let X, Y, Z be the resolved parts along the axes of the normal reaction and the two frictions at the point $\xi,\ \eta,\ \zeta$, and let the mass of the body be unity.   By Euler's equations we have
$$\left.\begin{array}{l} A\omega_1' - (B - C)\,\omega_2\omega_3 = \eta Z - \zeta Y \\ B\omega_2' - (C - A)\,\omega_3\omega_1 = \zeta X - \xi Z \\ C\omega_3' - (A - B)\,\omega_1\omega_2 = \xi Y - \eta X \end{array}\right\}\ \ldots\ldots\ldots\ldots\ldots\ldots\ldots\ldots (1),$$

where accents denote differentiations with regard to the time.

Also the equations of motion of the centre of gravity are by Art. 5,
$$\left.\begin{array}{l} u' - v\omega_3 + w\omega_2 = gp + X \\ v' - w\omega_1 + u\omega_3 = gq + Y \\ w' - u\omega_2 + v\omega_1 = gr + Z \end{array}\right\}\ \ldots\ldots\ldots\ldots\ldots\ldots\ldots (2).$$

Also since the line $(p,\ q,\ r)$ remains always vertical (Art. 18),
$$\left.\begin{array}{l} p' = q\omega_3 - r\omega_2 \\ q' = r\omega_1 - p\omega_3 \\ r' = p\omega_2 - q\omega_1 \end{array}\right\}\ \ldots\ldots\ldots\ldots\ldots\ldots\ldots (3).$$

Since the point $(\xi,\ \eta,\ \xi)$ which, for the moment, is fixed relatively to the moving axes is also, for the moment, fixed in space, we have by Art. 17,

$$\left.\begin{array}{l} U= u - \eta\omega_3+ \zeta\omega_2 =0 \\ V= v - \zeta\omega_1+ \xi\omega_3 =0 \\ W=w - \xi\omega_2+ \eta\omega_1 =0 \end{array}\right\}\dots\dots\dots\dots\dots\dots\dots\dots (4),$$

where U, V, W are the resolved parts of the velocity of the point of contact P in the positive directions of the axes.

246. *Secondly, let the plane be perfectly smooth.* The equations (1), (2), (3), apply equally to this case, but equations (4) are not true. Since the resultant of X, Y, Z is a reaction R normal to the fixed plane, we have

$$X= -pR, \qquad Y= -qR, \qquad Z= -rR\dots\dots\dots\dots\dots\dots (5).$$

The negative sign is prefixed to R because (p, q, r) are the direction-cosines of the outward direction of the normal, and it is clear that when these are taken positively, the components of R are all negative. If at any moment R vanishes and changes sign the body will leave the plane.

Since the velocity of G parallel to the fixed plane is constant in direction and magnitude, it will usually be more convenient to replace the equations (2) by the following single equation. Let GM be the perpendicular on the fixed plane and let $MG=z$, then $\qquad z'' = -g+R\dots\dots\dots\dots\dots\dots\dots\dots\dots\dots\dots\dots (6).$

It is necessary that the velocity of the point of contact resolved normal to the plane should be zero, this condition may be written in either of the equivalent forms $\qquad\qquad Up + Vq + Wr=0$

$$\left.\begin{array}{l} Up + Vq + Wr=0 \\ z' + (\eta\omega_3 - \zeta\omega_2)\, p + (\zeta\omega_1 - \xi\omega_3)\, q + (\xi\omega_2 - \eta\omega_1)\, r=0 \end{array}\right\}\dots\dots\dots (7).$$

247. *Thirdly, let the body slide on an imperfectly rough plane.* The equations (1), (2), (3) and (7) hold as before. If μ be the coefficient of friction the resultant of the forces X, Y, Z must make an angle $\tan^{-1}\mu$ with the normal at the point of contact, hence

$$\frac{(Xp + Yq + Zr)^2}{X^2+Y^2+Z^2} = \frac{1}{1+\mu^2}\dots\dots\dots\dots\dots\dots\dots\dots\dots\dots\dots (8).$$

Also since the resultant of (X, Y, Z), the normal at P and the direction of sliding must lie in one plane, we have the determinantal equation

$$X(qW - rV) + Y(rU - pW) + Z(pV - qU)=0\dots\dots\dots\dots\dots (9).$$

Since the friction must act *opposite* to the direction of sliding, we must have $XU + YV + ZW$ negative. When this vanishes and changes sign, the point of contact ceases to slide.

If the body start from rest we must use the method explained in Vol. i. Chap. iv. to determine whether the point of contact will begin to slide or not. The rule may be briefly stated as follows. Assume X, Y, Z to be the forces necessary to prevent sliding. Then since $u, v, w, \omega_1, \omega_2, \omega_3$ are all initially zero, we have by differentiating (4) and eliminating the differential coefficients of $u, v, w, \omega_1, \omega_2, \omega_3$ three linear equations to find X, Y, Z, in terms of the known initial values of (p, q, r) and (ξ, η, ζ). The point of contact will slide or not according as these values make the left-hand side of equation (8) less or greater than the right-hand side.

In this way when the point of contact is fixed for the moment *the equations* (1), (2), *and* (4) *are sufficient to find the initial values of* X, Y, Z, *i.e. the components of the reaction at the point of contact.* This is also the rule given in Vol. i. Chap. iv. under the heading *Initial Motions* to find the initial value of a reaction, viz. we differentiate the geometrical equations, and substitute from the dynamical equations. This seems the simplest method of proceeding, but we may also adopt either of the following methods.

The equations to find X, Y, Z may be obtained by treating the forces as if they were indefinitely small impulses. In the time dt, we may regard the body as acted on by an impulse gdt at G and a blow whose components are Xdt, Ydt, Zdt at P. It is shown in the chapter on Momentum in Vol. I. that we may consider these in succession. The effect of the first is to communicate to P a velocity gdt in a direction normal to the fixed plane and outwards. If P does not slide, the effect of the blow at P must be to destroy this velocity.

In the chapter on Momentum in Vol. I. certain formulæ have been deduced from the ordinary equations of impact by which we can find the resolved initial velocities of the point of application of any impulse. A geometrical representation of these formulæ is also given by the help of an ellipsoid, $E = $ constant, where E is the vis viva generated by the impulse. To avoid the repetition of this investigation we may use these formulæ to find X, Y, Z. We accordingly write $u_1 = pg$, $v_1 = qg$, $w_1 = rg$ and u_2, v_2, w_2 each equal to zero on the left-hand sides and (to suit the notation of this article) change p, q, r on the right-hand sides into ξ, η, ζ. Geometrically the point of contact will not slide if the diametral line of the fixed plane with regard to the ellipsoid called E makes a less angle with the normal than $\tan^{-1} \mu$.

In any of these cases when p, q, r have been found, the inclinations of the principal axes to the vertical are known. Their motion round the vertical may then be deduced by the rule given in Art. 19. When u, v, w and the motions of the axes have been found, the velocity of the centre of gravity resolved along any straight line fixed in space may be found by resolution.

248. Some integrals of these equations are supplied by the principles of angular momentum and vis viva. If the plane is perfectly smooth we have

$$A\omega_1 p + B\omega_2 q + C\omega_3 r = a,$$

$$A\omega_1^2 + B\omega_2^2 + C\omega_3^2 + (dz/dt)^2 = \beta - 2gz,$$

where a and β are two constants. If the plane is perfectly rough we have

$$A\omega_1^2 + B\omega_2^2 + C\omega_3^2 + u^2 + v^2 + w^2 = \beta - 2gz.$$

249. **Examples. Ex. 1.** A body rests with a plane face on an imperfectly rough horizontal plane whose coefficient of friction is μ. The centre of gravity of the body is vertically over the centre of gravity of the face, and the form of the face is such that the radius of gyration of the face about any straight line in its plane through its centre of gravity is γ. The body is now projected along the plane so that the initial velocity of its centre of gravity is v_0 and the initial rotation about a vertical axis through its centre of gravity is ω_0. If ω_0 be very small, prove that the centre of gravity moves in a straight line and that its velocity at the end of any time t is $v_0 - \mu gt$. If ω be the angular velocity at the same time, prove that $\frac{\gamma^2}{k^2} \log \frac{\omega}{\omega_0} = 1 - \frac{\mu gt}{v_0}$, where k is the radius of gyration of the body about a vertical through the centre of gravity. [*Poisson, Traité de Mécanique.*]

Ex. 2. A body of any form rests with a plane face in contact with a smooth fixed plane so that the perpendicular from the centre of gravity G on the plane falls within the face. If the body is then struck by a blow which passes through G, or begins to move from rest under the action of any finite forces whose resultant passes through G, prove that it will not turn over, but will begin to slide along the plane, even if the line of action of the force cuts the plane outside the base. [*Cournot.*]

Ex. 3. A heavy ellipsoid is placed on an inclined plane, touching it at a point

P whose co-ordinates referred to the principal diameters are (ξ, η, ζ). Deduce from Arts. 246 and 247 the initial values of the reaction at P when the plane is (1) perfectly rough, and (2) perfectly smooth. Thence deduce the initial direction of motion of the centre of gravity.

250. **Oscillations on a rough horizontal plane.** Whatever the shape of a body may be we may suppose it to be set in rotation about the normal at the point of contact with an angular velocity n. If the body continue to rotate about that normal as a permanent axis, the normal must be a principal axis at the point of contact, and yet it must pass through the centre of gravity. This cannot be unless the normal is a principal axis at the centre of gravity. If however $n=0$, this condition is not necessary. There are therefore two cases to be considered.

Case 1. *A body of any form is placed in equilibrium resting with the point C on a rough horizontal plane, with a principal axis at the centre of gravity vertical, and is then set in rotation with an angular velocity n about GC. A small disturbance being given to the body, it is required to find the motion.*

Case 2. *A body of any form is placed in equilibrium on a rough horizontal plane with the centre of gravity over the point of contact. A small disturbance being given to the body, it is required to find the motion.*

251. *Case* 1. Supposing the body not to depart far from its initial position, all the quantities $p, q, u, v, w, \omega_1, \omega_2$ are small and $r=1$ nearly. Hence, by (2), when we neglect the squares of small quantities, we see that X, Y are also small, and that $Z = -g$ nearly. It follows by (1) that ω_3 is constant and $\therefore = n$. Also ξ and η are small and $\zeta = h$ nearly, where h was the altitude of the centre of gravity above the horizontal plane before the motion was disturbed. The equation to the surface may, by Taylor's theorem, be written in the form

$$\zeta = h - \frac{1}{2}\left(\frac{\xi^2}{a} + \frac{2\xi\eta}{b} + \frac{\eta^2}{c}\right),$$

where (a, b, c) are some constants depending on the curvatures of the principal sections of the body at the point C.

The squares of all small quantities being neglected, the equations of Art. 245 become
$$A\omega_1' - (B - C)\,n\omega_2 = -g\eta - hY\Big\}$$
$$B\omega_2' - (C - A)\,n\omega_1 = hX + g\xi \quad\Big\},$$
$$u' - nv = gp + X, \qquad v' + nu = gq + Y,$$
$$p' = nq - \omega_2, \qquad q' = \omega_1 - np,$$
$$u - n\eta + h\omega_2 = 0, \qquad v - h\omega_1 + n\xi = 0,$$
$$p = \frac{\xi}{a} + \frac{\eta}{b}, \qquad q = \frac{\xi}{b} + \frac{\eta}{c}.$$

Eliminating $X, Y, u, v, \omega_1, \omega_2$ from these equations, we get, as in Art. 15,
$$(A + h^2)\,q'' + (A + B + 2h^2 - C)\,n\,p' - \{(B - C)\,n^2 + hg + h^2 n^2\}\,q = -(g + hn^2)\,\eta + hn\xi'$$
$$- (B + h^2)\,p'' + (A + B + 2h^2 - C)\,nq' + \{(A - C)\,n^2 + hg + h^2 n^2\}\,p = (g + hn^2)\,\xi + hn\eta'.$$

It will be found convenient to express ξ, η in terms of p, q. The right-hand sides of each of the equations will then take the form
$$Lp + Mq + L_1 p' + M_1 q'.$$

To solve the equations, we must then assume p, q to be of the form
$$p = P_0 \cos \lambda t + P_1 \sin \lambda t, \qquad q = Q_0 \cos \lambda t + Q_1 \sin \lambda t.$$

If the tangents to the lines of curvature of the moving body at C are parallel to the principal axes at the centre of gravity, these equations admit of considerable simplification. In that case the equation to the surface may be written in the form

$$\zeta = h - \frac{1}{2}\left(\frac{\xi^2}{a} + \frac{\eta^2}{c}\right),$$

where a and c are the radii of curvature of the lines of curvature. The right-hand sides of the equations then become respectively

$$- (g + hn^2)\, cq + hnap' \quad \text{and} \quad (g + hn^2)\, ap + hncq'.$$

To satisfy the equations, it will be sufficient to put

$$p = F \cos (\lambda t + f), \qquad q = G \sin (\lambda t + f).$$

This simplification is possible because we can see beforehand, that if we substitute these values, the first equation will contain only $\sin (\lambda t + f)$ and the second only $\cos (\lambda t + f)$. These trigonometrical terms may then be divided out of the equations leaving two relations between the constants F, G and λ. Eliminating the ratio F/G, we get the following quadratic to determine λ^2

$$[(A + h^2)\lambda^2 + \{B - C + h(h - c)\}\, n^2 + g(h - c)][(B + h^2)\lambda^2 + \{A - C + h(h - a)\}\, n^2 + g(h - a)]$$
$$= \lambda^2 n^2 \{A + B + 2h^2 - C - ha\}\{A + B + 2h^2 - C - hc\}.$$

If λ_1, λ_2 are the roots of this equation, the motion is represented by the equations

$$p = F_1 \cos (\lambda_1 t + f_1) + F_2 \cos (\lambda_2 t + f_2)$$
$$q = G_1 \sin (\lambda_1 t + f_1) + G_2 \sin (\lambda_2 t + f_2)$$

where G_1/F_1, G_2/F_2 are known functions of λ_1, λ_2 respectively, and F_1, F_2, f_1, f_2 are constants to be determined by the initial values of p, q, dp/dt, dq/dt.

In order that the motion may be stable, it is necessary that the roots of this quadratic should be real and positive. The conditions may be easily expressed.

252. Examples. Ex. 1. A solid of revolution is placed with its axis vertical on a perfectly rough horizontal plane and is set in rotation about its axis with an angular velocity n. If c be the radius of curvature at the vertex, h the altitude of the centre of gravity, k the radius of gyration about the axis, k' that about an axis through the vertex perpendicular to the axis of figure, show that the position of the body will be stable if $n > 2\, \dfrac{k'\sqrt{g\,(h - c)}}{k^2 + hc}$.

Ex. 2. An ellipsoid is placed with one of its vertices in contact with a smooth horizontal plane. What angular velocity of rotation must it have about the vertical axis in order that the equilibrium may be stable?

Result. Let a, b, c be the semi-axes, c the vertical axis, then the angular velocity must be greater than $\sqrt{\dfrac{5g}{c}} \cdot \dfrac{\sqrt{c^4 - a^4} + \sqrt{c^4 - b^4}}{a^2 + b^2}$. *[Puiseux.]*

Ex. 3. A solid of any form is placed in equilibrium with the point C on a smooth horizontal plane, a principal axis GC at the centre of gravity being vertical, and an angular velocity n is then communicated to it about GC. A small disturbance being given, show that the harmonic periods may be deduced from the quadratic

$$(A\lambda^2 + E)(B\lambda^2 + F) = (A + B - C)\, n^2 \lambda^2 + g^2 (\rho' - \rho)^2 \sin^2 \delta \cos^2 \delta,$$

where

$$E = (B - C)\, n^2 + g\, \{(h - \rho) \sin^2 \delta + (h - \rho') \cos^2 \delta\},$$
$$F = (A - C)\, n^2 + g\, \{(h - \rho) \cos^2 \delta + (h - \rho') \sin^2 \delta\}.$$

Also h is the altitude of the centre of gravity, ρ, ρ' are the principal radii of curvature at the vertex, and δ is the angle the principal axis GA makes with the plane of the section whose radius of curvature is ρ. *[Puiseux.]*

Ex. 4. A heavy rigid body with a plane base rests in equilibrium on the summit of a rough fixed sphere, the principal axis GC at the centre of gravity G being vertical. It is then set in rotation about the axis GC with an angular velocity n, it is required to determine the periods of the oscillations about this steady motion.

In the general case the result is rather long, but it is simplified when $A = B$. Let $(p, q, -c)$ be the co-ordinates of the point of contact referred to axes GA, GB, GC fixed in the body. Let a be the radius of the sphere and let the mass of the body be unity. We then have $p = P \sin \rho t$, $q = Q \cos \rho t$, where ρ is determined by

$$\{(A + c^2) \rho^2 - (g + cn^2) a\} (\rho^2 - n^2) + gc\rho^2 = \pm n\rho \{(C - A - c^2 + ca) (\rho^2 - n^2) + gc\}.$$

One factor is obviously $\rho \mp n$.

If (ξ, η) be the co-ordinates of the point of contact referred to axes fixed in space, we have $\xi = P \sin (\rho + n) t$, $\eta = Q \cos (\rho + n) t$.

Ex. 5. A solid cube is in neutral equilibrium on the summit of a rough fixed sphere of radius c. It is then set in rotation about a vertical axis through its centre of gravity with an angular velocity n. Prove that this state of steady motion will not be stable unless $n^2 > (55 + 7\sqrt{70}) \, g/4c$.

In this case the cubic obtained in the last question reduces to

$$\rho^3 + n\rho^2 - \tfrac{3}{5} n^2 \rho - \tfrac{3}{5} (n^2 + g/c) \, n = 0.$$

The roots are real if the condition given is satisfied.

253. *Case* 2. Returning now to the general problem enunciated in Art. 250, we proceed to discuss *the oscillations about equilibrium of a heavy body resting on a rough horizontal plane with the centre of gravity over the point of contact.*

Supposing the disturbance to be small, all the quantities ω_1, ω_2, ω_3, u, v, w are small. Hence, when we neglect the squares of small quantities, the equations (1) and (2) of Art. 245 become respectively,

$$A\omega_1' = \eta Z - \zeta Y, \qquad B\omega_2' = \zeta X - \xi Z, \qquad C\omega_3' = \xi Y - \eta X \dots \dots \dots \dots (\text{i}),$$

$$u' = gp + X, \qquad v' = gq + Y, \qquad w' = gr + Z \dots \dots \dots \dots \dots (\text{ii}).$$

Let ξ_0, η_0, ζ_0 be the co-ordinates of the point of contact in the position of equilibrium, and let $\xi = \xi_0 + \xi_1$, $\eta = \eta_0 + \eta_1$, $\zeta = \zeta_0 + \zeta_1$. Then in the small terms of equation (4) we may write ξ_0, η_0, ζ_0 for ξ, η, ζ. Hence, differentiating these and eliminating X, Y, Z, u, v, w by help of equations (i) and (ii), we get

$$(A + \eta_0^2 + \zeta_0^2) \omega_1' - \xi_0 \eta_0 \omega_2' - \xi_0 \zeta_0 \omega_3' = - g (\eta r - \zeta q) \dots \dots \dots \dots (\text{iii}),$$

and two similar equations.

Let p_0, q_0, r_0 be the values of p, q, r in the position of equilibrium. Then $\xi_0/p_0 = \eta_0/q_0 = \zeta_0/r_0 = \rho$, where ρ is the radius vector from G to the point of contact. Now in the small terms of equations (3) we may write ρp_0, ρq_0, ρr_0 for ξ_0, η_0, ζ_0. Hence equations (iii) become by substitution from the second and third of equations (3)

$$A\omega_1' = \eta_0 \rho r'' - \zeta_0 \rho q'' - g (\eta r - \zeta q) \dots \dots \dots \dots \dots (\text{iv}),$$

and two similar equations. At the time t let $p = p_0 + x$, $q = q_0 + y$, and $r = r_0 + z$.

Then since $(p_0 + x)^2 + (q_0 + y)^2 + (r_0 + z)^2 = 1$, we have $p_0 x + q_0 y + r_0 z = 0$. The form of the surface being known we can find x, y, z in terms of ξ_1, η_1, ζ_1, and thus express $\eta r - \zeta q$, $\zeta p - \xi r$, $\xi q - \eta p$ in the form $- g (\eta r - \zeta q) = Lx + My$.

The equations (iv) now become

$$A\omega_1' = \eta_0 \rho z'' - \zeta_0 \rho y'' + Lx + My \dots \dots \dots \dots \dots \dots (\text{v}),$$

and two similar equations.

Differentiating equations (3), and substituting for $d\omega_1/dt$, $d\omega_2/dt$, $d\omega_3/dt$, from (v), and for z and z'' from $p_0 x + q_0 y + r_0 z = 0$, we get equations of the form

$$\left. \begin{aligned} Fx'' + Gy'' &= Hx + Ky \\ F_1 x'' + G_1 y'' &= H_1 x + K_1 y \end{aligned} \right\} .$$

To solve these we put $x = P \cos (\lambda t + f)$, $y = Q \cos (\lambda t + f)$, substituting, and eliminating the ratios P/Q, we have the following quadratic to determine λ^2

$$\left| \begin{array}{cc} F\lambda^2 + H, & G\lambda^2 + K \\ F_1\lambda^2 + H_1, & G_1\lambda^2 + K_1 \end{array} \right| = 0 \quad\quad\quad\quad\text{(vi)}.$$

Thus, by virtue of the relation existing between x, y, z, each of these may be represented by an expression of the form

$$P_1 \cos (\lambda_1 t + f_1) + P_2 \cos (\lambda_2 t + f_2).$$

Substituting these values in equations (v) we see that ω_1, ω_2, ω_3 can each be represented by an expression

$$\Omega_1 + E_1 \cos (\lambda_1 t + f_1) + E_2 \cos (\lambda_2 t + f_2),$$

where E_1, E_2 are known functions of P_1, P_2...and λ_1, λ_2, but Ω_1, Ω_2, Ω_3 are small arbitrary quantities. By substituting in equations (3) and equating the coefficients of $\cos (\lambda_1 t + f_1)$ and $\cos (\lambda_2 t + f_2)$, we may find the values of E_1 and E_2 without difficulty. And we also see that we must have $\Omega_1/p_0 = \Omega_2/q_0 = \Omega_3/r_0$, so that, of the three Ω_1, Ω_2, Ω_3, only one is really arbitrary. We have therefore but five arbitrary constants, viz. P_1, P_2, f_1, f_2, and Ω_1. These are determined by the initial values of ω_1, ω_2, ω_3, x and y.

To find the motion of the principal axes round the vertical, let ϕ be the angle the plane containing GC and the vertical makes with the plane of AC. Then, by drawing a figure for the standard case in which p, q, r are all positive, it will be seen that, if μ be the rate at which GC goes round the vertical,

$$\mu\sqrt{1 - r^2} = \omega_1 \cos \phi + \omega_2 \sin \phi = (p_0\omega_1 + q_0\omega_2)/\sqrt{1 - r_0^2}.$$

Substituting for ω_1, ω_2, this takes the form

$$\mu = n_3 + N_1 \cos (\lambda_1 t + f_1) + N_2 \cos (\lambda_2 t + f_2),$$

where n_3, N_1, N_2 are all known constants.

In order that the equilibrium may be stable it is necessary that both the roots of the quadratic (vi) should be real and positive. These conditions may easily be expressed.

These conditions being supposed satisfied, the expressions for x, y, z will only contain periodical terms, and thus the inclinations of the principal axes to the vertical will not be sensibly altered. But the expressions for ω_1, ω_2, ω_3 *may* each contain a non-periodical term, and if so the rate at which the principal axes will go round the vertical will also contain non-periodical terms. The body therefore may gradually turn with a slow motion round the normal at the point of contact. The expressions for u, v, w will contain only periodic terms, so that the body will have no motion of translation in space.

Motion of a Rod.

254. When the body whose motion is to be determined is a rod, it is often more convenient to recur to the original equations of motion supplied by D'Alembert's principle. The equations of Lagrange may also be used with advantage. These methods will be illustrated by the following problem.

A uniform heavy rod, suspended from a fixed point O by a string, makes small oscillations about the vertical. Determine the motion.

Let O be taken as origin, and let the axis of z be measured vertically downwards; let $2a$ be the length of the rod, b the length of the string. Let (l, m, n) (p, q, r) be the direction-cosines of the string and rod. Then l, m, p, q are small quantities whose squares are to be neglected, and we may put n and r each equal to unity.

Let u be the distance of any element du of the rod from that extremity A of the rod to which the string is attached. Let (x, y, z) be the co-ordinates of the element du, then we have $\qquad x = bl + up, \quad y = bm + uq, \quad z = b + u \dots\dots\dots\dots\dots\dots\dots(1)$.

Let M be the mass of the rod, MT the tension of the string. The equations of motion of the centre of gravity will be

$$bl'' + ap'' = -Tl \qquad bm'' + aq'' = -Tm \qquad 0 = g - T \dots\dots\dots\dots(2),$$

where accents denote differential coefficients with regard to t.

By D'Alembert's principle the equation of moments round x will be

$$\Sigma du\,(yz'' - zy'') = \Sigma du\,(yZ - zY) = \Sigma du\,(yg).$$

By equations (1) this reduces to

$$\int_0^{2a} du\,\{-(b+u)\,(bm'' + uq'')\} = 2ag\,(bm + aq).$$

Integrating, we get

$$-6ab\,(bm'' + aq'') - 6ba^2m'' - 8a^3q'' = 6ag\,(bm + aq),$$

which, by equations (2), reduces to

$$3bm'' + 4aq'' = -3gq \dots\dots\dots\dots\dots\dots\dots\dots\dots(3).$$

Therefore by (2) and (3) the four equations of motion are

$$bl'' + ap'' = -gl, \qquad 3bl'' + 4ap'' = -3gp \dots\dots\dots\dots\dots\dots(4),$$

and two similar equations for m, q. These equations do not contain m or q, and on the other hand the equations to find m and q do not contain l or p. This shows that the oscillations in the plane xz are not affected by those in the perpendicular plane yz.

To solve these equations, put $\quad l = F \sin(\lambda t + a), \quad p = G \sin(\lambda t + a)$,

we get $\qquad b\lambda^2 F + a\lambda^2 G = gF, \qquad b\lambda^2 F + \tfrac{4}{3} a\lambda^2 G = gG$;

$$ab\lambda^4 - (4a + 3b)\,g\lambda^2 + 3g^2 = 0,$$

and the values of λ may be found from this equation.

255. In order to make a comparison of different methods, let us deduce the motion from Lagrange's equations. In this case we must determine the semi vis viva T true to the *squares* of the small quantities, p, q, l, m, we cannot therefore put $r = 1$, $n = 1$. Since $p^2 + q^2 + r^2 = 1$, $l^2 + m^2 + n^2 = 1$, we have

$$r = 1 - \tfrac{1}{2}(p^2 + q^2), \qquad n = 1 - \tfrac{1}{2}(l^2 + m^2),$$

we must therefore replace the third of equations (1) by

$$z = bn + ur = b + u - \tfrac{1}{2}b\,(l^2 + m^2) - \tfrac{1}{2}u\,(p^2 + q^2).$$

We have $\qquad \Sigma mx'^2 = \Sigma m\,(b^2l'^2 + 2bl'p'u + p'^2u^2) = M\,(b^2l'^2 + 2bl'p'a + \tfrac{4}{3}a^2p'^2)$.

The value of $\Sigma my'^2$ may be found in a similar manner. The value of $\Sigma mz'^2$ is of the fourth order and may be neglected. Hence the vis viva is

$$2T = b^2\,(l'^2 + m'^2) + 2ab\,(l'p' + m'q') + \tfrac{4}{3}a^2\,(p'^2 + q'^2).$$

Also we have $U = -\tfrac{1}{2}gb\,(l^2 + m^2) - \tfrac{1}{2}ga\,(p^2 + q^2) + \text{constant}$.

The equation $\dfrac{d}{dt}\dfrac{dT}{dl'} - \dfrac{dT}{dl} = \dfrac{dU}{dl}$ becomes $bl'' + ap'' = -gl$;

similarly we get $\qquad\qquad\qquad\qquad bl'' + \tfrac{4}{3}ap'' = -gp$.

These are the same equations which we deduced from D'Alembert's principle, and the solution may be continued as before.

EXAMPLES *.

1. A uniform rod, moveable about one extremity, moves in such a manner as to make always nearly the same angle a with the vertical; show that the time of a small oscillation is $2\pi\sqrt{\dfrac{2a}{3g} \cdot \dfrac{\cos a}{1+3\cos^2 a}}$, a being the length of the rod.

2. If a rough plane inclined at an angle a to the horizon be made to revolve with uniform angular velocity n about a normal Oz and a sphere be placed at rest upon it, show that the path in space of the centre will be a prolate, a common, or a curtate cycloid, according as the point at which the sphere is initially placed is without, upon, or within the circle whose equation is $x^2+y^2 = (35g \sin a/2n^2)\, x$, the axis Oy being horizontal.

   When the sphere is placed at rest on the moving plane, it should be noticed that a velocity is suddenly given to it by the impulsive frictions.

3. A circular disc capable of motion about a vertical axis through its centre perpendicular to its plane is set in motion with angular velocity Ω. A rough uniform sphere is gently placed on any point of the disc, not the centre, prove that the sphere will describe a circle on the disc, and that the disc will revolve with angular velocity $\dfrac{7Mk^2}{7Mk^2+2mr^2}\,\Omega$, where Mk^2 is the moment of inertia of the disc about its centre, m is the mass of the sphere and r the radius of the circle traced out.

4. A sphere is pressed between two perfectly rough parallel boards which are made to revolve with the uniform angular velocities Ω and Ω' about fixed axes perpendicular to their planes. Prove that the centre of the sphere describes a circle about an axis which is in the same plane as the axes of revolution of the boards and whose distances from these axes are inversely proportional to the angular velocities about them.

   Show that when the boards revolve about the same axis, their points of contact will trace on the sphere small circles, the tangents of whose angular radii will be $\dfrac{c}{a} \cdot \dfrac{\Omega \sim \Omega'}{\Omega + \Omega'}$, a being the radius of the sphere and c that of the circle described by its centre.   [Math. Tripos, 1861.

5. A perfectly rough circular cylinder is fixed with its axis horizontal. A sphere being placed on it in a position of unstable equilibrium is so projected that the centre begins to move with a velocity V parallel to the axis of the cylinder. It is then slightly disturbed in a direction perpendicular to the axis. If θ be the angle the radius through the point of contact makes with the vertical, prove that the velocity of the centre parallel to the axis at any time t is $V\cos\sqrt{\tfrac{2}{7}}\,\theta$ and that the sphere will leave the cylinder when $\cos\theta = \tfrac{10}{17}$.

6. A uniform sphere is placed in contact with the exterior surface of a perfectly rough cone. Its centre is acted on by a force the direction of which always meets the axis of the cone at right angles and the intensity of which varies inversely as the cube of the distance from that axis. Prove that if the sphere be properly started the path described by its centre will meet every generating line of the cone on which it lies in the same angle. See the *Solutions of Cambridge Problems* for 1860, page 92.

7. Every particle of a sphere of radius a, which is placed on a perfectly rough

---

* These Examples are taken from the Examination Papers which have been set in the University and in the Colleges.

sphere of radius c, is attracted to a centre of force on the surface of the fixed sphere with a force varying inversely as the square of the distance ; if it be placed at the extremity of the diameter through the centre of force and be set rotating about that diameter and then slightly displaced, determine its motion ; and show that when it leaves the fixed sphere the distance of its centre from the centre of force is a root of the equation $20x^3 - 13\,(2c+a)\,x^2 + 7a\,(2c+a)^2 = 0$. [Math. Tripos, 1867.

8. A perfectly rough plane revolves uniformly about a vertical axis in its own plane with an angular velocity n, a sphere being placed in contact with the plane rolls on it under the action of gravity, find the motion.

Take the axis of revolution as axis of z, and let the axis of x be fixed in the plane. Let a be the radius, m the mass of the sphere; F, F' the frictions resolved parallel to the axes of x and z, and R the normal reaction. The motions of the axes (Art. 5) are given by $\theta_1 = 0$, $\theta_2 = 0$, $\theta_3 = n$. The equations of motion (Arts. 5, 10) are

$$u = dx/dt - an, \qquad v = xn, \qquad w = dz/dt,$$

$$du/dt - vn = F/m, \qquad dv/dt + un = R/m, \qquad dw/dt = -g + F'/m,$$

$$d\omega_x/dt - n\omega_y = -F'a/k^2, \qquad d\omega_y/dt + n\omega_x = 0, \qquad d\omega_z/dt = Fa/k^2.$$

Since the point of contact has the same motion as the plane the geometrical equations are $u + a\omega_z = 0$, $w - a\omega_x = 0$. Solving these equations we find that the sphere will not fall down. If the sphere start from *relative* rest at a point in the axis of x, we have $n^2 z = -g \tan^2 i \{1 - \cos(nt \cos i)\}$, where $\sin i = \sqrt{\tfrac{5}{7}}$. The sphere will therefore never descend more than $5g/n^2$ below its original position.

9. A perfectly rough vertical plane revolves with a uniform angular velocity μ about an axis perpendicular to itself, and also with a uniform angular velocity Ω about a vertical axis in its own plane which meets the former axis. A heavy uniform sphere of radius c is placed in contact with the plane ; prove that the position of its centre at any time t, will be determined by the equations

$$7\xi'' - 5\Omega^2 \xi - 2\mu z' = 0, \qquad 7z''' + 2\Omega^2 z' + 2\mu\,(\xi'' + \Omega^2 \xi) = 0,$$

z denoting the distance of the centre from the horizontal plane through the horizontal axis of revolution, and ξ that from the plane through the two axes.

Prove also that $7u = 7c\Omega + 2\mu b$, $7v + 2\mu a = 0$, if a and b be the initial values of ξ and z, u and v those of $d\xi/dt$ and dz/dt.

10. A hoop $AGBF$ revolves about AB its diameter as a fixed vertical axis. GF is a horizontal diameter of the same circle, which is without mass and which is rigidly connected to the circle ; DC is a smaller concentric hoop which can turn freely about GF as diameter. If Ω, Ω', ω, ω', be the greatest and least angular velocities about AB, GF respectively, prove that $\Omega \cdot \Omega' = \omega^2 - \omega'^2$.

11. OA, OB, OC are the principal axes of a rigid body which is in motion about a fixed point O. The axis OC has a constant inclination a to a line OZ fixed in space, and revolves with uniform angular velocity Ω round it, and the axis OA always lies in the plane ZOC. Prove that the constraining couple has its axis coincident with OB, and that its moment is $-(A - C)\,\Omega^2 \sin a \cos a$.

12. A ring of wire, of radius c, rests on the top of a smooth fixed sphere of radius a, and is set rotating about the vertical diameter of the sphere with an angular velocity n. If the ring is slightly displaced, prove that the motion is unstable if $n^2 c^4$ is less than $2g\,(2a^2 - c^2)\sqrt{(a^2 - c^2)}$. [Math. Tripos, 1885.

Since the ring is rigidly connected with the centre of the fixed sphere it may be regarded as a top having that point for vertex. The condition of stability for a top has been shown in Art. 212 to be $C^2 n^2 > 4Agl$. Substituting the values of A and C for a ring the result follows at once.

13. A uniform right circular cone of mass M' is capable of turning freely about its vertex which is fixed above a rough horizontal table on which is a sphere of mass M and radius a. The cone rolls on the sphere whose centre is describing with velocity V a circle of radius b whose centre is vertically below the vertex of the cone ; prove that in the steady motion of the cone and the sphere, the angular velocity Ω of the cone about its axis is

$[nab \sin \beta \cos \beta - V (a \sin a \cos \beta \cos \gamma + b \cos a \sin \gamma - b \sin^2 \beta)]/[b (b + a \cos \beta) \sin \gamma]$,

and that the condition that such steady motion may be possible is

$$V^2 k_2^2 \cos a + \Omega V b k_1^2 < g h b^2,$$

where γ is the semi-vertical angle of the cone, a the inclination of its axis to the vertical, $\beta = a - \gamma$, and k_1, k_2 are the radii of gyration of the cone about its axis and a perpendicular to its axis through the vertex and h is the distance of the centre of gravity of the cone from its vertex, and n the spin of the sphere about the vertical.

[Math. Tripos, 1889.

14. A rough sphere of radius a and radius of gyration K capable of moving about its centre is initially at rest ; another sphere of $1/n$ the mass and of radius b and radius of gyration k is placed gently on it, having initially an angular velocity ω about the common normal which makes an acute angle a with the vertical drawn upwards. Prove that the second sphere will not roll off provided

$$\omega^2 > \frac{2\mu (a + b) g}{(3\mu + 1) b^2} \{(3\mu + 1)^2 - 4\mu^2 \cos^2 a\} \sec a, \text{ where } \mu = a^2/nK^2 + b^2/k^2.$$

[Math. Tripos, 1888.

CHAPTER VI.

Linear Differential Equations.

256. It has been shown in Chap. III. that the problem of determining the small oscillations of a system about a state of steady motion is really the same as that of solving a corresponding system of linear differential equations. In that chapter the forces were assumed to have a potential, so that the differential equations had a certain symmetry which simplified the solution. We now propose to remove this restriction. Taking the differential equations in their most general form, but still with constant coefficients, we shall briefly discuss any peculiarities of their solution which appear to have dynamical applications.

The chief object of this chapter is to determine the conditions that the undisturbed motion should be stable. This resolves itself into two questions (1) under what circumstances do positive powers of the time enter into the expressions for the co-ordinates, and what is the highest power which presents itself? (2) when the roots of the fundamental equation cannot be found, what conditions must be satisfied by the coefficients of that equation that stability may be assured? In order to make our remarks on these two questions intelligible it will be necessary to sum up a few propositions which belong rather to the theory of differential equations than to that of dynamics. The discussion of the first question begins therefore at Art. 268 though alluded to before that article. The second question will occupy the next section.

257. Following the same notation as in Art. 111, let θ, ϕ, &c. be the co-ordinates of the system. Let the system be moving in any known manner determined by $\theta = f(t)$, $\phi = F(t)$, &c. We now suppose the system to be slightly disturbed from this state of motion. To discover the subsequent motion we put $\theta = f(t) + x$, $\phi = F(t) + y$, &c. These quantities x, y, &c. are in the first instance very small because the disturbance is small. The quantities

x, y, &c. are said to be *small* when it is possible to choose some quantity numerically greater than all of them which is such that its square can be neglected. This quantity may be called the standard of reference for small quantities.

258.   To determine whether x, y, &c. remain small, we substitute these new values of θ, ϕ, &c. in the equations of motion. Assuming, for the moment, that x, y, &c. remain small we may neglect their squares, and thus the resulting equations will be linear. The coefficients of x, dx/dt, d^2x/dt^2, y, dy/dt &c. in these equations may be either constants or functions of the time. Following the definitions in Art. 111, the undisturbed motion in the former case is said to be steady.

259.   We propose to consider first the case in which the system depends on two independent co-ordinates or (as it is sometimes called) has two degrees of freedom. This is a case which occurs very frequently, and as the results are comparatively simple it seems worthy of a separate discussion. We shall then proceed to the general case in which the system has any number of co-ordinates.

260.   **Two degrees of freedom.**   The equations of motion of a dynamical system performing its natural oscillations with two degrees of freedom may be written

$$\left. \begin{aligned} A\frac{d^2x}{dt^2} + B\frac{dx}{dt} + Cx + A'\frac{d^2y}{dt^2} + B'\frac{dy}{dt} + C'y = 0 \\ E\frac{d^2x}{dt^2} + F\frac{dx}{dt} + Gx + E'\frac{d^2y}{dt^2} + F'\frac{dy}{dt} + G'y = 0 \end{aligned} \right\}.$$

To solve these equations we put

$$x = \left[A'\frac{d^2}{dt^2} + B'\frac{d}{dt} + C' \right] V \qquad y = -\left[A\frac{d^2}{dt^2} + B\frac{d}{dt} + C \right] V,$$

these suppositions evidently satisfying the first equation whatever V may be. Substituting in the second, and using the symbol δ to represent $\dfrac{d}{dt}$ for the sake of brevity we find

$$\begin{vmatrix} A\delta^2 + B\delta + C & A'\delta^2 + B'\delta + C' \\ E\delta^2 + F\delta + G & E'\delta^2 + F'\delta + G' \end{vmatrix} V = 0.$$

This is an equation to find V in terms of t. Since δ enters into the determinant in the fourth power, the value of V when found will contain four arbitrary constants. Thence we find both x and y by means of the formulæ given above. It will be observed that *these require no operation to be performed except differentiation*. Thus, no matter how complicated V may be, the values of x and y readily follow.

261.   Let $\Delta(\delta)$ represent the determinant which is the operator on V. Then making $\Delta(\delta) = 0$, we have a biquadratic to find δ.

If the roots of this biquadratic are m_1, m_2, m_3, m_4, we know by the rules for solving differential equations that

$$V = L_1 e^{m_1 t} + L_2 e^{m_2 t} + L_3 e^{m_3 t} + L_4 e^{m_4 t}$$

where L_1, L_2, L_3, L_4, are the four arbitrary constants.

If all the roots of the biquadratic are real and unequal, this is the proper expression to use for V. But it takes a variety of different forms when the biquadratic contains imaginary or equal roots. These however are described in the theory of differential equations, and will be summed up in Art. 264.

262. **Many degrees of freedom.** The equations which occur in Dynamics are in general all of the second order, but as this restriction is not necessary in what follows, we shall suppose the equations to contain differential coefficients of any order.

Let there be n dependent variables represented by x, y, z, &c. and one independent variable represented by t. If the symbol δ represent differentiation with regard to t, the n equations to find x, y, &c. may be written:

$$\left. \begin{array}{l} f_{11}(\delta)\, x + f_{12}(\delta)\, y + f_{13}(\delta)\, z + \ldots = 0 \\ f_{21}(\delta)\, x + f_{22}(\delta)\, y + f_{23}(\delta)\, z + \ldots = 0 \\ \qquad \ldots \qquad \ldots \qquad \ldots \qquad = 0 \end{array} \right\} \ldots\ldots\ldots\ldots(1).$$

To solve these, we use the analogy which exists between the rules for combining symbols of differentiation and those of common algebra. Omitting for the moment any one equation, say the first, and proceeding to solve the remaining $n - 1$ equations by the rules of common algebra, we find the ratios

$$\frac{1}{I_1(\delta)}\, x = \frac{1}{I_2(\delta)}\, y = \frac{1}{I_3(\delta)}\, z = \&c. = V \ldots\ldots\ldots\ldots(2),$$

where each of the equalities has been put equal to V. Here we have used the letter I to stand for the minors of the determinant

$$\Delta(\delta) = \begin{vmatrix} f_{11}(\delta), f_{12}(\delta), f_{13}(\delta), \ldots \\ f_{21}(\delta), f_{22}(\delta), f_{23}(\delta), \ldots \\ \ldots \quad \ldots \quad \ldots \quad \ldots \end{vmatrix} \ldots\ldots\ldots\ldots(3).$$

The suffix of the letter I indicates the number of the column in which the constituent of the omitted equation lies whose minor is required.

Substituting these values of x, y, z, &c. in the equation previously omitted, we obtain

$$\Delta(\delta)\, V = 0 \ldots\ldots\ldots\ldots\ldots(4).$$

This is an equation to determine a single quantity V as a function of t. We may call V the *type of the solution*. Supposing this equation to be solved by the usual rules, the values of x, y, z, &c. are found by equations (2). Thus we have

$$x = I_1(\delta)\, V, \quad y = I_2(\delta)\, V, \&c. \ldots\ldots\ldots\ldots(5).$$

These operators, $I_1(\delta)$, $I_2(\delta)$, &c., are all *integral and rational functions of* δ; so that when V is once known, all the other operations necessary for the complete solution of the equations are reduced to the one operation of continued differentiation.

263. This arrangement of the solution of the differential equations (1) has the advantage of expressing the results by means of integral and rational functions of the symbol δ. In practice, this will be found to introduce a great simplification into the solution. The type V can always be immediately written down by the usual rules for solving equation (4). It is sometimes very complicated, and in such cases it is very convenient to be able to deduce the forms of x, y, z, &c. without having to perform any inverse operation.

264. **Different types of the solution.** If the roots of the determinantal equation $\Delta(\delta) = 0$ be m_1, m_2, &c. the type V is known to be

$$V = L_1 e^{m_1 t} + L_2 e^{m_2 t} + \dots\dots\dots\dots\dots\dots\dots\dots\dots(6),$$

where L_1, L_2, &c. are arbitrary constants. When a pair of imaginary roots of the form $r \pm p\sqrt{-1}$ occurs we replace the two corresponding imaginary exponentials by the terms $\qquad V = e^{rt}(L \cos pt + M \sin pt) \dots\dots\dots\dots\dots(7).$
If equal roots occur, the value of V thus given has no longer the full number of constants. Supposing that we have a roots each equal to m, the type of the solution which depends on these roots is

$$V = (L_0 + L_1 t + \dots + L_{a-1} t^{a-1}) e^{mt} \dots\dots\dots\dots(8)$$

where the L's are a arbitrary constants. This may be put into the form

$$V = \left(L_0 + L_1 \frac{d}{dm} + \dots + L_{a-1} \frac{d^{a-1}}{dm^{a-1}} \right) e^{mt} \dots\dots\dots\dots(9).$$

If we have a equal pairs of imaginary roots of the form $r \pm p\sqrt{-1}$ we replace the a pairs of terms by

$$e^{rt}(L_0 \cos pt + M_0 \sin pt) + \frac{d}{dr} e^{rt}(L_1 \cos pt + M_1 \sin pt) + \text{\&c.} \dots\dots(10).$$

Here, if we please, we may replace the differentiation with regard to r by a differentiation with regard to p.

The peculiarity of the case of equal roots is the presence of terms containing some power of t as a factor. The occurrence of a equal roots in general indicates the presence of terms containing all the integral powers of t up to t^{a-1}.

265. In order to deduce the corresponding values of x, y, &c. from these types, we shall have, in the absence of equal roots, to operate with some integral and rational function of δ such as $I(\delta)$ on an exponential real or imaginary.

I. We have the theorem $\qquad I(\delta) e^{mt} = I(m) e^{mt}$,
so that when the roots of the equation $\Delta(\delta) = 0$ are all real and unequal we have immediately $\qquad x = L_1 I_1(m_1) e^{m_1 t} + L_2 I_1(m_2) e^{m_2 t} + \text{\&c.},$

$$y = L_1 I_2(m_1) e^{m_1 t} + L_2 I_2(m_2) e^{m_2 t} + \text{\&c.},$$

$$z = \text{\&c.}$$

II. If X be any function of t, we have the theorem $I(\delta) e^{rt} X = e^{rt} I(\delta + r) X$, so that when a pair of imaginary roots occurs, and we have to operate on the product of a real exponential and a sine or cosine, we can immediately remove the real exponential, and reduce the operator to that of continued differentiation of the sine or cosine.

III. We have the theorem $f(\delta^2) \sin mt = f(-m^2) \sin mt$.

Hence if we have to operate with $F(\delta)$, we arrange the operator in the form $\phi(\delta^2) + \delta\psi(\delta^2)$. We then have $F(\delta) \sin mt = \phi(-m^2) \sin mt + \psi(-m^2) m \cos mt$.

266. When the determinantal equation $\Delta(\delta) = 0$ has equal roots we have to operate on expressions which contain some powers of t. But since the operators d/dt and d/dm or d/dr are independent we may use the theorem

$$I(\delta) \frac{d^\kappa}{dm^\kappa} e^{mt} = \frac{d^\kappa}{dm^\kappa} \{I(m) e^{mt}\}.$$

Thus when the equation $\Delta(\delta) = 0$ has a roots each equal to m we may write the solution given by equations (5) and (9) of Arts. 262, 264 in the form

$$x = L_0 [I_1(m) e^{mt}] + L_1 \frac{d}{dm} [I_1(m) e^{mt}] + \ldots + L_{a-1} \frac{d^{a-1}}{dm^{a-1}} [I_1(m) e^{mt}],$$

$$y = L_0 [I_2(m) e^{mt}] + L_1 \frac{d}{dm} [I_2(m) e^{mt}] + \ldots + L_{a-1} \frac{d^{a-1}}{dm^{a-1}} [I_2(m) e^{mt}],$$

$$z = \&c.$$

267. Ex. 1. If there be two roots of the determinantal equation $\Delta(\delta) = 0$ each equal to m, show by an actual comparison of the several terms that we have the same solutions for x, y, &c. whether we use as operators the minors of the first or the minors of any other row of the determinant $\Delta(\delta)$.

Ex. 2. The values of x, y, &c. are obtained from V by operating with certain functions of δ, viz. $I_1(\delta)$, $I_2(\delta)$, &c. If instead of these operators we use $\mu I_1(\delta)$, $\mu I_2(\delta)$, &c. where μ is some function of δ such as $\mu = f(\delta)$, show that the effect is merely to alter the arbitrary constants L_0, L_1, &c. Thence show that the solutions are the same, whether there be equal roots or not, whatever set of first minors of $\Delta(\delta)$ are used as operators.

268. **An Indeterminate Case.** If the roots of the determinantal equation $\Delta(\delta) = 0$ are m_1, m_2, &c. we have shown that the values of x, y, &c. are given by

$$x = \Sigma L I_1(m) e^{mt}, \quad y = \Sigma L I_2(m) e^{mt}, \quad \&c.$$

But we see at once that there is a case of failure. If one of the roots of the equation $\Delta.(\delta) = 0$ makes all the minors, $I_1(m)$, $I_2(m)$, &c. equal to zero, the solution becomes incomplete, for then one of the constants called L disappears from the solution. If all the minors of only one row vanished, we could find the values of x, y, z, &c. by choosing as our operators the minors of some other row. But this cannot be done if all the minors of all the rows are zero.

269. We shall now prove that this indeterminate case cannot occur unless the determinantal equation $\Delta(\delta) = 0$ has equal roots. To show this, we differentiate equation (3) of Art. 262. We find

$$\frac{d\Delta(\delta)}{d\delta} = I_{11} \frac{df_{11}}{d\delta} + I_{12} \frac{df_{12}}{d\delta} + \&c. + I_{21} \frac{df_{21}}{d\delta} + \&c.$$

where the letter I stands for the minor of that constituent of the determinant $\Delta\,(\delta)$ which is indicated by the suffix. We notice that the right-hand side of this equation vanishes when all the first minors are zero. Thus the equation $\Delta\,(\delta) = 0$ must have at least two equal roots. In the same way, if the second minors are all zero also, any first minor has two equal roots, and therefore the original equation has three equal roots, and so on.

270. We may notice two obvious results. (1) If all the first minors of a determinant have a root a times, the determinant has the root $a + 1$ times at least. (2) If a determinant have r equal roots, and all its first, second, &c. minors vanish for these roots, then each of the first minors has the equal root $r - 1$ times, each of the second minors $r - 2$ times, and so on.

271. Let us consider, as an example, Lagrange's determinant to find the periods of the small oscillations of a system about a position of equilibrium, Art. 57. Suppose this determinant has two equal roots, then by Art. 266, we might expect that each co-ordinate of the system would contain a term of the form $(A + Bt)\,e^{mt}$. Thus the amplitude of the oscillation will contain powers of the time.

By Art. 61 we know that every first minor of Lagrange's determinant also contains this root, so that the solution given by Art. 266 fails. Accordingly we shall find in Art. 273 that the solution does not contain any powers of the time, but that the independent constants arrange themselves in another manner which may be conveniently represented by using a double type of solution. See also Art. 281.

272. We may now consider the following general problem :—

Let the determinant $\Delta\,(\delta)$ have α roots each equal to m. Let β of these roots make every first minor of $\Delta\,(\delta)$ equal to zero. Let γ of these last make every second minor equal to zero, and so on. It is required to state the general form of the solution and to explain how the α constants in that solution are to be found.

Solution with a single type. First, let us consider the α roots which are equal to m. It has been proved in Art. 266, that the part of the solution which depends on these may be written in the form

$$x = L_0[I_1(m)e^{mt}] + L_1\frac{d}{dm}[I_1(m)e^{mt}] + \ldots + L_{a-1}\frac{d^{a-1}}{dm^{a-1}}[I_1(m)\,e^{mt}],$$

with similar expressions for y, z, &c.

If these first minors are finite, these formulæ contain powers of t from t^0 to t^{a-1}, and thus supply the α constants which belong to the α equal roots. If the first minors have β roots equal to m, $I_1(m)$, $I_2(m)$, &c., and their differential coefficients up to the $(\beta - 1)$th are all zero. In this case the powers of t extend only to $t^{a-\beta-1}$, and thus these formulæ do not supply the full number of constants.

When all the first minors have the root α times and all the second minors have the root β times, we know by Art. 270 that $\alpha - \beta - 1$ cannot be negative.

273. Solution with a double type. To find the proper forms for x, y, &c. when the first minors are all zero, we return to the analogy between operations and quantities alluded to in Art. 262. We now reject any two of the equations (1), say the first two. Solving the remaining $n - 2$ equations we can express all the co-ordinates z, u &c. in terms of x and y, thus obtaining a series of equations of the form

$$z = \phi(\delta)\, x + \psi(\delta)\, y,$$

where the functional symbols are really second minors of the determinant $\Delta(\delta)$. We now substitute these expressions for z, u, &c. in the two omitted equations. These two equations will be satisfied provided x and y have any values which make $I(\delta)\, x = 0$ and $I(\delta)\, y = 0$, where $I(\delta)$ is any first minor of $\Delta(\delta)$.

We notice also that these two equations are satisfied by the *separate* parts of these values of z, u, &c. which arise from x and from y. We may therefore arrange the solution so as to find these two parts separately, and then finally add the results. The following arrangement will be found convenient in practice.

When the first minors are all zero, reject some one of the given differential equations (1), say the first. We have now $n - 1$ equations to determine the n co-ordinates. Putting $y = 0$ in these equations we find x, z, &c. in terms of a single type ξ, where ξ satisfies the equation $I_2(\delta)\, \xi = 0$. Here I_2 represents the minor of the second constituent of the first line of the determinant $\Delta(\delta)$. We write the solution thus found in the form

$$x = J_{21}(\delta)\, \xi, \qquad y = 0, \qquad z = J_{23}(\delta)\, \xi, \&c.$$

where the operators are the *second minors* of the constituents in the first two lines of $\Delta(\delta)$. Next, putting $x = 0$ instead of y in the equations after the first, we obtain another solution, by which x, z, &c. are expressed in terms of another single type η. Here η satisfies the equation $I_1(\delta)\, \eta = 0$, where I_1 is the minor of the first constituent of the first line of $\Delta(\delta)$. We write the solution thus found in the form

$$x = 0, \qquad y = J_{12}(\delta)\, \eta, \qquad z = J_{13}(\delta)\, \eta, \&c.$$

Adding these two solutions together, we have the following values of x, y, z, &c.

$$x = J_{21}(\delta)\, \xi, \quad y = J_{12}(\delta)\, \eta, \quad z = J_{23}(\delta)\, \xi + J_{13}(\delta)\, \eta, \&c.$$

These evidently satisfy all the equations except the one rejected. But this equation also is satisfied because by hypothesis we take those parts only of these solutions which make all the first minors equal to zero.

The types ξ, η are the same exponentials but with different constants, the operators also are different. Suppose for example the determinant $\Delta\,(\delta)$ had two roots only equal to m and that these make every first minor of $\Delta\,(\delta)$ equal to zero. The terms of x, y, z &c. which depend on the roots other than m are found each from its own exponential by the rule given in Art. 262 for a single type. The terms of x, y, z &c. which depend on the root m are found by putting $\xi = L_1 e^{mt}$, $\eta = L_2 e^{mt}$ where L_1 and L_2 are two different arbitrary constants. The portions of the solution due to these are respectively

$$x = L_1 J_{21}(m) e^{mt}, \qquad y = 0, \qquad z = L_1 J_{23}(m) e^{mt}, \&\text{c.}$$
$$x = 0, \qquad y = L_2 J_{12}(m) e^{mt}, \qquad z = L_2 J_{13}(m) e^{mt}, \&\text{c.}$$

where J_{ab} is a second minor which may be deduced from $\Delta\,(\delta)$ by rejecting the ath and bth columns and the two first rows, giving the second minor thus left its proper sign. The suffix 2 occurs in every J in the first line and the suffix 1 in every J in the second. The complete solution due to the root m is the sum of these two partial solutions. We notice that the two arbitrary constants L_1, L_2 so enter into the values of x, y, z &c. that the exponential e^{mt} is accompanied by two arbitrary constants instead of one and these are not separated by the presence of powers of t.

274. If the minors which the types ξ and η are to satisfy contain the root $\delta = m$, β times, we have therefore

$$\xi = (G_0 + G_1 t + \ldots + G_{\beta-1} t^{\beta-1})\, e^{mt},$$
$$\eta = (H_0 + H_1 t + \ldots + H_{\beta-1} t^{\beta-1})\, e^{mt}.$$

The corresponding values of x, y, &c. are found by substitution, and may be written in the form

$$x = G_0 [J_{21}(m)\, e^{mt}] + G_1 \frac{d}{dm}[J_{21}(m)\, e^{mt}] + \ldots + G_{\beta-1} \frac{d^{\beta-1}}{dm^{\beta-1}}[J_{21}(m)\, e^{mt}],$$

with similar expressions for y, &c.

The peculiarity of the solutions which are derived from the double type ξ, η is that the corresponding terms in the expressions for x and y have *independent constants*.

If the second minors which form the operators are all finite, these formulæ contain powers of t up to $t^{\beta-1}$ and supply 2β constants. But if these second minors contain γ roots equal to m, the powers of t extend only to $t^{\beta-\gamma-1}$, and thus the full number of constants has not been found.

275. **Solution with a triple type.** Thirdly, we have to find the solution when the second minors are zero as well as the first minors. In this case the solution just found becomes again insufficient. To determine the proper forms of x, y, z, &c. we now reject any three of the differential equations (1) of Art. 262, and proceed as before. We thus have $n-3$ equations to find the n co-ordinate. We see at once that we can express all the co-ordinates in terms of any three we please, say

x, y, z. We thus have three times as many arbitrary constants as there are roots equal to m.

In the same way as before we can express the solution in terms of a triple type ξ, η, ζ. Putting y and z equal to zero, we find the remaining co-ordinates, viz. x, u, &c. in terms of a single type ξ. Putting x and z equal to zero (instead of y and z) in these $n-3$ equations we obtain a second solution depending on another single type η. Lastly, putting x and y equal to zero we obtain a third solution depending on ζ. Adding together these three solutions we find that all the co-ordinates may be expressed by means of operators which are really third minors of the determinant Δ (δ). The subjects of operation are the three independent functions ξ, η, ζ. These are such that if I (δ) be any of the second minors of the constituents of the three omitted equations I (δ) $\xi = 0$, I (δ) $\eta = 0$, I (δ) $\zeta = 0$. If these contain the root $\delta = m$, γ times, each of the three ξ, η, ζ will be expressed by a series of the form $\qquad (K_0 + K_1 t + \ldots + K_{\gamma-1} t^{\gamma-1}) e^{mt}$, but with independent constants.

276. **The number of constants.** Each of the sets of values of x, y, &c. given in Arts. (272), (273), and (275) is, of course, a *solution*. The complete solution is really the sum of these partial solutions, provided it has the proper number of constants. We appear, however, to have too many constants. We must therefore examine these, and determine what terms are absolutely zero and what terms are repeated in the several partial solutions.

We begin with the solution derived from the type V, Art. (272), by the help of the first minors. Since the first minors have β roots each equal to m, the first β terms of each of the expressions for x, y, &c. are easily seen to be zero. Consider the solution derived from any term L_k, where k lies between $\beta - 1$ and 2β. In the case of the variables x and y they are expressions of the form

$$(A_0 + A_1 t + \ldots + A_{k-\beta} t^{k-\beta}) e^{mt}.$$

All these are evidently included amongst the terms derived from ξ, η by the help of the second minors. The corresponding terms in z, u, &c. must be related to the terms in x, y by the formula given in Art. (273), and are therefore also included in the series derived from ξ, η. Lastly, consider the solution derived from the terms from $L_{2\beta}$ to $L_{\alpha-1}$. They include powers of t from t^β to $t^{\alpha-1-\beta}$. These $\alpha - 2\beta$ terms are not included in the terms derived from ξ and η, and they supply $\alpha - 2\beta$ arbitrary constants.

Secondly, we turn our attention to the solution derived from the double type ξ, η by the help of the second minors (Arts. 273 and 274). Each of these second minors has γ roots each equal to m; hence, by the same reasoning as before, the first γ terms of the series for x and y are zero, and the highest power of t is $\beta - 1 - \gamma$ instead of $\beta - 1$. In consequence of this, the terms of the series derived from the single type V, and not included in those derived from the double type ξ, η, now extend their powers of t from $t^{\beta-\gamma}$ to $t^{\alpha-1-\beta}$. There are therefore $\alpha - 2\beta + \gamma$ such terms instead of $\alpha - 2\beta$.

The same reasoning applies to all the other partial solutions derived from the triple and higher types. *We therefore conclude that the partial solution derived from a single type by operating with the first minors of the first row of the fundamental determinant supplies $\alpha - 2\beta + \gamma$ terms not included in the solutions which follow. These supply as many arbitrary constants. The partial solution derived from a double type by operating with the second minors of the two first rows of the fundamental determinant supplies $\beta - 2\gamma + \delta$ terms not included in the solutions which follow.*

These supply twice as many constants. The partial solution derived from a triple type by operating with the third minors of the three first rows supplies $\gamma - 2\delta + \epsilon$ *terms and thrice as many constants, and so on.*

Thus suppose (for example) the fourth minors are not all zero; the number of constants supplied by each of the several partial solutions is indicated by the terms of the series $(a - 2\beta + \gamma) + 2 (\beta - 2\gamma + \delta) + 3 (\gamma - 2\delta) + 4\delta.$

If none of the terms of this series are negative, we have obtained a series of partial solutions containing the proper number of constants. This point we now proceed to discuss.

277. If a determinant contain the root just a times, if the first minors of the two first constituents of the two first rows contain the root just β times, if the second minor of these four constituents contain the root just γ times, then $a - 2\beta + \gamma$ is positive.

To prove this, let Δ be the determinant, I_1, I_2, J_1, J_2 the four first minors, Δ_2 the second minor. Then we know that $\Delta\Delta_2 = I_1 J_2 - I_2 J_1$. The left-hand side contains the root just $a + \gamma$ times, the right-hand side contains the root at least 2β times. Hence $a + \gamma - 2\beta$ is positive.

In the same way we may show, on similar suppositions, that $\beta - 2\gamma + \delta$ is positive, and so on.

278. Example. Solve the differential equations

$$\left.\begin{array}{c} (\delta - 1)^2 (\delta + 1)\, x - (\delta - 1)\, (\delta - 2)\, y + (\delta - 1)\, z = 0 \\ 3\, (\delta - 1)^2 x - (\delta - 1)\, (\delta - 3)\, y + 2\, (\delta - 1)\, z = 0 \\ (\delta - 1)^2 x + (\delta - 1)\, y + (\delta - 1)\, z = 0 \end{array}\right\}.$$

The fundamental determinant (Art. 262) is $\Delta (\delta) = - (\delta - 1)^6$. This determinant (Art. 271) has six equal roots ($a = 6$), every first minor has the root three times ($\beta = 3$), and every second minor has the root once ($\gamma = 1$). The part of the solution depending on a single type (Art. 276) will supply $a - 2\beta + \gamma$ (i.e. one) constants. These accompany the highest powers of t which occur in the type, one constant for each power (Art. 272). The part of the solution depending on a double type will supply $2 (\beta - 2\gamma)$ (i.e. two) constants. These accompany the highest powers of t which occur in *this* type, two constants to each power. The part of the solution depending on a triple type will supply 3γ (i.e. three) constants which again accompany the highest powers of t, three constants to each power. To obtain the full number of constants it is necessary in this example to retain only the one highest power of t which occurs in each type.

The single type is $\xi = (\&\text{c.} + A t^5)\, e^t$ by Art. 264. Taking the minors of the first row of $\Delta (\delta)$ we have by Art. 262 $x = - (\delta - 1)^3 \xi$, $y = - (\delta - 1)^3 \xi$, $z = \delta (\delta - 1)^3 \xi$.

To find the part of the solution which depends on a double type we reject the first equation (Art. 273). Putting $x = 0$ we find $y = (\delta - 1)\, \xi$, $z = - (\delta - 1)\, \xi$ where $(\delta - 1)^3 \xi = 0$. Putting $y = 0$ we find $x = (\delta - 1)\, \eta$, $z = - (\delta - 1)^2 \eta$ where $(\delta - 1)^3 \eta = 0$. The double type is therefore $\xi = (\&\text{c.} + Bt^2)\, e^t$, $\eta = (\&\text{c.} + Ct^2)\, e^t$. The values of the co-ordinates are $x = (\delta - 1)\, \eta$, $y = (\delta - 1)\, \xi$, $z = - (\delta - 1)\, \xi - (\delta - 1)^2 \eta$.

To find the part of the solution which depends on a triple type we reject the two first equations (Art. 275). The three partial solutions are then *first*, $x = 0$, $y = 0$, $z = De^t$; *secondly*, $x = 0$, $y = Ee^t$, $z = 0$, *thirdly*, $x = Fe^t$, $y = 0$, $z = 0$. The sum of these is the solution derived from a triple type.

Adding up the solutions which are derived from all the different types and simplifying the constants we have

$$x = (F + Ct + At^2)\, e^t, \qquad y = (E + Bt + At^2)\, e^t, \qquad z = \{D - Bt - A\, (t^2 + 2t)\}\, e^t.$$

279. Conversely, *suppose it is given that we have such a solution as that described in Art. 276, let us enquire what minors must be zero.*

Let it be given that the solution contains terms depending on a triple type containing $(\gamma - 1)$ powers of t accompanied by independent constants in some three co-ordinates. Putting any two of these co-ordinates equal to zero the differential equations are satisfied by a solution depending on a single type. Thus we have n equations containing $n - 2$ co-ordinates all satisfied by values of the co-ordinates which contain powers of t up to the $(\gamma - 1)$th. This shows that all the second minors which can be formed from these equations must be zero and each of these minors must contain the root γ times.

From this we infer by Art. 270 that every first minor must contain the root $\gamma + 1$ times. But let us suppose that the given solution contains also terms derived from a double type which have powers of t extending up to the $(\beta - \gamma - 1)$th with independent constants in some two of the co-ordinates. Reasoning in the same way as before, we see that every first minor must have the root $(\beta - \gamma - 1)$ times. These must be in addition to the $\gamma + 1$ roots already counted, because we may regard the given solutions derived from the double and triple types as solutions which depend on unequal roots and then make these roots become equal in the limit. It follows therefore that every first minor has the root β times.

We now infer by Art. 270 that the determinant (4) of Art. 261 must have the root $\beta - 1$ times. But if the given solution also contains terms derived from a single type with powers of t extending to the $(a - \beta - 1)$th, we deduce by the preceding reasoning that the determinant (4) must have the root a times.

280. We may notice as a corollary of this theory that the solution cannot contain terms in which the high powers of t depend on a larger type than the low powers of t. For example, if the term $t^n e^{mt}$ occur accompanied by k independent constants, this term must be part of a solution derived from a kth type. It follows that all the lower powers of t which multiply the same exponential will be part of the same type and must be accompanied by at least k independent constants.

281. **Condition that all powers of t are absent.** *In some dynamical problems it is well known that, though the fundamental determinant has a equal roots, yet there are no terms in the solution with powers of* t. *We may now determine the conditions that this may occur.*

We see by Art. 272 that, unless every first minor has the root $a - 1$ times at least, a solution can be deduced from the first minors which has some power of t greater than zero in the coefficient. Again, unless every second minor has the root $a - 2$ times at least, a solution can be deduced from the second minors with some power of t in the coefficient. *On the whole, we infer that when a equal roots occur in the determinant, and the terms in the solution with* t *as a factor are to be absent, it is necessary as well as sufficient that all the first, second, &c. minors up to the $(a - 1)$th should be zero.*

282. **Dynamical Meaning of the Types.** We shall now consider how the three different types of solution given in Art. 264 indicate different kinds of motion. Let us begin with a real root. In this case every co-ordinate has a term of the form $M e^{mt}$. *If m be positive* this term will become greater as time goes on, and the

system will therefore depart widely from its undisturbed state, and our equations will represent only the manner in which the system begins its travels. *If m be negative* this term will gradually dwindle away and the motion will finally depend on the other terms in the solution.

Similar remarks apply whenever we have a real exponential whether multiplied by a trigonometrical function or not. We may therefore state as a general principle, subject to some reservations in the case of equal roots which will be presently mentioned, that *the necessary and sufficient conditions of stability are that the real roots and the real parts of the imaginary roots should be all negative or zero.* A simple rule to determine whether this is the case or not will be given in another section of this chapter.

283. Effect of equal roots on stability. When there are equal roots in the determinantal equation we have seen that the solution in general has terms which contain powers of t as a factor. The important question for us to determine is the effect of these terms on the stability of the system. If m be positive the presence of a term $Mt^q e^{mt}$ will of course make the system unstable. But if m be negative, this term can never be numerically greater than $M\left(\dfrac{q}{em}\right)^q$. If m be very small the initial increase of the term may make the values of x, y, &c. become large, and the motion cannot be regarded as a small oscillation. But if the system be not so disturbed that $M\left(\dfrac{q}{em}\right)^q$ is large, the term will ultimately disappear and the motion may be regarded as stable. If m be wholly imaginary and equal to $n\sqrt{-1}$, this term will take the form $t^q \sin nt$ and will of course cause the system to be unstable.

Thus equal roots do not disturb the stability if their real parts are negative, but do render the system unstable if their real parts are zero or positive.

284. It is clear from this that the whole character of the motion depends on the nature of the roots of the determinantal equation $\Delta(\delta) = 0$. If we can solve this equation and find the roots, we of course know immediately the nature of the motion. But if this cannot be done we must have recourse to the theory of equations to determine whether the roots are real or imaginary, and whether any roots are equal or not. The theorems of Fourier and Sturm will be of use in the equations of the higher orders, but in many dynamical problems we have only to deal with two co-ordinates, and we have therefore to examine the roots of the biquadratic in Art. 260.

Rules by which the analysis of a biquadratic is made to depend on the solution of a cubic are given in most treatises on the theory of equations; but as this form is not convenient in practice, a short analysis will be given here for reference. The criteria of the nature of the roots of the biquadratic are very conveniently summed up in Art. 68 of the *Theory of Equations* by Burnside and Panton, 1886.

285.   **Analysis of a biquadratic.** Let the biquadratic be

$$ax^4 + 4bx^3 + 6cx^2 + 4dx + e = 0,$$

so that the invariants are $I = ae - 4bd + 3c^2$, and $J = ace + 2bcd - ad^2 - eb^2 - c^3$. This last may also be written as a determinant. It will generally be found convenient to clear the equation of the second term. Let the equation so transformed be

$$a\xi^4 - 2aH\xi^2 + aG\xi - aF = 0,$$

where $a^2H = 3\,(b^2 - ac)$ and $a^3G = 4\,(2b^3 - 3abc + a^2d)$. By using the invariants or by actual transformation it is easy to see that

$$I = \tfrac{1}{3}a^2H - a^2F \text{ and } J = \tfrac{4}{27}a^3H^3 - \tfrac{1}{16}a^3G^2 - \tfrac{1}{3}aIH.$$

Let Δ be the discriminant, i.e. $\Delta = I^3 - 27J^2$, then it is proved in all books on the theory of equations that if Δ is negative and not zero, the biquadratic has two real and two imaginary roots. If Δ is positive and not zero the roots are either all real or all imaginary.

Usually we can distinguish between these two cases by ascertaining if the biquadratic has or has not a real root. Thus if a and e have opposite signs, one root is real and therefore all the roots are real. In any case we can use the following criterion. Having cleared the given biquadratic of the second term we may write the resulting equation in the form $(\xi^2 - H)^2 + G\xi = K$.

If S_n be the arithmetic mean of the nth powers of the roots, we have by Newton's theorem on the sums of powers, $S_1 = 0$, $S_2 = H$, $4S_3 = -3G$ and $K = S_4 - S_2^2$. If all the roots are real we have S_2 positive and by a known theorem in "inequalities" S_4 is greater than S_2^2. Hence H and K are both positive. If all the roots are imaginary, let them be $r \pm p\sqrt{-1}$ and $-r \pm q\sqrt{-1}$. Then

$$H = S_2 = r^2 - \tfrac{1}{2}(p^2 + q^2),$$
$$K = S_4 - S_2^2 = \tfrac{1}{4}(p^2 - q^2)^2 - 2r^2(p^2 + q^2).$$

If H is positive or zero we see that K is negative. The criterion may therefore be stated thus. *If H and K are both positive the four roots are real. If either is negative or zero the four roots are imaginary.*

If the discriminant Δ is zero but I and J not zero, it is known that the biquadratic has two roots equal. If two of the roots are real and equal and the other two imaginary we see (by putting $q = 0$) that if H is positive or zero, K must be negative. The criterion therefore is, *If H and K are both positive all the roots are real, if H or K is negative or zero, two roots are real and two are imaginary.* If G is zero, there are then two pairs of equal roots. In this case K is zero and these roots are all real if H is positive, all imaginary if H is negative.

Lastly let *the discriminant Δ be zero and also both I and J zero.* The biquadratic has three roots equal and therefore all the roots are equal. If H is also zero the four roots are all equal and real.

Ex.   If the discriminant of a biquadratic be positive, clear the equation of the term containing the third power in the usual manner, and then arbitrarily erase the term containing the first power. If both the roots of the quadratic thus formed be real and the sum of the roots be positive, then all the four roots of the biquadratic are real. If either contingency fail the four roots are imaginary.

Conditions of Stability.

286. It has been shown that the determination of the oscillation of a system can be reduced to the solution of a certain
determinantal equation, which has been represented in Art. 262,
by $\Delta = f(\delta) = 0$. In many cases it is impracticable to solve this
equation and therefore the motion cannot be properly found. If
however we only wish to ascertain whether the position of equilibrium or the steady motion about which the system is in oscillation
is stable or unstable we may proceed without solving the equation.

It is clear from Art. 282 that the conditions of stability are
that the real roots and the real parts of the imaginary roots should
all be negative. It is now proposed to investigate a method to
decide whether the roots are of this character or not.

287. Taking first the case of a biquadratic; let the equation
to be considered be

$$f(z) = az^4 + bz^3 + cz^2 + dz + e = 0,$$

where we have written z for δ. Let us form that symmetrical
function of the roots which is the product of the sums of the roots
taken two and two. If this be called X/a^3, we find*

$$X = bcd - ad^2 - eb^2 = \tfrac{1}{2} \begin{vmatrix} 2a & b & c \\ b & 0 & d \\ c & d & 2e \end{vmatrix}.$$

It will be convenient to consider first the case in which X is
finite. Suppose we know the roots to be imaginary, say $\alpha \pm p\sqrt{-1}$,
and $\beta \pm q\sqrt{-1}$. Then

$$X/a^3 = 4\alpha\beta \{(\alpha + \beta)^2 + (p + q)^2\} \{(\alpha + \beta)^2 + (p - q)^2\}.$$

---

* The value of X may be found in several ways more or less elementary. If
we substitute $z = E \pm Z$ in the given biquadratic and equate to zero the even and
odd powers of Z, we have

$$aZ^4 + (6aE^2 + 3bE + c) Z^2 + aE^4 + bE^3 + cE^2 + dE + e = 0 \Big\}$$
$$(4aE + b) Z^3 + (4aE^3 + 3bE^2 + 2cE + d) Z = 0 \Big\} .$$

Rejecting the root $Z = 0$ and eliminating Z we have

$$64a^3E^6 + \ldots\ldots + bcd - ad^2 - eb^2 = 0,$$

where only the first and last terms of the equation are retained, the others not
being required for our present purpose. Since $z = E \pm Z$ it is clear that each value
of E is the arithmetic mean of two values of z. We have an equation of the sixth
degree to find E because there are six ways of combining the four roots of the
biquadratic two and two. The product of the roots of the equation in E may be
deduced in the usual manner from the first and last terms, and thence the value
of X is seen to be that given in the text.

If we eliminated E we should obtain an equation in Z whose roots are the
arithmetic means of the differences of the roots of the given equation taken two
and two. If we put $4Z^2 = \zeta$, we obtain by an easy process the equation whose roots
are the squares of the differences of the roots of the given equation $f(z=0)$.

Thus, $\alpha\beta$ always takes the sign of X/a, and $\alpha + \beta$ always takes the sign of $-b/a$. The signs of both α and β can therefore be determined; and if a, b, X have the same sign, the real parts of the roots are all negative.

Suppose, next, that two of the roots are real and two imaginary. Writing $q'\sqrt{-1}$ for q, so that the roots are $\alpha \pm p\sqrt{-1}$ and $\beta \pm q'$, we find

$$X/a^3 = 4\alpha\beta\left\{[(\alpha + \beta)^2 + p^2 - q'^2]^2 + 4p^2 q'^2\right\}.$$

Just as before, $\alpha\beta$ takes the sign of X/a, and $\alpha + \beta$ takes the sign of $-b/a$. Also, $\beta^2 - q'^2$ takes the sign of the last term e/a of the biquadratic. This determines whether β is numerically greater or less than q'. If, then, a, b, e, and X have the same sign, the real roots and the real parts of the imaginary roots are all negative.

Lastly, suppose the roots to be all real. Then, if all the coefficients are positive, we know, by Descartes' rule, that the roots must be all negative, and the coefficients cannot be all positive unless all the roots are negative. In this case, since X is the product of the sums of the roots taken two and two, it is clear that X/a will be positive.

Whatever the nature of the roots may be, yet if the real roots and the real parts of the imaginary roots are negative, the biquadratic must be the product of quadratic factors all whose terms are positive. It is therefore necessary for stability that every coefficient of the biquadratic should have the same sign. It is also clear that no coefficient of the equation can be zero unless either some real root is zero or two of the imaginary roots are equal and opposite.

Summing up the several results which have just been proved, we conclude that *if* X *and* e *are finite, the necessary and sufficient conditions that the real roots and the real parts of the imaginary roots should be negative are that every coefficient of the biquadratic and also* X *should have the same sign.*

288.   The case in which $X = 0$ does not present any difficulty. It follows from the definition of X, that if X vanishes two of the roots must be equal with opposite signs, and conversely if two roots are equal with opposite signs X must vanish. Writing $-z$ for z in the biquadratic and subtracting the result thus obtained from the original equation we find $bz^3 + dz = 0$. The equal and opposite roots are therefore given by $z = \pm\sqrt{-d/b}$. If b and d have opposite signs these roots are real, one being positive and one negative. If b and d have the same sign, they are a pair of imaginary roots with the real parts zero.

The sum of the other two roots is equal to $-b/a$ and their product is be/ad. We therefore conclude that *if* X $= 0$, *the real roots and the real parts of the imaginary roots will be negative or zero, if every coefficient of the biquadratic is finite and has the same sign.*

289. If either a or e vanishes, the biquadratic reduces to a cubic, see note to Art. 105. Putting e zero, we have

$$X/a^3d = bc - ad.$$

If the coefficients have all the same sign it is easy to see that it is necessary for stability that $bc - ad$ should be positive or zero.

If a and e be not zero and one of the two b, d vanish, the other must vanish also, for otherwise X could not have the same sign as a. In this case X vanishes, and the biquadratic reduces to the quadratic $\qquad az^4 + cz^2 + e = 0.$

As this equation admits of an easy solution, no difficulty can arise in practice from this case. It is necessary for stability that the roots of the quadratic should be real and negative. The conditions for this are, *firstly* the coefficients a, c, e, must all have the same sign, *secondly* that $c^2 > 4ae$.

290. **Equation of the nth degree.** When the degree of the equation is higher than a biquadratic the conditions of stability become more numerous. A very simple rule will now be proved by which these conditions can be calculated as quickly as they can be written down. Besides this we propose to give an extension of this rule by which *we may determine how many roots there are, real or imaginary, which have their real parts positive.* If there be no such roots the conditions of stability are supposed to be satisfied. The number of roots with their real parts equal to zero is also found.

291. To discover this rule we have recourse to a theorem of Cauchy. Let $z = x + y\sqrt{-1}$ be any root, and let us regard x and y as co-ordinates of a point referred to rectangular axes. Substitute for z and let

$$f(z) = P + Q\sqrt{-1}.$$

Let any point whose co-ordinates are such that P and Q both vanish be called a radical point. Describe any contour, and let a point move round this contour in the positive direction, and notice how often P/Q passes through the value zero and changes its sign. Suppose it changes a times from $+$ to $-$ and β times from $-$ to $+$. Then Cauchy asserts that the number of radical points within the contour is $\frac{1}{2}(a - \beta)$. It is however necessary that no radical point should lie on the contour.

Let us choose as our contour the infinite semicircle which bounds space on the positive side of the axis of y. Let us first travel from $y = -\infty$ to $y = +\infty$ along the circumference. If $\qquad f(z) = p_0 z^n + p_1 z^{n-1} + \dots + p_n \dots$ (1),

we have, changing to polar co-ordinates,

$$f(z) = p_0 r^n (\cos n\theta + \sin n\theta \sqrt{-1}) + \dots$$

Hence $\qquad \begin{aligned} P &= p_0 r^n \cos n\theta + p_1 r^{n-1} \cos(n-1)\theta + \dots \\ Q &= p_0 r^n \sin n\theta + p_1 r^{n-1} \sin(n-1)\theta + \dots \end{aligned} \Big\}$ .... (2).

In the limit since r is infinite $P/Q = \cot n\theta$.

$$P/Q \text{ vanishes when } \theta = \pm \frac{1}{n}\frac{\pi}{2}, \quad \pm \frac{3}{n}\frac{\pi}{2}, \quad \pm \frac{5}{n}\frac{\pi}{2} \dots \dots \dots \dots (A).$$

$$P/Q \text{ is infinite when } \theta = 0, \qquad \pm \frac{2}{n}\frac{\pi}{2}, \quad \pm \frac{4}{n}\frac{\pi}{2} \dots \dots \dots \dots (B).$$

The values of θ in series (B), it will be noticed, *separate* those in series (A).

When θ is small and very little greater than zero, P/Q is positive and therefore changes sign from $+$ to $-$ at every one of the values of θ in series (A). If therefore n be even there will be n changes of sign.

If n be odd there will be $n-1$ changes of sign excluding $\theta = \pm \frac{1}{2}\pi$, in this case P/Q is positive when θ is a little less than $\frac{1}{2}\pi$ and negative when θ is a little greater than $\frac{1}{2}\pi$, but this result will not be wanted in the sequel.

Let us now travel along the axis of y, still in the positive direction round the contour, viz. from $y = +\infty$ to $y = -\infty$. Substituting $z = x + y\sqrt{-1}$ in (1) and remembering that $x = 0$ along the axis of y, we have, *when n is even,*

$$\left. \begin{aligned} P &= p_n - p_{n-2}y^2 + p_{n-4}y^4 - \dots + (-1)^{\frac{1}{2}n}p_0 y^n \\ Q &= p_{n-1}y - p_{n-3}y^3 + \dots \qquad - (-1)^{\frac{1}{2}n}p_1 y^{n-1} \end{aligned} \right\} \dots \dots \dots \dots (3).$$

$$\therefore -\frac{P}{Q} = \frac{p_0 y^n - p_3 y^{n-2} + \dots}{p_1 y^{n-1} - p_3 y^{n-3} + \dots} \dots \dots \dots \dots (4).$$

Let e be the excess of the number of changes of sign from $-$ to $+$ over that from $+$ to $-$ in *this* expression as we travel from $y = +\infty$ to $y = -\infty$, then by Cauchy's theorem the whole number of radical points on the positive side of the axis of y is $\frac{1}{2}(n+e)$. This of course expresses the number of roots which have their real parts positive.

292. To count these changes of sign we use Sturm's theorem. Taking

$$\left. \begin{aligned} f_1(y) &= p_0 y^n - p_2 y^{n-2} + \dots \\ f_2(y) &= p_1 y^{n-1} - p_3 y^{n-3} + \dots \end{aligned} \right\} \dots \dots \dots \dots (5)$$

we perform the process of finding the greatest common measure of $f_1(y)$ and $f_2(y)$, changing the sign of each remainder as it is obtained. Let the series of modified remainders thus obtained be $f_3(y)$, $f_4(y)$, &c. Then, as in Sturm's theorem, we may show that when any one of these functions vanishes the two on each side have opposite signs. It also follows that no two successive functions can vanish unless $f_1(y)$ and $f_2(y)$ have a common factor. This exception will be considered presently.

Taking then the functions $f_1(y)$, $f_2(y)$, &c., using them, as in Sturm's theorem, we see that no change of sign can be lost or gained except at *one* end of the series. Now the last is a constant and cannot change sign, hence changes of sign can be gained or lost only by the vanishing of the function $f_1(y)$ at the beginning of the series.

Consider now the beginning of the series of functions $f_1(y)$, $f_2(y)$, &c., and using them in Sturm's manner let y proceed from $+\infty$ to $-\infty$. We see that a change of sign is lost when the first two change from unlike to like signs, i.e. when the ratio of $f_1(y)$ to $f_2(y)$ changes from $-$ to $+$. In the same way a change of sign is gained when the ratio changes from $+$ to $-$. Hence e is equal to the number of variations or changes of sign lost in the series as we travel from $y = +\infty$ to $y = -\infty$.

293. When $y = \pm\infty$ we need only consider the coefficients of the highest powers in the series of functions $f_1(y)$, $f_2(y)$, &c. Let these coefficients when y is positive be called p_0, p_1, q_3, q_4,   &c.

When y is negative the signs, since n is even, will be indicated by

$$p_0, \quad -p_1, \quad q_3, \quad -q_4, \quad \&c.$$

Then we have just proved that e is equal to the number of variations or changes of sign lost as we proceed from the first series to the second.

294. If every term of the series $p_0, p_1, q_3,$ &c. have the same sign, it is evident that n changes of sign will be gained and therefore $e = -n$; and e cannot $= -n$ unless all these terms have the same sign. In this case there will be no radical point on the positive side of the axis of y. We therefore infer the following theorem. *The necessary and sufficient conditions that the real part of every root of the equation $f(z) = 0$ should be negative are that all the coefficients of the highest powers in the series $f_1(y), f_2(y),$ &c. should have the same sign**.

295. Suppose next that these coefficients do not all have the same sign. The degree of the equation being n, there are $n+1$ functions in the series $f_1(y), f_2(y),$ &c., and therefore on the whole there are n variations and permanencies. Let there be k variations and $n-k$ permanencies of sign. Now every permanency in the series $y = +\infty$ changes into a variation in the series $y = -\infty$, and every variation into a permanency. It follows that there will be $n-k$ variations and k permanencies

---

* As these are the conditions of stability in dynamics (Art. 282) it is worth while to give a short summary of the argument as adapted to this special case. Putting $z = x + yi$, let $f(z) = P + Qi$. Regarding P and Q as functions of x and y, let us trace the curves $P = 0$, $Q = 0$; it is evident that each intersection corresponds to a root of $f(z) = 0$. The polar forms of these curves are given in equations (2) of Art. 291. The P curve has evidently n asymptotes whose directions are given by $\cos n\theta = 0$, the Q curve has also n asymptotes but these are given by $\sin n\theta = 0$.

We shall first show that the conditions given in Art. 294 are necessary, if there is to be no radical point on the positive side of the axis of y. Draw a circle of infinite radius, and let it cut the asymptotes of the P curve in $P_1, P_2 \ldots P_n$ and the asymptotes of the Q curve in $Q_1, Q_2 \ldots Q_n$. These points alternate with each other. Taking only those points which lie on the positive side of the axis of y, the P and Q curves may be said to begin at these infinitely distant points and passing towards the negative side of the axis of y are not to intersect each other on the positive side of that axis. The branches of the two curves must therefore remain alternate with each other throughout the space on the positive side of the axis of y. Their points of intersection with the axis of y must also be alternate. If we put $x = 0$, in the equations $P = 0$, $Q = 0$ we have $f_1(y) = 0$, $f_2(y) = 0$ (Art. 292), and these equations must therefore be such that their roots are real and that the roots of each must separate or lie between the roots of the other. It is then pointed out in Art. 292, that the conditions that the roots of one equation should separate those of the other may practically be found by Sturm's theorem.

Conversely, we may deduce from Cauchy's theorem that the conditions given in Art. 292 are sufficient. For suppose the intersections of the P and Q curves with the axis of y are known to be alternate. It is evident that as we travel round the contour formed by the infinite semicircle which bounds space on the positive side of the axis of y, we pass over each P branch and each Q branch twice, crossing each in one direction on the semicircle and in the opposite direction on the axis of y. In Art. 293 the consequent changes of sign of P/Q are counted and it is shown that the changes of sign balance each other. It follows by Cauchy's theorem that there is no radical point within the contour.

in this second series. Hence the number e of variations lost in proceeding from the first to the second series is $2k - n$. But the number of radical points on the positive side of the axis of y has been proved to be $=\frac{1}{2}(n+e)$; substituting for e, this becomes equal to k. We therefore infer the following theorem. *If we form the series of coefficients of the highest powers of the functions $f_1(y)$, $f_2(y)$, &c., every variation of sign implies one radical point within the positive contour, and therefore one root with its real part positive.*

296. We require some rule to construct the series of coefficients with facility. If we perform the process of Greatest Common Measure on the functions $f_1(y)$, $f_2(y)$ changing the signs of the remainders, we find that the first three functions are

$$f_1(y) = p_0 y^n - p_2 y^{n-2} + p_4 y^{n-4} - \&c.,$$

$$f_2(y) = p_1 y^{n-1} - p_3 y^{n-3} + p_5 y^{n-5} - \&c.,$$

$$f_3(y) = \frac{p_1 p_2 - p_0 p_3}{p_1} y^{n-2} - \frac{p_1 p_4 - p_0 p_5}{p_1} y^{n-4} + \&c.$$

Thus the *coefficients of $f_3(y)$ may be obtained from those of $f_1(y)$ and $f_2(y)$ by a simple cross-multiplication, and may therefore be written down by inspection.* The coefficients of $f_4(y)$ may be derived from those of $f_2(y)$ and $f_3(y)$ by a similar cross-multiplication and so on. These successive functions may be called the *subsidiary functions.*

297. **First form of the Rule.** Summing up the preceding arguments, we have the following rule. The equation being

$$f(z) = p_0 z^n + p_1 z^{n-1} + p_2 z^{n-2} + \ldots$$

arrange the coefficients in two rows thus

$$
\begin{array}{cccc}
p_0, & p_2, & p_4, & \&c. \\
p_1, & p_3, & p_5, & \&c.
\end{array}
$$

Form a new row by cross-multiplication in the following manner

$$\frac{p_1 p_2 - p_0 p_3}{p_1}, \quad \frac{p_1 p_4 - p_0 p_5}{p_1}, \quad \&c.$$

Form a fourth row by operating on these two last rows by a similar cross-multiplication. Proceeding thus the number of terms in each row will gradually decrease, and we stop only when no term is left. *Then in order that there may be no roots whose real parts are positive it is necessary and sufficient that the terms in the first column should be all of one sign. If they be not all of one sign, the number of variations of sign is equal to the number of roots with their real parts positive.*

The terms which constitute the first column may be called the *test functions.*

As in forming these rows we only want their signs, we may multiply or divide any one by any positive quantity which may be convenient. We may thus often avoid complicated fractions.

298. **Equations of an odd degree.** In order to simplify the argument we have supposed the degree of the equation to be even. If n be odd, let as before

$$f(z) = p_0 z^n + p_1 z^{n-1} + \ldots + p_n.$$

We may regard this equation as the limit of

$$p_0 z^{n+1} + p_1 z^n + \ldots + p_n z + p_n h = 0.$$

If h be positive and indefinitely small the additional root of this equation is real and negative, and ultimately equal to $-h$. Those roots also of the two equations which lie within the positive contour are ultimately the same.

Since $n+1$ is even we may apply to this equation the preceding rule. The two first rows are

$$p_0, \quad p_2 \text{ &c.,} \quad p_{n-1}, \quad p_n h,$$
$$p_1, \quad p_3 \text{ &c.,} \quad p_n.$$

We easily see by calculating a few rows that none of the coefficients in the subsequent rows contain h as a factor except the extreme coefficients on the right-hand side. Hence in the general case all the test functions, except the two last, remain finite when h is put equal to zero ; and therefore have the same sign as if the rows had been calculated before the addition of the final term $p_n h$. The last two coefficients in the first column, when only the principal power of h is retained, are p_n and $p_n h$. But since h is positive there can be no variation of sign in this sequence. We may therefore omit this final term $p_n h$ altogether as giving nothing to the number of variations of sign. The result is that the *rule to calculate the number of roots whose real parts are positive is the same whether the degree of the equation is even or odd*.

299. Simplification of the rule when tests of stability only are required. In a dynamical point of view it is generally more important to determine the conditions of stability than to count how many times those conditions are broken. *If we only want to discover these conditions we may in forming the successive subsidiary functions by the rule of cross-multiplication omit the divisor at every stage provided p_0 be made positive to begin with*, for this divisor being one of the test functions must in every case be positive.

Supposing the conditions of stability to be satisfied we see by reference to Art. 292 that the proper number of variations cannot be lost at the beginning of the series unless *the roots of the equation $f_1 (y)$ are all real and the roots of $f_2 (y)$ separate the roots of $f_1 (y)$ and therefore are all real also*. Then because when a subsidiary function vanishes the two on each side have opposite signs it follows that the *roots of $f_3 (y)$ are real and separate those of $f_2 (y)$ and so on*.

Supposing the roots of the equation $f(z) = 0$ to have their real parts negative, the real quadratic factors made up of those roots must have their terms positive. Thus every term of the equation $f(z) = 0$ must be positive. It follows from the definition of the functions $f_1 (y)$ and $f_2 (y)$ in Art. 292 that the signs of their terms are alternately positive and negative, and since their roots are real every one of those roots is positive. Hence all the subsequent auxiliary functions $f_3 (y), f_4 (y)$, &c. have their roots real and positive. The signs therefore of all their terms are alternately positive and negative, and by Art. 297 the coefficient of the highest power is in every case positive.

In this way we are led to an extension of the theorem in Art. 297. Supposing p_0 to have been made positive, we see by the preceding reasoning that though it is necessary and sufficient that all the terms in the first column should be positive, yet *it is also true that the terms in every column must be positive. Hence as we perform the process indicated in that article we may stop as soon as we find any negative term*, and conclude at once that $f(z)$ has some roots with their real parts negative.

300. Ex. 1. Express the condition that the real roots and the real parts of the imaginary roots of the cubic $z^3 + p_1 z^2 + p_2 z + p_3 = 0$ should be all negative.
By Art. 296
$$f_1 (y) = y^3 - p_2 y,$$
$$f_2 (y) = p_1 y^2 - p_3.$$
Using the method of cross-multiplication given in Art. 297 and omitting the divisors as shown in Art. 299 we have
$$f_3 (y) = (p_1 p_2 - p_3) y, \qquad f_4 (y) = (p_1 p_2 - p_3) p_3.$$
The necessary conditions are that p_1, $p_1 p_2 - p_3$, and p_3 should be all positive.

We have retained the powers of y in order to separate the terms, and also the negative signs in the second column, but both these are unnecessary and in accordance with Art. 297 might have been omitted. In both this and the next example all the numerical calculations are shown.

Ex. 2. Express the corresponding conditions for the biquadratic

$$z^4 + p_1 z^3 + p_2 z^2 + p_3 z + p_4 = 0,$$

$$f_1(y) = y^4 \qquad\qquad\qquad -p_2 y^2 + p_4,$$
$$f_2(y) = p_1 y^3 \qquad\qquad\qquad -p_3 y,$$
$$f_3(y) = (p_1 p_2 - p_3) y^2 \qquad\qquad -p_1 p_4,$$
$$f_4(y) = \{(p_1 p_2 - p_3) p_3 - p_1{}^2 p_4\} y,$$
$$f_5(y) = \{(p_1 p_2 - p_3) p_3 - p_1{}^2 p_4\} p_1 p_4.$$

The conditions are that p_1, $p_1 p_2 - p_3$, $(p_1 p_2 - p_3) p_3 - p_1{}^2 p_4$ and p_4 should be all positive. These are evidently equivalent to the conditions given in Art. 287.

301. **Second Form of the rule.** When the degree of the equation is very considerable there is some labour in the application of the rule given in Art. 297. The objection is that we only want the terms in the first column and to obtain these we have to write down all the other columns. *We shall now investigate a method of obtaining each term in the first column from the one above it without the necessity of writing down any expression except the one required.*

We notice that each function is obtained from the one above it by the same process. Now the three first functions are written down in Art. 297. The first and second lines will be changed into the second and third by writing for

$$p_0, \qquad p_1, \qquad p_2, \qquad p_3, \ \&c.$$

the values $\left. p_1, \ p_2 - \dfrac{p_0 p_3}{p_1}, \ p_3, \ p_4 - \dfrac{p_0 p_5}{p_1}, \ \&c. \right\} \quad \dots\dots\dots\dots\dots(A).$

We therefore infer the following rule. *To form the test functions of Art.* 297 *we write down the first, viz.* p_0; *the second may be obtained from the first and the third from the second and so on by changing each letter as indicated in the schedule* A *just above.*

In these changes we always increase the suffix, hence we may write zero for any letter as soon as its suffix becomes greater than the degree of the equation.

We thus form the test functions, each from the preceding, and we stop as soon as we have obtained the proper number, viz. (counting p_0 as one test function) one more than the degree of the equation.

302. Example. *Express the test functions for the quintic*

$$f(z) = p_0 z^5 + p_1 z^4 + p_2 z^3 + p_3 z^2 + p_4 z + p_5 = 0.$$

Here we notice that p_6, p_7, &c. are all zero, so that any term which has the factor p_5 will become zero in the next test function. Following the rule the six test functions

are $p_0, \qquad p_1, \qquad p_2 - \dfrac{p_0 p_3}{p_1},$

$$p_3 - \frac{p_1 (p_1 p_4 - p_0 p_5)}{p_1 p_2 - p_0 p_3}, \qquad p_4 - \frac{p_0 p_5}{p_1} - \frac{(p_1 p_2 - p_0 p_3)^2 p_5}{p_1 p_3 (p_1 p_2 - p_0 p_3) - p_1{}^2 (p_1 p_4 - p_0 p_5)},$$

and lastly, p_5.

If we regard z as of one dimension in space it is clear that the dimensions of the several coefficients p_0, p_1, &c. are indicated by their suffixes. Hence we may test the correctness of our arithmetical processes by counting the dimensions of the several terms in each of the test functions.

303. When any test function vanishes this process causes an infinite term to appear in the next function. In such a case we may replace the vanishing function by an infinitely small quantity a and then proceed as before. Thus suppose $p_1 = 0$, writing a for p_1 the six functions become p_0, a, $-p_0 p_3 / a$, p_3, $p_4 - p_2 p_5 / p_3 + p_0 p_5^2 / p_3^2$, p_5. Consider the first four of these functions; the signs of p_0 and p_3 being given, it is easy to see by trial that there will be the same number of variations of sign whether we regard a as positive or negative. Thus if p_0 and p_3 have the same sign, the middle terms have always opposite signs and there will be just two variations; if p_0 and p_3 have opposite signs, the middle terms are both positive or both negative and there will be just one variation.

304. **Vanishing of a Subsidiary function.** In the preceding theory two reservations have been made.

1. In applying Cauchy's theorem it has been assumed that there were no radical points on the axis of y.

2. It has been assumed that P and Q have no common factor. In this case as we continue the process of finding the greatest common measure in order to construct the subsidiary functions $f_3(y)$, &c. we arrive at a function which is this greatest common measure and the next function is absolutely zero. Thus we are warned of the presence of common factors by the absolute vanishing of one of the subsidiary functions.

It is clear that if $f(z) = 0$ have two roots which are equal and opposite, the even and odd powers of z must separately vanish. It follows from the definition in Art. 292 that $f_1(y)$ and $f_2(y)$ will have these roots common to each. The greatest common measure of $f_1(y)$ and $f_2(y)$ must therefore contain as factors all the roots of $f(z)$ which are equal and opposite. *Conversely*, the greatest common measure of $f_1(y)$ and $f_2(y)$ is necessarily a function of y which contains only even powers of y*, and if it be equated to zero, its roots are necessarily equal and opposite. These roots must obviously satisfy $f(z) = 0$.

Now if any radical point lie on the axis of y, $f(z)$ must have roots of the form $\pm k \sqrt{-1}$ and these are equal and opposite. The two reserved cases therefore are included in the one case in which $f_1(y)$ and $f_2(y)$ have common factors.

305. Let the greatest common measure of $f_1(y)$ and $f_2(y)$ be $\psi(y^2)$. If then we put $f(z) = \psi(-z^2) \phi(z)$, the function $\phi(z)$ is such that no two of its roots are equal and opposite, and to this function we may therefore apply Cauchy's theorem without fear of failure. By Art. 295, the number of roots of $\phi(z)$ which have their real parts positive is equal to the number of variations of sign in the coefficients of the highest powers of the subsidiary functions of $\phi(z)$. But, since $\psi(-z^2)$ is real when we write $z = y\sqrt{-1}$, the subsidiary functions of $\phi(z)$ become, when each is multiplied by $\psi(y^2)$, the subsidiary functions of $f(z)$. The presence of this common factor will not affect the number of variations of signs in the series. Suppose then we agree to omit the consideration of the factors of $\psi(-z^2)$, we may test the positions of the remaining radical points by discussing either of the functions $f(z)$ or $\phi(z)$.

We may therefore make the following addition to the rule given in Art. 297. *If we apply that rule, using only the subsidiary functions which do not wholly vanish, we obtain the number of roots which have their real parts positive, but excluding those roots which are in pairs equal and opposite to each other.*

---

* If $p_n = 0$, we have an additional root, viz. $z = 0$, which is not included in this remark. But this root may be either divided out of the equation $f(z) = 0$, or it may be included in the following reasoning as a part of the function $\phi(z)$.

These omitted roots are of course given by equating to zero, the last subsidiary function which does not wholly vanish. Putting $y\sqrt{-1}=z$ we may deduce the corresponding roots of the original equation.

It will be seen that for every pair of imaginary roots of y there will be one value of z which has its real part positive, and for every pair of real roots of y there will be two values of z of the form $\pm k\sqrt{-1}$. The former indicate an unstable, the latter a stable motion according to the rule of Art. 283.

306. Usually we may best find the nature of these roots by solving the equation formed by equating to zero the last subsidiary function. But if this be troublesome we may conveniently use Sturm's theorem. Since the powers of y in any subsidiary function decrease two at a time we may effect Sturm's process of finding the greatest common measure exactly as described in Art. 297. We may also show by the same kind of reasoning as in Art. 295, that for every variation of sign when $y=+\infty$ in Sturm's functions there will be a pair of imaginary values of y. We may thus make a second addition to the rule given in Art. 297.

In forming the successive subsidiary functions, as soon as we arrive at one which wholly vanishes, we write instead of it the differential coefficient of the last which does not vanish and proceed to form the succeeding functions by the same rule as before. Every variation of sign in the first column will then indicate one root with its real part positive. The remaining roots will have their real parts negative or zero.

307. **Equal Roots.** We know by Art. 283 that whether a single root of the form $a+b\sqrt{-1}$ indicate stability or instability, several equal roots will indicate the same, except when $a=0$. In this latter case while solitary roots of the form $\pm b\sqrt{-1}$ imply stability, several equal roots indicate instability. It is therefore generally important to determine if the roots of the latter form are repeated or not.

When the equal roots are of the first form and there happen to be no others equal and opposite to them, their number is fully counted in using Cauchy's theorem. When the equal roots are of the second form, i.e. $\pm b\sqrt{-1}$, they appear in the common factor $\psi(-z^2)$. If we can solve the equation $\psi(-z^2)=0$, we know at once whether the repeated roots are of the first or second forms. If we analyse the equation by Sturm's theorem (Art. 306) and stop as usual at the first Sturmian function which does not vanish, we must remember that these equal roots will be counted as if they were one root. The last Sturmian function which does not vanish gives by its factors the sets of equal roots with a loss of one root in each set. If we differentiate this function and continue the process described in Art. 297, we are really applying Sturm's theorem anew to this function, and will arrive at another Sturmian function containing the sets of equal roots with a loss of two of each set. Thus by continuing the process the number of repetitions may be counted.

Numerical Examples. Determine how many roots of the equation

$$z^{10}+z^9-z^8-2z^7+z^6+3z^5+z^4-2z^3-z^2+z+1=0$$

have their real parts positive.

Forming the first two rows by the rule of Art. 297 we have

$$
\begin{array}{lrrrrr}
y^{10} & 1, & -1, & 1, & 1, & -1, & 1, \\
y^9 & 1, & -2, & 3, & -2, & 1, &
\end{array}
$$

where we have written on the left-hand side the highest power of each subsidiary function, and have omitted the negative signs given in the second, fourth and sixth columns of Art. 292. We may notice that the presence of negative terms shows that the equation indicates an unstable motion (Art. 299). Hence *if we merely wish to determine the question of stability or instability the process terminates at the first negative sign.* To illustrate the other rules we continue as follows.

Operating by the rule of Art. 297 we have

$$y^8 \qquad 1, \qquad -2, \qquad 3, \qquad -2, \qquad 1.$$

These are the same as the figures in the last line, hence the next subsidiary function will wholly vanish. Therefore $\psi(-z^2) = z^8 - 2z^6 + 3z^4 - 2z^2 + 1$. By Art. 306 we replace the next function by the differential coefficient

$$y^7 \quad \begin{cases} 8, & -12, & 12, & -4, \text{ divide by } 4, \\ 2, & -3, & 3, & -1, \end{cases}$$

$$y^6 \quad \begin{cases} -\tfrac{1}{2}, & \tfrac{3}{2}, & -\tfrac{3}{2}, & 1, \text{ multiply by } 2, \\ -1, & 3, & -3, & 2, \end{cases}$$

$$y^5 \quad \begin{cases} 3, & -3, & 3, \text{ divide by } 3, \\ 1, & -1, & 1, \end{cases}$$

$$y^4 \quad \begin{cases} 2, & -2, & 2, \text{ divide by } 2, \\ 1, & -1, & 1. \end{cases}$$

Here again the next function vanishes. There are therefore equal roots given by $z^4 - z^2 + 1 = 0$. The nature of these roots may be found by solving this equation. Disregarding this, we may (Art. 307) replace the next function by the differential coefficient

$$y^3 \quad \begin{cases} 4, & -2, \text{ divide by } 2, \\ 2, & -1, \end{cases}$$

$$y^2 \qquad -1, \qquad 2, \text{ after multiplication by } 2,$$

$$y \qquad 3,$$

$$y^0 \qquad 2.$$

Looking at the first column, we see that there are four changes of sign. Hence there are four roots whose real parts are positive. We verify this by remarking that the given equation may be written in the form $(z^4 - z^2 + 1)^2 (z^2 + z + 1) = 0$. In this example we have exhibited all the numerical calculations.

Ex. 2. Show that the roots of the equations

$$z^4 + 2z^3 + z^2 + 1 = 0,$$

$$z^8 + 2z^7 + 4z^6 + 4z^5 + 6z^4 + 6z^3 + 7z^2 + 4z + 2 = 0,$$

do not satisfy the conditions of stability.

Ex. Show that the roots of the equations

$$z^4 + 3z^3 + 5z^2 + 4z + 2 = 0,$$

$$z^6 + z^5 + 6z^4 + 5z^3 + 11z^2 + 6z + 6 = 0,$$

do satisfy the conditions of stability.

The conditions of stability given in this section are taken from the third chapter of the author's essay on *Stability of Motion*. Other methods of testing the roots of the equation $f(z) = 0$ are given in the second chapter of that essay. The conditions for a biquadratic were read before the Mathematical Society in 1874. The theory of linear differential equations with especial reference to the indeterminate case is abridged from a paper by the author in the *Proceedings of the Mathematical Society*, 1883.

CHAPTER VII.

Free Oscillations.

308. THE difference between free and forced vibrations will be explained in the next section of this chapter. The following rough distinction will be sufficient for our present purpose. When the forces which act on a system depend only on the deviations of the several particles from their undisturbed motion, every term in the equations of motion, as explained in Art. 257, will contain the first powers of the co-ordinates. The equations of motion will then take the form given to them in Art. 310 of this chapter. The oscillations of such a system are called its *free oscillations.*

Besides these forces we may have others due to external causes which may be functions of the time, and may not vanish when the system is placed in its undisturbed position. Such forces are usually written on the right-hand side of the equations of motion, to intimate that their effects must be calculated by different rules from the former forces. The oscillations produced by these forces are called *forced oscillations.*

309. **Introductory summary.** The propositions in this section are constructed for the purpose of examining the small oscillations of a system which depends on many co-ordinates. But as they are of general application they are here presented in a form which is purely mathematical. No reference is made to any dynamical principle and when dynamical terms are used it is only for the sake of explanation.

We begin by taking the equations of the second order with n dependent variables in their most general forms, though such general forms do not occur in dynamics. Two typical equations are then deduced, and from these, the chief propositions in the section are derived.

The first step usually taken in solving simultaneous equations is to form a certain determinant (Art. 262). The general form of the solution and the stability of the resulting motion depend on the roots of this determinant. If as explained in Art. 282 the real parts of the roots are positive the motion is unstable. Two propositions are shown to follow immediately from the typical equations. If three functions, here called A, B, C, are one-signed it is shown (1) that, however general the equations may be, the real roots of the determinant cannot be positive, (2) that, if the equations have that simpler character which occurs in dynamics, the real part of every imaginary root is negative.

When we apply our equations to the case of a system oscillating about a position of equilibrium we see that the function A corresponds to half the vis viva, B to the dissipation function, and C to the potential of the forces of restitution.

The first of these propositions has been established by Lagrange and Sir W. Thomson when the equations represent the oscillations of a system about a position of equilibrium. The second is to be found in the author's essay on the *Stability of Motion* but expressed in a different form. It is also given in the last edition of Thomson and Tait's *Natural Philosophy.* The reader is also referred to a paper by the author read in April 1883 before the Mathematical Society of London.

310. The roots of the fundamental determinant. Let there be any number of dependent variables x, y, z, &c., to be found in terms of t, by means of as many differential equations of the second order with constant coefficients. Whatever these equations may be, they may be very conveniently written in the form

$$(A_{11}\delta^2 + B_{11}\delta + C_{11})\, x + \left(\begin{array}{c} A_{12}\delta^2 + B_{12}\delta + C_{12} \\ + D_{12}\delta^2 + E_{12}\delta + F_{12} \end{array} \right) y + \left(\begin{array}{c} A_{13}\delta^2 + B_{13}\delta + C_{13} \\ + D_{13}\delta^2 + E_{13}\delta + F_{13} \end{array} \right) z + \&c. = 0,$$

$$\left(\begin{array}{c} A_{12}\delta^2 + B_{12}\delta + C_{12} \\ - D_{12}\delta^2 - E_{12}\delta - F_{12} \end{array} \right) x + (A_{22}\delta^2 + B_{22}\delta + C_{22})\, y + \left(\begin{array}{c} A_{23}\delta^2 + B_{23}\delta + C_{23} \\ + D_{23}\delta^2 + E_{23}\delta + F_{23} \end{array} \right) z + \&c. = 0,$$

$$\left(\begin{array}{c} A_{13}\delta^2 + B_{13}\delta + C_{13} \\ - D_{13}\delta^2 - E_{13}\delta - F_{13} \end{array} \right) x + \left(\begin{array}{c} A_{23}\delta^2 + B_{23}\delta + C_{23} \\ - D_{23}\delta^2 - E_{23}\delta - F_{23} \end{array} \right) y + (A_{33}\delta^2 + B_{33}\delta + C_{33})\, z + \&c. = 0,$$

$$\&c. \qquad + \qquad \&c. \qquad\qquad + \&c. = 0,$$

where the symbol δ represents differentiation with regard to t, and the order of suffixes is immaterial, so that $A_{12} = A_{21}$, and so on.

We see here two sets of terms, (1) those which depend on the letters A, B, C, and which by themselves constitute a symmetrical determinant; (2) those which depend on the letters D, E, F, and which by themselves constitute a skew determinant.

311. For the reasons given in Chap. IX. of Vol. I., we may call the terms which depend on the letter A the *effective forces*, those which depend on the letter B the *forces of resistance*, those on C the *forces of restitution*. It generally happens that the terms which depend on the letters D and F are absent. The terms which depend on the letter E occur when we consider the oscillations about a state of motion, Chap. III., Art. 112. These we shall call the *centrifugal forces.*

If we write A, B, C for the three functions

$$A = \tfrac{1}{2}A_{11}x^2 + A_{12}xy + \tfrac{1}{2}A_{22}y^2 + \ldots\ldots,$$
$$B = \tfrac{1}{2}B_{11}x^2 + B_{12}xy + \tfrac{1}{2}B_{22}y^2 + \ldots\ldots,$$
$$C = \tfrac{1}{2}C_{11}x^2 + C_{12}xy + \tfrac{1}{2}C_{22}y^2 + \ldots\ldots,$$

the terms in the several equations which arise from A, B, C may be written

$$\delta^2 \frac{dA}{dx} + \delta \frac{dB}{dx} + \frac{dC}{dx}, \quad \delta^2 \frac{dA}{dy} + \delta \frac{dB}{dy} + \frac{dC}{dy}, \quad \&c.$$

Hence A, B, C *may be called respectively the potentials of the effective forces, the forces of resistance, and the forces of resti-tution.*

312.   When we compare the equations of motion with those given by Lagrange for the oscillations about a position of equilibrium (Chap. II.), we see that the function A cannot be otherwise than positive.   So also these oscillations are stable if the function C be always positive.

Thus, it frequently occurs that the three functions A, B, C, or some of them, are such that they *keep one sign whatever real quantities* we write for x, y, z, &c., *and do not become zero except when x, y, &c. are all zero.*   Such functions will be referred to as *one-signed quadrics.*

313.   The method of solving the differential equations in Art. 310 has been explained in Chap. VI.   Let m_1, m_2, &c., be the roots of the fundamental determinant, which we need not here write down.   This determinant is the same as that represented by the symbol $\Delta(\delta)$ in Art. 262.   Let us suppose that these roots are unequal, the case of equal roots being regarded as a limiting case of unequal roots.   The solution may be written thus:—

$$x = x_1 e^{m_1 t} + x_2 e^{m_2 t} + \dots$$
$$y = y_1 e^{m_1 t} + y_2 e^{m_2 t} + \dots, \qquad dx/dt = x_1' e^{m_1 t} + x_2' e^{m_2 t} + \dots$$
$$z = \&c. \qquad dy/dt = y_1' e^{m_1 t} + y_2' e^{m_2 t} + \dots,$$
$$\&c. = \&c.$$

where $x_1' = x_1 m_1$, $y_1' = y_1 m_1$, &c., $x_2' = x_2 m_2$, &c.

Here x_1, y_1, z_1, &c. contain as a common factor one constant of integration, x_2, y_2, &c. another constant, and so on.   The forms of these constants are not wanted here.   It is enough that we should remember that the *coefficients which belong to a real exponential are themselves real.*   On the other hand, if m_1, m_2, be a pair of imaginary roots, the coefficients (x_1, x_2), &c., take the form $P \pm Q \sqrt{-1}$.

314.   **The first equation.**   If we substitute the first terms of each of these values of x, y, z, &c., in the equations of Art. 310, we obtain a set of equations which differs from those only in having m_1 written for δ, and x_1, y_1, &c. for x, y, &c.   Multiply these respectively by x_1, y_1, &c., and add the results together; we have

$$(A_{11}x_1^2 + 2A_{12}x_1y_1 + \&c.) m_1^2 + (B_{11}x_1^2 + 2B_{12}x_1y_1 + \&c.) m_1$$
$$+ (C_{11}x_1^2 + 2C_{12}x_1y_1 + \&c.) = 0.$$

It should be noticed that *the terms which depend on the letters D, E, F have altogether disappeared from this equation.*

It should also be noticed that *the coefficients of the powers of m are twice the functions A, B, C with x_1, y_1, &c. written for x, y, &c.*

315. Prop. I. **On real roots.**—We may immediately deduce the three following theorems:—

(1) *If the potentials* A, B, C *are either zero or one-signed functions, and if all three have the same sign, the fundamental determinant cannot have a real positive root.*

For if m_1 were real, the coefficients x_1, y_1, &c. would be real. We should thus have the sum of three positive quantities equal to zero.

(2) *If there are no forces of resistance, i.e. if the term* B *is absent, and if the potentials* A *and* C *are one-signed and have the same sign, the fundamental determinant cannot have a real root, positive or negative.*

(3) *If* A, B, C *are one-signed functions, but if the sign of* B *is opposite to that of* A *and* C, *the fundamental determinant cannot have a negative root.*

These propositions are true, whether there are any terms in the differential equations which depend on the functions D, E, F *or not.*

We may also notice that, *unless the potential* C *can vanish for some real values of the co-ordinates other than zero, the fundamental determinant cannot have a root equal to zero.* If, for example, the co-ordinate x is absent from C (Art. 98), then C vanishes when the other co-ordinates are zero and x is finite. In this case m_1 can be equal to zero. If the forces depending on B are also absent the determinant will have two roots equal to zero.

When two zero roots occur terms such as $nt + a$ must be added to some of the expressions for the co-ordinates given in Art. 313. Unless the initial conditions are such as to make the constants n and a equal to zero, these terms should be included in the expressions $\theta = f(t)$, $\phi = F(t)$, &c., which as explained in Art. 257 give the steady motion. The presence of these terms thus indicates a slight change in the steady motion about which the system has been supposed to oscillate.

316. **The two equations.** Exactly as in Art. 314, let us again substitute the first term of each of the values of x, y, &c. in the equations of motion. But let us now multiply these by x_2, y_2, &c., and add the results. We thus obtain

$$[A_{11}x_1x_2 + A_{12}(x_1y_2 + x_2y_1) + A_{23}(y_1z_2 + y_2z_1) + \&c.] m_1{}^2$$
$$+ [B_{11}x_1x_2 + \&c.] m_1 + [C_{11}x_1x_2 + \&c.]$$
$$= [D_{12}(x_1y_2 - x_2y_1) + D_{23}(y_1z_2 - y_2z_1) + \&c.] m_1{}^2$$
$$+ [E_{12}(x_1y_2 - x_2y_1) + \&c.] m_1 + [F_{12}(x_1y_2 - x_2y_1) + \&c.].$$

To bring this equation within bounds, we must use some notation to shorten the coefficients. Let us represent the halves of these series by their first terms, omitting suffixes to A, B, &c. We may therefore write the equation in the form

$$A(x_1x_2) m_1{}^2 + B(x_1x_2) m_1 + C(x_1x_2)$$
$$= D(x_1y_2) m_1{}^2 + E(x_1y_2) m_1 + F(x_1y_2).$$

In the same way we have

$$A\,(x_1 x_2)\,m_2^2 + B\,(x_1 x_2)\,m_2 + C\,(x_1 x_2)$$
$$= -\,D\,(x_1 y_2)\,m_2^2 - E\,(x_1 y_2)\,m_2 - F\,(x_1 y_2).$$

Also we deduce from these the two equations

$$\left.\begin{aligned}
A\,(x_1 x_1)\,m_1^2 + B\,(x_1 x_1)\,m_1 + C\,(x_1 x_1) &= 0 \\
A\,(x_2 x_2)\,m_2^2 + B\,(x_2 x_2)\,m_2 + C\,(x_2 x_2) &= 0
\end{aligned}\right\}.$$

The first of these is the same as that already found in Art. 314.

Here we may notice that the functions $A\,(xx)$, $B\,(xx)$, $C\,(xx)$ are really the same as those we have already more simply denoted by A, B, C. We also notice that $D\,(x_1 y_1) = 0$, $E\,(x_1 y_1) = 0$, and $F\,(x_1 y_1) = 0$.

317. Let us now suppose that there is a pair of imaginary roots in the fundamental determinant of the form $m_1 = r + p\sqrt{-1}$, $m_2 = r - p\sqrt{-1}$. The values of x, y, &c., given in Art. 313, become

$$x = (x_1 + x_2)\,e^{rt}\cos pt + (x_1 - x_2)\sqrt{-1}\,e^{rt}\sin pt + \&c.,$$
$$y = (y_1 + y_2)\,e^{rt}\cos pt + (y_1 - y_2)\sqrt{-1}\,e^{rt}\sin pt + \&c.,$$

which may be conveniently abbreviated into

$$\left.\begin{aligned}
x &= X_1 e^{rt}\cos pt + X_2 e^{rt}\sin pt + x_3 e^{m_3 t} + \cdots \\
y &= Y_1 e^{rt}\cos pt + Y_2 e^{rt}\sin pt + y_3 e^{m_3 t} + \cdots \\
z &= \&c.
\end{aligned}\right\}.$$

If $X_1' = rX_1 + pX_2$ and $X_2' = -pX_1 + rX_2$, &c.,

$$\left.\begin{aligned}
dx/dt &= X_1'\,e^{rt}\cos pt + X_2'\,e^{rt}\sin pt + x_3'\,e^{m_3 t} + \cdots \\
dy/dt &= Y_1'\,e^{rt}\cos pt + Y_2'\,e^{rt}\sin pt + y_3'\,e^{m_3 t} + \cdots \\
\&c. &= \&c.
\end{aligned}\right\}.$$

318. Returning now to the two first equations of Art. 316, let us divide them by m_1 and m_2 respectively. If we first add and then subtract the results, we have

$$A\,(x_1 x_2)\,r + B\,(x_1 x_2) + C\,(x_1 x_2)\frac{r}{r^2 + p^2} = \left\{ D\,(x_1 y_2)\,p - F\,(x_1 y_2)\frac{p}{r^2 + p^2} \right\}\sqrt{-1},$$

$$A\,(x_1 x_2)\,p - C\,(x_1 x_2)\frac{p}{r^2 + p^2} = \left\{ D\,(x_1 y_2)\,r + E\,(x_1 y_2) + F\,(x_1 y_2)\frac{r}{r^2 + p^2} \right\}\frac{1}{\sqrt{-1}}.$$

By substitution, we find that

$$\left.\begin{aligned}
4A\,(x_1 x_2) &= A\,(X_1 X_1) + A\,(X_2 X_2) \\
-\,2D\,(x_1 y_2)\sqrt{-1} &= D\,(X_1 Y_2)
\end{aligned}\right\},$$

with similar results for the other letters. We also infer from these equations that if A *is a one-signed function*, $A\,(x_1 x_2)$ *is not only real, but has always the same sign as* A. Similar remarks apply to the functions B and C.

If the functions D, E, F are absent, the two first equations of this Article reduce to

$$\left. \begin{array}{l} A\,(x_1x_2)\,2r + B\,(x_1x_2) = 0 \\ - A\,(x_1x_2)\,(r^2 + p^2) + C\,(x_1x_2) = 0 \end{array} \right\},$$

except when $p = 0$, *i.e.*, except when the roots (which we have supposed imaginary) are real.

These equations may be conveniently written

$$r = -\tfrac{1}{2}\frac{B\,(X_1X_1) + B\,(X_2X_2)}{A\,(X_1X_1) + A\,(X_2X_2)}, \qquad r^2 + p^2 = \frac{C\,(X_1X_1) + C\,(X_2X_2)}{A\,(X_1X_1) + A\,(X_2X_2)},$$

thus giving r and p when the amplitudes of the oscillations are known.

319. PROP. II. **On imaginary roots.**—We may immediately deduce the following theorem from the equations of Art. 318.

(1) *Let the fundamental determinant be symmetrical, i.e., let the functions* D, E, F *be all absent. Let the potentials* A *and* B *be one-signed and have the same sign (whether* C *be a one-signed function or not). Then the real part* r *of every imaginary root must be negative and not zero. But if the potential* B *is absent, then the real part of every imaginary root is zero.*

If the potentials A *and* C *are one-signed and have opposite signs, there can be no imaginary roots.*

These results follow by simply looking at the two last equations of Art. 318.

(2) *If the terms depending on* D *and* F *are absent from the equations, whether the terms depending on* E *are present or not, and if the three potential functions* A, B, C *are all one-signed and have the same sign, then the real part* r *of every imaginary root is negative, and not zero. But if the forces of resistance, i.e.* B, *are also absent, then the real part of every imaginary root is zero.*

(3) *If the terms depending on* D *and* E *are absent, but not* necessarily those depending on F, and if A, B, C are all one-signed and have the same sign, then the real part r of every imaginary root must be negative, or, if positive, must be less than p.

320. Ex. 1. If A is a one-signed function prove that $\{A\,(x_1x_2)\}^2$ is always less than the product $A\,(x_1x_1) \cdot A\,(x_2x_2)$.

Ex. 2. If $\Delta\,(m)$ is the determinant of motion, $\Delta_1\,(m)$ the minor of its leading constituent, x_1y_1, &c. the minors of the first row, and m any quantity not necessarily a root of $\Delta\,(m)$, prove the identity

$$A\,(x_1x_1)\,m^2 + B\,(x_1x_1)\,m + C\,(x_1x_1) = \Delta\,(m)\,\Delta_1\,(m).$$

Ex. 3. If m_1, m_2 are any two quantities, not necessarily roots of the determinant $\Delta\,(m)$, prove that

$$\left. \begin{array}{l} A\,(x_1x_2)\,m_1{}^2 + B\,(x_1x_2)\,m_1 + C\,(x_1x_2) \\ - D\,(x_1y_2)\,m_1{}^2 - E\,(x_1y_2)\,m_1 - F\,(x_1y_2) \end{array} \right\} = \Delta\,(m_1)\,\Delta_1\,(m_2).$$

Ex. 4. If the determinant is symmetrical, and if the potentials A and C are one-signed and have opposite signs, then, whatever sign the potential B may have, the roots of the determinant are all real.

Ex. 5. If the terms depending on F and E are absent, but not necessarily those depending on D, and if the three potentials A, B, C are all one-signed and have the same sign, then the real part r of every imaginary root must be negative or if positive less than p.

321. **Effect of the forces of resistance on oscillations about a position of equilibrium.** Let a system be oscillating about its position of equilibrium under no forces of resistance, so that the functions B, D, E, F are all zero. We also suppose the functions A and C to be one-signed and to have the same sign.

By referring to the equations of motion in Art. 310 we see at once that the determinant of the motion viz. $\Delta(\delta)$ contains only even powers of δ. This determinant is of course the same as the Lagrangian determinant discussed in Chap. II. It follows either from Chap. II. or from Arts. 315 and 319 of this chapter that all the roots of the equation $\Delta(\delta) = 0$ are of the form $\pm p\sqrt{-1}$. Any co-ordinate therefore is represented by a series of the form

$$x = X_1 \cos pt + X_2 \sin pt + \ldots\ldots$$

Let now some small forces of resistance act on the system, we therefore introduce into the equations of motion the terms which depend on the function B. The forces thus introduced are supposed to be so small that we may reject the squares of the coefficients of the function B. We represent this by supposing every coefficient to contain a factor κ whose square can be neglected. It is the effect of these additional forces on the former motion which we wish to discover.

Referring again to the equations of motion in Art. 310, let $\Delta_1(\delta)$, $\Delta_2(\delta)$ be the determinants of motion before and after the introduction of these forces of resistance. The determinantal equation therefore becomes

$$\Delta_2(\delta) = \Delta_1(\delta) + B_{11}\delta I_{11}(\delta) + \&c. = 0,$$

where the symbol I indicates the minors of the constituents of $\Delta_1(\delta)$ as explained in Chap. VI.

This equation may be written in the form $\Delta_1(\delta) + \kappa\delta\phi(\delta) = 0$, where $\phi(\delta)$ contains only even powers of δ. Since $p\sqrt{-1}$ is a root of $\Delta_1(\delta) = 0$, we let the corresponding root of this new equation be $p\sqrt{-1} + r$ where r is a small quantity, real or imaginary, whose square can be neglected. We find by Taylor's theorem

$$\Delta_1'(p\sqrt{-1})\,r + \kappa p\sqrt{-1}\,\phi(p\sqrt{-1}) = 0.$$

Hence, since $\Delta_1'(\delta)$ contains only odd powers of δ, it follows that r is necessarily real.

We have thus proved that the correction to any root of the determinantal equation when we introduce the resistances is necessarily real. This means that the correction to the imaginary part of the root depends on the *square* of the resistances. *The addition* r *to the real part of the root introduces a real exponential factor* e^{rt} *into the amplitude of any oscillation.* The addition to the imaginary part alters the period of the oscillation (Art. 317). *Thus the periods of the oscillations are affected only by the squares of small quantities when we introduce the resisting forces.*

322. The series for any co-ordinate now takes the form (see Art. 317)

$$x = X_1 e^{rt} \cos pt + X_2 e^{rt} \sin pt + \ldots$$

where p is the same as before and, by Art. 319, r is negative. With the same given initial values of x, y, &c, dx/dt, dy/dt, &c. the coefficients X_1, &c. are changed only by terms which contain the factor κ, and being themselves small, these changes may be neglected.

The value of r may be deduced from the expressions given at the end of Art. 318. If the forces of resistance were zero, the real exponentials would be absent and the ratios X_1/X_2, Y_1/Y_2 would all be equal. With small forces of resistance these ratios differ from each other by quantities which contain the small factor κ. It follows that the ratios $B\,(X_1 X_1)/A\,(X_1 X_1)$ and $B\,(X_2 X_2)/A\,(X_2 X_2)$ are also equal when we reject the *square* of the small quantity. The expression for r therefore reduces to the simple form

$$r = -\tfrac{1}{2}\frac{B\,(X_1 X_1)}{A\,(X_1 X_1)} = -\tfrac{1}{2}\frac{B_{11}X_1{}^2 + 2B_{12}X_1 Y_1 + \ldots}{A_{11}X_1{}^2 + 2A_{12}X_1 Y_1 + \ldots}.$$

Translating this formula into English we see by Art. 73 that *the numerical value of r, for any one principal oscillation, is one half the ratio of the mean value of the dissipation function to the mean value of the kinetic energy for that oscillation.*

Forced Oscillations.

323. We may suppose a system to be moving in a given state of motion defined, as explained in Art. 257, by the co-ordinates $\theta = \theta_0$, $\phi = \phi_0$, &c. where θ_0, ϕ_0, &c. are known functions of the time. This motion we shall call sometimes the undisturbed motion and sometimes the steady motion. If the system be now disturbed in any manner, we write $\theta = \theta_0 + x$, $\phi = \phi_0 + y$, &c. where x, y, &c. are so small that we may reject their squares. This disturbance may have been made by some small impulse and the system may then have been left to oscillate about the undisturbed motion.

We may also have continuous forces acting on the system tending to make it oscillate about the undisturbed motion. As the object of our enquiry is the *oscillation* of a system, we shall suppose that these forces when they exist are periodic. If $f\,(t)$ represents any one we may suppose this function to be expanded by the known processes of Trigonometry in a series of multiple angles; thus

$$f\,(t) = Pe^{-\kappa t}\sin\,(\lambda t + \alpha) + P'e^{-\kappa' t}\sin\,(\lambda' t + \alpha') + \&c.$$

Each of these terms is called a *disturbing force*. The coefficient of the trigonometrical factor of any term is called the *magnitude* or *amplitude* of that term. The angle $\lambda t + \alpha$ is called sometimes the *phase* and sometimes the *argument*.

It frequently happens that the real exponentials are absent from the expression for the force. This case will therefore be more particularly considered in what follows. When we wish to call attention to the absence of the real exponential, the disturbing force is often called a *permanent force*. When the real exponential is present with a negative index, we may call the force *evanescent*.

Sometimes instead of the force being given, some point of the system is compelled to oscillate in a given manner. We then have some given relation between the co-ordinates of the system of the form $ax + by + cz + \&c. = Ge^{-gt}\sin\,(\nu t + \gamma)$

where a, b, c &c., G, g &c. are given constants. There may also be several similar relations between some or all the co-ordinates. In such a case we suppose these given relations to be included amongst the differential equations, though they cannot be derived from a Lagrangian function as in Art. 111. The method of finding the corresponding forced vibration given in Art. 326, will then still be applicable.

324. The general equations of motion of the second order are given in Art. 310, but in dynamics the terms which depend on the functions D and F are in general absent. The mode in which these are formed when the resisting forces are absent is explained in Art. 111. Including these resistances we may suppose that the equations of motion take the form

$$(A_{11}\delta^2 + B_{11}\delta + C_{11})\, x + \left(\begin{matrix} A_{12}\delta^2 + B_{12}\delta + C_{12} \\ + E_{12}\delta \end{matrix}\right) y + \ldots = P e^{-\kappa t} \sin(\lambda t + \alpha)$$

$$\left(\begin{matrix} A_{12}\delta^2 + B_{12}\delta + C_{12} \\ - E_{12}\delta \end{matrix}\right) x + (A_{22}\delta^2 + B_{22}\delta + C_{22})\, y + \ldots = Q e^{-\kappa' t} \sin(\mu t + \beta)$$

$$\&c. = \&c.$$

where we have written on the right-hand side only one disturbing force in each equation as a specimen.

For the sake of brevity, it will be found convenient to distinguish the equation in which any disturbing force occurs by some simple phrase. The first equation is obtained from Lagrange's equations (Art. 111) by differentiating with regard to θ or x. The second by differentiating with regard to ϕ or y. The force on the right-hand side of the first equation may therefore be said to *act directly* on the co-ordinate x and *indirectly* on y, z, &c. So the force on the right-hand side of the second equation acts directly on the co-ordinate y and indirectly on x, z, &c.

325. **Forced and Free Oscillations.** It is proved in the theory of differential equations that the solution of these equations leads to an expression for each of the co-ordinates which contains two sets of terms. The first set is called a *particular integral* and consists of any solution obtained by any process however restricted. The second set is called the *complementary function* and represents the value of the co-ordinate when all the disturbing forces on the right-hand side are omitted. The complementary function is therefore the same as the solution found and discussed in the first section of this chapter.

The complementary functions in the expressions for the co-ordinates give the oscillations of the system about the undisturbed motion when not influenced by any disturbing forces. These integrals are therefore said to constitute *the natural or free vibrations of the system*. The particular integrals in the several co-ordinates which indicate the effects of any disturbing force are called *the forced vibrations or oscillations due to that force*.

According to this definition any particular integral may be

14—2

taken to represent the forced vibration. But in practice there is one particular integral which is more convenient than any other. What this is will be made clear by the next proposition.

A free oscillation does not necessarily mean a principal oscillation though it is sometimes used in that sense (Arts. 53 and 116). Any motion represented by any number of terms selected from the complementary function will be a free motion. The word "free" is meant to be a contrast to the word "forced."

The term "Complementary Function" is used by Liouville in Vol. 13 of the *Journal Polytechnique*, 1832 in an article on fractional differential coefficients. It is also used in *Gregory's Examples*, 1841. The distinction of Waves into "free" and "forced" may be found in *Airy's Tides and Waves*, published in the Encyclopædia Metropolitana, 1842.

326. To find the Forced Vibration. To find a particular integral for any force $Pe^{-\kappa t} \sin(\lambda t + \alpha)$ we follow the methods already explained in Chap. VI. If $\Delta(\delta)$ be the determinant of the motion and $I_1(\delta)$, $I_2(\delta)$, &c. be the minors of the first, second, &c. terms in that row of $\Delta(\delta)$ which corresponds to the equation in which the force occurs, we have

$$x = \frac{I_1(\delta)}{\Delta(\delta)} Pe^{-\kappa t} \sin(\lambda t + \alpha), \quad y = \frac{I_2(\delta)}{\Delta(\delta)} Pe^{-\kappa t} \sin(\lambda t + \alpha), \quad z = \&c.$$

We shall now prove that these operators will lead to two trigonometrical terms in each of the co-ordinates. These two terms constitute the forced vibration in that co-ordinate.

327. To perform the operations indicated by these functions of δ, we use the following simple rule. *To perform the operation* $F(\delta) = \dfrac{I(\delta)}{\Delta(\delta)}$ *on* $Pe^{-\kappa t} \dfrac{\sin}{\cos}(\lambda t + a)$ *we write* $\delta = -\kappa + \lambda\sqrt{-1}$ *and reduce the operator to the form* $L + M\sqrt{-1}$. *The required result is then* $Pe^{-\kappa t}\left(L + M\dfrac{\delta}{\lambda}\right)\dfrac{\sin}{\cos}(\lambda t + a)$.

To prove this rule, we notice that by Art. 265 $F(\delta) e^{mt} = (L + M\sqrt{-1}) e^{mt}$ where $m = -\kappa + \lambda\sqrt{-1}$. If we now replace the imaginary part of the exponential by its trigonometrical value, and equate the real and imaginary parts on each side of the equation, the result follows at once.

328. Ex. If the determinant $\Delta(\delta)$ have a roots each equal to m, i.e. $-\kappa + \lambda\sqrt{-1}$, the result assumes an infinite form. Prove that in this case the operator may be replaced by $\{t^a I(\delta) + at^{a-1}I'(\delta) + \ldots + I^a(\delta)\}/\Delta^a(\delta)$, where the coefficients follow the binomial law, and $\Delta^a(\delta)$, &c. have been written to express the ath differential coefficient of $\Delta(\delta)$, &c. Every one of these operations may now be performed by the rule given in the last article.

To prove this, we replace the root m by $m + h$ where h is to be afterwards put equal to zero. We then find

$$\frac{I(\delta)}{\Delta(\delta)} e^{mt} = \left\{I(m) e^{mt} + \ldots + \frac{d^a}{dm^a}(I(m) e^{mt}) \frac{h^a}{La}\right\} \bigg/ \left\{\Delta^a(m) \frac{h^a}{La}\right\}.$$

The first a terms of this series in each co-ordinate, though infinite, may be absorbed into the complementary function, see Art. 266. The solution is therefore expressed by the $(a+1)$th term. This, by the theorem of Leibnitz to find the ath differential coefficient of a product, reduces to the operator given above.

329. Ex. A particle, say the earth, describes a nearly circular orbit about a centre of force whose attraction varies inversely as the square of the distance. It is also acted on by two disturbing forces represented by $P \sin \lambda t$ and $Q \sin \lambda t$ acting respectively along and perpendicular to the radius vector. If the polar co-ordinates r, θ be given by $r = a + x$, $\theta = nt + y$, prove that the equations of motion are

$$\left. \begin{aligned} (\delta^2 - 3n^2)\, x - 2an\delta y &= P \sin \lambda t \\ 2n\delta x + a\delta^2 y &= Q \sin \lambda t \end{aligned} \right\},$$

show also that the forced vibrations are given by

$$x = \frac{P}{n^2 - \lambda^2} \sin \lambda t - \frac{2nQ}{\lambda\,(n^2 - \lambda^2)} \cos \lambda t, \qquad y = \frac{2nP}{a\lambda\,(n^2 - \lambda^2)} \cos \lambda t + \frac{(3n^2 + \lambda^2)\,Q}{a\lambda^2\,(n^2 - \lambda^2)} \sin \lambda t.$$

330. **Smooth and Tremulous Motion.** We have supposed the system to be capable of moving in some state of steady motion, just as a hoop rolls on the ground in a vertical plane. But owing to some small disturbances the system really oscillates on each side of this steady motion, the amount of disturbance being always represented for each co-ordinate by the sum of the natural and forced oscillations. When the period of one of these is small the system rapidly changes from one side to the other of its mean or steady motion. The mean motion then appears to the eye to be *tremulous*. When the periods of all the oscillations are very long the changes from one side of the mean motion to the other takes place so slowly that it is hardly perceived to be an oscillation. The mean motion is then said to be *smooth*.

331. **Disappearance of the Free Vibrations.** When a system is set in vibration by any continuous permanent disturbing force, we have seen that two kinds of vibration are excited in the system, viz. the free and the forced vibrations. If there are no forces of resistance both these continue to coexist throughout the motion. But if there are any forces of resistance an exponential is introduced into the free vibration which causes its amplitude to decrease continually so that finally the free vibration becomes insensible (Art. 319). The amplitude however of the forced vibration is not similarly decreased. Thus the oscillation of the system is ultimately independent of the initial conditions and depends only on the forced vibrations. The forced vibration produced by a permanent disturbing force is therefore sometimes called the *permanent vibration*.

332. *It is sometimes important to compare the rates at which the different free oscillations tend to become extinct under the influence of the resisting forces.* It is clear that this depends on the magnitude of the negative quantity r in the exponential factor e^{rt} introduced by these resistances. Since this factor is not necessarily the same in all the terms, it follows that all the free

vibrations do not diminish at the same rate. Some may become insensible before the magnitudes of others have been much impaired.

When the initial amplitudes of any one principal oscillation are known in all the co-ordinates, the value of r for that oscillation can be deduced from the equations given in Art. 318. But when the system is oscillating about a position of equilibrium and the forces of resistance are small the expression for r takes the very simple form given in Art. 322. If X_1, Y_1, &c. be the amplitudes in the co-ordinates x, y, &c. of any one free principal oscillation, this expression is

$$r = -\tfrac{1}{2}\frac{B_{11}X_1{}^2 + 2B_{12}X_1Y_1 + \ldots}{A_{11}X_1{}^2 + 2A_{12}X_1Y_1 + \ldots},$$

where the vis viva and twice the dissipation function are given by

$$2A = A_{11}x'^2 + 2A_{12}x'y' + \ldots, \quad 2B = B_{11}x'^2 + 2B_{12}x'y' + \ldots.$$

The use of this expression for r will be best shown by a few examples.

333.   Ex. 1.   Let us regard a homogeneous tight chain as constructed of a series of equal very small particles, each of mass m, connected by very short strings each of length l and without mass. Let x, y, &c. be the displacements of the particles of such a string vibrating, say, transversely. Then the vis viva is given by $\Sigma m x'^2$. Suppose the resistance of the atmosphere to be represented by a *retarding force on each particle which varies as its actual velocity.* Prove that the dissipation function B may be represented by $2B = \Sigma \kappa m x'^2$. Taking κ to be the same for all the particles it immediately follows that $r = -\tfrac{1}{2}\kappa$, *so that the proportional effect of the resistance of the air on all the free vibrations is the same.*

Ex. 2.   If the particles of the chain vibrate longitudinally instead of transversely the effects of the resistance of the air will be less than before while the effects of viscosity or imperfect elasticity will be more apparent. Let us suppose that these may be represented by a series of forces resisting compression or extension between adjacent particles, *each force being proportional to the relative velocities of the two particles between which it acts and reacts.* Prove that the dissipation function B may be represented by $2B = \Sigma \kappa m (x' - y')^2$.

Speaking in general terms, we infer that r is greatest for that kind of oscillation in which the differences of the amplitudes of the oscillations of adjacent particles are greatest. Oscillations of this kind disappear the soonest, while those in which the adjacent particles move nearly together may remain perceptible for a long time after. This is sometimes briefly expressed by saying that the *effect of viscosity is to extinguish the shorter waves before the longer ones.*

Ex. 3.   If the co-ordinates are so chosen that the dissipation function and the vis viva take the forms $2B = B_{11}x'^2 + B_{22}y'^2 + \ldots$ $2T = A_{11}x'^2 + A_{22}y'^2 + \ldots$ then the value of r for every principal vibration lies between the greatest and least of the fractions $B_{11}/2A_{11}$, $B_{22}/2A_{22}$, &c. It may be noticed that *these limits are independent of the force function and are therefore the same whatever the forces may be.*

Ex. 4.   The membrane which forms a drum-head vibrates transversely when struck. If the resistance of the air be slight and vary as the actual velocity of each particle, show that all the free vibrations have the same real exponential factor.

Ex. 5.   When successive notes are sounded on a musical instrument the terminal motion of one note is the initial motion of the next. Explain why each note is not sensibly affected by the preceding one. See Art. 331.

334. **Herschel's Theorem on the period of the Forced Vibration.** On comparing the terms in Art. 327 which constitute the forced vibration with that which forms the disturbing force, we notice that the period of the forced vibration is the same as that of the force to which it is due. Thus *if any periodical cause of disturbance act on a system of vibrating particles the forced vibrations follow the period of the exciting cause.* This important theorem is due to Sir J. Herschel, who first enunciated it in his Theory of Sound (*Encyc. Met.* 1830). His demonstration however is totally different from that given here.

More generally, the disturbing force and the resulting forced vibration have not only the same period, but have the same real exponential also. Thus, when the fundamental determinant has no equal roots the two have the same general form or type. A permanent force produces a permanent vibration, an evanescent vibration follows only from an evanescent force.

In the proof of this theorem we have assumed that the system of vibrating particles is such that the squares of the displacements can be neglected.

The theorem also only applies to the forced vibrations. If therefore we wish to apply Herschel's theorem to the actual visible motion, a time sufficient to allow the free vibrations to die away, must have elapsed since the initial motion. See Art. 331.

335. As an example of this principle we may notice that when a sounding body (such as a drum) excites vibrations in the air, the period or pitch of the sound produced in the air and in the ear is the same as that of the sounding body.

336. As another example we may take one given by Herschel. Let a ray of light fall on a refracting substance like glass. The vibrations of the incident light must excite vibrations inside the glass. These last as long as the exciting cause continues and therefore constitute the forced vibration. The period of the refracted light is, by Herschel's theorem, the same as that of the incident light.

There are however some exceptions to this result. Thus in the *Phil. Trans.* for 1852 Sir G. Stokes has pointed out that light beyond the ultra violet by passing through certain substances may have its period so lengthened as to become visible. And Prof. Tyndall by means of the ultra red rays heated a platinum foil to incandescence and thus so shortened the periods that the vibrations became visible. See his *Rede Lecture*, 1865.

To understand the cause of these exceptions we must remember that *the forces of restitution have been taken proportional to the first power of the displacements,* i.e. only the first powers of x, y, &c. have been retained. Now the molecules of a body may be compounded of smaller atoms closely packed together. When the oscillations under consideration are such that only the molecules move amongst each other these displacements may be so small compared with the distances of the molecules from each other that the force of restitution $f(\xi)$, due to a displacement ξ of any molecule, may be replaced by the first power which occurs in McLaurin's expansion. But when the oscillations are such that the closely packed atoms of each molecule move amongst each other, the force of restitution may no longer vary as the first power of the displacement. Thus the equations of Art. 324 may apply to the former but not to the latter kind of motion. The reader will find more complete explanations in Sir G. Stokes' paper, see pages 549, 550.

It is obvious that the motion may be very different from that described above when the squares and cubes of the small quantities cannot be rejected. This will be especially noticeable when the terms of the first order are absent. An elementary example is given in Vol. I. Art. 450, where an oscillation leading to the differential equation $\delta^2\theta + a\theta^3 = 0$ is discussed. It is shown that the period, so far from being constant, varies inversely as the arc of vibration. If we represent a disturbing force by the term $P \sin \lambda t$ on the right-hand side of this equation, it is clear that the equation cannot be satisfied by a term of the form $\theta = Q \sin \lambda t$, so that the period of the forced oscillation is not the same as that of the force.

337. How a disturbing force is magnified. In dynamical problems as they occur in nature we often have a system oscillating freely about some mean position and acted on by a crowd of small forces which tend to disturb this motion. Some of these forces are very small in magnitude, others are greater. May we reject the small ones as compared with the greater? The number of forces is perhaps too large for us to consider the effects of each. It is evident that we require some rule to guide us in choosing those forces which produce the most important effects. For instance in the Planetary Theory each planet is pulled about by an innumerable number of causes of disturbance. It would be impossible to determine the actual motion without some principle to enable us to reject those forces which produce insensible disturbance.

Let a system be acted on by two permanent disturbing forces which we may represent by the two terms $P \sin(\lambda t + \alpha)$ and $Q \sin(\mu t + \beta)$ both placed in the first equation of Art. 324. The corresponding forced vibrations in the co-ordinate x are given by

$$x = \frac{I(\delta)}{\Delta(\delta)} P \sin(\lambda t + \alpha) + \frac{I(\delta)}{\Delta(\delta)} Q \sin(\mu t + \beta),$$

where $I(\delta)$ is the minor of the x term in the first line of the determinant $\Delta(\delta)$. These coefficients contain the operator δ and their magnitudes will therefore depend on λ and μ. We therefore infer that the *effects of different permanent disturbing forces acting under similar conditions on the same co-ordinate are not simply proportional to their respective magnitudes but depend on their periods.*

338. Without however restricting ourselves to permanent disturbing forces, let us consider the forced vibration produced by the disturbing force $Pe^{-\kappa t} \sin \lambda t$. Writing as before (Art. 327) $m = -\kappa + \lambda\sqrt{-1}$, the resulting forced vibration is the coefficient of $\sqrt{-1}$ in $\quad \dfrac{I(\delta)}{\Delta(\delta)} Pe^{mt} = P \dfrac{I(m)}{\Delta(m)} e^{mt}.$

If m is nearly equal to a root of $\Delta(\delta) = 0$ the denominator of this expression is very small. But the types of the free vibrations are given by $\Delta\,(m) = 0$ as shown in Art. 262. We therefore infer that *a disturbing force whose period and real exponential are nearly the same as those of any one free vibration produces a large forced vibration.*

339. Usually a disturbing force is of the permanent type $P \sin (\lambda t + \alpha)$. If there were no forces of resistance there would be free permanent oscillations in the system of the form $A \sin (pt + \beta)$, and we have just seen that, if λ were nearly equal to any value of p, the disturbing force would produce a magnified forced oscillation. But the resisting forces introduce real exponentials as factors of the free vibrations, (Art. 319). Thus the type of the disturbing force is no longer the same as that of any free vibration. We conclude that *one effect of the resistances on a disturbing permanent force, which would otherwise produce a magnified forced oscillation, is to modify that oscillation and keep it within bounds.*

340. As a simple example of this dynamical principle, *let us consider how easily a heavy swing can be set into violent oscillation by a series of little pushes and pulls if properly timed.* If we push when the swing is receding and pull when it is approaching us, the swing is continually accelerated and the arc of oscillation will be greater and greater at each succeeding swing. Such a series of alternations of push and pull is practically what we have called a permanent disturbing force whose period is the same as that of the free vibration of the swing. But if the period is very unequal to that of the free vibration, though a few pushes and pulls may increase the arc of vibration, yet a time comes when the effect is reversed. The force acts opposite to the motion of the swing and the oscillations will decrease just as they before increased.

It is well known that when a piano string is exposed to the air and is acted on by vibrations in that medium, the string will sometimes appear to be unaffected by the motion and at other times will sound a note. The reason is that though the string is always set in motion, yet, unless the aërial impulses on it are *properly timed*, the motion produced is too slight to be sensible. If however one of the existing notes in the air has the same period as one of those of the string, the pressure of the air on the string, like the impulses on the pendulum described above, will continually tend to increase the motion.

On the other hand the intensity of this particular note in the air is weakened by the amount communicated to the string while the intensities of the other aërial notes appear to be unaffected. Thus a piano string, or any vibrating body, will absorb or extract from the surrounding medium the same notes which it would produce in the air if independently set in motion. Sir G. Stokes uses this theory to explain on dynamical principles the discovery of Foucault and Kirchhoff, that a body may be at the same time a source of light giving out rays of a definite period and an absorbing medium extinguishing rays of the same period which traverse it. [*Phil. Mag.*, March, 1860.]

341. *We may take a second example from the rolling of ships at sea.* The ship has its own natural vibration together with that forced one which follows the oscillation of the waves. If the periods of these synchronise the rolling of the ship may become very great. Mr White in his *Manual of Naval Architecture* men-

tions several interesting examples of this. After noticing how some vessels are made to roll heavily by an almost imperceptible swell, he mentions the case of the Achilles, a vessel of great reputation for steadiness, which rolled more heavily off Portland in an almost dead calm than it did off the coast of Ireland in very heavy weather. Again in the cruise of the combined squadrons in 1871, though the Monarch far surpassed most of the vessels present in steadiness when the weather was heavy, there was one occasion (possibly owing to a near agreement between the natural period of this ship and the period of the waves) when the ship rolled more heavily in a long swell than some of the most notorious heavy rollers.

342. A good use of this principle was made by Capt. Kater in his experiments to determine the length of the seconds' pendulum. *It was important to determine if the support of his pendulum was perfectly firm.* He had recourse to a delicate and simple instrument invented by Mr Hardy a clockmaker, the sensibility of which is such that had the slighest motion taken place in the support it must have been instantly detected. The instrument consists of a steel wire, the lower part of which is inserted in the piece of brass which forms its support, and is flattened so as to form a delicate spring. On the wire a small weight slides by means of which it may be made to vibrate in the *same time* as the pendulum to which it is to be applied as a test. When thus adjusted it is placed on the material to which the pendulum is attached, and should this not be perfectly firm, the motion will be communicated to the wire, which in a little time will accompany the pendulum on its vibrations. This ingenious contrivance appeared fully adequate to the purpose for which it was employed, and afforded a satisfactory proof of the stability of the point of suspension. See *Phil. Trans.* 1818.

343. It has been shown in Art. 338 that a disturbing force may produce a large vibration in x if its period is such that the denominator $\Delta (\delta)$ is small. But this result is affected by the operator $I (\delta)$ which occurs in the numerator. If for instance the result of the operation of the minor $I (\delta)$ is zero, the forced vibration disappears.

Now these minors are just the operators used in finding the free vibrations. Thus in Art. 262, we have $x = I (\delta)$ [type].

If then any one of the free vibrations is absent from one of the co-ordinates though present in the others, then a disturbing force of nearly the same period does not produce a large forced vibration in that co-ordinate. We infer that a *disturbing force can produce a large forced vibration in any co-ordinate only if there be in that co-ordinate a free vibration of nearly the same period and containing nearly the same real exponential.*

344. If the force is nearly equal to $Pe^{-\kappa t} \sin (\lambda t + a)$, it may occur that the determinant $\Delta (\delta)$ has a roots equal to $-\kappa + \lambda \sqrt{-1}$, while the minor $I (\delta)$ has none of them. Referring to the expressions for the forced vibrations in the co-ordinates x, y, &c. given in Art. 326, we see that in this case the forced vibration is divided a times by a small quantity and is said to be magnified a times. But if the minor $I (\delta)$ has β of these roots, the forced vibration is magnified $a - \beta$ times. By reference to Art. 272 we see that the co-ordinate x has in this case powers of t up to the $(a - \beta - 1)^{\text{th}}$ in the coefficients of its free vibration. We infer that *the forced vibration in any co-ordinate is magnified once more than the highest power of t which occurs in that co-ordinate in connexion with the free vibrations of nearly the same period.*

345. As an example let us consider the case of a planet describing a circle about

the sun considered as fixed in the centre ; the radius vector r is then equal to a constant and the longitude $\theta = nt + \epsilon$. If slightly disturbed and acted on only by the attraction of the sun the planet describes an ellipse of small eccentricity e. The consequent changes in the radius vector and longitude are small and these changes may be represented by what we have called x and y. From the theory of elliptic motion we know these are approximately

$$x = a - ae \cos (nt + a),$$
$$y = bt + c + 2e \sin (nt + a),$$

where a, b, c are small quantities and $2\pi/n$ is the period of the planet. These are of course the free vibrations. Comparing these with the type $\sin (\lambda t + a)$ we see that two free vibrations occur in x, viz. $\lambda = n$ and $\lambda = 0$. There are three free vibrations in the expression for y, viz. $\lambda = n$ and two equal values of λ each zero. These equal values introduce the terms with powers of t as explained in Art. 266.

We infer that any small permanent periodical force produces a magnified disturbance both in the radius vector and longitude of a planet, if its period is nearly equal to that of the planet or is very long. Since there are two equal free periods in the longitude whose type is $\lambda = 0$ and only one in the radius vector, those small disturbing forces whose periods are very long are twice magnified in their effects on the longitude and once magnified in the radius vector. If any such forces as these act on the planet it is necessary to examine into their effects. Small disturbing forces, whose magnitudes are less than the standard of small quantities to be retained, may be disregarded only if their periods are different from those just indicated.

These rules are used in the Lunar and Planetary Theories to assist us in estimating the values of the disturbing forces. They enable us to separate from the crowd of small forces those which can produce sensible effects on the motions of the planets, see Art. 337.

346. How a disturbing force is diminished. Let us resume the expression given in Art. 326 for the forced vibration due to a continuous disturbing force. We remark in the first place that the denominator of the coefficient contains higher powers of λ than the numerator. To show this it may be sufficient to notice that the determinant of the motion $\Delta (\delta)$ has two powers of δ more than any of its minors. We therefore infer that, in the limit, when λ is very great, i.e. *when the period of the disturbing force is much smaller than that of any free oscillation, the forced vibration produced is in general insignificant.*

347. When the type of a continuous disturbing force $f(t)$ which acts directly on the co-ordinate x is such that it satisfies the differential equation $I_1 (\delta) f(t) = 0$, we remark in the second place that the forced oscillation in the co-ordinate x wholly vanishes. Now $I_1 (\delta) = 0$ is the determinantal equation whose roots give the free vibration when the co-ordinate x is constrained to be zero. We infer that when *the type of a disturbing force which acts directly on any co-ordinate x is nearly the same as any one of the modes of free vibration when x is constrained to be zero, then the forced vibration in x is very small.* See Art. 343.

348. Ex. A tight string, whose extremities A and B are fixed, is acted on

tranversely at any point C by a permanent disturbing force. If the period of the force is equal to any one of the periods of a string stretched with the same tension but whose length is either AC or CB, show that the forced vibration does not disturb the point C. If the strings AC, CB have no free period in common, show that one string is not moved by the forced vibration.

We may also deduce this result from some elementary considerations. Let the string be held at rest at C and let the part AC be set in motion, CB being at rest. The pressure at C when resolved perpendicular to the string will represent a permanent disturbing force whose period is equal to that of any one of the free vibrations of AC. Replacing the pressure by the disturbing force we have AC in vibration and CB at rest.

349. How an Impulse is diminished. When a system of machinery is moving in some state of steady stable motion it may be liable to disturbance from any sudden jerks whose effects it may be important to diminish as much as possible. Let us consider briefly what means we have to abate an impulse.

When the jerk has completed its work and has ceased to act, the system is displaced from its proper state of motion. It now begins to oscillate about this state. Thus one effect of the jerk is to introduce a new set of free oscillations. If there be any forces of resistance these free vibrations will begin to fade away and the system will tend to assume a state of steady motion. *One method of correcting the effects of a disturbing impulse is therefore to increase the resisting forces.*

The resistances which are thus intentionally introduced into the machinery should be properly arranged. They should be such as not to affect the steady motion, but to begin to act only when the machine deviates from its intended course. An example of this has been given in Art. 105, where the motion of the governor was discussed.

350. The actual effect of a jerk X on any co-ordinate such as x is easily deduced from the equations of Art. 118. If Δ be the discriminant of the quadric A where $2A = A_{11}x^2 + 2A_{12}xy + \ldots\ldots$ and I_{11} the minor of the constituent A_{11}, we have

$$\delta x_1 - \delta x_0 = (I_{11}/\Delta)\, X.$$

If then it is important to lessen the effects of the impulse X, we may make some addition to the machine or modify the arrangement of its parts so as to increase the discriminant Δ as compared with I as much as possible.

If the function A is a positive one-signed function, its discriminant Δ is positive. We may then show, as in the next article, that the ratio of I_{11} to Δ is in general *decreased* by the addition of the square of any linear function of x, y, &c. to the function A. Now the quadric function A with accented co-ordinates is part of the expression for the vis viva (Art. 111) and is always a positive function. Hence if any addition is made to the vis viva the corresponding addition to this function is also positive and may be expressed as the sum of a number of squares of linear functions. *We may therefore in general weaken the direct effects of jerks on a system by increasing the vis viva.*

The usual method of effecting this is to attach a fly-wheel to the machine. The vis viva of a rotating body is $Mk^2\omega^2$, where Mk^2 is the moment of inertia of the body about the axis and ω is the angular velocity. The advantage of using a wheel is that with a given quantity, of additional matter, the additional terms may be increased to any extent by increasing the radius of gyration.

351. **Ex. 1.** If the co-ordinates be so chosen that the square factor added to the quadric $2A$ is of the form μy^2, where y is any co-ordinate other than x, show that the ratio I_{11}/Δ becomes $(I_{11} + \mu \Delta_2)/(\Delta + \mu I_{22})$, where Δ_2 is the second minor formed by omitting the first two rows and columns, and the suffix of each I indicates as usual the constituent of which that I is the minor. Show also that the second ratio is less than the first by $I_{12}^2 \mu/\Delta (\Delta + \mu I_{22})$. Show also that *this difference is positive or zero and has a finite limit when μ is infinite.*

Ex. 2. If the square factor added to the quadric $2A$ be $\mu (ax + by + cz + ...)^2$, show that the direct effect of an impulse represented by X on the co-ordinate x is *not altered* by this addition to the inertia if $a^2 I_{11}^2 + 2ab I_{11} I_{12} + b^2 I_{12}^2 + ... = 0$.

352. **The interval at which any phase of effect follows the same phase of cause.** Any disturbing force tends alternately to increase and decrease the deviation of the system from its undisturbed position, but it is not necessarily true that this deviation actually increases when the force urges an increase or decreases when the force urges a decrease. To examine into this point we notice that by Art. 326 the forced vibration produced by a disturbing force $Pe^{-\kappa t} \sin (\lambda t + \alpha)$ is

$$Pe^{-\kappa t} \{ L \sin (\lambda t + \alpha) + M \cos (\lambda t + \alpha) \}$$
$$= P \sqrt{L^2 + M^2} e^{-\kappa t} \sin (\lambda t + \alpha + \tan^{-1} M/L).$$

In this transformation it is clear that if the square root in the coefficient be regarded as positive, the angle added to the phase must be such that its sine has the same sign as M and its cosine the same sign as L. The consequence is that all the possible values of the change of phase differ by multiples of 2π.

Comparing the expression for the forced vibration with that for the disturbing force we see that their maxima do not occur simultaneously. The maximum of the oscillation occurs later than the maximum of the force by an interval equal to $-(1/\lambda) \tan^{-1} (M/L)$. In the same way *every phase of the oscillation follows the corresponding phase in the force after the same interval.*

The change of phase in any co-ordinate thus depends on the values of L and M for that co-ordinate. These are easily found by the rule given in Art. 327, where it is shown that if we write $\delta = -\kappa + \lambda \sqrt{-1}$ in the operator $I(\delta)/\Delta(\delta)$ for that co-ordinate the result is $L + M \sqrt{-1}$.

353. If the disturbing force is permanent, i.e. is of the form $P \sin (\lambda t + \alpha)$, and if the forces of resistance are neglected, the determinant $\Delta (\delta)$ contains only even powers of δ. We infer therefore from Art. 326 that *if the minor $I (\delta)$ also contains only even powers of δ, the phase of the forced oscillation is the same as that of the force or is greater by π. If the minor $I (\delta)$ contains only odd powers, the phase of the oscillation is greater than that of the force by $\pm \frac{1}{2}\pi$.*

If we consider the *direct* effect of a force on any co-ordinate the minor $I (\delta)$

contains only even powers of δ, as well as the determinant $\Delta(\delta)$. If the centrifugal forces are absent as when the system oscillates about a position of equilibrium, every minor contains only even powers of δ. *In these cases the forced vibration is simply a multiple positive or negative of the disturbing force without further change of phase.*

354. Ex. A particle describes a nearly circular orbit about a centre of force which attracts according to the Newtonian law, and is acted on by a permanent disturbing force along the radius vector. Show that the particle at any moment is inside the mean circular orbit when the force acts outwards and outside when the force acts inwards, provided the period of the force is less than that of the particle in its undisturbed orbit round the centre of the force. But the reverse of this is the case if the period of the disturbing force is greater than that of the particle. Would there be a similar distinction of cases if the centre of force attracted according to some inverse power greater than 3 ? See Art. 329.

Second approximations.

355. When we try to find the oscillations of a dynamical system we generally proceed by continued approximations. We first reject all the squares of the small quantities and thus obtain a set of linear differential equations. Solving these we substitute the results in the terms of the second order and treat these functions of t as disturbing forces. Their corresponding forced vibrations are then found. The operation may be repeated for a third approximation and so on.

It has been shown in Art. 337 that when the forces of resistance are small, a permanent disturbing force whose period is nearly equal to that of any one of the free vibrations produces a *magnified forced vibration*. It follows that a small force of proper period which would appear in the differential equations only when we include terms of (say) the third order may produce oscillations in the co-ordinates which are of the second or first order.

If therefore we wish to have our results correct to any given order it will be necessary to retain, for examination, those periodic terms of higher orders in the differential equations whose periods are nearly equal to any of the free vibrations.

We also see the importance of proceeding to higher approximation. These small terms which produce such large forced vibrations may not make their appearance until the terms of the higher orders are examined. Thus some important oscillations may be missed if we stopped at a first approximation.

356. When we substitute our first approximation in the terms of the higher orders it sometimes happens that permanent disturbing forces make their appearance *whose periods are exactly the same as those of some of the free vibrations included in the first approximation*. When this occurs, it has been shown in Art. 328 that the forced vibration changes its character. The solution now contains terms with powers of t as factors. These terms (not being balanced by the proper exponential factors, Art. 283) will become large, so that the system will depart widely from the state indicated by the approximate solution.

This is another way of saying that what we have taken as our first approximation is not sufficiently near to the truth to serve as an approximation. In most dynamical problems the disturbing forces are given as functions of the co-ordinates and then by the approximate solution expressed as functions of the time. Thus the expressions for the forces themselves are only approximations. It *may* therefore happen that if we can obtain a more correct first approximation to the motion the small terms which indicate such a large departure from the first approximation may not make their appearance.

To find a sufficiently correct first approximation to the motion it may not be enough to take the solution of the differential equations when all the terms of the higher orders are neglected. We must include in these differential equations all those small terms of the higher orders which materially affect the motion. The solution of these modified equations (if one can be found) is to be taken as our first approximation.

Let us repeat the argument in a slightly different form. The first approximation comprises all the largest terms in the expressions for the co-ordinates and may generally be taken to represent the visible motion of the system. If now a disturbing force, such as that we have just described, act on the system, it greatly modifies the visible motion and in turn its own period is modified by the change of motion. Thus the system takes up some new state of steady motion with oscillations about that steady motion. This obliges us to abandon the former first approximation in order to use one which may be a permanent representation of the new visible motion.

When we examine this new first approximation, as in the following examples, we find that it sometimes has the same general character as the former, but with the important exception that the free vibration whose period was the same as that of the force has been modified. We therefore infer that *when a small disturbing force is wholly or in part a function of the co-ordinates and has the same period as a free oscillation of the system, it may have the effect of removing that type of free oscillation from the system and replacing it by some other type of a different period.*

357. Before proceeding to the general theory we shall illustrate the method of proceeding by a simple example.

A particle oscillates in a straight line about a centre of force whose attraction at a distance x is represented by $p^2x + \beta x^3$. Find the time of a small oscillation.

The equation of motion is clearly

$$\ddot{x} + p^2 x = -\beta x^3 \dots \dots \dots (1),$$

where dots represent differentiations with regard to t.

As a first approximation we reject the term on the right-hand side as being of the third order of small quantities. We then find

$$x = M \sin(pt + a) \dots \dots \dots (2).$$

Proceeding to a second approximation we substitute this in the term previously rejected. We have $\quad \ddot{x} + p^2 x = -\tfrac{1}{4}\beta M^3 \{3 \sin(pt + a) - \sin 3(pt + a)\} \dots \dots (3).$

The first trigonometrical term on the right-hand side has the same period as the oscillation which represents the first approximation and therefore modifies that approximation (Art. 356). To include its effects we must alter equation (2). This modified solution when substituted in the differential equation must make the left-hand side, not equal to zero as before, but equal to a very small quantity, viz. the small disturbing force. As a trial solution we shall therefore retain the same general form. The letters M and a, being undetermined, will still serve for general symbols, but we shall replace p by $p + \mu$ where μ is some small quantity to be

determined by the disturbing force. We shall therefore write the first approximation in the form
$$x = M \sin\{(p+\mu)\,t+a\} \quad\text{...........................(4)}.$$
Proceeding to a second approximation we have
$$\ddot{x} + p^2 x = -\tfrac{3}{4}\beta M^3 \sin\{(p+\mu)\,t+a\}.$$
If our correction is successful, this equation must be satisfied by our amended first approximation. Substituting we find the equation is satisfied provided
$$M\{-(p+\mu)^2 + p^2\} = -\tfrac{3}{4}\beta M^3, \quad \therefore\ \mu = \tfrac{3}{8}\frac{\beta}{p}M^2 \text{ nearly}.$$

Thus the oscillations of the particle about the centre of force are very nearly represented by equation (4). *The effect of the disturbing force $-\beta x^3$ is to shorten the time of oscillation by a quantity which depends on the square of the arc.*

358.   If the force of attraction had been $p^2x + \beta\,(dx/dt)^3$ instead of that given above, we may show that this process would have failed.

Taking the first approximation as before and substituting in the differential equation we obtain
$$\ddot{x} + p^2 x = -\tfrac{1}{4}\beta M^3\{3\cos(pt+a) + \cos 3\,(pt+a)\}.$$
Neglecting the second trigonometrical term as before, let us try to include the other in our first approximation. Taking the amended form (4) and substituting we find that we should have
$$M\{-(p+\mu)^2 + p^2\}\sin\{(p+\mu)\,t+a\} = -\tfrac{3}{4}\beta M^3\cos\{(p+\mu)\,t+a\}.$$
But this equation cannot be satisfied by any constant value of μ. *The effect of this disturbing force is therefore not merely to alter the time of oscillation.*

359.   **Ex.**   A particle describes a nearly circular orbit about a centre of force whose attraction at a distance r is represented by $\mu\,(u^2+\beta u^n)$ where u is the reciprocal of r. If β is very small show that the path is nearly represented by
$$u = a\{1 + e\cos(c\theta - a)\},$$
where
$$c = 1 - \tfrac{1}{2}\beta a^{n-2}\,(n-2)\{1 + \tfrac{1}{8}\,(n-3)\,(n-4)\,e^2 + \&c.\},$$
provided the square of β can be neglected. This example is a modification of a case which occurs in the Lunar Theory.

360.   **General Theory.**   Having illustrated the method of treating the terms of the higher orders by several examples, we shall now consider the subject more generally. *Our object is to so modify the first approximate solution as to include in it (when such a thing is possible) the effects of small forces whose periods are the same as those of the free vibrations* (Art. 356). The general result arrived at will be given in the summary at the end of the argument.

We shall suppose the left-hand sides of the differential equations to contain all the first powers of the small co-ordinates x, y, z, &c. These therefore take the form given in Art. 324 or more generally Art. 262. The disturbing forces are placed on the right hand sides and contain powers and products higher than the first of the co-ordinates x, y, &c., and their differential coefficients. Thus all these disturbing forces would be neglected if we took into account only the terms of the first order. We shall also suppose that these disturbing forces are not explicit functions of the time. If this condition is not satisfied, the following analysis must be slightly modified.

361.   To avoid a complication of symbols let us resume the exponential values of the sine and cosine. Let then the first approximation obtained by neglecting in the differential equation all terms beyond the first order be
$$x = M_1 e^{m_1 t} + M_2 e^{m_2 t} + \dots, \quad y = N_1 e^{m_1 t} + N_2 e^{m_2 t} + \dots, \quad \&c. = \&c., \quad\text{.........(1)},$$

where m_1, m_2, &c. are the roots real or imaginary of the determinant Δ (δ) (Art. 262). On proceeding to a second approximation we substitute these values of x, y, &c. in the several small terms which were before neglected. Taking some term which contains the products and powers of the variables the result of the substitution produces disturbing forces of the form

$$\Sigma P e^{(f m_1 + g m_2 + \ldots) t} \quad \ldots\ldots\ldots\ldots\ldots\ldots\ldots\ldots\ldots\ldots\ldots\ldots(2),$$

where the order of the term is $f + g + \ldots$ If these quantities f, g, &c. are such that any number of relations hold of the form

$$f m_1 + g m_2 + \ldots = m_1 \quad \ldots\ldots\ldots\ldots\ldots\ldots\ldots\ldots\ldots\ldots\ldots(3),$$

there are just so many of these disturbing forces which take the type $P e^{m_1 t}$. The forced vibrations derived from these are obtained by using the operator $I(\delta)/\Delta(\delta)$ and are evidently infinite. To include these in the first approximation we replace the equations (1) by

$$x = M_1 e^{n_1 t} + M_2 e^{n_2 t} + \ldots \quad y = N_1 e^{n_1 t} + N_2 e^{n_2 t} + \ldots \quad \text{&c.} = \text{&c.} \quad \ldots\ldots\ldots(4),$$

where the M's N's, &c. are not necessarily the same as before, and each n only differs slightly from the corresponding m. Substituting as before we of course obtain a disturbing force of the form (2) but with n's written for the m's. If we assume the same relations to hold as before between the exponents, viz.

$$f n_1 + g n_2 + \ldots = n_1 \ldots\ldots\ldots\ldots\ldots\ldots\ldots\ldots\ldots\ldots\ldots(5),$$

this force takes the type $P e^{n_1 t}$. There may also be other relations similar to (5) but with n_2 or n_3, &c. written for n_1 on the right-hand side and these introduce other disturbing forces whose effects have also to be included in the new first approximation.

Including these forces we may write the differential equations in the form

$$\left.\begin{array}{l} f_{11}(\delta)\, x + f_{12}(\delta)\, y + \ldots = P_1 e^{n_1 t} + P_2 e^{n_2 t} + \ldots \\ f_{21}(\delta)\, x + f_{22}(\delta)\, y + \ldots = Q_1 e^{n_1 t} + Q_2 e^{n_2 t} + \ldots \\ \quad\quad \text{&c.} = \text{&c.} \end{array}\right\} \quad \ldots\ldots\ldots\ldots\ldots(6),$$

where the functional symbols $f_{11}(\delta)$ &c. have been used for the sake of brevity. If we have been successful in including the effects of these disturbing forces in our new first approximation, these differential equations must be satisfied by the values of x, y, &c. given in (4). Substituting we have

$$\left.\begin{array}{l} f_{11}(n_1)\, M_1 + f_{12}(n_1)\, N_1 + \ldots = P_1 \\ f_{21}(n_1)\, M_1 + f_{22}(n_1)\, N_1 + \ldots = Q_1 \\ \quad\quad \text{&c.} = \text{&c.} \end{array}\right\} \quad \ldots\ldots\ldots\ldots\ldots\ldots\ldots(7),$$

with similar equations for each of the other disturbing forces.

In these equations the M's are to be regarded as arbitrary, their values being reserved to satisfy the initial conditions of the motion. Our object is to find the values of the remaining coefficients, viz., the N's and also the values of the n's in terms of the M's. *These values of the n's must also satisfy the relations* (5). *Supposing this test to be satisfied* we have found values of the co-ordinates which satisfy the differential equations to the first order, and include the disturbing forces which appeared to threaten the stability of the system.

362.   The forces P, Q, &c. may each consist of several terms of different orders of smallness. But the lowest is supposed to be of a higher order than the coefficients M, N &c. Taking only the lowest powers which occur in P, Q &c., we may easily find a first approximation to the values of n_1, n_2 &c. Solving the equations (7) we find

$$M_1 \Delta(n_1) = P_1 I_{11}(n_1) + Q_1 I_{21}(n_1) + \text{&c.},$$

where $I_{11}(n)$ &c. are as usual the minors of the determinant $\Delta(n)$. Let $n_1 = m_1 + \mu_1$, $n_2 = m_2 + \mu_2$ &c. Since all the terms on the right-hand side are smaller than M_1 we may in these terms write $n_1 = m_1$, $n_2 = m_2$ &c. Remembering that $\Delta(m_1) = 0$, we have

$$M_1 \frac{d\Delta(m_1)}{dm_1} \mu_1 = P_1 I_{11}(m_1) + Q_1 I_{21}(m_1) + \&c. \dots\dots\dots\dots\dots(8).$$

In the same way we have

$$M_2 \frac{d\Delta(m_2)}{dm_2} \mu_2 = P_2 I_{11}(m_2) + Q_2 I_{21}(m_2) + \&c.$$

The forces P_1 &c. are functions of M_1, N_1 &c., M_2, N_2 &c. But looking at equations (7) we see that the ratios of M_1, N_1 &c., differ from the ratios of the minors $I_{11}(m_1)$, $I_{12}(m_1)$ &c. by quantities of the order P/M. We may therefore in calculating the values of P_1 &c., substitute for N_1 &c., N_2 &c. by the help of these ratios. Thus the right-hand sides of the equations (8) are all known functions of the arbitrary M's and of the roots of the determinantal equation $\Delta(\delta) = 0$.

The quantities f, g &c. are usually positive integers. In this case the orders of the quantities P &c. are not less than $f + g + \&c.$ It follows that the corrections μ_1, μ_2 &c. are of the order $f + g + \&c. - 1$ at least.

363. Summary of results. We may embody the results of equations (8) in a rule.

Taking the first approximation viz. $x = M_1 e^{m_1 t} + \&c.$ found by rejecting all terms of the higher orders in the differential equations, we proceed to a second approximation. Suppose that in consequence of some relations such as

$$fm_1 + gm_2 + \&c. = m_1,$$

we arrive at disturbing forces $P_1 e^{m_1 t}$, $P_2 e^{m_2 t}$ &c. These would produce infinite terms in the co-ordinate x, if we employed the operators $I(\delta)/\Delta(\delta)$, &c. as usual (Art. 326). Instead of these let us employ the operators $I(\delta)/\Delta'(\delta)$, &c. simply replacing $\Delta(\delta)$ by $\Delta'(\delta)$. Let the result be $x = He^{m_1 t} + Ke^{m_2 t} + \&c.$, where H and K contain powers of M_1, M_2, &c. above the first. Then the effects of these disturbing forces may be taken account of to the next approximation by replacing the first approximation by $x = M_1 e^{(m_1 + \mu_1)t} + M_2 e^{(m_2 + \mu_2)t}$ where $\mu_1 = H/M_1$, $\mu_2 = K/M_2$ &c., provided that these new indices satisfy the relations $f\mu_1 + g\mu_2 + \&c. = \mu_1$, &c.

Supposing this condition to be satisfied, we see that a disturbing force of the same type and period as a free vibration has the effect of removing that type from the system and replacing it by some other type of vibration which is more and more remote from the original type the greater the amplitude of the vibration.

364. Examples. *A pendulum swings in a very rare medium, resisting partly as the velocity and partly as the square of the velocity, to find the motion.*

Let θ be the angle the straight line joining the point O of support to the centre of gravity G of the pendulum makes with the vertical. Let $g = ln^2$ where l is the length of the simple equivalent pendulum. Then the equation of motion is

$$\ddot{\theta} + n^2 \sin\theta = -2\kappa\dot{\theta} - \mu\dot{\theta}^2 \dots\dots\dots\dots\dots\dots\dots(1),$$

where 2κ and μ are the coefficients of the resistance divided by the moment of inertia of the pendulum about the axis of suspension. Since θ is small we may write the equation in the form

$$\ddot{\theta} + n^2\theta = -2\kappa\dot{\theta} - \mu\dot{\theta}^2 + \tfrac{1}{6}n^2\theta^3 - \dots$$

Since κ and θ are very small, we might at first suppose that it would be sufficient as a first approximation to reject all the terms on the right-hand side.

This gives $\theta = a \sin nt$, the origin of measurement of t being so chosen that t and θ vanish together. If we substitute this in the small terms we get

$$\ddot{\theta} + n^2\theta = -2\kappa n \cdot a \cos nt + \tfrac{1}{3} n^2 a^3 \sin nt + \&c.,$$

which gives $\qquad \theta = a \sin nt - \kappa a t \sin nt + \tfrac{1}{16} n a^3 t \cos nt + \&c.$

These additional terms contain t as a factor, and show that our first approximation was not sufficiently near the truth to represent the motion except for a short time. To obtain a sufficiently near first approximation we must include in it the small term $2\kappa\, d\theta/dt$ (Art. 356). We have therefore

$$\ddot{\theta} + 2\kappa\dot{\theta} + n^2\theta = 0.$$

This gives $\theta = a e^{-\kappa t} \sin mt$, where for the sake of brevity we have put $n^2 - \kappa^2 = m^2$.

In our second approximation we reject all terms of the order a^3 or $a^2\kappa$ unless they are such that after integration they rise in importance in the manner explained in Art. 344. We thus get

$$\ddot{\theta} + 2\kappa\dot{\theta} + n^2\theta = -\tfrac{1}{2}\mu a^2 m^2 e^{-2\kappa t}(1 + \cos 2mt) + \tfrac{1}{24} n^2 a^3 e^{-3\kappa t}(3 \sin mt - \sin 3mt)$$
$$- \tfrac{1}{2}\mu a^2 \kappa e^{-2\kappa t}(-\kappa + \kappa \cos 2mt + 2m \sin 2mt),$$

where all the terms on the right-hand side after the first are of the third order, and are to be rejected unless they rise in importance. To solve this, let us first consider the general case

$$\ddot{\theta} + 2\kappa\dot{\theta} + n^2\theta = e^{-p\kappa t} \cdot (A \sin rmt + B \cos rmt).$$

Put $\theta = e^{-p\kappa t}(L \sin rmt + M \cos rmt)$. Substituting we get

$$\left.\begin{array}{l} L\{(p-1)^2\kappa^2 + m^2(1-r^2)\} + 2(p-1)\kappa rmM = A \\ M\{(p-1)^2\kappa^2 + m^2(1-r^2)\} - 2(p-1)\kappa rmL = B \end{array}\right\}.$$

Now κ is very small; if then r be not equal to unity, we have $L = \dfrac{A}{m^2(1-r^2)}$, $M = \dfrac{B}{m^2(1-r^2)}$ nearly; but if $r = 1$, we have $L = \dfrac{-B}{2(p-1)\kappa m}$, $M = \dfrac{A}{2(p-1)\kappa m}$ nearly. The case of $p = 1$ does not occur in our problem. It appears that those terms only in the differential equation which have $r = 1$ give rise to terms in the value of x which have the small quantity κ in the denominator. Hence in the differential equation the only term of the third order which should be retained is the first. We thus find, putting successively $r = 0$, $r = 2$, $r = 1$,

$$\theta = a e^{-\kappa t}\sin mt - \frac{\mu a^2}{2}e^{-2\kappa t} + \frac{\mu a^2}{6}e^{-2\kappa t}\cos 2mt + \frac{n^2 a^3}{32\kappa m}e^{-3\kappa t}\cos mt.$$

This equation determines the motion only during any one swing of the pendulum; when the pendulum turns to go back μ changes sign. Let us suppose the pendulum to be moving from left to right, and let us find the lengths of the arcs of descent and ascent. To do this, we must put $d\theta/dt = 0$. Let the equation be written in the form $\theta = f(t)$, then if we neglect all the small terms, $d\theta/dt$ vanishes when $mt = \pm\tfrac{1}{2}\pi$, say when $t = \pm T$. Put then $t = -T + x$ where x is a small quantity, we have

$$f'(t) = f'(-T) + f''(-T)x = 0.$$

Now

$$f'(t) = ae^{-\kappa t}(m \cos mt - \kappa \sin mt) - \frac{\mu a^2}{2}e^{-2\kappa t}\left(-2\kappa + \frac{2\kappa}{3}\cos 2mt + \frac{2m}{3}\sin 2mt\right)$$
$$+ \frac{n^2 a^2}{32\kappa m}e^{-3\kappa t}(-m \sin mt - 3\kappa \cos mt).$$

A sufficiently near approximation to the value of $f''(t)$ may be found by differentiating the first term of $f'(t)$. We thus find $m^2 x = -\kappa - \tfrac{4}{3}\mu a \kappa - \tfrac{1}{32}n^2 a^2/\kappa$;

the second of these terms being smaller than the other two might be neglected. We also find as the arc of descent

$$\theta = f(-T) + f'(-T) x = -\{ae^{\kappa T} + \tfrac{2}{3}\mu a^2 e^{2\kappa T} - mx (\kappa a e^{\kappa T} + \tfrac{1}{32} n^2 a^3 e^{6\kappa T}/\kappa m)\}.$$

To find the arc of ascent we put $t = T + y$. This gives $m^2 y = -\kappa - \tfrac{1}{32} n^2 a^2/\kappa$ and the arc of ascent is

$$\theta = a e^{-\kappa T} - \tfrac{2}{3}\mu a^2 e^{-2\kappa T} - my (\kappa a e^{-2\kappa T} + \tfrac{1}{32} n^2 a^3 e^{-3\kappa T}/\kappa m).$$

In these expressions for the arcs of descent and ascent the terms containing x and y are very small, and assuming κ not to be extremely small, these terms will be neglected *.

Now a is different for every swing of the pendulum, we must therefore eliminate a. Let u_n and u_{n+1} be two successive arcs of descent and ascent, and let $\lambda = e^{-\kappa T}$, so that λ is a little less than unity. Then we have

$$u_n = a \frac{1}{\lambda} + \frac{2}{3}\mu a^2 \frac{1}{\lambda^2}, \qquad u_{n+1} = a\lambda - \frac{2}{3}\mu a^2 \lambda^2;$$

eliminating a we have very nearly $\dfrac{1}{u_{n+1}} + \dfrac{1}{c} = \dfrac{1}{\lambda^2}\left(\dfrac{1}{u_n} + \dfrac{1}{c}\right)$,

where $c = \dfrac{3}{2\mu} \dfrac{1-\lambda^2}{1+\lambda^2} = \dfrac{3\kappa\pi}{4\mu m}$ nearly, and $T = \dfrac{\pi}{2m}$.

The successive arcs are, therefore, such that $1/u_n + 1/c$ is the general term of a geometrical series whose ratio is $e^{\kappa\pi/m}$. The ratio of any arc u_n to the following arc

u_{n+1} is

$$\frac{u_n}{u_{n+1}} = e^{2\kappa T} + \frac{u_n}{c}(e^{2\kappa T} - 1),$$

which continually decreases with the arc. In any series of oscillations the ratio is at first greater and afterwards less than its mean value. This result is found to agree with experiment.

To find the time of oscillation. Let t_1, t_2 be the times at which the pendulum is at the extreme left and right of its arc of oscillation. Then

$$mt_1 = -\frac{\pi}{2} - \frac{\kappa}{m} - \frac{n^2 a^2}{32 m\kappa}, \qquad mt_2 = \frac{\pi}{2} - \frac{\kappa}{m} - \frac{n^2 a^2}{32 m\kappa}.$$

The time of oscillation from one extreme position to the other is $t_2 - t_1$ which is equal to π/m. This result is independent of the arc, so that the time of oscillation remains constant throughout the motion. The time is however not exactly the same as in vacuo, but is a little longer; the difference depending on the *square* of the small quantity κ. See Art. 321.

Ex. 2. A rigid body is suspended by two equal and parallel threads attached to it at two points symmetrically situated with respect to a principal axis through the centre of gravity which is vertical, and being turned round that axis through a small angle is left to perform small *finite* oscillations. Investigate the reduction to infinitely small oscillations. [Smith's Prize.]

---

* If these terms are not neglected the equation connecting the successive arcs of descent and ascent becomes $\dfrac{1}{u_n} - \dfrac{\lambda^2}{u_{n+1}} = -\dfrac{2}{3}\mu(1+\lambda^2) + \dfrac{n^2 x}{32\kappa m}\dfrac{1-\lambda^4}{\lambda}$. Now $1-\lambda^4 = \dfrac{2\kappa\pi}{m}$ nearly, so that this additional term is very small compared with that retained.

CHAPTER VIII.

Method of Isolation.

365. OUR object in this chapter may be very briefly stated. Given any number of simultaneous differential equations with constant coefficients, it is known that the dependent variables $x, y, z,$ &c. can be expressed in terms of the independent variable t, by means of a series of exponentials real or imaginary. Let one of these exponentials be $x = Me^{mt}$, then M is a function of the initial values of the variables x, y, &c. and of their differential coefficients. It is here proposed to exhibit this function. Thus without solving the equations, any one term of the solution, if its exponent be known, can be separated from the others and have its value written down, without finding those other terms.

When the differential equations are not of a high order we can generally solve the determinantal equation and find all the possible values of m. In such a case it is merely a question of algebra to find the constants in terms of the initial values of the variables. We may, however, effect this more briefly and simply by using the rule here given. Sometimes it is impossible to solve the determinantal equation. We may find one or more roots, but the rest remain unknown. In such a case we could not proceed by the processes of common algebra, for the equations cannot be written down. *Our object is to find the constants which accompany these known terms without the knowledge of the remaining ones.*

This method is very simple and easy of application when the exponential to be separated from the others is connected with a solitary root of the fundamental determinant. But it may be used even though the root is repeated several times. The complication arises from the fact that the exponential is then accompanied by as many constants as there are equal roots. Each of these requires a separate operation to find its value.

The method is generally applicable whatever be the order of the equations, but there is considerable simplification when the order is not higher than the second. This is of course the most interesting case, as the equations may then be such as occur in dynamics.

In some cases the rule can be put into another form, which may possibly be thought simpler. In these cases it takes the form of the Method of Multipliers. When the number of dependent variables is infinite, we have an example in Fourier's rule to expand any function in a series of sines or cosines.

366. **The Determinant of Isolation.** Resuming the notation of Art. 262, we let the n equations to find x, y, z, &c. be written in the form

$$\left.\begin{aligned} f_{11}(\delta)\,x + f_{12}(\delta)\,y + f_{13}(\delta)\,z + \ldots &= 0 \\ f_{21}(\delta)\,x + f_{22}(\delta)\,y + f_{23}(\delta)\,z + \ldots &= 0 \\ \ldots\ldots\ldots\ldots\ldots\ldots\ldots\ldots\ldots\ldots\ldots &= 0 \end{aligned}\right\},$$

where δ as before stands for d/dt. In dynamical applications these functions of δ are all of the second degree, but at present we make no restriction of that kind.

To solve these we proceed as explained in Art. 262 and form the determinant

$$\Delta(\delta) = \begin{vmatrix} f_{11}(\delta), & f_{12}(\delta), & f_{13}(\delta) \ldots \\ f_{21}(\delta), & f_{22}(\delta), & f_{23}(\delta) \ldots \\ \ldots\ldots\ldots\ldots\ldots\ldots\ldots\ldots \end{vmatrix}.$$

If we equate this determinant to zero, we have an equation to find δ. Let its roots be m, m_2, &c. omitting the suffix of the first for the sake of brevity. Then we know that

$$x = M e^{mt} + M_2 e^{m_2 t} + \ldots.$$

It is our present object to find any one of these coefficients, say M, without finding any of the others.

To effect this we deduce from the determinant $\Delta(\delta)$ another determinant, which we write

$$\Pi(m) = \begin{vmatrix} \dfrac{f_{11}\delta - f_{11}m}{\delta - m}\,x + \dfrac{f_{12}\delta - f_{12}m}{\delta - m}\,y + \&c., f_{12}(m), f_{13}(m)\,\&c. \\[2ex] \dfrac{f_{21}\delta - f_{21}m}{\delta - m}\,x + \dfrac{f_{22}\delta - f_{22}m}{\delta - m}\,y + \&c., f_{22}(m), f_{23}(m)\,\&c. \\[2ex] \ldots\ldots\ldots\ldots\ldots\ldots\ldots\ldots\ldots\ldots\ldots\ldots\ldots\ldots \end{vmatrix}.$$

We form this determinant by the following rule. *Erase any column of the determinant $\Delta(\delta)$, say the first column. To replace it we divide the first equation by $\delta - m$, and rejecting the remainder place the quotient in the first row of the erased column. We divide the second equation by $\delta - m$ and place the quotient in the second row, and so on. Finally we put $\delta = m$ in the remaining columns.*

If we erase the second column of the determinant $\Delta(\delta)$ or $\Delta(m)$ we obtain a slightly different determinant, which we may write $\Pi_2(m)$, the suffix indicating which column of $\Delta(m)$ we erase.

The determinant $\Pi(m)$ is evidently a function of x, y, &c., δx, δy, &c., $\delta^2 x$, $\delta^2 y$, &c., up to one less than the highest power δ in the given differential equations. For all these we write their given initial values. We then have

$$M = \frac{\Pi(m)}{\Delta'(m)},$$

where $\Delta'(m)$ means as usual the differential coefficient of $\Delta(m)$ with regard to m. In the same way if Ne^{mt} be the corresponding term in the value of y, we have $N = \dfrac{\Pi_2(m)}{\Delta'(m)}$, and so on.

367. Examples. Before proceeding to the demonstration of this theorem let us consider some examples.

Ex. 1.   Taking the equations
$$\begin{aligned}(\delta^2 - 4\delta)\, x - (\delta - 1)\, y &= 0 \\ (\delta + 6)\, x + (\delta^2 - \delta)\, y &= 0\end{aligned}\bigg\},$$
we see that the fundamental determinant

$$\Delta(m) = \begin{vmatrix} m^2 - 4m, & -(m-1) \\ m+6, & m^2 - m \end{vmatrix} = m^4 - 5m^3 + 5m^2 + 5m - 6.$$

Equating this to zero, we find that one value of m is $m = -1$. Let us find the coefficient of e^{-t} in the value of x.

Dividing the equations by $\delta + 1$ and rejecting the remainders, we form at once the second determinant, viz.
$$\Pi(m) = \begin{vmatrix} (\delta - 5)\, x - y, & 2 \\ x + (\delta - 2)\, y, & 2 \end{vmatrix},$$
the second column being obtained by putting $m = -1$ in the second column of $\Delta(m)$. Expanding, and noticing that $\Delta'(m) = -24$ when $m = -1$, we find

$$-12M = \delta x - \delta y - 6x + y,$$

where M is the required coefficient. Here x, y, δx, δy are supposed to have their known initial values.

We may show in the same way that there is a term $M'e^{2t}$ in the value of x where $-3M' = 2\delta x + \delta y - 3x - y$.

Ex. 2.   Let us take another example, in which the differential coefficients rise to a higher order, but let us still restrict ourselves to two dependent variables to save space. Taking the equations
$$\begin{aligned}(\delta^3 + 2\delta^2 + \delta + 1)\, x + (\delta^3 + 2\delta + 1)\, y &= 0 \\ (\delta^2 + 2\delta + 2)\, x + (\delta^4 + \delta + 2)\, y &= 0\end{aligned}\bigg\},$$
we see by inspection that the determinantal equation is satisfied by $m = 1$. Thus $x = Me^t$ is a part of the solution. Let it be required to find M when the initial values of δx, $\delta^2 x$, δy, $\delta^2 y$, $\delta^3 y$ are all zero, and the initial values of x and y unity. Constructing the function Π by dividing each equation by $\delta - 1$, and putting $\delta = 0$ as we proceed, we have
$$\Pi(m) = \begin{vmatrix} 4x + 3y, & 4 \\ 3x + 2y, & 4 \end{vmatrix} = M\Delta'(m).$$
But, differentiating the determinant without expanding it, and putting $m = 1$, we have $\Delta'(m) = 16$. Hence, putting x and y each equal to unity, we immediately find $M = \frac{1}{2}$.

368. *We now proceed to the proof of the rule.*

Let p be some quantity which we shall write for m in the definition of the determinant $\Pi\,(m)$ in order to call attention to the fact that p is not necessarily a root of $\Delta\,(\delta) = 0$.

Taking the general expression for the determinant $\Pi\,(p)$ given in Art. 366, we may resolve it into the difference of two determinants, the first rows of each of which may be written as follows.

$$\Pi\,(p) = \frac{1}{\delta - p} \begin{vmatrix} f_{11}(\delta)\,x + f_{12}(\delta)\,y + \&c., & f_{12}(\delta), & \&c. \end{vmatrix}$$

$$-\frac{1}{\delta - p} \begin{vmatrix} f_{11}(p)\,x + f_{12}(p)\,y + \&c., & f_{12}(p), & \&c. \end{vmatrix}$$

Consider the first determinant, the first column is occupied by the functions which form the differential equations. Hence this determinant vanishes whenever x, y, &c. have values which satisfy the differential equations.

Consider the second determinant, it may be made into the sum of as many determinants as there are terms in the leading constituent. All these determinants have two columns the same except the first determinant. This first determinant is clearly $\Delta\,(p)\,x$. It immediately follows that

$$(\delta - p)\,\Pi\,(p) = -\,\Delta\,(p)\,x.$$

Solving this linear differential equation in the usual way, we have

$$\Pi\,(p) + \Delta\,(p)e^{pt}\int_0^t e^{-pt}x\,dt = Ce^{pt} \,\ldots\ldots\ldots\ldots\,(1).$$

Here p is any quantity at our disposal and x, y, &c. have any values which satisfy the differential equations

To find the value of the constant C, we put $t = 0$. The second term on the left-hand side is then zero because the limits coincide. It follows that C is the value of $\Pi\,(p)$ when we write for x, y, &c., δx, δy, &c. their initial values.

Since p is arbitrary we may differentiate the equation partially with respect to p. Differentiating and putting $p = m$, where m is a *solitary root* of the equation $\Delta\,(p) = 0$, we find

$$\frac{d\Pi\,(m)}{dm} + \Delta'\,(m)\,e^{mt}\int_0^t e^{-mt}x\,dt = Cte^{mt} + \frac{dC}{dp}\,e^{mt}.$$

Let us now substitute $x = Me^{mt} + M_2e^{m_2t} + \&c.$ with the corresponding values of y, z, &c. in the left-hand side of this equation and let us search for terms of the form te^{mt}. The operator $d\Pi\,(m)/dm$ is a linear function of x, y, &c., δx, &c., and can clearly give rise to no term of the required form. The remaining portion of the left-hand side gives only the single term $\Delta'\,(m)\,Mte^{mt}$ of the required form. Equating this to the corresponding term on the right-hand side we have $\Delta'\,(m)\,M = C$. Since C is the initial value of $\Pi\,(p)$, this equation is exactly equivalent to that given in Art. 366.

369. **On Repeated Roots.** When the root $p = m$ is a *repeated root* of the equation $\Delta(p) = 0$, the demonstration just given no longer applies. Since p is arbitrary we may differentiate the equation (1) as often as we please, and after each differentiation we may write $p = m$. Since $\Delta(m) = 0$, $\Delta'(m) = 0$ &c. the successive left-hand sides reduce to $\Pi(m)$, $d\Pi(m)/dm$, &c. On the successive right-hand sides we have only terms which contain the exponential e^{mt}.

It follows that if $\Delta(p) = 0$ have a roots each equal to m, the operators

$$\Pi(m), \quad \frac{d\Pi(m)}{dm}, \quad \frac{d^2\Pi(m)}{dm^2}, \ldots\ldots \frac{d^{a-1}\Pi(m)}{dm^{a-1}},$$

all produce zero when we substitute for x, y, &c. any solutions of the differential equations which do not contain the exponential e^{mt}.

Thus it appears that if we calculate the results of these operations by substituting the particular parts of the values of x, y, &c. which depend on the root m of the equation $\Delta(\delta) = 0$, the results will be general, i.e., will be the same as if we had substituted the complete values of x, y, &c.

Without using any further rule, therefore, we may find the a constants which depend on the repeated root $p = m$ by substituting in these a operators the particular terms in x, y, &c. which contain the exponential e^{mt}. Thus we obtain a expressions for the operators which contain the a constants. At the same time the values of the operators themselves may be found by giving the variables x, y, &c. their initial values.

This, however, requires that we should use *all* the co-ordinates, but if we wish to find the values of the constants which occur in one co-ordinate only, we may use the results of the following theorem.

370. *It is required to find in terms of the initial conditions the values of the constants which enter into the expression for any one of the co-ordinates when the fundamental determinant* $\Delta(p)$ *has* a *roots each equal to* m.

In this case the value of x contains powers of t, but how many will depend on whether the minors of the determinant $\Delta(\delta)$ are zero or not. Since, however the highest power of t cannot exceed $a - 1$ we may take as the general value of x

$$x = \left(M_0 + M_1 t + \ldots + \frac{M_{a-1} t^{a-1}}{L(a-1)} \right) e^{mt} + \Sigma N t^i e^{qt},$$

where the terms included in the Σ stand for those portions of the value of x which do not depend on the root m and $L(a-1) = 1 \cdot 2 \cdot 3 \ldots (a-1)$. There are similar expressions for y, z, &c. also containing powers of t not higher than the $(a-1)^{th}$, but it will be unnecessary to write these down.

We now proceed to differentiate equation (1) of Art. 368 r times with regard to p, and after substitution for x, y, &c., we shall search for the terms containing $t^\kappa e^{mt}$ where r and κ are any integers we may find convenient to use. The rth differential coefficient is clearly

$$\frac{d^r \Pi(p)}{dp^r} + \frac{d^r \Delta(p) P}{dp^r} = \frac{d^r}{dp^r} C e^{pt} \quad \ldots\ldots\ldots\ldots\ldots\ldots (2),$$

where $P = e^{pt} \int_0^t e^{-pt} x \, dt$.

We notice that the first of the two terms on the left-hand side is a linear function of x, y, &c. and their differential coefficients with regard to t. Hence no term of the form searched for can enter unless with powers of t less than a. If then we restrict ourselves to values of κ greater than $a - 1$, we may pay no further attention to this term.

This second term on the left-hand side of (2) may by Leibnitz's theorem be

written $\quad \Delta^r (p) P + r\Delta^{r-1} (p) \dfrac{dP}{dp} + \dots + \dfrac{L(r)}{L(a)\,L(r-a)} \Delta^a (p) \dfrac{d^{r-a}P}{dp^{r-a}}$.

In this series all the differential coefficients of $\Delta(p)$ below the ath have been omitted because the equation $\Delta(p)=0$ has been supposed to have a roots each equal to m.

If we substitute in the expression for P any such term as $Nt^i e^{qt}$ we find after integration only one term which is free from the exponential e^{qt}, and this one term is of the form He^{pt}. Hence d^sP/dp^s contains no power of t higher than the sth. In this series therefore, when we put $p=m$ and search for the terms of the form $t^\kappa e^{mt}$, if we restrict ourselves to values of κ greater than $r-a$, we may pay no further attention to such terms as $Nt^i e^{qt}$.

We have next to find the value of d^sP/dp^s when we substitute for x any term of

the form $\dfrac{M_{\kappa-1}}{L(\kappa-1)} t^{\kappa-1} e^{mt}$. Now whatever x may be we have

$$\frac{d^sP}{dp^s} = \frac{d^s}{dp^s} \frac{1}{\delta-p} x = \frac{Ls}{(\delta-p)^{s+1}} x = Ls\, e^{pt} \delta^{-s-1} (e^{-pt} x),$$

where $Ls=1\,.\,2\,.\,3\dots s$ as usual. Substituting for x and writing $p=m$, we may effect the integrations represented by δ^{-s} without difficulty. The exponential

disappears and we find at once $\dfrac{d^sP}{dp^s} = \dfrac{Ls}{L(\kappa+s)} M_{\kappa-1} t^{\kappa+s} e^{mt}$.

No correction is necessary to the integration since this vanishes with t.

Supposing then κ to be greater than both $a-1$ and $r-a$ we find for the coefficient of $t^\kappa e^{mt}$ on the left-hand side of the equation (2)

$$\frac{1}{L\kappa} \left\{ \Delta^r(m) M_{\kappa-1} + r\Delta^{r-1}(m) M_{\kappa-2} + \frac{r(r-1)}{1\,.\,2} \Delta^{r-2}(m) M_{\kappa-3} + \&c. \right\}.$$

On the right-hand side we find the coefficient of $t^\kappa e^{mt}$ to be $\dfrac{Lr}{L\kappa L(r-\kappa)} \cdot \dfrac{d^{r-\kappa}C}{dm^{r-\kappa}}$.

Equating these two we have

$$\frac{\Delta^r(m)}{Lr} M_{\kappa-1} + \frac{\Delta^{r-1}(m)}{L(r-1)} M_{\kappa-2} + \dots + \frac{\Delta^a(m)}{La} M_{\kappa-r+a-1} = \frac{1}{L(r-\kappa)} \frac{d^{r-\kappa}C}{dm^{r-\kappa}}.$$

Since the letter C stands for the initial value of $\Pi(m)$, it will be more convenient to replace it by the latter symbol, with the understanding *that all the coordinates have their initial values.*

Since κ must be greater than $a-1$ and $M_a=0$, the only useful value of κ is $\kappa=a$. Since κ must be greater than $r-a$, the only possible values of r are $r=a$, $a+1, \dots 2a-1$. Writing these in succession for r, we obtain

$$\frac{\Delta^a}{La} M_{a-1} = \Pi(m),$$

$$\frac{\Delta^{a+1}}{L(a+1)} M_{a-1} + \frac{\Delta^a}{La} M_{a-2} = \frac{d\Pi(m)}{dm},$$

$$\frac{\Delta^{a+2}}{L(a+2)} M_{a-1} + \frac{\Delta^{a+1}}{L(a+1)} M_{a-2} + \frac{\Delta^a}{La} M_{a-3} = \frac{1}{1\,.\,2} \frac{d^2\Pi(m)}{dm^2},$$

$$\&c. = \&c.$$

$$\frac{\Delta^{2a-1}}{L(2a-1)} M_{a-1} + \&c. + \frac{\Delta^{a+1}}{L(a+1)} M_1 + \frac{\Delta^a}{La} M_0 = \frac{1}{L(a-1)} \frac{d^{a-1}\Pi(m)}{dm^{a-1}}.$$

We have here just the right number of equations to find the a arbitrary constants which occur in the value of x without requiring the corresponding values of the other co-ordinates.

If all the first minors of the determinant $\Delta(\delta)$ have β roots equal to m, the first β operators on the right-hand side vanish whatever x, y, &c. may be. In this case therefore the coefficients $M_{a-1}...M_{a-\beta}$ are all zero. Thus the expression for x (as already explained in Art. 272) loses β of its highest powers of t.

In the same way we may find the constants which occur in y by using the operator called Π_2 in Art. 366 instead of Π.

371. **Another form of the determinant.** There is another form in which the operator $\Pi(m)$ can be written and which is particularly useful when the differential equations are of the second order. Returning to the proof given in Art. 368, we see that the determinant $\Pi(p)$ may be written as the difference between two determinants, the second of which is zero when $\Delta(p) = 0$. Looking at the first determinant, we may divide all the constituents of the first column by any power of δ we please, provided we finally multiply the determinant by the same power of δ. But these constituents are the functions which form the differential equations. We may therefore modify the rule given in Art. 366 as follows. *First divide the equations by any power of δ we please. Then form $\Pi(m)$ from these modified equations by the rule already given in Art. 366 and finally multiply the constituents of the first column by the same power of δ.* If this modified operator be called $\Pi'(m)$, we see that $\Pi(m)$ and $\Pi'(m)$ differ by some multiple of $\Delta(m)$. If $\Delta(\delta) = 0$ have α roots each equal to m, it follows that all the differential coefficients of $\Pi(m)$ and $\Pi'(m)$ up to the $(\alpha-1)$th are equal each to each.

372. Thus let the equations be

$$(A_{11}\delta^2 + B_{11}\delta + C_{11})\,x + (A_{12}\delta^2 + B_{12}\delta + C_{12})\,y = 0$$
$$(A_{21}\delta^2 + B_{21}\delta + C_{21})\,x + (A_{22}\delta^2 + B_{22}\delta + C_{22})\,y = 0$$

taking only two variables to shorten the results. We divide each equation by δ, then to form $\Pi(m)$ we divide by $\delta - m$ and reject the remainders. Finally we multiply again by δ. We thus have

$$\Pi(m) = \begin{vmatrix} A_{11}\delta x + A_{12}\delta y - \dfrac{C_{11}x + C_{12}y}{m}, & A_{12}m^2 + B_{12}m + C_{12} \\[2ex] A_{21}\delta x + A_{22}\delta y - \dfrac{C_{21}x + C_{22}y}{m}, & A_{22}m^2 + B_{22}m + C_{22} \end{vmatrix}.$$

In this form the constituents of the first column (when the equations are of the second degree) *may be written down by copying them from the equation.*

The advantage of this form is that the forces of resistance which depend on the potential B (Art. 311) have disappeared from the symbol $\Pi(m)$. It also leads to the method of multipliers to be explained in the next section.

373.   Ex. 1.   Let the equations be $(\delta^2 - 3\delta + 2)\,x + (\delta - 1)\,y = 0$
$- (\delta - 1)\,x + (\delta^2 - 5\delta + 4)\,y = 0$.

The fundamental determinant is

$$\Delta(m) = \begin{vmatrix} m^2 - 3m + 2 & m - 1 \\ -(m-1) & m^2 - 5m + 4 \end{vmatrix} = (m-1)^2 (m-3)^2$$

The equation $\Delta(m) = 0$ has therefore two roots each equal to 2 and the corresponding terms in the value of x will be $x = (M_0 + M_1 t)\, e^{3t}$.

It is required to find M_0 and M_1 in terms of the initial values of the co-ordinates.

We form the operator $\Pi(m)$ by the rule given in Art. 372, copying the columns from the equations given above

$$\Pi(m) = \begin{vmatrix} \delta x - \dfrac{2x - y}{m}, & m - 1 \\[2mm] \delta y - \dfrac{x + 4y}{m}, & m^2 - 5m + 4 \end{vmatrix} = (m-1)\left\{ (m-4)\,\delta x - \delta y - \dfrac{2m - 9}{m}\,x + y \right\}.$$

This gives when $m = 3$, $\Pi(m) = -2\{\delta x + \delta y - x - y\}$, $\dfrac{d\Pi(m)}{dm} = \delta x - \delta y - x + y$.

Also when $m = 3$ we have $\Delta(m) = 0$, $\Delta'(m) = 0$, $\Delta''(m) = 8$, $\Delta'''(m) = 24$. Hence by the rule given in Art. 370 $4M_1 = -2(\delta x + \delta y - x - y)$
$4(M_1 + M_0) = \delta x - \delta y - x + y$,

where the quantities on the right-hand side have their initial values.

Ex. 2.   Let the equations be $(\delta^2 - 2\delta)\,x - y = 0$
$(2\delta - 1)\,x + \delta^2 y = 0$.

Find the constants in $x = (M_0 + M_1 t + \tfrac{1}{2} M_2 t^2)\, e^t$.

The result is $2M_2 = \delta x + \delta y + x + y$, $2M_1 + M_2 = 2\delta x - x + y$, $2M_0 + M_1 = \delta x + x$.

374.   The following examples illustrate the application of the preceding theorems when the differential equation has but one dependent variable.

Ex. 1.   The differential equation $(\delta^3 - 2\delta^2 - \delta + 2)\,x = 0$ is satisfied by $x = Me^t$. If the initial values of x, δx, $\delta^2 x$ are a, a', a'', prove that $2M = 2a + a' - a''$.

Ex. 2.   Let the differential equation be $f(\delta)\,x = 0$ and let $f(\delta)$ contain only even powers of δ. If the terms of the solution depending on the pair of solitary roots $m = \pm\sqrt{-1}$ of $f(m) = 0$ be $x = F\cos kt + G\sin kt$, prove that

$$\frac{F}{2}\frac{f'(m)}{m} = \frac{f(\delta)}{\delta^2 + k^2}\,x \quad \text{and} \quad \frac{G}{2}\frac{f'(m)}{m} = \frac{f(\delta)}{\delta^2 + k^2}\frac{\delta x}{k}.$$

Ex. 3.   Let $A_n \delta^n x + \ldots + A_1 \delta x + A_0 x = 0$ be a differential equation. Representing this by $f(\delta)\,x = 0$, let m be a real solitary root of $f(\delta) = 0$, and let Me^{mt} be the corresponding term in the value of x. Prove that a superior limit to the value of $Mf'(m)$ is the sum of those terms in the series $A_n \delta^{n-1} x + \ldots + A_2 \delta x + A_1$ which have the same sign as $f'(m)$. Here of course x, δx, &c. are all supposed to have their known initial values.

375.   The following examples indicate another method of investigating the theorems of this section.

Ex. 1.   Let the first minors of the determinant $\Delta(\delta)$ be represented by the letter I, the suffix indicating the constituent of which it is the minor. If q be any root of $\Delta(\delta) = 0$ we know that a solution of the differential equations is

$$x = G I_{11}(q)\, e^{qt}, \qquad y = G I_{12}(q)\, e^{qt}, \qquad z = \&\text{c.},$$

where G is an arbitrary constant. Let us however suppose that q is unrestricted in value and is not necessarily a root of $\Delta(\delta)=0$. Prove that the result of the substitution of these values of x, y, &c. in $\Pi(p)$ is

$$\Pi(p) = G e^{qt} \frac{\Delta(q) I_{11}(p) - \Delta(p) I_{11}(q)}{q-p},$$

where p also is unrestricted in value.

This result may be proved by resolving $\Pi(p)$ into the difference between two determinants as in Art. 368, and then substituting in each.

Ex. 2. Deduce from the last example that if p and q be unequal solitary roots of $\Delta(\delta)=0$, then $\Pi(p)=0$. But if p and q be the same solitary root then

$$\Pi(p) = G I_{11}(p) \Delta'(p) e^{pt}.$$

Ex. 3. If the equation $\Delta(\delta)=0$ have β roots each equal to q, the form of the solution is indicated by $x = G_0 I_{11}(q) e^{qt} + \ldots + G_{\beta-1} (d/dq)^{\beta-1} \{I_{11}(q) e^{qt}\}$, with similar expressions for the other co-ordinates. If the equation $\Delta(\delta)=0$ have also a roots each equal to p, prove that the result of the substitution of these values of the co-ordinates in any one of the determinants $\Pi(p)$, $(d/dp)\,\Pi(p)\ldots$ $(d/dp)^{a-1}\Pi(p)$ is zero if p and q be unequal. If p and q be equal, we obtain the results given in Art. 366.

This may be proved by using Leibnitz's theorem to differentiate the equation of Ex. 1, i times with regard to p, and j times with regard to q, where i is less than a and j than β.

Ex. 4. When all the first minors of $\Delta(\delta)$ vanish for any particular value of δ, the solution depends on a double type ξ, η so that $x = J_{12}(\delta)\,\xi$, $y = J_{12}(\delta)\,\eta$ &c. where $J_{12}(\delta)$ is the second minor of $\Delta(\delta)$ formed by omitting the first two rows and columns as in Art. 273. Prove that if we write $\xi = G e^{qt}$, $\eta = H e^{qt}$, where G and H are two arbitrary constants which run through all the values of the other co-ordinates, then

$$\Pi(p) = G\,\frac{e^{qt}}{q-p}\left\{ \begin{vmatrix} I_{11}(p), & I_{12}(q) \\ I_{21}(p), & I_{22}(q) \end{vmatrix} - J_{12}(q)\,\Delta(p) \right\} - H\,\frac{e^{qt}}{q-p} \begin{vmatrix} I_{11}(p), & I_{11}(q) \\ I_{21}(p), & I_{21}(q) \end{vmatrix}.$$

Here p and q are unrestricted in value and do not necessarily satisfy $\Delta(\delta)=0$.

Ex. 5. Deduce from the result of Ex. 4, that if $\Delta(\delta)$ have two roots each equal to m one of which makes all the first minors zero, so that $x = M e^{mt}$, $y = N e^{mt}$ are parts of the solution where M, N are independent constants, then

$$\tfrac{1}{2}\Delta''(m)\,M = \frac{d\Pi}{dm}, \qquad \tfrac{1}{2}\Delta''(m)\,N = \frac{d\Pi_2}{dm},$$

where Π_2 is obtained from $\Delta(m)$ by erasing the second column instead of the first (see Art. 366). Here the co-ordinates on the right-hand side are supposed to have their initial values.

Ex. 6. Let the equation $\Delta(\delta)=0$ have a roots each equal to m, and let all the first minors have β roots also equal to m. Let us form from $\Pi(m)$ a new determinant $\Pi'(m)$ by omitting any row we please and any column except the first. Prove that if we substitute in the determinants $(d/dm)\,\Pi'(m)$, &c. $(d/dm)^{\beta-1}\Pi'(m)$ any values of the co-ordinates which satisfy the differential equations and which do not involve the exponential e^{mt}, the results are all zero.

Method of Multipliers.

376. In the last section we showed how the constant belonging to any one oscillation could be determined when the differential equations were of any order. We now propose to consider what

simplifications can be made in the rule when the differential equations are of the second order and of that simpler kind which usually occurs in dynamics.

Referring to Art. 310, we find the equations of the second order written at length. But forms so general as these seldom make their appearance. The two most important problems which occur in dynamics are those in which we have—

(1) Oscillations about a position of equilibrium, whether with forces of resistance or not.

(2) Oscillations about a state of steady motion.

In the first of these cases the terms depending on D, E, F are absent from the equations so that the fundamental determinant is therefore symmetrical. In the second the terms depending on D and F are absent, but those depending on the centrifugal forces E are present. In this case the forces of resistance B are generally absent.

377. We may therefore simplify these equations of motion and write them in the form

$$\left.\begin{array}{l} (A_{11}\delta^2 + B_{11}\delta + C_{11})\,x + \left(\begin{array}{c} A_{12}\delta^2 + B_{12}\delta + C_{12} \\ + E_{12}\delta \end{array}\right) y + \&\mathrm{c.} = 0, \\[1.2em] \left(\begin{array}{c} A_{12}\delta^2 + B_{12}\delta + C_{12} \\ - E_{12}\delta \end{array}\right) x + (A_{22}\delta^2 + B_{22}\delta + C_{22})\,y + \&\mathrm{c.} = 0, \\[1.2em] \qquad \&\mathrm{c.} \qquad\qquad + \&\mathrm{c.} \qquad\quad + \&\mathrm{c.} = 0. \end{array}\right\}$$

The solution of these equations has been already expressed in Arts. 313 and 317 in the following forms. If m_1, m_2, &c. be real roots of the fundamental determinant, we have

$$\left.\begin{array}{l} x = x_1 e^{m_1 t} + x_2 e^{m_2 t} + \&\mathrm{c.} \\ y = y_1 e^{m_1 t} + y_2 e^{m_2 t} + \&\mathrm{c.} \\ \&\mathrm{c.} = \&\mathrm{c.} \end{array}\right\} \qquad \left.\begin{array}{l} dx/dt = x_1{}' e^{m_1 t} + x_2{}' {}^{m_2 t} + \&\mathrm{c.} \\ dy/dt = y_1{}' e^{m_1 t} + y_2{}' {}^{m_2 t} + \&\mathrm{c.} \\ \&\mathrm{c.} = \&\mathrm{c.} \end{array}\right\}$$

Here x_1, y_1, z_1, &c., $x_1{}'$, $y_1{}'$, &c. contain as a common factor one constant of integration, x_2, y_2, &c., $x_2{}'$, $y_2{}'$, &c., another constant and so on. These are the constants called L_1, L_2 &c., in Arts. 261, 268 &c. Also $x_1{}' = x_1 m_1$, $y_1{}' = y_1 m_1$ and so on.

378. If there be a pair of imaginary roots in the fundamental determinant of the form $m_1 = r + p\sqrt{-1}$, $m_2 = r - p\sqrt{-1}$, the preceding solution takes the form

$$\left.\begin{array}{l} x = X_1 e^{rt}\cos pt + X_2 e^{rt}\sin pt + x_3 e^{m_3 t} + \&\mathrm{c.} \\ y = Y_1 e^{rt}\cos pt + Y_2 e^{rt}\sin pt + y_3 e^{m_3 t} + \&\mathrm{c.} \\ \&\mathrm{c.} = \&\mathrm{c.} \end{array}\right\}$$

$$\left.\begin{array}{l} dx/dt = X_1{}' e^{rt}\cos pt + X_2{}' e^{rt}\sin pt + x_3{}' e^{m_3 t} + \&\mathrm{c.} \\ dy/dt = Y_1{}' e^{rt}\cos pt + Y_2{}' e^{rt}\sin pt + y_3{}' e^{m_3 t} + \&\mathrm{c.} \\ \&\mathrm{c.} = \&\mathrm{c.} \end{array}\right\}$$

where $X_1 = x_1 + x_2$, $X_2 = (x_1 - x_2) \sqrt{-1}$ and $X_1' = rX_2 + pX_1$
$X_2' = -pX_1 + rX_2$. There are of course similar expressions for
the Y's, &c. Here we notice that all the coefficients in the first
two columns are linear functions of two constants of integration,
the coefficients of the third column are multiples of a third
constant and so on.

379. If we examine the form of the solution given in the
last article we see that the columns are arranged according to
the roots of the fundamental determinant. Each column contains
one or two arbitrary constants which have to be determined from
the initial values of x, y, &c. If the whole solution is known
we may therefore find the constants by common algebra, though
if there are many unknown constants the process may be very
long. But if the whole solution is not known the processes of
common algebra fail.

380. Thus suppose we have found only one root of the fundamental determinant, then we know the terms which occur in one
column only. The other columns depend on the other roots
which have not yet been investigated. We may yet wish to find
the value of the constant which occurs in this column in terms of
the initial values of the variables. We should then be able to
find the magnitude of any one oscillation without finding the others.

To effect this we use the *method of multipliers,* our object is to
find some multipliers for the equations which express the values
of x, y, &c., dx/dt, dy/dt, &c. such that on adding together the
products all the columns will disappear except the one we wish
to retain. Supposing this done we have one equation containing
the constant to be found and the initial values of x, y, &c. This
equation will be sufficient to determine the value of the constant.

There is this point of difference between the method of isolation and that of
multipliers. In the former we find the constant connected with any one term in
any column without caring for the other terms in that or any other column. In
the latter we require to use all the terms in that column to find the one constant.
In the former method we isolate any one term, in the latter we isolate any one
column.

381. The proper multipliers may be deduced from the determinant Π (m).
Taking the form given in Art. 371 as the best adapted for equations of the second
order, we have by expansion

$$\Pi\,(m) = Px + Qy + \&\text{c.} + P'\delta x + Q'\delta y + \&\text{c.},$$

where P, Q, &c. stand for the coefficients in the expanded determinant. Now it has
been proved in Art. 369 that Π (m) is zero when we write for x, y, &c., the terms of
any column of the solution in Art. 377 depending on a root other than m. It
follows at once that the proper multipliers to separate the column depending on the
root m from the other columns are P, Q, &c., P', Q', &c.

These multipliers are really determinants, and when there are many co-ordinates
it may be very troublesome to calculate their values. The coefficients of the

column which is to be separated from the others are also determinants. Both these sets of determinants are connected with the minors of the fundamental determinant; the former with the minors of some column, the latter with the minors of some row. When the differential equations are of the simpler kind which occurs in dynamics (Art. 377), the fundamental determinant has a certain symmetry about the leading diagonal. In this case the two sets of determinants are connected together so that the required multipliers can be expressed as simple functions of the coefficients of the column we wish to separate.

Instead of making the transformation from one set of determinants to the other, it will be simpler to adopt an independent mode of proof. The required multipliers follow at once from the two equations which have been made the foundation of the theorems in the first section of Chap. VII. (see Art. 316). As the equations now under consideration are simpler than those treated of in the section just referred to, the proofs of these two theorems will be briefly summed up in the next article. The definitions of the functions A, B, C (Art. 311) will also be adapted to the special use which we now intend to make of them.

382. If we substitute the terms in the first column of the expressions for x, y, &c. given in Art. 377 in the differential equations we obtain a set of equations which differs from the differential equations only in having m_1 written for δ and x_1, y_1, &c. for x, y, &c. First multiply these respectively by x_1, y_1, &c. and add the results together, the sum may be briefly written,

$$A\,(x_1 x_1)\,m_1^2 + B\,(x_1 x_1)\,m_1 + C\,(x_1 x_1) = 0.$$

Next, multiply these respectively by x_2, y_2, &c. and add the results together. The sum may be briefly written

$$A\,(x_1 x_2)\,m_1^2 + B\,(x_1 x_2)\,m_1 + C\,(x_1 x_2) = E\,(x_1 y_2)\,m_1.$$

The functional symbols A, B, C when not followed by the subject of the functions all represent functions of the co-ordinates x, y, z, &c. which have been defined in Art. 311. Thus

$$A = \tfrac{1}{2} A_{11} x^2 + A_{12} xy + \tfrac{1}{2} A_{22} y^2 + \ldots,$$
$$B = \tfrac{1}{2} B_{11} x^2 + B_{12} xy + \tfrac{1}{2} B_{22} y^2 + \ldots,$$
$$C = \tfrac{1}{2} C_{11} x^2 + C_{12} xy + \tfrac{1}{2} C_{22} y^2 + \ldots.$$

When the differential equations are given the following rule to find A, B, C will be useful:—*Multiply the equations by* x, y, z, &c. *and add the products, treating the operator* δ *as an algebraic factor. The halves of the coefficients of the powers of* δ *are the functions* A, B, C.

When we wish to substitute for the variables x, y, z, &c. any quantities we affix as usual those quantities to the functional symbol and write

$$A\,(x_1 x_1) = \tfrac{1}{2} A_{11} x_1^2 + A_{12} x_1 y_1 + \tfrac{1}{2} A_{22} y_1^2 + \ldots,$$

with similar expressions for $B\,(x_1 x_1)$ and $C\,(x_1 x_1)$.

We then generalize these expressions and for the sake of brevity write

$$A\,(x_1 x_2) = \tfrac{1}{2} A_{11} x_1 x_2 + \tfrac{1}{2} A_{12}(x_1 y_2 + x_2 y_1) + \tfrac{1}{2} A_{22} y_1 y_2 + \ldots.$$

383.   PROP. A.—*To determine the multipliers when the funda-mental determinant is symmetrical and the forces of resistance not absent.*

Let $m_1 m_2$ be any two roots of this determinant. Then, by Art. (382), since the terms depending on E are absent,

$$\left. \begin{array}{l} A\,(x_1 x_2)\,m_1^2 + B\,(x_1 x_2)\,m_1 + C\,(x_1 x_2) = 0 \\ A\,(x_1 x_2)\,m_2^2 + B\,(x_1 x_2)\,m_2 + C\,(x_1 x_2) = 0 \end{array} \right\} \dotsc\dotsc\dotsc(1).$$

Eliminating B and C in turn from these equations, we have

$$\left. \begin{array}{l} A\,(x_1 x_2)\,m_1 m_2 = C\,(x_1 x_2) \\ -\,A\,(x_1 x_2)\,(m_1 + m_2) = B\,(x_1 x_2) \end{array} \right\} \dotsc\dotsc\dotsc\dotsc(2),$$

except when m_1 and m_2 are the same root.

Either of these equations may be used to find the required multipliers. *We thus find two sets of multipliers.* We shall choose the first equation, as giving the simpler results.

If there be a pair of imaginary roots in the fundamental de-terminant, say $m_1 = r + p\sqrt{-1}$, $m_2 = r - p\sqrt{-1}$, and if m_3 be any other root, the first of equations (2) gives

$$\left. \begin{array}{l} A\,(x_1 x_3)\,(r + p\sqrt{-1})\,m_3 = C\,(x_1 x_3) \\ A\,(x_2 x_3)\,(r - p\sqrt{-1})\,m_3 = C\,(x_2 x_3) \end{array} \right\} \dotsc\dotsc\dotsc(3).$$

Remembering that A and C are linear functions, we see that these give by addition and subtraction

$$\left. \begin{array}{l} A\,(X_1' x_3)\,m_3 = C\,(X_1 x_3) \\ A\,(X_2' x_3)\,m_3 = C\,(X_2 x_3) \end{array} \right\} \dotsc\dotsc\dotsc\dotsc(4),$$

where X_1, X_1'; X_2, X_2' have the meaning given to them in Art. 378.

The function $A\,(x_1 x_2)$ may obviously be deduced from the potential $A\,(x_1 x_1)$ by the process

$$2A\,(x_1 x_2) = x_2 \frac{dA\,(x_1 x_1)}{dx_1} + y_2 \frac{dA\,(x_1 x_1)}{dy_1} + \dots,$$

where of course $A\,(x_1 x_1)$ (Art. 382) represents the value of $A\,(xx)$, or A when x_1, y_1, &c. have been written for x, y, &c. The functions B and C may be treated in a similar manner.

We may now immediately deduce the proper multipliers.

Taking the solutions written down in Art. 377, let us multiply the expressions for x, y, &c. by $- dC/dx$, $- dC/dy$, &c., after writing x_1, y_1, &c. in these multipliers for $x. y$, &c.; also let us multiply the expressions for dx/dt, &c. by dA/dx, &c., after writing x_1', y_1', &c., for x, y, &c., in these multipliers. Finally, let us add the products; then, by virtue of the first of equations (2), the sum of every column except the first is zero.

If we have imaginary roots in the fundamental determinant, we take the solution given in Art. 378. Treating it in the same way, we see by equations (4) that all the columns disappear except the two first. Repeating the process for the second column, we again find that all the columns except the two first disappear.

384. *The rule may be summed up as follows :—*

Let the fundamental determinant be symmetrical, and the forces of resistance not absent. Let it be required to separate by the method of multipliers any given column from the others. *The proper multipliers for the co-ordinates are the values of dC/dx, dC/dy, &c., after we have substituted for x, y, &c., in these multipliers the corresponding coefficients in the column we wish to preserve. The proper multipliers for the velocities are the values of − dA/dx, − dA/dy, &c., after we have substituted for x, y, &c. in these multipliers the corresponding coefficients in the column of velocities we wish to preserve. Finally, we add the products together.*

In this way we can find an equation connecting the initial values of the co-ordinates with the constant which accompanies any one column. Since these initial values are arbitrary, neither side of this equation can wholly vanish unless all the multipliers themselves vanish. Hence the coefficient of the exponential on the right-hand side cannot be zero, except in this one case.

The multipliers cannot all vanish unless the quadric functions C and A also vanish for some *finite values* of the co-ordinates. In dynamics the function A is such a function of the co-ordinates as the vis viva is of the velocities. It is therefore impossible that A could vanish for any finite values of the co-ordinates.

385. **Example.** Let us consider the equations

$$(\delta^2 + \delta + 1)\, x + \tfrac{1}{2}(\delta - \tfrac{3}{2})\, y = 0 \\ \tfrac{1}{2}(\delta - \tfrac{3}{2})\, x + (\delta^2 - \delta + \tfrac{1}{4})\, y = 0 \Big\} .$$

It is easily seen that the determinant of the solution reduces to $m^4 - \tfrac{5}{16} = 0$. We therefore have, if m now stand for $\tfrac{1}{2}\sqrt[4]{5}$,

$$x = x_1 e^{mt} + x_2 e^{-mt} + X_3 \cos mt + X_4 \sin mt \\ y = y_1 e^{mt} + y_2 e^{-mt} + Y_3 \cos mt + Y_4 \sin mt \Big\} ,$$

$$dx/dt = mx_1 e^{mt} - mx_2 e^{-mt} + mX_4 \cos mt - mX_3 \sin mt \\ dy/dt = my_1 e^{mt} - my_2 e^{-mt} + mY_4 \cos mt - mY_3 \sin mt \Big\} .$$

Also multiplying the equations by x and y, and taking the halves of the coefficients of the powers of δ, we have

$$A = \tfrac{1}{2}(x^2 + y^2), \qquad C = \tfrac{1}{2}x^2 - \tfrac{3}{4}xy + \tfrac{1}{8}y^2.$$

Suppose we wish to find the coefficients x_1, y_1 in terms of the initial conditions. Following the rule, we multiply x and y by the differential coefficients of C after we have written x_1, y_1 for x, y in the multipliers. We multiply the velocities by minus the differential coefficients of A, writing in the multipliers mx_1 and my_1 for x and y. Finally, we add the results. Thus we have

$$\begin{aligned} x(x_1 - \tfrac{3}{4}y_1) + y(-\tfrac{3}{4}x_1 + \tfrac{1}{4}y_1) \\ -\frac{dx}{dt}mx_1 - \frac{dy}{dt}my_1 \end{aligned} \Bigg\} = \begin{Bmatrix} x_1^2 - \tfrac{3}{2}x_1 y_1 + \tfrac{1}{4}y_1^2 \\ -m^2(x_1^2 + y_1^2) \end{Bmatrix} e^{mt}.$$

Putting $t=0$, and giving x, y and their velocities their known initial values, we have one equation to find the constants x_1, y_1. Their ratio,

$$\frac{y_1}{x_1} = -\frac{m^2+m+1}{\frac{1}{2}(m-\frac{3}{2})}, \qquad m=\frac{1}{2}\sqrt[4]{5},$$

being known from the first equation, we easily find both x_1 and y_1.

If we wish to find the coefficients of the trigonometrical terms, we use two sets of multipliers, because the two imaginary exponentials have become mixed up together in the trigonometrical term; or we may replace them by their imaginary exponentials, and find the coefficients of either by one set of multipliers. Taking the first alternative, one set of multipliers will be respectively

$$X_3 - \tfrac{3}{4}Y_3, \quad -\tfrac{3}{4}X_3 + \tfrac{1}{4}Y_3, \quad -mX_4, \quad -mY_4.$$

The other set will be $X_4 - \tfrac{3}{4}Y_4, \quad -\tfrac{3}{4}X_4 + \tfrac{1}{4}Y_4, \quad +mX_3, \quad +mY_3.$

386.   PROP. B.—*To determine the multipliers when the fundamental determinant is symmetrical and the forces of resistance are absent.*

This proposition is really included in the last. But as the absence of the function B introduces great simplification, it is worth while to consider this case separately.

Since the forces of resistance are absent, only even powers of δ enter into the equations. Hence for every root of the fundamental determinant there is another equal in magnitude but contrary in sign. If A and C are one-signed functions, and have the same sign, these roots are of the form $\pm p\sqrt{-1}$. Choosing this as the type, we may write the equations of Art. 378 in the form

$$x = X_1 \cos pt + X_2 \sin pt + x_3 e^{m_3 t} + \dots \qquad \&c. = \&c.,$$

$$dx/dt = X_1' \cos pt + X_2' \sin pt + x_3' e^{m_3 t} + \dots \qquad \&c. = \&c.$$

Here, unless there are equal roots, we have

$$\frac{X_2}{X_1} = \frac{Y_2}{Y_1} = \&c. = \frac{X_1'}{-X_2'} = \frac{Y_1'}{-Y_2'} = \&c. = H,$$

because the ratios of the coefficients of any exponential are expressed by the minors of the fundamental determinant, and these, containing only even powers of m, are the same when the exponents are equal in magnitude but contrary in sign.

Here H will stand for the constant in the second column on the right-hand side of the equations, the constant in the first column being included as a factor in X_1, Y_1, &c., X_2', Y_2', &c.

Since the function B is zero, the equations (2) of Art. 383 reduce to $A(x_1 x_2) = 0, \qquad C(x_1 x_2) = 0,$

except when $m_1 = \pm m_2$. For a pair of imaginary roots such as $m_1 = r + p\sqrt{-1}$, $m_2 = r - p\sqrt{-1}$, combined with a third root m_3, we have (exactly as in that article)

$$\left. \begin{array}{l} A(X_1 x_3) = 0 \\ A(X_2 x_3) = 0 \end{array} \right\}, \qquad \left. \begin{array}{l} C(X_1 x_3) = 0 \\ C(X_2 x_3) = 0 \end{array} \right\}.$$

387.   We may use either the function A or the function C to

supply the proper multipliers. *We thus find two sets of multipliers. Which we should choose depends on the forms of* A *and* C.

If either of these functions contain only the squares of the co-ordinates, i.e. if it be of the form

$$ax^2 + by^2 + cz^2 + \ldots,$$

it is clear that its differential coefficients will be much simpler than if the terms containing the products of the co-ordinates were also present. The multipliers are indicated by these differential coefficients, and will therefore also be simpler. That function is therefore to be chosen which has the fewest terms containing the products of the co-ordinates.

Choosing the function A, we have the following rule to find the multipliers. Let it be required to separate from the others any particular oscillation—say the two columns containing the phase pt. *The proper multipliers for the co-ordinates* x, y, &c. *are the values of* $\dfrac{dA}{dx}$, $\dfrac{dA}{dy}$, &c., *after we have substituted for* x, y, &c. *in these multipliers the coefficients of either of the columns containing the phase* pt. *Adding these products, we have one equation from which all the oscillations except the one to be preserved have disappeared. The same multipliers may now be used for the velocities, and thus by a second addition we obtain another equation of the same kind.*

The two equations thus obtained may be written thus:—

$$x \frac{dA\,(X_1 X_1)}{dX_1} + \&c. = 2A\,(X_1 X_1)\,\{\cos pt + H \sin pt\},$$

$$\frac{dx}{dt} \frac{dA\,(X_1 X_1)}{dX_1} + \&c. = 2A\,(X_1 X_1)\,\{Hp \cos pt - p \sin pt\}.$$

Putting $t = 0$ either before or after using the multipliers, we have two equations to determine H and the other constant included in X_1, Y_1, &c.

388. A rule to find the functions A and C when the differential equations are known has already been given in Art. 382. But in using Lagrange's method it is sometimes more convenient to refer to the expression for the vis viva and the force function from which these equations have been derived. Referring to Vol. I. we see that the vis viva is

$$2T = A_{11} x'^2 + 2A_{12} x' y' + \ldots$$

Thus the function A is derived from T by merely dropping the accents from the co-ordinates. The function C is of course the same as the function $U_0 - U$ defined in Vol. I.

389. PROP. C.—*To determine the multipliers when the forces of resistance are absent but the determinant is skewed by the centrifugal forces.*

Referring to the equations of motion in Art. 377, we form the determinant which we have called the fundamental determinant. It is unnecessary to write this determinant, as its form is evident from the merest inspection of the equations. It is also given at length in Art. 112.

If in this determinant we write $-\delta$ for δ, the rows of the new determinant are the same as the columns of the old, so that the determinant is unaltered. When expanded, the determinant will contain only even powers of δ, and therefore its roots enter in pairs. We shall therefore take as our standard form of solution, instead of that in Art. 378, the expressions

$$\left.\begin{aligned} x &= X_1 \cos pt + X_2 \sin pt + x_3 e^{m_3 t} + \dots \\ y &= Y_1 \cos pt + Y_2 \sin pt + y_3 e^{m_3 t} + \dots \\ \&\text{c.} &= \&\text{c.} \end{aligned}\right\} \dots\dots\dots(1);$$

$$\left.\begin{aligned} dx/dt &= X_1' \cos pt + X_2' \sin pt + x_3' e^{m_3 t} + \dots \\ dy/dt &= Y_1' \cos pt + Y_2' \sin pt + y_3' e^{m_3 t} + \dots \\ \&\text{c.} &= \&\text{c.} \end{aligned}\right\} \dots\dots\dots(2).$$

Here the first two columns represent the most common form of a principal oscillation, and the third column represents any other form. When the centrifugal forces (i.e. the terms depending on E) are present, the minors of the fundamental determinant do not contain only even powers of δ. It follows that the coefficients in the second column do not necessarily bear a uniform ratio to those in the first column.

Since the function B is absent, we have by Art. 382, the equations

$$\left.\begin{aligned} A\,(x_1 x_2)\,m_1 + C\,(x_1 x_2)\,\frac{1}{m_1} &= E\,(x_1 y_2) \\ A\,(x_1 x_2)\,m_2 + C\,(x_1 x_2)\,\frac{1}{m_2} &= -E\,(x_1 y_2) \end{aligned}\right\} \dots\dots\dots(3).$$

Adding these to eliminate the functional symbol E, we find

$$A\,(x_1 x_2)\,m_1 m_2 + C\,(x_1 x_2) = 0 \dots\dots\dots\dots\dots(4),$$

except when $m_1 = -m_2$.

We notice also that, by Art. 382,

$$\left.\begin{aligned} A\,(x_1 x_1)\,m_1^2 + C\,(x_1 x_1) &= 0 \\ A\,(x_2 x_2)\,m_2^2 + C\,(x_2 x_2) &= 0 \end{aligned}\right\} \dots\dots\dots\dots(5).$$

We might also eliminate the function A or C from the equations (3) instead of the function E, *and in each case we may deduce a rule to find the multipliers; but the simplest rule is found by eliminating the function E.*

The formula (4) resembles that used in Art. 383, and there called (2), except in the sign of A. Proceeding therefore exactly as in that article, we shall deduce the corresponding rule for the multipliers.

Instead of equations (3) of Art. 383, we now have (since $r = 0$)

$$\left. \begin{array}{l} A\,(x_1 x_3)\,p m_3 \sqrt{-1} + C\,(x_1 x_3) = 0 \\ -\,A\,(x_2 x_3)\,p m_3 \sqrt{-1} + C\,(x_2 x_3) = 0 \end{array} \right\} \quad\quad (6).$$

Remembering that A and C are linear functions of the letters of any one suffix, these give by addition and subtraction

$$\left. \begin{array}{l} A\,(X_1' x_3)\,m_3 + C\,(X_1 x_3) = 0 \\ A\,(X_2' x_3)\,m_3 + C\,(X_2 x_3) = 0 \end{array} \right\} \quad\quad (7).$$

where as before $X = x_1 + x_2$, $X_2 = (x_1 - x_2)\sqrt{-1}$, $X_1' = p X_2$, $X_2' = -p X_1$.

Also writing $m_1 = p\sqrt{-1}$, $m_2 = -p\sqrt{-1}$ in equations (5), we find by subtraction $A\,(X_1' X_2') + C\,(X_1 X_2) = 0$ ....(8).

390. From these formulæ we now deduce the following rule to find the multipliers.

Let the forces of resistance be absent, and let the fundamental determinant be skewed by the centrifugal forces only. Let it be required to separate any principal oscillation from the others. *Selecting one of the two columns which form the oscillation, the proper multipliers for the co-ordinates x, y, &c. are the values of* $\dfrac{dC}{dx}$, $\dfrac{dC}{dy}$, *&c., after we have substituted for x, y, &c. in these multipliers the corresponding coefficients in the column selected. The proper multipliers for the velocities are the values of* $\dfrac{dA}{dx}$, $\dfrac{dA}{dy}$, *&c., after we have substituted for x, y, &c. in these multipliers the coefficients corresponding to these velocities in the column selected. Finally, we add all these products together. We then repeat the process with the coefficients of the other of the two columns which form the oscillation.*

By virtue of equations (5) and (8) it will be found that in each of these processes every column *except one* will disappear from the final summation. But we may notice a curious difference between the columns which contain real exponentials and those which contain trigonometrical expressions. If we operate with the coefficients of one of the former introduced into the multipliers, it is the *companion column which does not disappear;* but if we operate with the coefficients of one of the latter, it is the *column whose coefficients we have used which does not disappear.*

391. **Example.** Consider the equations $\left. \begin{array}{l} (\delta^2 - 8)\,x + \sqrt{6}\,\delta y = 0 \\ -\sqrt{6}\,\delta x + (\delta^2 + 2)\,y = 0 \end{array} \right\}$.

It is easily seen that the fundamental determinant reduces to $m^4 - 16 = 0$. Hence

we have
$$\left. \begin{array}{l} x = X_1 \cos 2t + X_2 \sin 2t + x_3 e^{2t} + x_4 e^{-2t} \\ y = Y_1 \cos 2t + Y_2 \sin 2t + y_3 e^{2t} + y_4 e^{-2t} \end{array} \right\},$$
$$\left. \begin{array}{l} dx/dt = 2X_2 \cos 2t - 2X_1 \sin 2t + 2x_3 e^{2t} - 2x_4 e^{-2t} \\ dy/dt = 2Y_2 \cos 2t - 2Y_1 \sin 2t + 2y_3 e^{2t} - 2y_4 e^{-2t} \end{array} \right\} ;$$

where
$$\left. \begin{array}{l} 2x_3 = \sqrt{6}\,y_3 \\ 2x_4 = -\sqrt{6}\,y_4 \end{array} \right\}, \quad\quad \left. \begin{array}{l} Y_1 = -\sqrt{6}\,X_2 \\ Y_2 = \sqrt{6}\,X_1 \end{array} \right\} .$$

Also multiplying the equations (Art. 382) by x, y, adding and taking the halves of the coefficients of the powers of δ,

$$A = \tfrac{1}{2}(x^2 + y^2), \qquad C = \tfrac{1}{2}(-8x^2 + 2y^2).$$

The proper multipliers are indicated (Art. 390) by the formula

$$x\,\frac{dC}{dx} + y\,\frac{dC}{dy} + \frac{dx}{dt}\frac{dA}{dx} + \frac{dy}{dt}\frac{dA}{dy}.$$

Now
$$\frac{dC}{dx} = -8x, \quad \frac{dC}{dy} = 2y, \quad \frac{dA}{dx} = x, \quad \frac{dA}{dy} = y.$$

Having chosen the column whose coefficients are to be used in the multipliers, we see by Art. 390 that the proper multiplier for the first equation is minus eight times the coefficient of the column in that equation; the proper multiplier for the second equation is twice the coefficient in that equation; the proper multipliers for the third and fourth equations are the coefficients themselves in those equations.

Suppose first we wish to find x_4, y_4, then, because the fourth column contains a *real* exponential, we operate with the coefficients of the companion column.

The multipliers are therefore
$$\frac{dC}{dx} = -8x_3, \quad \frac{dC}{dy} = 2y_3, \quad \frac{dA}{dx} = 2x_3, \quad \frac{dA}{dy} = 2y_3.$$

Hence we find
$$-8x_3 x + 2y_3 y + 2x_3 \frac{dx}{dt} + 2y_3 \frac{dy}{dt} = 16 y_3 y_4 e^{-2t};$$

substituting for x_3 in terms of y_3 and putting $t = 0$, we find

$$-4\sqrt{6}\,x + 2y + \sqrt{6}\,\frac{dx}{dt} + 2\frac{dy}{dt} = 16 y_4,$$

which determines y_4 in terms of the initial values of the co-ordinates and their initial velocities.

Suppose next we wish to find X_1, X_2. Taking the coefficients of the first column, the multipliers are
$$\frac{dC}{dx} = -8X_1, \quad \frac{dC}{dy} = 2Y_1, \quad \frac{dA}{dx} = 2X_2, \quad \frac{dA}{dy} = 2Y_2.$$

Since these columns contain trigonometrical expressions, we know that when we operate with the coefficients of either column in the multipliers, the other column disappears. Hence, paying no attention to any column except the first, we have

$$-8X_1 x + 2Y_1 y + 2X_2\,dx/dt + 2Y_2\,dy/dt = 16\,(X_1^2 + X_2^2)\cos 2t;$$

substituting for Y_1 and Y_2 and putting $t = 0$, we find

$$-8X_1 x - 2\sqrt{6}X_2 x + 2X_2 dx/dt + 2\sqrt{6}X_1\,dy/dt = 16\,(X_1^2 + X_2^2).$$

Operating in the same way with the coefficients of the second column, we have

$$-8X_2 x + 2Y_2 y - 2X_1\,dx/dt - 2Y_1\,dy/dt = 16\,(X_1^2 + X_2^2)\sin 2t;$$

substituting as before, we have

$$-8X_2 x + 2\sqrt{6}X_1 y - 2X_1\,dx/dt + 2\sqrt{6}X_2\,dy/dt = 0.$$

These equations determine X_1 and X_2 in terms of the initial values of x, y, and their differential coefficients.

392. PROP. D.—*To consider the effect of equal roots on the rules already given.*

When there are equal roots in the fundamental determinant, we require only some slight modification of our rules. Referring to the general solution exhibited in Art. 377, let us suppose, for example, that there are three roots equal to m_1. Regarding these as the limits of the unequal roots, m_1, $m_1 + h$, $m_1 + k$, we may write that solution in the form

$$x = x_1 e^{m_1 t} + G \frac{d}{dm_1}(x_1 e^{m_1 t}) + H \frac{d^2}{dm_1{}^2}(x_1 e^{m_1 t}) + x_4 e^{m_4 t} + \ldots$$

$$y = y_1 e^{m_1 t} + G \frac{d}{dm_1}(y_1 e^{m_1 t}) + H \frac{d^2}{dm_1{}^2}(y_1 e^{m_1 t}) + y_4 e^{m_4 t} + \ldots$$

&c. = &c.,

$$\frac{dx}{dt} = x_1' e^{m_1 t} + \frac{d}{dm_1}(x_1' e^{m_1 t}) + \frac{d^2}{dm_1{}^2}(x_1' e^{m_1 t}) + x_4' e^{m_4 t} + \ldots$$

&c. = &c. ;

where $x_1' = x_1 m_1$, $x_4' = x_4 m_4$, &c., and G, H are the two constants in addition to the one included in x_1, y_1, &c.

Two questions now present themselves :—(1) When we use certain multipliers to separate a column which depends on a solitary root such as m_4, will the columns which depend on other equal roots such as m_1 (and therefore contain powers of t as factors) still disappear ?

(2) What multipliers must we use to separate the three columns which depend on the three equal roots from the remaining columns ?

393. Taking the first of these questions, suppose we wish to separate the fourth column of the equations of Art. 392 from the others. Let us use the same multipliers as if there were no equal roots. It is obvious that, since the three first columns disappear in the general case in which h and k have any values, these columns must also disappear when h and k are indefinitely small. *We therefore infer that any column which depends on a solitary root may be separated by the same rules as before.*

As an example, take the rule given in Prop. A, Art. 383. To separate the fourth column, we multiply the equations by

$$dC(x_4 x_4)/dx_4, \quad \&c., \quad -dA(x_4' x_4')/dx_4', \quad \&c.,$$

and add the products. Since the three first columns must disappear, we have

$$C(x_1 x_4) - A(x_1' x_4') = 0$$
$$C\left(\frac{dx_1}{dm_1} x_4\right) - A\left(\frac{dx_1'}{dm_1} x_4'\right) = 0$$
$$C\left(\frac{d^2 x_1}{dm_1{}^2} x_4\right) - A\left(\frac{d^2 x_1'}{dm_1{}^2} x_4'\right) = 0$$

The last two of these equations also follow from the first by an evident process.

394. Taking the second question, we wish to find what multipliers will separate the three first columns from the others.

But these are supplied by the equations just written down. Since m_4 is any other root, and

$$2C\,(x_1x_4) = \frac{dC}{dx_1}\,x_4 + \frac{dC}{dy_1}\,y_4 + \dots\,,$$

we have merely to use the multipliers indicated by the coefficients of x_4, y_4, &c. in these equations. The rule may be enunciated as follows :—

Multiply the equations by the proper factors for the first column, treating x_1, y_1, &c., $x_1{}'$, $y_1{}'$, &c. as the coefficients, and add the products. We thus have one of the three required equations. Multiply the equations by the proper factors for the second column as if $\dfrac{dx_1}{dm_1}$, $\dfrac{dy_1}{dm_1}$, &c., $\dfrac{dx_1{}'}{dm_1}$, &c. were the coefficients, and add the products. We thus obtain the second equation. Lastly, multiply the equation by the proper factors for the third column as if $\dfrac{d^2x_1}{dm_1{}^2}$, &c., $\dfrac{d^2x_1{}'}{dm_1{}^2}$, &c., were the coefficients, and add the products. We thus have, on the whole, three equations to find the three constants which enter into the three first columns.

The proper factors just mentioned are those calculated from the coefficients by the rules of Prop. A or Prop. C.

395. In some cases of equal roots it is known that some of the terms with t as a factor fail to introduce themselves into the solution. The number of constants is then made up by a greater indeterminateness in the coefficients which accompany the exponential. Regarding these equal roots as the limits of unequal roots, as in Art. 393, it follows that we can still use the same rules to find the multipliers. We arrange our solution in columns with one constant in each column. Then using the proper multipliers, as described above, we can separate any solitary root at once. To determine the constants which accompany the equal roots, we shall require as many sets of multipliers as there are columns with that root or its companion root.

396. **Example.** Consider the equations $\left.\begin{array}{l}(\delta^2-1)\,x+y+z=0\\x+(\delta^2-1)\,y+z=0\\x+y+(\delta^2-1)\,z=0\end{array}\right\}$.

It is easily seen that the fundamental determinant reduces to $(m^2-2)^2\,(m^2+1)=0$. Putting $a=\sqrt{2}$, we write the solution in the form

$$\left.\begin{array}{l}x=\quad Ee^{at}\qquad+Ge^{-at}\qquad\qquad+K\sin t+L\cos t\\y=\qquad\quad+Fe^{at}\qquad\qquad+He^{-at}+K\sin t+L\cos t\\z=-Ee^{at}-Fe^{at}-Ge^{-at}-He^{-at}+K\sin t+L\cos t\end{array}\right\},$$

where E, F, G, H, K, L are the six constants to be determined.

Looking at the equations to be solved, we see that the potential functions A and C are given by
$$\left.\begin{array}{l}2C=-x^2-y^2-z^2+2xy+2yz+2zx\\2A=\quad x^2+y^2+z^2\end{array}\right\}.$$

Following the rule indicated in Art. 387, we choose the function A to operate with, because this function will supply the simplest multipliers. The proper multipliers will therefore be $\qquad dA/dx = x, \quad dA/dy = y, \quad dA/dz = z,$
where we write for x, y, z the coefficients of the column under consideration. The proper multipliers are therefore the coefficients of the columns in succession.

Suppose we wish to find K and L. The coefficients in either of these two columns are all equal. The multipliers are therefore equal. We therefore obtain, by adding the equations and putting $t = 0$,

$$x + y + z = 3L.$$

Treating the differential coefficients in the same way (Art. 387), we have

$$\delta x + \delta y + \delta z = 3K.$$

If we wish to find the four constants E, F, G, H which are all connected with the companion roots $\pm a$, we must find four equations. According to the rule, the multipliers are the coefficients of the several columns. We thus obtain, when $t = 0$,

$$\left. \begin{aligned} Ex + 0y - Ez &= E\,(2E + 2G + F + H) \\ 0x + Fy - Fz &= F\,(E + G + 2F + 2H) \end{aligned} \right\}\,,$$
$$\left. \begin{aligned} E\delta x + 0\delta y - E\delta z &= Ea\,(2E - 2G + F - H) \\ 0\delta x + F\delta y - F\delta z &= Fa\,(E - G + 2F - 2H) \end{aligned} \right\}\,.$$

This simple and obvious example sufficiently illustrates the method of proceeding when the proper multipliers could not be otherwise found.

397. Ex. If the differential equations are such that the fundamental determinant is symmetrical about the leading diagonal whether the forces of resistance are present or not, we have by Art. 262, $x_1/I_{11}\,(m_1) = y_1/I_{12}\,(m_1) = \&\text{c.} = G$, where G is an arbitrary constant. There will be similar equations for the other roots of the fundamental determinant. Thence show that the operator $\Pi\,(m)$ on expansion takes the form

$$G\Pi\,(m) = \frac{dA\,(x_1x_1)}{dx_1}\,\delta x + \frac{dA\,(x_1x_1)}{dy_1}\,\delta y + \&\text{c.} - \frac{1}{m_1}\frac{dC\,(x_1x_1)}{dx_1}\,x - \frac{1}{m_1}\frac{dC\,(x_1x_1)}{dy_1}\,y - \&\text{c.}$$

Thence deduce the forms of the multipliers given in Prop. A, Art. 383.

Fourier's Rule.

398. Of the two important problems which occur in dynamics (Art. 376) the most common is that in which the system is oscillating about a position of equilibrium free from any forces of resistance. This of course, is Lagrange's problem and the solution has been discussed in Chapter II.

It often happens that the co-ordinates chosen are such that the vis viva $2T$ can be written in the form

$$2T = x'^2 + y'^2 + \ldots$$

without any terms containing the products of the velocities. In other cases when the vis viva contains products, it may happen that the force function U can be written in the form

$$2U = x^2 + y^2 + \ldots$$

without any terms containing the products of the co-ordinates.

In either of these two cases if we follow the same line of argument as in Art. 386 we arrive at a simple rule. Taking the first case, Lagrange's equations are

$$\left.\begin{aligned} \delta^2 x + C_{11}x + C_{12}y + \ldots &= 0 \\ \delta^2 y + C_{12}x + C_{22}y + \ldots &= 0 \\ \&\text{c.} &= 0 \end{aligned}\right\} \ldots\ldots\ldots\ldots\ldots(1),$$

As in Art. 386 the solutions of these may be written in the form

$$\left.\begin{aligned} x &= X_1 \cos pt + X_2 \sin pt + X_3 \cos qt + X_4 \sin qt + \&\text{c.} \\ y &= Y_1 \cos pt + Y_2 \sin pt + Y_3 \cos qt + Y_4 \sin qt + \&\text{c.} \\ \&\text{c.} &= \&\text{c.} \end{aligned}\right\} \ldots(2),$$

where the coefficients of any one column are in the ratio of the minors of Lagrange's determinant and are therefore known multiples of the same undetermined constant; see Vol. I. Art. 457. The constants in the several columns are those represented in Art. 53 of this volume by $L_1 \cos \alpha_1$, $L_1 \sin \alpha_1$; $L_2 \cos \alpha_2$, $L_2 \sin \alpha_2$; &c. respectively. Our object is to find these constants.

Since the equations (1) are analytically satisfied by the values of x, y, &c. expressed by any one column, let us substitute for x, y, &c. the terms in the first column and multiply the resulting equations by X_3, Y_3, &c. respectively. Adding these results we find, after division by $\cos pt$,

$$p^2(X_1X_3 + Y_1Y_3 + \ldots) = C_{11}X_1X_3 + C_{12}(X_1Y_3 + X_3Y_1) + \&\text{c.}$$

Since the right-hand side is a symmetrical function of the coefficients of the first and third columns, we have

$$p^2(X_1X_3 + \&\text{c.}) = q^2(X_1X_3 + \&\text{c.}).$$

It immediately follows that unless $p = \pm q$ we must have

$$X_1X_3 + Y_1Y_3 + \&\text{c.} = 0 \ldots\ldots\ldots\ldots\ldots\ldots(3).$$

An exactly similar proof applies in the case in which the products are absent from the force function.

In either of these cases any column, say the first, may be separated by using as multipliers the coefficients X_1, Y_1, &c. of that column. Putting $t = 0$, so that the co-ordinates x, y, &c. have their initial values, the second, fourth, and all the even columns disappear from (2). Then multiplying by X_1, Y_1, &c. we have

$$xX_1 + yY_1 + \&\text{c.} = X_1^2 + Y_1^2 + \&\text{c.} \ldots\ldots\ldots\ldots(4).$$

In the same way by differentiating the equations (2) we turn the sines into cosines so that the first, third and all the odd columns disappear when $t = 0$. Multiplying by X_2, Y_2, &c. we have

$$(dx/dt)X_2 + (dy/dt)Y_2 + \&\text{c.} = p(X_2^2 + Y_2^2 + \&\text{c.}) \ldots(5).$$

We therefore have two equations to find the two constants which accompany the principal oscillation whose period is $2\pi/p$. These may be put into the form of a rule which when applied to some problems in heat or sound is usually called *Fourier's Rule.*

This may be stated as follows. *Multiply each co-ordinate by the coefficient of the cosine in the column we wish to separate, add the results together and put $t = 0$. All the other columns will disappear from this sum, leaving one equation to find the constant of integration which accompanies that cosine.*

To find the constant of integration which accompanies the sine which occurs in any column, we differentiate the co-ordinates and thus turn sines into cosines. Repeating the same process as before we have an equation to find the constant. These rules are simple corollaries from that given in Art. 387.

399. It sometimes happens that the vis viva $2T$ can be written in the form $2T = m_1 x'^2 + m_2 y'^2 + \ldots$
where m_1, m_2, &c. are the constants connected with the co-ordinates x, y, &c. In such a case the rule requires only a slight modification. By the same reasoning as before, we show that

$$m_1 X_1 X_3 + m_2 Y_1 Y_3 + \ldots = 0.$$

Thus the multipliers necessary to separate the first column of the values of x, y, &c. from the other columns are $m_1 X_1$, $m_2 Y_1$, &c. It will often happen that the coefficients m_1, m_2, &c. are the masses of some particles connected with the co-ordinates x, y, &c. Using this phraseology we have the following rule. *To separate any column we multiply the co-ordinates of the several particles as before by the coefficients in that column and by the masses of the several particles. We then add these results and proceed as before.*

400. *The investigation we have here given of Fourier's rule is purely analytical. All we have assumed is that the values of x, y, &c. satisfy certain differential equations. But we may also give a physical meaning to the process and show that we have really been using the principle of Virtual Velocities.*

It has been shown in the first volume that that general principle may be analytically represented by the equation

$$\left(\frac{d}{dt}\frac{dT}{dx'} - \frac{dU}{dx}\right)\xi + \left(\frac{d}{dt}\frac{dT}{dy'} - \frac{dU}{dy}\right)\eta + \&c. = 0,$$

where ξ, η, &c. are any small arbitrary variations of the co-ordinates x, y, &c. consistent with the geometrical conditions.

Let us suppose the system to be performing any principal oscillation, say the one represented by the first column in the values of x, y, &c. Let us take as the arbitrary variation of the co-ordinates, a displacement along any other principal oscillation, say the one represented by the third column in the expressions for x, y, &c. This variation is consistent with the geometrical conditions since the two oscillations might coexist in the same motion.

In this case ξ, η, &c. are proportional to X_3, Y_3, &c. After substituting for x, y, &c. their values as given by the terms in the first column and dividing by $\cos pt$, the equation becomes

$$-p^2(X_1X_3 + Y_1Y_3 + \ldots) = C_{11}X_1X_3 + C_{12}(X_1Y_3 + X_3Y_1) + \&c.$$

Since the right-hand side is a symmetrical function of the co-efficients of the first and third columns, we immediately have, as before, $X_1X_3 + Y_1Y_3 + \ldots = 0,$

except when p and q are numerically equal.

Lagrange shows how to find the constants of integration in certain cases in Sect. VI. of the second part of his *Mécanique Analytique*. Poisson devotes Chapters VII. and VIII. of his *Théorie de la Chaleur* to an explanation of the method of expressing arbitrary functions in a series of sines and cosines. Another treatment of Fourier's rule is given in Arts. 93 and 94 of Lord Rayleigh's *Theory of Sound*.

The reader may consult two papers by the author on the several subjects discussed in this Chapter. The first is in No. 75 of the *Quarterly Journal of Pure and Applied Mathematics*, 1883. The second may be found in the *Proceedings of the London Mathematical Society* for the same year. The solutions also of many of the examples given in this Chapter may be found in these two papers.

CHAPTER IX.

Solution of Problems.

401. IN the first section of this chapter we propose, by the consideration of some examples, to show how the Calculus of Finite Differences may be applied to the solution of dynamical problems. In the second section we shall examine a few remarkable points in the theory of such oscillations.

The calculus of finite differences may be used when the system contains a great many oscillatory bodies arranged in some order. Perhaps there are so many that to write down all their equations of motion individually would be impossible. If however there be a sufficient amount of similarity between the motions of successive bodies taken in order, it may be possible by writing down a few equations of differences to include all the equations of motion. To show how this can be done we shall begin with the following problem.

402. **Oscillations of a chain of particles connected by strings.** Ex. *A string of length $(n+1)\,l$, an insensible mass, stretched between two fixed points with a force T, is loaded at intervals l with n equal masses m not under the influence of gravity and is slightly disturbed ; if $T/lm = c^2$, prove that the periodic times of the simple transversal vibrations which in general coexist are given by the formula $(\pi/c)\ \mathrm{cosec}\ i\pi/2\,(n+1)$ on putting in succession $i = 1, 2, 3\ldots n$.*

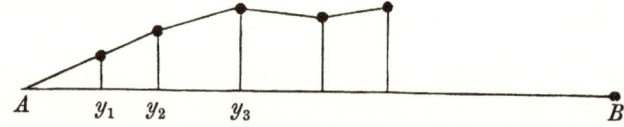

Let A, B be the fixed points; y_1, y_2,...y_n the ordinates at time t of the n particles. The motion of the particles parallel to AB is of the second order, and hence the tensions of all the strings must be equal, and in the small terms we may put this tension equal to T. Consider the motion of the particle whose ordinate is y_k. The equation of motion * is

$$m\frac{d^2y_k}{dt^2} = \frac{y_{k+1}-y_k}{l}T - \frac{y_k - y_{k-1}}{l}T;$$

$$\therefore \frac{d^2y_k}{dt^2} = c^2(y_{k+1} - 2y_k + y_{k-1}) \dots \dots \dots (1).$$

Now the motion of each particle is vibratory, we may therefore expand y_k in a series of the form

$$y_k = \Sigma L \sin(pt + \omega) \dots \dots \dots (2),$$

where Σ implies summation for all values of p.

As there may be a term of the argument pt in every y, let L_1, L_2,... be their respective coefficients. Then substituting, we have

$$L_{k+1} - 2L_k + L_{k-1} = -\frac{p^2}{c^2}L_k \dots \dots \dots (3).$$

To solve this linear equation of differences we follow the usual rule. Putting $L_k = A a^k$, where A and a are two constants, we get after substitution and reduction $a - 2 + 1/a = -(p/c)^2$, or

$$\sqrt{a} - \frac{1}{\sqrt{a}} = \frac{p}{c}\sqrt{-1}, \text{ and } \sqrt{a} + \frac{1}{\sqrt{a}} = \pm 2\left\{1 - \left(\frac{p}{2c}\right)^2\right\}^{\frac{1}{2}};$$

$$\therefore \sqrt{a} = \pm\left\{1 - \left(\frac{p}{2c}\right)^2\right\}^{\frac{1}{2}} + \frac{p}{2c}\sqrt{-1}.$$

Let these values of a be called α and β, then

$$L_k = A\alpha^k + B\beta^k$$

is a solution, and since it contains two arbitrary constants it is the general solution.

The constants A, B, α, β are the same for all the particles, but not necessarily the same for all the trigonometrical terms defined by the different values of p. When we wish to discuss the properties of any particular A and B we write as a suffix the letter p by which they are distinguished.

---

* This equation might also be deduced from Lagrange's general equations of motion. If U be the force function, the position of equilibrium being the position of reference, we have $2U = -\frac{T}{l}y_1{}^2 - \frac{T}{l}(y_2 - y_1)^2 - \&c. - \frac{T}{l}(y_n - y_{n-1})^2 - \frac{T}{l}y_n{}^2.$

The vis viva is evidently $my_1'^2 + my_2'^2 + \dots + my_n'^2.$

Substituting these in Lagrange's equations of motion we obtain the equations represented by (1).

This problem is discussed by Lagrange in his *Mécanique Analytique*. He deduces the solution from his own equations of motion. He also determines the oscillations of an inextensible string charged with any number of weights and suspended by both ends or by one only. Though several solutions of these problems had been given before his time, he considers that they were all more or less incomplete.

The term distinguished by $p = 0$ requires some further consideration. In this term the two values of a viz. α and β are each equal to unity, and the solution of equation (3) loses one of its arbitrary constants. But this defect is easily cured by following the usual rules for treating equations of differences. Just as in differential equations, when t is the independent variable, the presence of equal roots indicates that there are powers of t in the solution, (Art. 266), so in equations of differences powers of the independent variable k make their appearance under similar circumstances. We therefore have

$$L_k = A_0 + B_0 k.$$

The term distinguished by $p = 2c$ also presents some peculiarity. In this term the two values of a are each equal to -1. We have therefore $\qquad L_k = (A_{2c} + B_{2c}k)(-1)^k.$

Summing up, the solution of equation (1) may be written at length

$$y_k = A_0 + B_0 k + (A_{2c} + B_{2c}k)(-1)^k \sin(2ct + \omega_{2c})$$
$$+ \Sigma (A_p \alpha^k + B_p \beta^k) \sin(pt + \omega_p) \dots\dots\dots (4),$$

where the Σ implies summation for all existing values of p. We know from the theory of equations of differences that the first four terms in this expression are really included in the last as the limiting case of the terms distinguished by $p = 0$ and $p = 2c$. Unless therefore we wish to call attention to these terms, they may be omitted in the expression for y_k.

403. The equation (1) represents the motion of every particle except the first and last. In order that it may represent these also it is necessary to suppose that y_0 and y_{n+1} are both zero though there are no particles corresponding to the values of k equal to 0 and $n + 1$. With this understanding the solution (4) represents the motion of every particle from $k = 1$ to $k = n$.

404. Since $y = 0$ when $k = 0$ for all values of t every term in the series (4) must vanish; $\therefore A_0 = 0$, $A_{2c} = 0$ and $A_p + B_p = 0$. Also $y = 0$ when $k = n + 1$ for all values of t, $\therefore B_0 = 0$, $B_{2c} = 0$ and $A_p \alpha^{n+1} + B_p \beta^{n+1} = 0$. These equations give $\alpha^{n+1} = \beta^{n+1}$. If p be greater than $2c$ the ratio of α to β is real and different from unity. Hence we must have p less than $2c$. Let then

$$p/2c = \sin\theta, \quad \therefore a = \cos 2\theta \pm \sin 2\theta \sqrt{-1}.$$

Hence by what has been proved before

$$(\cos 2\theta + \sin 2\theta \sqrt{-1})^{n+1} = (\cos 2\theta - \sin 2\theta \sqrt{-1})^{n+1};$$
$$\therefore \sin 2(n+1)\theta = 0; \quad \therefore \theta = i\pi/2(n+1),$$

and the complete period of any term is $P = 2\pi/p = \pi c/\sin\theta$. The letter i indicates any integer, but since $p = 2c \sin\theta$, we see that it is necessary to consider only the integers from $i = 1$ to $i = n$. The values $i = 0$ and $i = n + 1$ are excluded because they make $p = 0$ and $p = 2c$ which have been already taken account of.

The periods thus determined are those of the principal oscillations. Taking any one of these values of p^2, the corresponding values of $y_1, y_2, \ldots y_n$ are given by the equation (4) which reduces to

$$y_k = C \sin 2k\theta \sin (pt + \omega).$$

The oscillations indicated by the several values of p are very different from each other. When θ has its least value, the sign of $\sin 2k\theta$ is the same for all values of k from $k = 1$ to n, so that the chain oscillates in the form of a single loop. When θ has its next least value the first half of the terms $y_1, y_2, \ldots$ have the same sign and this sign is opposite to that of the second half, so that the chain always oscillates in the form of a double loop. When θ has its next value the chain oscillates with three loops and so on. The several kinds of motion are easily distinguished from each other by tracing the curves whose ordinate is y_k and abscissa k, the time t having any given value. They also follow at once from *Sturm's Theorems* given a little further on, where it is proved that similar distinctions exist whenever the connected system of particles is such that the equation of differences takes a certain standard form.

405. In forming the differential equation (1) we have supposed the distance l between any two successive particles to be unaltered. This will practically be the case if $y_k - y_{k-1}$ is small compared with the distance l. This limitation however does not prevent us from enquiring what would be the effect of reducing the masses of all the particles and placing them proportionally closer, so that the total mass per unit of length is unaltered. The restriction is that the inclinations of the strings must still be sufficiently small. The interest of this change is that the closer the particles are placed the more nearly does the system approach to that of a uniform string stretched between the two fixed points A and B.

Let us represent by ρ the mass per unit of length, then $c^2 l^2 = Tl/m = T/\rho$. Put $a = cl$, then a is equal to the square root of the ratio of the tension to the mass of a unit of length. Thus a is unaltered by any of these changes of the particles.

If the length of the string AB be L we have $L = (n + 1) l$. If n be very great we find $p = 2c \sin \theta = a \, i\pi/L$ very nearly.

Thus the notes sounded by a string loaded with small particles at short intervals are such that their periods are given by $P = 2L/ai$. The note given by $i = 1$ is called the fundamental note, those given by the higher integer values of i are called the harmonics.

406. **Determination of Constants.** If we express a and β in terms of θ and substitute these in equation (4) we find the typical equation

$$y_k = \Sigma E_i \sin 2k\theta \cos (2ct \sin \theta) + \Sigma F_i \sin 2k\theta \sin (2ct \sin \theta) \ldots\ldots\ldots (5),$$

where E_i and F_i have been written for $2A_p \sin \omega_p \sqrt{-1}$ and $2A_p \cos \omega_p \sqrt{-1}$. As before $\theta = i\pi/2 \, (n+1)$ and the symbol Σ implies summation for all values of i from $i=1$ to $i=n$. This equation has n terms and thus we have $2n$ arbitrary constants, viz. $E_1, E_2 ... E_n$ and $F_1, F_2 ... F_n$. These have to be determined from the known initial values of the n co-ordinates $y_1, y_2 ... y_n$ and of their initial velocities $y_1', y_2' ... y_n'$.

Since k may have any value from $k=1$ to $k=n$ the typical equation (5) represents as many equations as there are particles. We may imagine there to be written down, one under another, exactly as described in Chap. VIII. Art. 379. To find the constant E_i which runs through all the terms in any one column we use the multiplier to separate that column from the others. To find this multiplier we write down the vis viva of the system which in our case is $2T = \Sigma m y_k'^2$. According to the rule given in Chap. VIII. Art. 387 or Art. 399, the proper multiplier for the equation giving y_k is found by differentiating T with regard to y_k' and substituting for y_k' the coefficient of the oscillation we wish to separate. The differentiation in our case is $m y_k'$. The proper multipliers to separate the two columns distinguished by any value of i are therefore $m E_i \sin 2k\theta$ and $m F_i \sin 2k\theta$. Thus we find after division by common factors

$$\left. \begin{array}{l} \Sigma \left\{ y_k \sin 2k\theta \right\} = \tfrac{1}{2} E_i \, (n+1) \\ \Sigma \left\{ y_k' \sin 2k\theta \right\} = \tfrac{1}{2} F_i \, (n+1) \, 2c \sin \theta \end{array} \right\} .$$

Here we have written on the right-hand side for $\Sigma \, (\sin 2k\theta)^2$ its value $\tfrac{1}{2} \, (n+1)$ which is easily found by ordinary trigonometrical processes.

These equations determine the values of E_i and F_i for any particular value of i. On the left-hand side the co-ordinates $y_1, y_2,$ &c. and the velocities $y_1', y_2',$ &c. are supposed to have their initial values, and the symbol Σ implies summation for all values of k from $k=1$ to $k=n$, the value of i included in θ being given.

407. **Ex. 1.** A string of length $2\,(n+1)\,l$ is stretched between two fixed points A and B as before and loaded with $2n+1$ particles at distances apart each equal to l. Taking the origin at the middle particle, let the particles from $k = -\epsilon$ to $k = +\epsilon$ be initially displaced so that $y_k = C \sin k\pi/\epsilon$. Let all the other particles be in their undisturbed positions in the straight line AB, so that $y_k = 0$ for all values of k not comprised between the limits $\pm \, \epsilon$. Let also the system start from rest. Then by proceeding as explained in the last article, we find that the motion is given by

$$y_k = \Sigma E_i \sin 2k\theta \cos (2ct \sin \theta),$$

where $\qquad \theta = \dfrac{i\pi}{2\,(n+1)}, \qquad E_i = \dfrac{C \cos i\pi}{2\,(n+1)} \dfrac{\sin 2\epsilon\theta \sin \pi/\epsilon}{\sin^2 \pi/2\epsilon - \sin^2 \theta}.$

Ex. 2. A string of length $(n+1)\,l$ is stretched between two fixed points A and B and loaded with n particles at distances each equal to l. The extremity A, defined by $k=0$, is suddenly moved a small space equal to y_0 at right angles to the original position of the string and is there kept fixed. The motion of the kth particle is given by $\qquad y_k = y_0 \left(1 - \dfrac{k}{n+1} \right) - \Sigma \dfrac{y_0}{n+1} \cot \theta \sin 2k\theta \cos (2ct \sin \theta),$

where $\theta = i\pi/2 \, (n+1)$, and the symbol Σ implies summation for all values of i from $i=1$ to n.

To prove this we have the following conditions; (1) for *all* values of t we have $y_k = y_0$ when $k=0$, and $y_k = 0$ when $k=n+1$. These give $B_0 = y_0$ and $A_0 \,(n+1) = -y_0$, (2) when $t=0$ we have $y_k = 0$ for all values of k except $k=0$.

408. **Agitation of one extremity.** When one extremity of the string of particles is agitated according to any given law,

a slight modification of the solution given in Art. 402 will enable us to find the motion. *Let us suppose that the extremity* A, *defined by* $k = 0$, *is agitated so that its motion is continuously given by* $y_0 = C \sin \mu t$; *it is required to find the motion of the particles.* We may notice that it is sufficient for our present purpose that the law of agitation, however complicated, can be represented by a finite series of terms of this form. The resultant motion of any particle is then found by compounding together the motions due to the several terms of the series.

The motion of the string of particles may be regarded as made up of two separate oscillatory motions. There are (1) the forced oscillation whose period is the same as that of the agitating force, and (2) the free oscillations whose periods are the same as those found in Art. 404 when the two extremities of the string were fixed. Our present object is to find the former of these.

Proceeding as before, we have by equation (4)

$$y_k = A_0 + B_0 k + (A_{2c} + B_{2c}k)(-1)^k \sin(2ct + \omega_{2c}) + \Sigma (A_p \alpha^k + B_p \beta^k) \sin(pt + \omega_p).$$

Since $y_k = C \sin \mu t$ when $k = 0$ we have $p = \mu$, $\omega_p = 0$ in the forced vibration. Also unless $\mu = 0$ or $2c$ we have $A_0 = 0$, $A_{2c} = 0$. Again, $y_k = 0$ when $k = n + 1$, hence $B_0 = 0$, $B_{2c} = 0$ and the forced vibration is given by

$$A_\mu + B_\mu = C, \quad A_\mu \alpha^{n+1} + B_\mu \beta^{n+1} = 0,$$

where α and β are the two values of a given by

$$\sqrt{a} = \pm \left\{1 - \left(\frac{\mu}{2c}\right)^2\right\}^{\frac{1}{2}} + \frac{\mu}{2c}\sqrt{-1}.$$

409. *If μ be greater than* $2c$, let $\mu = 2c/\sin\phi$, and all possible cases are included if we suppose ϕ to lie between 0 and $\frac{1}{2}\pi$, so that $\tan\frac{1}{2}\phi$ is less than unity. Making the necessary substitutions we find for the forced oscillation

$$y_k = \frac{(\tan\frac{1}{2}\phi)^{2(n+1-k)} - (\cot\frac{1}{2}\phi)^{2(n+1-k)}}{(\tan\frac{1}{2}\phi)^{2(n+1)} - (\cot\frac{1}{2}\phi)^{2(n+1)}} \cdot (-1)^k \, C \sin \mu t \dots (1).$$

If the string is very long we have n infinite, and this expression takes the simpler form

$$y_k = (\tan\frac{1}{2}\phi)^{2k}(-1)^k \, C \sin \mu t \dots\dots\dots\dots\dots(2).$$

The first of these two expressions applies to a finite string of particles and is clearly made up of two expressions like the latter, the coefficients being such that the displacements of A and B are respectively $C \sin \mu t$ and zero. The motion has therefore been analysed as the resultant of two motions each of which is represented by equation (2).

410. *If μ be less than* $2c$, let $\mu = 2c \sin\psi$, the forced vibration then becomes

$$y_k = \frac{\sin 2(n+1-k)\psi}{\sin 2(n+1)\psi} C \sin \mu t \dots\dots\dots\dots (3).$$

17—2

This can be written in the form

$$y_k = \frac{C \cos \left[\mu t - 2\left(n+1-k\right)\psi\right]}{2 \sin 2\left(n+1\right)\psi} - \frac{C \cos \left[\mu t + 2\left(n+1-k\right)\psi\right]}{2 \sin 2\left(n+1\right)\psi} \quad \ldots (4).$$

Taking the first of these two terms by itself, we see that after a time T given by $\mu T = 2\psi$, the term is unaltered if we write $k-1$ for k. This term therefore represents a wave which travels the space between one particle and the next in the time T. In the same way the second term represents a wave which travels with the same velocity in the opposite direction.

We may notice that the denominator of either of the terms in (4) is very small when μ is nearly equal to $2c \sin i\pi/2\,(n+1)$, i.e. the forced vibration is magnified when the period of the agitating force is nearly equal to one of the periods of the free vibrations of the string, both ends being fixed.

411. Two kinds of possible motion. Attention should be particularly directed to the great difference between the two kinds of oscillatory motions. If the period of the agitating force, viz. $2\pi/\mu$ is long enough to make $\mu < 2c$, the forced oscillation transmitted to the string of particles is formed by the superposition of two waves which travel in opposite directions without change of magnitude. Thus the particles near the further extremity B of the string may be as greatly agitated as those near the point of application of the force. Suppose $\psi = \pi/2q$, where q is some integer, then by (3) every qth particle counting from the further extremity B is permanently at rest and *forms a node*. The strings of particles between these successive nodes form equal loops which are alternately on one side and the other of the straight line AB.

Let us now compare this state of motion with that which results from the agitating force when its period is so short that $\mu > 2c$. In this case no motion in the nature of a wave is transmitted along the string. Taking the case of a very long string, the particles are alternately on opposite sides of AB, while their displacements form a series in geometrical progression. Thus the displacements of the particles are less and less the more remote they are from the agitating force.

412. The transition from the one kind of motion to the other is easily understood by supposing the period of the agitating force to grow gradually less and less until it passes the critical value. It is clear that $\sin \psi$ will increase, but it cannot become greater than unity. The number of particles, viz. $q-1$, between two successive nodes decreases and finally vanishes when $\psi = \frac{1}{2}\pi$. But since no further decrease is possible the motion changes its character.

The expressions (1) and (3) both assume the form $0/0$ when $\phi = \psi = \frac{1}{2}\pi$. The motion in the transitional state may be deduced

from either of these expressions by the usual rules in the differential calculus. But we see independently by Art. 402 that it is given by $\quad y_k = (A + Bk)(-1)^k \sin 2ct.$
Since $y_k = C \sin 2ct$ when $k = 0$ and $y_k = 0$ when $k = n + 1$, we easily find $\quad y_k = \{1 - k/(n+1)\}(-1)^k C \sin 2ct.$

413. Discontinuous agitating force. When the agitation communicated to the extremity A is not continuous, but acts for a short time only, the resulting motion may be found by the method of the superposition of small motions.

Thus if the extremity A be suddenly moved at the time $t = 0$ a short distance y_0 at right angles to AB, the resulting motion has been found in Ex. 2, Art. 407. Let us represent this motion by $y_k = y_0 f(k, t)$. After a time $t = u$ has elapsed, let the extremity A receive another displacement Y_0, the rest of the string being undisturbed. If we superimpose these two motions we obtain

$$y_k = y_0 f(k, t) + Y_0 f(k, t - u).$$

At the time $t = u$, the second function and its differential coefficient with regard to t both vanish for all values of k from $k = 1$ to $k = n + 1$. Thus the initial conditions of motion at this time are expressed by the first function. This equation therefore represents the motion produced by these two disturbances for all time from $t = u$ to $t = \infty$.

Generalizing this, we see that if the extremity A be moved according to any law say $y_0 = F(t)$ for a time extending from $t = 0$ to $t = \gamma$, then the motion of the string

is given by $\quad y_k = \int_0^\gamma F'(u) f(k, t - u) \, du$

for all time extending from $t = \gamma$ to $t = \infty$.

Since the agitating force ceases to act after the time $t = \gamma$ it is clear that the motion of the string after this time is made up of the free vibrations belonging to a string of particles having each end fixed. Accordingly, if we substitute for the function $f(k, t - u)$ its value given in Art. 407, we see that this expression for y_k consists of n oscillations whose periods are the same as those already found in Art. 404. Their phases and magnitudes depend on the action of the agitating force.

414. Ex. Let the extremity A of the string of particles already described be moved so that $y_0 = C \sin \mu t$ for a time extending from $t = 0$ to $t = \pi/\mu$. Supposing the extremities to remain at rest for all subsequent time, prove that the motion of the kth particle is given by

$$y_k = \Sigma \frac{4C\mu \cos\theta \sin 2k\theta}{n+1} \cdot \frac{\sin\left[2c\sin\theta\left(t - \dfrac{\pi}{2\mu}\right)\right]\cos\left[\dfrac{c\pi}{\mu}\sin\theta\right]}{\mu^2 - 4c^2\sin^2\theta},$$

where $\theta = i\pi/2(n+1)$ and the Σ implies summation for all integer values of i from $i = 1$ to $n + 1$.

If the string is very long, n is infinite and we may write $d\theta = \pi/2(n+1)$. The expression then becomes

$$y_k = \frac{8c\mu}{\pi}\int_0^\infty d\theta \cos\theta \sin 2k\theta \sin\left\{2c\sin\theta\left(t - \frac{\pi}{2\mu}\right)\right\}\frac{\cos(c\pi\sin\theta/\mu)}{\mu^2 - 4c^2\sin^2\theta}.$$

The subject of integration is not infinite when $\sin\theta = \mu/2c$, for the last factor then becomes $\pi/4\mu^2$.

415. Analysis by Waves. There is another method of arranging the solution of the equation of motion given in Art. 402 which has the advantage of enabling us to analyse the motion by waves instead of by Lagrangian elements, see Art. 85. Writing δ for d/dt as usual the equation of motion becomes

$$y_{k+1} - 2y_k + y_{k-1} = \frac{\delta^2}{c^2} y_k \dots\dots\dots\dots\dots\dots (1).$$

Treating the operator on the right-hand side as a constant, we proceed to solve the equation of differences in the manner already explained in Art. 402. The two constants A and B are now functions of t. Hence if we put

$$\Omega = \left\{1 + \left(\frac{\delta}{2c}\right)^2\right\}^{\frac{1}{2}} - \frac{\delta}{2c} \qquad \text{we have} \qquad y_k = \Omega^{2k} f(t) + \Omega^{-2k} F(t) \dots\dots(2).$$

This is a symbolical solution of the equation of differences with its two arbitrary functions $f(t)$ and $F(t)$. When the forms of these functions are given, the operation represented by Ω can be performed and a solution of the equations of differences will be found.

416. To obtain one interpretation of this symbolical solution let us suppose that the functions $f(t)$ and $F(t)$ can be expressed in a series whose general term is $A \cos(2c \sin \theta t + \omega)$, where θ is the parameter whose value distinguishes any term of the series from another. All cases are clearly included if we suppose θ to lie between the limits 0 and $\frac{1}{2}\pi$.

Since the radical in the operator Ω contains only even powers of δ, we obtain the result of its operation by writing $-(2c \sin \theta)^2$ for δ^2, see Art. 265. We therefore find

$$\Omega \cos(2c \sin \theta t + \omega) = \cos(2c \sin \theta t + \omega - \theta).$$

Repeating this process $2k$ times we have

$$y_k = \Sigma A \cos(2c \sin \theta t + \omega - 2k\theta) + \Sigma B \cos(2c \sin \theta t + \omega + 2k\theta).$$

If we take by itself any one term of the first series we see that if we write for k, $k+1$ and for t, $t+T$, where T is given by $c \sin \theta T = \theta$, the term is unaltered. Hence (exactly as in Art. 87) any one term represents a wave which travels the space between one particle and the next in the time T. In the same way the corresponding term of the second series represents a wave which travels in the opposite direction with the same velocity. See Art. 410.

Each term of either series represents a wave. Each wave travels with a uniform velocity but the different waves have different velocities. Consider the wave defined by any given value of θ, and let $a = cl$. If v be the velocity, λ the length of the wave measured from ridge to ridge, and P the period of oscillation of any one particle, we have

$$v = a\frac{\sin \theta}{\theta}, \qquad \lambda = \frac{\pi l}{\theta}, \qquad P = \frac{\pi l}{a \sin \theta}.$$

Since θ lies between 0 and $\frac{1}{2}\pi$, we see that the velocities of all these waves lie between a and $2a/\pi$; the length of every wave is greater than $2l$; the period of oscillation of every particle is greater than $\pi l/a$. The longer the waves are the more nearly do they travel with the same velocity.

If we suppose l to decrease the particles become closer together, and if each particle have proportionally less mass the quantity a is unchanged. Considering then all waves whose lengths have a given inferior limit, we see that *the closer the particles are together, the mass of a unit of length being unchanged, the more nearly do waves of all lengths travel with the same velocity.*

Other interpretations of the symbolical solution given in Art. 415 may be obtained by substituting other forms for the arbitrary functions $f(t)$ and $F(t)$. Thus we may have $\qquad y_k = \Omega^{2k} \frac{1}{2} C e^{\mu t \sqrt{-1}} + \Omega^{-2k} \frac{1}{2} C e^{-\mu t \sqrt{-1}}.$

If μ be greater than $2c$ we may introduce the subsidiary angle ϕ as in Art. 409. This expression then reduces to $\qquad y_k = (-1)^k (\tan \frac{1}{2}\phi)^{2k} C \cos \mu t.$

417. Ex. If we write $x = kl$ and make the interval l between the particles indefinitely small, the operation represented by Ω^{2k} takes the singular form 1^{∞}. Show by finding the limit in the usual manner that $\Omega^{2k} = e^{-(x/a)\delta}$ and thence deduce

$$y_x = f(-x/a + t) + F(x/a + t).$$

418. Examples. Ex. 1. A long row of particles, each of mass m, is placed on a smooth horizontal table. Each is connected with the two adjacent ones by similar light elastic stretched strings of natural length l. They receive small longitudinal disturbances such that each of them proceeds to perform a harmonic oscillation: prove that there will be two waves of vibrations in opposite directions with the

same velocity, viz. $l' \sqrt{\dfrac{E}{ml}} \dfrac{q}{\pi} \sin \dfrac{\pi}{q}$, where l' is the average distance between two

successive particles, q the number of intervals between two particles in the same phase, and E the modulus of elasticity. [Math. Tripos, 1873.

Ex. 2. A light elastic string of length nl and coefficient of elasticity E is loaded with n particles each of mass m ranged at intervals l along it, beginning at one extremity. If it be suspended by the other prove that the periods of its vertical

oscillations are given by the formula $\pi \sqrt{\dfrac{lm}{E}} \operatorname{cosec} \dfrac{2i+1}{2n+1} \dfrac{\pi}{2}$, where $i = 0, 1, 2 \ldots n-1$

successively. Hence show that the periods of the vertical oscillations of a heavy

elastic string are given by the formula $\dfrac{4}{2i+1} \sqrt{\dfrac{ML}{E}}$, where L is the length of the

string, M its mass, and i is zero or any positive integer. [Math. Tripos, 1871.

Ex. 3. A railway engine is drawing a train of equal carriages connected by spring couplings of strength μ, and the driving power is so adjusted that the velocity is $A + B \sin qt$. Show that if $q^2\{(M + 4m)b^2 + 4mk^2\}$ be nearly equal to $2\mu b^2$ the couplings will probably break, M being the mass of a carriage which is supported on four equal wheels of mass m, radius b and radius of gyration k. Are there any other values of q for which the couplings will probably break? [Coll. Exam. 1880.

Ex. 4. Equal uniform rods, n in number, and each of mass m, are smoothly hinged together at their ends and are suspended by light elastic strings which are fastened to the joints and the free ends. The other extremities of the strings are attached to $n+1$ points in a horizontal line whose distance apart is equal to the length of a rod. The strings are all of a natural length l and modulus E, except the extreme ones whose modulus is $\frac{1}{2}E$. The system rests in equilibrium under the action of gravity and the rods are in a horizontal straight line and all the strings vertical. Show that the periods of the small co-existent oscillations about this

position of equilibrium are $\dfrac{2\pi}{\sqrt{3E}} \left\{ ml \left(2 + \cos \dfrac{i\pi}{n} \right) \right\}^{\frac{1}{2}}$, where i is zero or any integer,

the joints and ends being supposed to move approximately in vertical straight lines. [Coll. Exam. 1881.

Ex. 5. A number of uniform circular discs of radius a but of any masses are freely moveable in a vertical plane about their centres which are fixed in a horizontal line at distances $4a$ apart. A fine rough string of indefinite length having two equal particles of mass m at its extremities is laid over these circles, and uniform circular discs each of radius a and mass $2m$ are laid on the string so as to hang between the other circles, the parts of the string not in contact with a circle being vertical. Show that if the system be in motion under the action of gravity, all its parts will move uniformly so long as the centres of all the discs $2m$ are below the line of fixed centres. [Coll. Exam. 1880.

It will be seen on writing down a few of the equations of motion that both the dynamical and geometrical equations are all linear with constant coefficients. When this is the case the reactions are all constant, being independent both of the time and of the initial conditions, see Vol. I. Chap. IV. Arts. 135—136. The system is initially in equilibrium and only moves because it is disturbed; hence the reactions throughout the motion retain their equilibrium values. The tension therefore of every portion of the string is equal to mg. It easily follows that the motion is uniform.

Ex. 6. From the same sheet of indefinitely thin metal of uniform width are made n cylinders of radii $a_1, a_2, \ldots a_n$ (in descending order of magnitude). They are placed one inside the other, and the whole are then placed inside a fixed cylinder of radius a whose axis is horizontal, so that the axes of all the cylinders are parallel. Show that if ω_r be the angle turned through by the cylinder of radius a_r, and if M_r denote the sum $a_n + a_{n-1} + \ldots a_r$, the equations giving the small motions of the system are of the form $2a_r^2 (d^2\omega_r/dt^2) + g (M_r\chi_r - M_{r-1}\chi_{r+1}) = 0$,

where $a_r (\omega_r + \chi_r) = a_{r-1} (\omega_{r-1} + \chi_r)$.   [Coll. Exam. 1880.

419. Oscillations of a chain made of rods or gyrostats* connected by strings. Ex. 1. The links of a chain are alternately uniform rods each of length $2a$, and inelastic strings each of length $2l$; the number of rods being equal to that of the strings. The system is stretched with the rods and strings in one straight line, the extremity of the first string being attached to a fixed point A and the extremity of the last rod to another fixed point B. The system being slightly displaced in one plane, it is required to find the small oscillations.

Let n be the number of rods, $y_1, y_2 \ldots y_n$ the ordinates of their centres of gravity; $q_1, q_2 \ldots q_n$ their inclinations to AB. Let $s_1, s_2 \ldots s_n$ be the inclinations of the strings to the same straight line. Let m be the mass of each rod, mA the moment of inertia about the centre of gravity. Let mT be the tension of the chain.

The equations of motion of the kth rod are

$$y_k'' = T (s_{k+1} - s_k) \dots\dots\dots\dots\dots\dots\dots\dots\dots\dots\dots\dots\dots\dots(1),$$
$$A q_k'' = Ta (s_k + s_{k+1} - 2q_k) \dots\dots\dots\dots\dots\dots\dots\dots\dots(2),$$

where accents denote differential coefficients with regard to the time. Besides these we have the geometrical equation

$$y_{k+1} - y_k = a (q_k + q_{k+1}) + 2ls_{k+1}\dots\dots\dots\dots\dots\dots\dots\dots\dots(3).$$

These equations, when solved, give the motion of the chain however long it may be. We have to find a solution adapted to the condition that at two points A and B

$$y_0 + aq_0 = 0, \qquad y_n + aq_n = 0 \dots\dots\dots\dots\dots\dots\dots\dots(4),$$

throughout the motion. These being satisfied we may suppose the points A and B to be fixed and all the chain except the portion between A and B removed.

---

* In April 1875, Sir W. Thomson made a communication to the London Mathematical Society on vibrations and waves in a stretched uniform chain of symmetrical gyrostats connected together by universal flexure joints; see questions 4 and 5. In the Mathematical Tripos 1889, Part II. Prof. Burnside set a question on the motion of an endless train of waves on a chain of gyrostats connected by ball and socket joints; see question 6. Questions 1, 2, 3 of the above series have been constructed with the view of showing how the conditions at the extremities of a finite chain of connected rigid bodies are to be treated.

To solve these we use the method already explained in Art. 402. We put

$$y_k = Y\rho^k \sin(pt+a), \qquad q_k = Q\rho^k \sin(pt+a), \qquad s_k = S\rho^k \sin(pt+a).$$

Substituting, the equations (1), (2), (3) become

$$-p^2 Y = T(\rho - 1) S \dots\dots\dots\dots\dots\dots\dots\dots\dots\dots\dots\dots\dots(5),$$

$$-(Ap^2 - 2Ta) Q = Ta(\rho + 1) S \dots\dots\dots\dots\dots\dots\dots\dots\dots(6),$$

$$Y(\rho - 1) = a(\rho + 1) Q + 2l\rho S \dots\dots\dots\dots\dots\dots\dots\dots(7).$$

Eliminating the ratios Y, Q, S by a determinant we find

$$(\rho^2 + 1)\{Ap^2 - 2Ta - a^2 p^2\} - 2\rho\left\{(Ap^2 - 2Ta)\left(1 - \frac{lp^2}{T}\right) + a^2 p^2\right\} = 0 \dots\dots(8).$$

For each value of p we have a quadratic to find ρ whose roots ρ, ρ_1 are such that $\rho\rho_1 = 1$. Putting therefore $\phi = pt + a$, we have

$$\left.\begin{array}{l} s_k = (S\rho^k + S_1\rho_1{}^k) \sin\phi \\[2mm] y_k = (Y\rho^k + Y_1\rho_1{}^k) \sin\phi = -\dfrac{T}{p^2}\{S(\rho - 1)\rho^k + S_1(\rho_1 - 1)\rho_1{}^k\} \sin\phi \\[2mm] q_k = (Q\rho^k + Q_1\rho_1{}^k) \sin\phi = -\dfrac{Ta}{Ap^2 - 2Ta}\{S(\rho + 1)\rho^k + S_1(\rho_1 + 1)\rho_1{}^k\} \sin\phi \end{array}\right\} \dots(9).$$

Referring to equations (4) we find by putting $k = 0$ and $k = n$

$$(Y + Qa) + (Y_1 + Q_1 a) = 0, \qquad (Y + Qa)\rho^n + (Y_1 + Q_1 a)\rho_1{}^n = 0 \dots\dots\dots(10).$$

These show that either

$$\rho^n = \rho_1{}^n \dots\dots\dots(11), \qquad \text{or both } Y + Qa = 0, \qquad Y_1 + Q_1 a = 0 \dots\dots\dots(12).$$

Taking first the alternative (11) we see that, since $\rho\rho_1 = 1$, we may put

$$\rho = \cos\theta + \sin\theta \sqrt{-1}, \qquad \sin n\theta = 0 \dots\dots\dots\dots\dots\dots\dots(13).$$

Since $\rho^2 + 1 = 2\rho\cos\theta$, the determinantal equation (8) becomes

$$(Ap^2 - 2Ta)\{lp^2 - T(1 - \cos\theta)\} - Ta^2 p^2(1 + \cos\theta) = 0 \dots\dots\dots\dots\dots(14).$$

This quadratic gives two positive values of p^2, separated by $p^2 = (1 - \cos\theta)T/l$. The values of $\cos\theta$ are given by $\cos\theta = \cos i\pi/n$, where i has all integer values from $i = 1$ to $i = n - 1$. The values $i = 0$ and $i = n$ are excluded because they make $\rho = \rho_1$ and when this happens the solution (9) changes its character and contains integer powers of k.

Considering next the second alternative (12), we find by putting $Y = -aQ$ in (5) and (7)

$$\{p^2 l + T(\rho - 1)\} S = 0 \dots\dots\dots\dots\dots\dots\dots\dots\dots\dots(15),$$

with a similar equation obtained by writing ρ_1 and S_1 for ρ and S. We thus find from (15) and (8) that

$$\rho = 1 - lp^2/T, \qquad (A + a^2)lp^2 = 2Ta(a + l) \dots\dots\dots\dots\dots\dots(16).$$

Since ρ_1 is the reciprocal of ρ and cannot also have the same value as ρ, we must have $S_1 = 0$. Substituting these values of ρ and S_1 in (9), the solution adapted to the second alternative has been found.

The peculiarity of the motion given by the second alternative is that $y_k + aq_k = 0$ for all values of k, so that *the second extremity of each rod is at rest throughout the motion.*

We have yet to examine the portion of the solution due to the equal roots of equation (8). Since $\rho\rho_1 = 1$, these are $\rho = \pm 1$. In this case we have

$$y = (Y_1 + Y_2 k)(\pm 1)^k \sin(pt + a),$$

with similar expressions for q and s obtained by writing Q_1, Q_2 and S_1, S_2 for Y_1, Y_2. The relations between these six coefficients may be found by substituting in the equations of motion and equating to zero the several powers of k. Also equations (4) give $Y_1 + aQ_1 = 0$, $Y_2 + aQ_2 = 0$. These eight equations cannot be satisfied by finite values of the coefficients except in one case which is included in (16) by putting $\rho = -1$ and $A = al$. We therefore infer that when the extremities A and B of the chain are fixed, terms with k as a factor do not appear in the solution.

The system has $3n$ co-ordinates, viz. $y_1...y_n$, $q_1...q_n$, $s_1...s_n$ and $n-1$ geometrical equations given by (3) with two more given by (4). By Lagrange's rule for the oscillations of a system about a position of equilibrium we should have $2n-1$ values of p^2. *Of these periods $2(n-1)$ are given by the $n-1$ values of $\cos\theta = i\pi/n$, each value leading to a quadratic for p^2 with unequal roots, viz. equation (14). One more period is given by equation (16).*

Ex. 2. The links of a chain are alternately uniform rods each of length $2a$, and inelastic strings each of length $2l$, the number of the rods being equal to that of the strings. Each rod has attached to its middle point a fly wheel which rotates freely in a plane perpendicular to the rod. The system is stretched with the rods and strings in one straight line, the extremity of a string being attached to a fixed point A and the extremity of the last rod to another fixed point B. The system being slightly displaced it is required to find the small oscillations.

In consequence of the presence of the fly wheels the motion cannot be analysed into two independent oscillations in perpendicular planes. It is therefore necessary to treat the problem as one in three dimensions.

Let AB be the axis of z, and let the axes of x and y be fixed in space. Let (x_k, y_k) be the co-ordinates of the centre of gravity of the kth rod, $(p_k, q_k, 1)$ its direction cosines, $(r_k, s_k, 1)$ those of the preceding string. Let the mass of each rod and fly wheel be m, let mC, mA be the moments of inertia about the rod and a perpendicular to it at the centre of gravity. Let n be the angular velocity of any fly wheel about its axis, then n is constant throughout the motion. Let mT be the tension. Let ν be the number of rods.

The equations of motion of the kth rod are

$$x_k'' = T\,(r_{k+1} - r_k), \qquad y_k'' = T\,(s_{k+1} - s_k), \quad ....(1),$$

$$\left. \begin{aligned} -Aq_k'' + Cnp_k' &= -Ta\,(s_k + s_{k+1} - 2q_k) \\ Ap_k'' + Cnq_k' &= Ta\,(r_k + r_{k+1} - 2p_k) \end{aligned} \right\} \quad ....(2).$$

Besides these we have the geometrical equations

$$\left. \begin{aligned} x_{k+1} - x_k &= a\,(p_k + p_{k+1}) + 2lr_{k+1} \\ y_{k+1} - y_k &= a\,(q_k + q_{k+1}) + 2ls_{k+1} \end{aligned} \right\} \quad ....(3).$$

There are also the conditions at the ends A and B of the chain

$$\left. \begin{aligned} x_0 + ap_0 &= 0 \\ y_0 + aq_0 &= 0 \end{aligned} \right\} , \qquad \left. \begin{aligned} x_\nu + ap_\nu &= 0 \\ y_\nu + aq_\nu &= 0 \end{aligned} \right\} \quad ....(4).$$

In these equations accents denote differentiation with regard to the time.

The equations (2) may be obtained by the rule given in Vol. i. Art. 265, viz. the angular momentum of a uniaxal body about any line through its centre of gravity is the same as that of two particles of equal mass, viz. $\frac{1}{2}m$ placed on the axis at a distance $b = \sqrt{A/m}$ from the centre of gravity together with the angular momentum Cn about the axis. We therefore have

$$h_x = m\,(\eta\zeta' - \zeta\eta') + mCnp, \qquad h_y = m\,(\zeta\xi' - \xi\zeta') + mCnq,$$

where (ξ, η, ζ) are the co-ordinates of either particle referred to the centre of gravity as origin. In our case $\xi = bp$, $\eta = bq$, $\zeta = b$. The equations of motion are then given by $dh_x/dt = L$ &c. see Vol. i. Art. 261. The moments on the right hand sides are formed by the usual rules of statics, viz. $L = \Sigma\,(yZ - zY)$ &c. Another method of forming these equations is given in Art. 15 of this volume.

To solve these equations we proceed as in the last example. We put

$$x = X\rho^k \sin \phi, \qquad p = P\rho^k \sin \phi, \qquad r = R\rho^k \sin \phi,$$

$$y = Y\rho^k \cos \phi, \qquad q = Q\rho^k \cos \phi, \qquad s = S\rho^k \cos \phi,$$

where $\phi = pt + \alpha$. Substituting in the equations (1), (2), (3) and eliminating the ratios of X, Y, P, Q, R, S, we find

$$(\rho^2 + 1) \{Ap^2 + Cnp - 2aT - a^2p^2\} = 2\rho \left\{ (Ap^2 + Cnp - 2aT)\left(1 - \frac{lp^2}{T} \right) + a^2p^3 \right\} \dots(8).$$

Since this equation gives two values of ρ for each value of p, it follows that each term in $\sin \theta$ or $\cos \theta$ is accompanied by *two* exponents. Let ρ, ρ_1 be the roots of equations (8), then $\rho\rho_1 = 1$.

Substituting next in equations (4), we find that there are two alternatives, viz. (1) $\rho^n = \rho_1^n$ or (2) both $X + aP = 0$, $Y + aQ = 0$.

Taking the first alternative we find as before that, since $\rho\rho_1 = 1$,

$$\rho = \cos \theta + \sin \theta \sqrt{-1}, \qquad \sin n\theta = 0 \dots\dots\dots\dots\dots\dots\dots\dots\dots(13).$$

The determinantal equation (8) then becomes

$$\{Ap^2 + Cnp - 2aT\} \{lp^2 - T(1 - \cos \theta)\} - Ta^2p^2(1 + \cos \theta) = 0 \dots\dots\dots(14).$$

This biquadratic leads to two real positive and two real negative values of p, each pair of values being separated by a root of the quadratic $lp^2 = T(1 - \cos \theta)$. The values of $\cos \theta$ are given by $\cos \theta = \cos i\pi/\nu$ where i has all integer values from $i = 1$ to $i = \nu - 1$, and ν is the number of rods.

Considering next the second alternative, we find by treating equations (1) and (3) exactly as in the last example

$$\left. \begin{array}{l} \rho = 1 - lp^2/T \\ (Ap^2 + Cnp + a^2p^2)\, l = 2Ta(a + l) \end{array} \right\} \dots\dots\dots\dots\dots\dots\dots\dots\dots(16).$$

The peculiarity of this motion is that *one extremity of every rod is at rest throughout the motion.*

The system has 6ν co-ordinates and $2(\nu - 1) + 4$ geometrical conditions, we therefore should have $2(2\nu - 1)$ values of p, Art. 111. *Of these periods $4(\nu - 1)$ are given by the $\nu - 1$ values of $\cos \theta$, each value leading to a biquadratic with unequal roots. Two more periods are given by the quadratic* (16).

Ex. 3. The links of a chain are formed of heavy uniform rods each of length $2a$ freely hinged together at their extremities. These are stretched out in a horizontal straight line with one end of the chain hinged to a point fixed in space. If the system starts from rest, show that the initial reaction at the kth hinge is

$$\frac{(-1)^k mg}{2\sqrt{3}} \cdot \frac{(2 + \sqrt{3})^{n+1-k} - (2 - \sqrt{3})^{n+1-k}}{(2 + \sqrt{3})^n + (2 - \sqrt{3})^n} .$$

If the links are made of rods with rotating fly wheels, such that the moment of inertia of each link about a perpendicular axis through its centre of gravity is $\frac{1}{3}ma^2$, show that the initial reactions at the hinges are also given by the above formula.

Ex. 4. A chain consists of alternate gyrostats each of length $2a$ and massless connecting links each of length $2l$, the connection being by universal flexure joints at the ends of the axis of each gyrostat. A finite length of such a chain being placed with its links forming an open plane polygon with its extremities A, B held fixed by universal flexure joints, the system is so set in motion that it rotates with angular velocity μ round AB as if it were a rigid polygon. It is required to form the equations of steady motion.

A gyrostat is a rapidly rotating fly wheel, angular velocity n, pivoted without friction on a stiff moveable framework or within a containing case.

[Math. Soc. 1875.

Taking AB as the axis of z, let the plane xz rotate round AB with angular velocity μ so that it always contains the chain. Let p_k, s_k be the inclinations of the kth rod and string to AB. Let mP be the resolved tension parallel to AB, which is therefore the same for every rod. The required equations are then

$$x_{k+1} - x_k = a\,(\sin p_{k+1} + \sin p_k) + 2l \sin s_{k+1} \qquad - \mu^2 x_k = P\,(\tan s_{k+1} - \tan s_k),$$

$$\mu\,\{ -C_2\mu\,(1 - \cos p_k) + C_1 n\}\sin p_k - A\mu^2\sin p_k \cos p_k$$
$$= Pa\,\{(\tan s_{k+1} + \tan s_k)\cos p_k - 2\sin p_k\},$$

where mC_1 and mC_2 are the moments of inertia of the fly wheel and the case about the axis, and mA that of both about a perpendicular axis.

To obtain the equation of moments, we notice that by the geometry of the universal joint each gyrostatic link moves as if its axis were produced to and joined to the fixed axis AB by a universal flexure joint. Thus each case has an angular velocity $-\mu$ about its axis and an angular velocity $+\mu$ about a parallel to AB drawn through its centre of gravity, Art. 33. By resolutions we find the angular momenta about the axes of C and A and thence the angular momenta about the co-ordinate axes x, y, z. Substituting in the equations of Art. 10 and remembering that in steady motion the angular momenta are constant we obtain the three equations of moments. Two are identically satisfied and the third is given above.

Ex. 5.　Supposing the polygon in the last question to be so nearly straight that the cubes of p and s can be neglected, show that the centres of gravity of the gyrostats lie on the harmonic curve $x = A\cos(\theta z/b) + B\sin(\theta z/b)$, where $b = 2a + 2l$ and θ is given by　　$(C_1 \mu n - A\mu^2 + 2Pa)(1 - \cos\theta - l\mu^2/P) = \mu^2 a^2\,(1 + \cos\theta).$

If the polygon, instead of being fixed at A and B, is produced indefinitely in each direction in the form of the above curve, then in the time π/μ the polygon makes a half turn round the axis of z and the harmonic curve appears to advance a distance $\pi b/\theta$ along that axis. Thus the velocity V of propagation is given by $V = \mu b/\theta$.

[Math. Soc. 1875.

Ex. 6.　A chain, whose tension is T, consists of alternate links of lengths $2a$ and $2b$ connected by smooth ball-and-socket joints; those of length $2a$ being massless connecting rods and the others symmetric gyrostats. The mass of each gyrostat is unity and its moments of inertia about its axis and a perpendicular to it are C and A, while its angular velocity about its axis is ω. Investigate the general equations for the small motions of such a chain; and show that an endless train of waves of period $2\pi/p$ will be propagated along it with velocity V given by the equations

$$C^2\omega^2 p^2 = \left[Ap^2 - 2bT - \frac{b^2 p^2\,(x+1)^2}{x^2 - 2\,(1 - ap^2/T)\,x + 1}\right]^2, \text{ and } x^2 - 2x\cos\frac{2(a+b)p}{V} + 1 = 0.$$

[Math. Tripos, 1889.

Ex. 7.　Equal balls, n in number, connected by flexible springs, are constrained to move in a circular groove into which the springs are also placed, the system of balls and springs forming a closed chain. If the mass of the springs be very small compared with that of the balls, and if the distance between the balls measured along the circular groove is initially equal to the unstretched length of any one of the springs, prove that the times of vibration of the system are $\pi\,(m/\mu)^{\frac{1}{2}}\operatorname{cosec} i\pi/n$ where m is the mass of one of the balls, μ the force required to increase the length of any one of the springs by unity and i an integer which may have any value from 1 to n. With what physical problem does this coincide when n is infinite and what are then the times of vibration?   [Math. Tripos, 1887.

Ex. 8. $2n$ equal uniform rods each of mass m are hinged together and are held so that they are alternately vertical and horizontal, thus forming a figure resembling a set of steps, each vertical rod being lower than the preceding one; the highest rod is horizontal and is capable of turning freely round its end which is fixed; prove that, when the rods are let go, the horizontal component X_{2r} and the vertical component Y_{2r} of the initial action between the $2r$th and the $2r+1$ the rods are given by $X_{2r} = B(-5 + 2\sqrt{6})^r + C(-5 - 2\sqrt{6})^r,$

$$Y_{2r} = B'(-5 + 2\sqrt{6})^r + C'(-5 - 2\sqrt{6})^r,$$

the constants B, C, B', C' being determined by the equations, $X_{2n} = 0$, $Y_{2n} = 0$, $X_2 + 2X_0 = 0$, $2Y_2 + 16Y_0 - 5mg = 0$.   [Math. Tripos, 1889.

420. Network of Particles.   Let columns of threads in one plane be cut at right angles by rows of threads. Let a particle of mass m be attached to them at each intersection. Let the interval between two adjacent columns be l and the interval between two adjacent rows be l'. Let the tensions of the rows and columns be respectively T and T'. Let the particles vibrate perpendicularly to the plane of the threads, and let the whole system be removed from the action of gravity.

Ex. 1.   If w be the displacement of the particle in the hth column and kth row and $T/ml = c^2$, $T'/ml' = c'^2$, prove that the equation of motion is

$$d^2w/dt^2 = c^2(w_{h+1} - 2w_h + w_{h-1}) + c'^2(w_{k+1} - 2w_k + w_{k-1}).$$

Ex. 2.   Prove that the motion of the particles may be represented by the series whose general term is

$$w = \Sigma\{a^h(Ab^k + Bb^{-k}) + a^{-h}(A'b^k + B'b^{-k})\}\sin pt \quad \dots\dots\dots\dots (1),$$

where the Σ implies summation for all values of a and b connected by the equation

$$-p^2 = c^2\left(a - 2 + \frac{1}{a}\right) + c'^2\left(b - 2 + \frac{1}{b}\right).$$

Show that if a and b are both real, one at least is negative. Show also that if the circumstances of the problem permit $b = \pm 1$ the corresponding coefficient of $\sin pt$ becomes $(\pm 1)^k\{a^h(A + Bk) + a^{-h}(A' + B'k)\} \quad \dots\dots \dots\dots\dots\dots (2).$
If a and b are both $= \pm 1$ the corresponding coefficient is

$$(\pm 1)^h(\pm 1)^k(A + Bh + Ck + Dhk)\dots\dots\dots\dots\dots\dots\dots(3).$$

What is the general form of the solution, when one of the two a and b is imaginary and the other real? When both are imaginary with unity for modulus, show that
$$\left.\begin{array}{l} w = \Sigma P \sin(pt - 2h\theta - 2k\phi) \\ p^2 = c^2(2\sin\theta)^2 + c'^2(2\sin\phi)^2 \end{array}\right\} \quad \dots\dots\dots\dots\dots\dots\dots (4).$$

Ex. 3.   Show that the solution (4) of the last example represents a wave motion. If λ be the length of the wave, v its velocity, and α the angle the direction in which it travels makes with the rows of thread, prove that

$$\lambda\theta = \pi l\cos\alpha, \qquad \lambda\phi = \pi l'\sin\alpha, \qquad v^2(\pi/\lambda)^2 = c^2\sin^2\theta + c'^2\sin^2\phi.$$

Ex. 4.   If the network is so constituted that $cl = c'l'$, prove that there are two directions in which a wave of given length travels with the greatest velocity, and that in these cases the fronts are the diagonals of the openings between the threads. The two directions of least velocity are those in which the fronts are along the threads.

Ex. 5.   If $cl = c'l'$ and if the intervals between the threads are very small, prove that the network becomes a membrane which is equally stretched in all directions. In this case waves of all finite length and all directions of front travel with the same velocity.

Ex. 6.   A network, otherwise infinite, is bounded by a rod which runs along the diagonals of the openings. The rod is agitated according to the law $w = P \sin pt$. Prove that two distinct motions result according as the period of agitation is greater or less than $\pi/(c^2 + c'^2)^{\frac{1}{2}}$. In the former case waves travel over the network, in the latter the motion resembles that described in Art. 411.

421. Network with Quadrilateral openings.   To bring these particles into order we regard them as arranged in rows and columns, as in rectangular networks, though these are no longer straight lines. If the network be so stretched that the tension of every thread is proportional to the length of the thread along which it acts, the ratio being equal to c^2, the equation of motion may be proved to be

$$\delta^2 w_{hk} = c^2 \left(\Delta^2 w_{h-1, k} + \Delta'^2 w_{h, k-1} \right),$$

where Δ operates on h and Δ' on k. This is exactly the same equation as that which determines the motion of a rectangular network when $c = c'$. Thus the motions of the two networks will be the same when the central and boundary conditions are made to correspond.

In this way we may deduce the motion of one kind of network from another just as in Hydrodynamics we change one fluid motion into another by the method of conjugate functions.

Ex. 1.   Show that the geometrical peculiarity of this quadrilateral network is that each particle is the centre of gravity of the four adjacent particles to which it is connected by strings.

Ex. 2.   If (x, y) be the Cartesian co-ordinates of the particle (hk), prove that x and y both satisfy the equation of differences $\Delta^2 x_{h-1, k} + \Delta'^2 x_{h, k-1} = 0$. Show also that the values of x and y may be written in the compendious form

$$x + y\sqrt{-1} = \Sigma A e^{2ah + 2\beta k \sqrt{-1}}, \qquad \tfrac{1}{2}(e^a - e^{-a}) = \pm \sin\beta.$$

Other forms of the solution may be deduced as in Art. 420. For example, we may have $x = A + Bh + Ck + Dhk$.

In all these solutions the directions of the threads which form the sides of the quadrilateral openings are defined (1) by making h constant and k variable, (2) by making k constant and h variable. Thus taking a single exponential, we find $x = A e^{2ah} \cos 2\beta k$, $y = A e^{2ah} \sin 2\beta k$. These lead to $x^2 + y^2 = A^2 e^{4ah}$, $y/x = \tan 2\beta k$. The quadrilateral openings are therefore formed by concentric circles and radii vectores from their centre.

Ex. 3.   When the openings of the network are indefinitely small, the result of the last example becomes $x + y\sqrt{-1} = f(h + k\sqrt{-1})$, so that that result may be regarded as an extension to Finite Differences of the theory of conjugate functions.

Ex. 4.   If in Ex. (2) the values of h and k are not restricted to be integral, prove that $\Delta x_{h-\frac{1}{2}, k} = \pm \Delta' y_{h, k-\frac{1}{2}}$, $\Delta' x_{h, k-\frac{1}{2}} = \mp \Delta y_{h-\frac{1}{2}, k}$.
The analogy of these results to some well-known theorems in conjugate functions is obvious.

Ex. 5.   The Cartesian co-ordinates of the particles of a triangular network are given by $x = h$, $y = hk$, where h, k are any integers. The equations to the three fixed boundaries are $x = n$, $y = 0$, $y = n'x$. Following the rule given in Ex. 2, show that the quadrilateral openings are formed by radii vectores from the origin and ordinates parallel to the axis of y. Prove that the period of vibration, viz. $2\pi/p$, is given by $p^2/c^2 = \sin^2(i\pi/2n) + \sin^2(i\pi/2n')$.

Theory of Equations of Differences.

422. General Equations of Motion. Let a series of n particles of masses m_1, m_2... be arranged in a straight row at intervals equal to l_1, l_2... and be in equilibrium under the action of external forces and their mutual attractions. Let these particles be now displaced from their positions of equilibrium either all at right angles to the axis of the row, or all along its length. Let the displacements at the time t be y_1, y_2...y_n. Our object is to find these y's as functions of the time.

The forces which act on the particles are of several kinds. (1) There are the external forces of restitution which are functions of the displacements of the particle acted on from its position of equilibrium. These must supply terms to the force function of the form $-\frac{1}{2}\Sigma a_k y_k^2$; all the higher powers of the displacements being rejected. (2) There are the forces of restitution which depend on the action of the adjacent particles on each side of the particle under consideration. These must supply terms to the force function which contain squares of the y's and products of y's with adjacent suffixes. But since $2y_k y_{k+1} = y_k^2 + y_{k+1}^2 - (y_{k+1} - y_k)^2$, the only additional terms thus introduced into the force function will be of the form $-\frac{1}{2}\Sigma b_k (y_{k+1} - y_k)^2$. (3) There are the forces of restitution which depend on the action of the two adjacent particles on each side of the particle under consideration. These supply terms to the force function containing squares and products of y's whose suffixes differ at most by 2. But since $2y_k y_{k+2} = (y_{k+2} - 2y_{k+1} + y_k)^2 + \&c.$, where the &c. indicates squares of y's and products of y's whose suffixes differ by unity, it is clear that the only additional terms introduced into the force function are of the form　　$-\frac{1}{2}\Sigma c_k (y_{k+2} - 2y_{k+1} + y_k)^2$.

The forces which depend on the action of the three adjacent particles may be treated in the same way.

Besides these forces there may be some external *forces of constraint* acting on the two extremities of the row. These are functions respectively of y_1 and y_n and therefore supply terms to the force function of the form $-\frac{1}{2}\lambda y_1^2$ and $-\frac{1}{2}\mu y_n^2$. If the forces of constraint act on the two last particles at each end we must add to these the terms $-\frac{1}{2}\lambda_2 (y_2 - y_1)^2$ and $-\frac{1}{2}\mu_{n-1} (y_n - y_{n-1})^2$.

Let U be the force function and let the position of equilibrium be the position of reference. To simplify the argument let us in the first instance restrict ourselves to the following terms

$$2U = -\lambda y_1^2 - \mu y_n^2 - \Sigma a_k y_k^2 - \Sigma b_k (y_{k+1} - y_k)^2.$$

If $2T$ be the vis viva, we have $2T = \Sigma m_k y_k'^2$.

The Lagrangian equations of motion may therefore be written in the typical form

$$m_k y_k'' = -a_k y_k + [b_k (y_{k+1} - y_k) - b_{k-1} (y_k - y_{k-1})],$$
$$= -a_k y_k + \Delta (b_{k-1} \Delta y_{k-1}),$$

where Δ has the usual meaning given to it in the calculus of differences.

The case in which $a = 0$ and b is a constant has been solved in Art. 402.

423. The Boundary Conditions. This typical equation represents the motion of all the particles except the first and last. It does not include the case $k = 1$, because the term $-b_0 (y_1 - y_0)^2$ is missing from $2U$, and the term $-\lambda y_1^2$ has not been taken account of. If the differential coefficients of these with regard to y_1 were equal, the errors would correct each other. This gives

$$b_0 (y_1 - y_0) = \lambda y_1.$$

Treating the other extremity in the same way, we find

$$-b_n (y_{n+1} - y_n) = \mu y_n.$$

There are no particles corresponding to the values $k=0$ and $k=n+1$, but the n equations of motion corresponding to $k=1$ to $k=n$ are all truly represented by the same equation of differences if we suppose y_0 and y_{n+1} to stand for their values as given by these two conditions.

424.   In the same way we may show that, if we take the more general value for U, viz.

$$2U = -\lambda_1 y_1^2 - \lambda_2 (\Delta y_1)^2 - \mu_n y_n^2 - \mu_{n-1} (\Delta y_{n-1})^2$$
$$- \Sigma a_k y_k^2 - \Sigma b_k (\Delta y_k)^2 - \Sigma c_k (\Delta^2 y_k)^2,$$

the typical equation of motion becomes

$$m_k y_k'' = -a_k y_k + \Delta (b_{k-1} \Delta y_{k-1}) - \Delta^2 (c_{k-2} \Delta^2 y_{k-2}).$$

The terminal conditions at one extremity are

$$b_0 \Delta y_0 - \Delta (c_{-1} \Delta^2 y_{-1}) = \lambda_1 y_1, \qquad -c_0 \Delta^2 y_0 = \lambda_2 \Delta y_1.$$

There are similar conditions at the other extremity.

425.   **Method of Solution.**   To solve the typical equation of motion

$$m_k y_k'' = -a_k y_k + \Delta (b_{k-1} \Delta y_{k-1}),$$

we follow the method of Lagrange.   To find a *principal oscillation* we put

$$y_k = L_k \sin (pt + \omega).$$

We thus have $a_k L_k - \Delta (b_{k-1} \Delta L_{k-1}) = p^2 m_k L_k.$

This equation can also be written in the form

$$b_k L_{k+1} = (a_k + b_{k-1} + b_k - p^2 m_k) L_k - b_{k-1} L_{k-1}.$$

If we wrote down at length the n equations given by $k=1, 2 \dots n$ we could by successive substitutions express the value of L_k as a linear function of L_0 and L_1. But since the ratio of L_0 to L_1 is given by one of the equations at the limits, we can find L_k in the form $L_k = C\phi (k, p)$, where C is either L_0 or L_1 at our pleasure or any function of L_0 and L_1.   See Art. 423.

If we make a few of the substitutions indicated it will be at once evident that $\phi (k, p)$ is an integral rational function of p^2 of the $(k-1)^{th}$ degree.   We must now substitute this result in the equation of condition at the other limit.   We thus have after division by C $b_n \{\phi (n+1, p) - \phi (n, p)\} + \mu \phi (n, p) = 0,$

This equation will be shortly represented by $\psi (p) = 0$.   We may notice that this reasoning is perfectly general, so that no value of L_k not included in this solution can satisfy the equation of differences.

This process is strictly Lagrange's method of finding the principal oscillations, and the final equation $\psi (p) = 0$ is merely Lagrange's determinantal equation in an expanded form.   Accordingly we see that it is an equation of the nth degree to find the n values of p^2.

But if n be considerable this method of elimination cannot always be employed. The Calculus of Finite Differences sometimes enables us (as in Art. 402) to arrive at a solution in a simpler manner.   But whatever method is adopted the solution obtained, whether partial or complete, must be included in that indicated above.

426.   If the given function b_k is such that $b_0 = 0, b_n = 0$ and λ, μ are also zero, there are no conditions at the limits.   In this case the equation of differences defined by $k=0$ only contains L_1 and L_2, the term $-b_0 (y_1 - y_0)$ being now absent.   This equation therefore determines the ratio of L_1 to L_0 and the argument proceeds as before.

It is however more convenient to regard this case as included in the former with the condition that y_0, y_1, y_{n-1}, y_n are not to be infinite.   With this proviso the terms $-b_0 (y_1 - y_0)$ and $b_n (y_{n+1} - y_n)$ cannot become finite.

427. The corresponding Differential Equation. The limiting case of this equation of differences is peculiarly interesting. Let us make all the intervals l_1, l_2, &c. between the particles equal to each other and each equal to l; and let us write $x = kl$. Then in the limit when l is indefinitely small we have $dx = l$, and all the various functions of k may therefore be regarded as continuous functions of x. Writing $m_k = m_x dx$, $a_k = a_x dx$, and $b_k = b_x/dx$ the equation of differences becomes in the limit

$$a_x y_x - \frac{d}{dx}\left(b_x \frac{dy}{dx} \right) = p^2 m_x y_x.$$

This equation is to hold for all values of x between certain limits, say $x = 0$ to $x = L$. The conditions at the limits are

$$x = 0, \quad b_x \frac{dy}{dx} = \lambda y, \qquad x = L, \quad -b_x \frac{dy}{dx} = \mu y.$$

In the same way we may find the differential equation which corresponds to the equation of differences given in Art. 424.

In this equation it is not necessary to suppose y to be small, for since the equation is linear we may multiply y by any constant quantity we please. It is necessary however that all the functions and as many of their differential coefficients as enter into the equation should be finite.

Suppose that the function $b_x = 0$ at each limit and that λ and μ are both zero. The conditions at the limit disappear for a differential equation of the second order. We thus have no equation to find p. But in the following theorems, the condition that the solutions chosen for y must be finite between the limits remains in full force. In some cases this one condition will limit the values of p.

428. Ex. If the differential equation is $-\frac{d}{dx}\left\{ (1-x^2)\frac{dy}{dx} \right\} = p^2 y$ and the limits are $x = 0$ and $x = 1$, show that no solution can be finite at both limits unless $p^2 = i(i+1)$ where i is any positive integer.

429. This equation of differences and its limiting case the differential equation are of considerable importance in other besides dynamical investigations. It is therefore useful to notice that though the equation presented itself with a dynamical meaning, yet the results in this section are perfectly general. We may regard the equations of motion as simply so many differential equations to find y_1, y_2, &c. derived, as explained in Chap. VII., from the two auxiliary functions A and C, the other auxiliary functions, B, D, E, F being all zero. The functions A and C are here called T and $-U$ and the symbol m is here replaced by $p\sqrt{-1}$.

430. Three Propositions. We immediately infer the following theorems concerning the values of p.

Prop. 1. If the function m_k or m_x is positive between the limits, the function T is a one-signed positive function. It therefore follows from Art. 319, *that all the values of p^2 are real.*

This also follows from the theorem that all the roots of Lagrange's determinant are real *.

---

* Another proof that all the values of p^2 are real is given by Poisson in Art. 90 of his *Théorie Mathématique de la Chaleur*. He there shows that if p^2 could have a pair of imaginary values of the form $f \pm g\sqrt{-1}$, the integral $\int_0^L m_x X_x Y_x dx$ could not be zero (see Art. 432). The argument is as follows. Since, by Art. 435, L_x is a function of p^2, it follows that the corresponding values of X_x and Y_x may be written $F \pm G\sqrt{-1}$. This leads to the result $\int_0^L m_x (F^2 + G^2)\, dx = 0$, which is an

431. Prop. 2. If the functions a_k, b_k, &c. or a_x, b_x, &c. as well as m_k or m_x are positive between the limits, and if λ, μ are also positive, the function $C = -U$ is a one-signed positive function. It therefore follows from Art. 315, *that all the values of p^2 are positive.*

This also follows from the theorem in Vol. I. that when the force function U is a maximum in the position of equilibrium, that position of equilibrium is stable.

432. Prop. 3. Let p and q be two unequal possible values of the parameter p, and let the corresponding solutions be indicated by the typical equations

$$y_k = X_k \sin pt, \text{ and } y_k = Y_k \sin qt.$$

Then we may use the method of multipliers as explained in Chap. VIII. Art. 399, *and assert that* $\Sigma m_k X_k Y_k = m_1 X_1 Y_1 + \ldots + m_n X_n Y_n = 0.$
In the case of the differential equation this becomes $\int_0^L m_x X_x Y_x \, dx = 0.$

By referring to the standard example Art. 402 we may perceive the separate uses of these three propositions. The values of p^2 there found are all real and positive and the third proposition was used in Art. 406 to determine the constants of integration when the initial conditions are known.

433. Sturm's Theorems. Restricting ourselves to the case in which the equation of differences has the form

$$a_k y_k - \Delta \left(b_{k-1} \Delta y_{k-1}\right) = p^2 m_k y_k,$$

let us compare the different kinds of motion indicated by different values of p^2.

In order to realize the motions of the several particles more easily, let an ordinate be drawn perpendicular to the length of the row at the position of each particle when in equilibrium. Let the length of this ordinate be equal to the displacement of that particle at the time t. The curve traced out by the extremities of these ordinates will exhibit to the eye the nature of the motion. The intersections of this curve with the axis of the row are called *nodes*, the maxima and minima ordinates are called *loops*.

In the example of Art. 402 these ordinates are the actual displacements of the several particles. In the general case we are now considering this curve is merely a conventional method of exhibiting to the eye the varying state of the system but in that particular case it is suggested by the visible motion.

Let all the possible values of p be arranged in ascending order beginning with the least.

In the solution given by the least value of p, it will be shown that *at any one moment all these ordinates have the same sign.* Thus throughout the motion the indicating curve forms an arc with a single loop which oscillates from one side to the other of the axis of x.

In the solution given by the next smallest value of p, it will be shown that *at any instant there is one change of sign among the ordinates, as we travel from one extremity of the row to the other.* Thus throughout the motion the indicating curve forms a double arc with two loops separated by a node.

In the solution given by the third smallest root there are at any instant two changes of sign among the ordinates. Thus the indicating curve forms three loops separated by two nodes, and so on through all the values of p.

---

impossible equation if m_x keep one sign between the limits. Poisson applies his argument to the case of a differential equation of the second order, but it may evidently be extended to the general case of a differential equation or an equation of differences of any order.

In all these cases the nodes which belong to any value of p are separated by the nodes which belong to the next value of p in the series.

434. The Lemma. To prove these theorems we require the following lemma. Let p and q be two values of p, and let the corresponding motions be given by $y_k = X_k \sin pt$ and $y_k = Y_k \sin qt$. We have therefore

$$\left. \begin{array}{l} a_k X_k - \Delta(b_{k-1}\Delta X_{k-1}) = p^2 m_k X_k \\ a_k Y_k - \Delta(b_{k-1}\Delta Y_{k-1}) = q^2 m_k Y_k \end{array} \right\}.$$

Eliminating the function a_k we find

$$(q^2 - p^2) m_k X_k Y_k = b_k (X_{k+1} Y_k - X_k Y_{k+1}) - b_{k-1}(X_k Y_{k-1} - X_{k-1} Y_k).$$

This gives by summation from $k = a$, to $k = k$

$$(q^2 - p^2)\{m_a X_a Y_a + \ldots + m_k X_k Y_k\} = b_k(X_{k+1} Y_k - X_k Y_{k+1}) - b_{a-1}(X_a Y_{a-1} - X_{a-1} Y_a).$$

The right-hand side may also be written

$$b_k(Y_k \Delta X_k - X_k \Delta Y_k) - b_{a-1}(Y_{a-1}\Delta X_{a-1} - X_{a-1}\Delta Y_{a-1}).$$

In the limiting case in which the equation of differences becomes the differential equation (Art. 427), this lemma takes the form

$$(q^2 - p^2)\int_0^L m X Y\, dx = \left[b_x \left(Y \frac{dX}{dx} - X \frac{dY}{dx} \right) \right]_0^L.$$

435. Cor. 1. Consider the full series of values $X_1, X_2 \ldots X_n$ arranged in order. We have ranges of positive and negative values succeeding each other. Let $X_a \ldots X_k$ be one of these ranges in which all the constituents have one sign, while those on each side, viz. X_{a-1} and X_{k+1}, have the opposite sign. *We shall prove that if $q > p$ there is one change of sign at least in the corresponding range of Y's extending from Y_{a-1} to Y_{k+1} both inclusive.*

For if possible let all these Y's have one sign, then every one of the four terms on the right-hand side of the equality in the lemma has the sign opposite to that of the product $X_k Y_k$. Hence the lemma could not be true.

We have made no assumption about the function of a_k, but b_k and m_k have been supposed to have the same sign, and to keep that sign from one limit to the other.

436. Cor. 2. Consider next a double range of values, say $X_a \ldots X_\beta \ldots X_k$, such that all the constituents from X_a to $X_{\beta-1}$ have one sign, say negative, and from X_β to X_k have the other sign, while (to make the double range complete) X_{a-1} and X_{k+1} have opposite signs to their adjacent constituents. *Then by Cor. 1, if $q > p$, Y must change sign between Y_{a-1} and Y_β and also between $Y_{\beta-1}$ and Y_{k+1}. We shall now prove that a single change of sign between $Y_{\beta-1}$ and Y_β will not suffice for both these requirements.*

For if it did, the products $X_a Y_a, \ldots, X_k Y_k$ would all have the same sign : but every one of the four terms on the right-hand side of the equality in the lemma has the sign opposite to that of the product $X Y_k$. Thus again the lemma could not be true.

In the same way if we consider a triple range of values $X_a \ldots X_\beta \ldots X_\gamma \ldots X_k$ so that X changes sign twice as k varies from one limit to the other, then, by Cor. 1, Y must change sign between Y_{a-1} and Y_β, $Y_{\beta-1}$ and Y_γ, $Y_{\gamma-1}$ and Y_{k+1}. But it follows exactly as before that two changes of sign of Y will not suffice for all three requirements.

437. Cor. 3. Consider the range of values $X_1, X_2 \ldots X_k$ all of one sign beginning at one extremity of the complete series and such that X_{k+1} has the opposite sign. *We shall prove that if $q > p$ there is one change of sign at least in the corresponding range of Y's extending from Y_1 to Y_{k+1}.*

In this case the range begins at one extremity, we have therefore the conditions $b_0 (X_1 - X_0) = \lambda X_1$ and $b_0 (Y_1 - Y_0) = \lambda Y_1$ which hold at that extremity. The equality in the lemma becomes therefore

$$(q^2 - p^2) (m_1 X_1 Y_1 + \ldots m_k X_k Y_k) = b_k (X_{k+1} Y_k - X_k Y_{k+1}).$$

If then all the Y's from Y_1 to Y_{k+1} had the same sign, every term on the left-hand side would have the same sign, and the two terms on the right-hand side would have the opposite sign, and thus the equality could not exist.

Similar remarks apply to a range terminating at the other extremity.

438. Cor. 4. Lastly consider all the n series $X_1 \ldots X_n$, $Y_1 \ldots Y_n$, &c., &c., corresponding to the n values of p, q, &c. arranged in order of magnitude beginning at the least. By the preceding corollaries, each of these series must have at least one more change of sign than any series before it. As there are but n terms in each series, the last, i.e. the nth, can have but $n - 1$ changes of sign. *Hence the first series has no changes of sign, the second has one change, the third has only two and so on. Also the changes of sign in each series alternate, in the manner already explained, with the changes of sign in any series next to it.*

439. It should be noticed that in Cor. 1 and 2 no use has been made of the conditions at the limits. In these propositions therefore p and q are any arbitrary quantities except that q must be greater than p. In Cor. 3 the conditions at one limit are introduced, so that all three corollaries are true if only $X_1/X_0 = Y_1/Y_0$ at one limit. Finally in Cor. 4 the conditions at both limits are supposed to be satisfied and therefore p and q must now be different roots of the equation represented in Art. 425 by $\psi (p) = 0$.

440. The fourth proposition. *To show that no two values of p^2 are equal.* Let us suppose that the conditions of constraint at one limit are satisfied as in Cor. 3. We may therefore write the lemma of Art. 434 in the form

$$(q^2 - p^2) \Sigma m X Y = b_n (X_{n+1} Y - X_n Y_{n+1}),$$

where the summation extends from $k = 1$ to $k = n$. Since p and q are now arbitrary quantities we may put $q^2 = p^2 + dp^2$. We therefore have to the first order of small quantities $\qquad dp^2 \Sigma m X^2 = b_n (X_{n+1} dX_n - X_n dX_{n+1}).$

This equation may be written in the form

$$\Sigma m X^2 = \frac{dX_n}{dp^2} \{ b_n (X_{n+1} - X_n) + \mu X_n \} - X_n \frac{d}{dp^2} \{ b_n (X_{n+1} - X_n) + \mu X_n \}.$$

But the quantity in brackets is the left-hand side of the equation $\psi (p) = 0$ arrived at in Art. 425 as the equation to find all the possible values of p when the conditions of constraint at both extremities are taken account of. We therefore infer that

$$\Sigma m X^2 = \frac{dX_n}{dp^2} \psi (p) - X_n \frac{d\psi (p)}{dp^2}.$$

It immediately follows from this equation that no value of p can make both $\psi (p) = 0$ and $\psi' (p) = 0$. *The equation $\psi (p) = 0$ cannot therefore have equal roots.*

441. Ex. 1. If n particles of any masses at any intervals are arranged in a straight row, as already explained, and oscillate transversely with the motion indicated by any one value of the parameter p, prove that the straight line joining any two particles cuts the axis of the row in a point which is fixed throughout the motion.

Ex. 2. If $y_k = X_k \sin pt$ represent the principal oscillation corresponding to the value p, prove that

$$p^2 \Sigma m_k X_k^2 = \Sigma a_k X_k^2 + \Sigma b_k (X_{k+1} - X_k)^2 + \lambda X_1^2 + \mu X_1^2.$$

The two first Σ's imply summation extending from $k=1$ to $k=n$, and the third from $k=1$ to $k=n-1$.

Ex. 3. If a_k, b_k and m_k are all positive and $2\pi/p$ is the longest period of a principal oscillation, prove that p^2 is less than the greatest value of $(a_k+b_k+b_{k-1})/m_k$ and greater than the least value of a_k/m_k.

If $2\pi/p$ is the shortest period of a principal oscillation, prove that p^2 is greater than the least value of $(a_k+b_k+b_{k-1})/m_k$ and less than the greatest value of $(a_k+2b_k+2b_{k-1})/m_k$. In this example b_0 and b_n are to be taken equal respectively to λ and μ.

Ex. 4. If the function a_k and b_k keep one and the same sign or are zero, show that no value of p can be zero unless λ and μ are both zero.

Ex. 5. Let $y_k=X_k \sin pt$, $y_k=Y_k \sin pt$ represent two principal oscillatory motions such that q is greater than p. If a range of values be taken, say $X_a...X_k$, which are all of one sign and such that X_k is at a loop and that a node lies between X_{a-1} and X_a, prove that either a node or a loop lies within the range $Y_{a-1}...Y_k$.

Thence show that either a node or a loop of the shorter-timed oscillation must lie within (or at the boundaries of) the space joining any node to any loop of the longer-timed oscillation.

Ex. 6. In the equation $P\dfrac{d^2y}{dx^2}+Q\dfrac{dy}{dx}+Ry=pSy$, where P, Q, R, S are given functions of x, let $y=X$ and $y=Y$ be two solutions corresponding to different values of p, and let μ be the integrating factor of the first two terms on the left-hand side. Prove that $\int \mu SXYdx=0$ for any limits between which X, Y and their differential coefficients are finite, provided that at each limit either

$$P=0 \text{ or } \frac{dY}{dx}\Big/Y = \frac{dX}{dx}\Big/X.$$

Ex. 7. Let additional external forces be applied to the system (Art. 422) so that a_k is changed to a_k' where $a_k'-a_k$ is positive between the limits $k=1$ and $k=n$, then if m_k is also positive prove that every value of p^2 is increased. On the other hand, if the inertia is increased so that m_k becomes m_k', then, if both $m_k'-m_k$ and m_k are positive between the limits, prove that all the values of p^2 are decreased.

These results follow from Art. 76 and Art. 77, Ex. 1. They may also be deduced from the lemma.

Ex. 8. Let the equation of motion of a dynamical system be

$$a_x y_x - \frac{d}{dx}\left(b_x \frac{dy}{dx}\right)=p^2 m_x y_x,$$

where the values of p^2 are deduced from the conditions at $x=0$ and $x=L$ given in Art. 427. Let some change be made in the system so that a_x is altered to a_x' where $a_x'-a_x$ is positive for all values of x between the limits. Then if m_x be also positive between the limits, prove that the values of p^2 are also increased.

The differential equation of the second order mentioned in Art. 427 is discussed by C. Sturm in the first volume of *Liouville's Journal*. He there establishes the theorems given in Art. 433 which we have called after his name. The extension of these to equations of finite differences will be found in a paper by the author in the eleventh volume of the *Proceedings of the Mathematical Society*, 1880. The theorems on a network of particles are taken from a paper by the author in the fifteenth volume of the same *Proceedings*, 1884.

CHAPTER X.

Principles of Least Action and Varying Action.

442. Two fundamental equations. Let $(q_1, q_2, q_3, \&c.)$ be the co-ordinates of a system of bodies, and let q stand for any one of these. Let $2T$ be the vis viva of the whole system and U the force-function, and let $L = T + U$. As before let accents denote differential coefficients with regard to the time.

Let us imagine the system to be moving in some manner, which we will call the actual motion or course. Then q_1, q_2, &c. are all functions of t, and it is generally our object to find the form of these functions. Let us suppose the system to move in some slightly different manner, i.e. let q_1, q_2, &c. be functions of t slightly different from their actual forms. Let us call the motion thus represented a neighbouring motion or course. We may pass, in our minds, from the actual motion to any neighbouring motion by the process called *variation* in the calculus of that name. By the fundamental theorem in that calculus

$$\delta \int_{t_0}^{t_1} L \, dt = \left[L \delta t \right]_{t_0}^{t_1} + \int_{t_0}^{t_1} \Sigma \left(\frac{dL}{dq} - \frac{d}{dt} \frac{dL}{dq'} \right) (\delta q - q' \delta t) \, dt + \left[\Sigma \frac{dL}{dq} (\delta q - q' \delta t) \right]_{t_0}^{t_1},$$

where the letter Σ implies summation for all the co-ordinates q_1, q_2, &c. and it is implied by the square brackets that the terms outside the integral sign are to be taken between limits.

The co-ordinates being independent of each other, each separate term under the integral sign vanishes by Lagrange's equations, and we have therefore

$$\delta \int_{t_0}^{t_1} L \, dt = \left[\left(L - \Sigma \frac{dT}{dq'} q' \right) dt + \Sigma \frac{dT}{dq'} \delta q \right]_{t_0}^{t_1}$$

$$= \left[- H \delta t + \Sigma \frac{dT}{dq'} \delta q \right]_{t_0}^{t_1},$$

where H is the reciprocal function of L, as explained in the first volume of this treatise.

The integral $\int_{t_0}^{t_1} L dt$ has been called by Sir W. R. Hamilton the *principal function*, and is usually represented by the letter S.

If the geometrical equations do not contain the time explicitly, T will be a quadratic homogeneous function of the velocities; we have therefore $\Sigma (dT/dq') q' = 2T$. In this case $H = T - U$. The equation of vis viva will now hold and therefore $T - U = h$, where h is a constant which represents the energy of the system. The Hamiltonian equation just proved now takes the simpler form

$$\delta S = \delta \int_{t_0}^{t_1} L dt = - h (\delta t_1 - \delta t_0) + \left[\Sigma \frac{dT}{dq'} \delta q \right]_{t_0}^{t_1}.$$

443. Other functions may be used instead of S. Let us put

$$V = S + [Ht]_{t_0}^{t_1}, \quad \therefore \ \delta V = \delta S + [H \delta t + t \delta H]_{t_0}^{t_1}$$

$$\therefore \ \delta V = \left[t \delta H + \Sigma \frac{\delta T}{dq'} \delta q \right]_{t_0}^{t_1}.$$

The function V is called the *characteristic function*.

444. If the geometrical equations do not contain the time explicitly, we have $H = h$, where h is a constant which may be used to represent the whole energy of the system. In this case

$$V = S + h (t_1 - t_0) = \int_{t_0}^{t_1} (T + U) \, dt + \int_{t_0}^{t_1} (T - U) \, dt,$$

$$\therefore \ V = 2 \int_{t_0}^{t_1} T dt.$$

The function V therefore expresses the whole accumulation of the vis viva, i.e. the *action* of the system in passing from its position at the time t_0 to its position at the time t_1.

For the sake of simplicity it will be generally assumed in this section that the geometrical equations do not contain the time explicitly.

445. In the proof of these theorems we have supposed that all the forces are conservative. If in addition to the impressed forces there are any reactions, such as rolling friction, which cannot be taken account of by reducing the number of independent co-ordinates, we must use Lagrange's equation in the form

$$\frac{d}{dt} \frac{dL}{dq'} - \frac{dL}{dq} = P,$$

where, as explained in Vol. I., $P \delta q$ is the virtual moment of these reactions corresponding to a displacement δq. In this case the quantity under the integral sign will not vanish unless the variations are such that

$$\Sigma P (\delta q - q' \delta t) = 0.$$

Now q being the value of any co-ordinate in the actual motion at the time t, $q + \delta q$ is its value in a neighbouring motion at the time $t + \delta t$. But $q' \delta t$ is the change of q in the time δt, hence $q + \delta q - q' \delta t$ is the value of the co-ordinate in the neighbouring motion at the time t. The neighbouring motions must therefore be such that the virtual moment of the reactions corresponding to a displacement of the system from any position in the actual motion into its position in a neighbouring motion at the same time is zero. With this restriction on the variations, the two equations which express the variations of S and V will still be true.

446. Another Proof. We may also establish these theorems without the use of Lagrange's equations. Let x, y, z be the Cartesian co-ordinates of any particle, and let m be the mass of this particle. Let U be such a function that dU/dx, dU/dy, dU/dz are the components of the impressed forces on this particle in the directions of the axes. We may write mX, mY, mZ as usual for these components. Then
$$L = T + U = \tfrac{1}{2}\Sigma m\,(x'^2 + y'^2 + z'^2) + U.$$
By the fundamental theorem in the Calculus of Variations, we have
$$\delta \int_{t_0}^{t_1} L\,dt = \Big[L\delta t \Big]_{t_0}^{t_1} + \Big[\Sigma\, \frac{dL}{dx'}(\delta x - x'\delta t) \Big]_{t_0}^{t_1} + \int_{t_0}^{t_1} \Sigma \left(\frac{dL}{dx} - \frac{d}{dt}\frac{dL}{dx'} \right)(\delta x - x'\delta t)\,dt,$$
where the variations δx, &c. are connected together by the geometrical relations of the system. If we substitute for L and remember that T is a homogeneous quadratic function of x', y', z', this becomes
$$\delta \int_{t_0}^{t_1} L\,dt = \Big[(U - T)\,\delta t + \Sigma m x'\delta x \Big]_{t_0}^{t_1} + \int_{t_0}^{t_1} \Sigma m\,(X - x'')\,(\delta x - x'\delta t)\,dt.$$

Now $\delta x - x'\delta t$ is the projection on the axis of x of the displacement of the particle m from its position in the actual motion at the time t to its position in a neighbouring motion at the *same time*. Hence the part under the integral sign vanishes by the principle of virtual velocities.

The term $\Sigma m x'\delta x$ is clearly the virtual moment of the momenta. If the coordinates be expressed as functions of any independent quantities q_1, q_2, &c., it has been proved in the first volume that this is equal to $\Sigma\,(dT/dq')\,\delta q$. Putting $T - U = H$ we have as before
$$\delta \int_{t_0}^{t_1} L\,dt = \Big[-H\delta t + \Sigma\,(dT/dq')\,\delta q \Big]_{t_0}^{t_1}.$$

447. Principle of Least Action. Let us call the positions of the system at the times t_0 and t_1 the initial and terminal positions. *Let us suppose these fixed so that the actual motion and all its neighbouring motions are to have the same initial and terminal positions.* In this case δq vanishes at each limit and the two fundamental equations giving the values of δS and δV take the simpler forms
$$\delta S = \delta \int_{t_0}^{t_1} L\,dt = -h\,(\delta t_1 - \delta t_0), \qquad \delta V = 2\delta \int_{t_0}^{t_1} T\,dt = (t_1 - t_0)\,\delta h,$$
where it has been supposed that the geometrical equations do not contain the time explicitly.

If the time of transit of the system from its initial to its terminal position is also given, we have $\delta t_1 = \delta t_0$, and therefore
$$\delta \int_{t_0}^{t_1} L\,dt = 0.$$

If the constant h is given, or which is the same thing, *if the energy of the system is given,* we have $\delta h = 0$, and therefore
$$\delta \int_{t_0}^{t_1} T\,dt = 0.$$

448.   Since $\delta V = 0$, it follows that for the actual motion V is a maximum or minimum, or at least that the change it undergoes in passing to any neighbouring motion is of the second order of small quantities. It cannot be a maximum since by causing the bodies to take circuitous paths we may make V as large as we please. Again, since the vis viva cannot be negative there must be some mode of motion from one given position to another for which the action is the least possible. When therefore the equations supplied by the Calculus of Variations lead to but one possible motion that motion must make V a minimum. But when there are several possible modes of motion, though none can be a maximum some may be neither maxima nor minima. With this understanding we may infer the two following theorems.

449. *Let any two positions of a dynamical system be given, the actual motion is such that $\int T dt$ is less than if the system were constrained, without violating any geometrical conditions, to move in some other manner from the one position to the other with the same energy; these other motions being such that, throughout, T is the same function of the co-ordinates and their differential coefficients.* This particular inference from the general equations in Art. 447 is usually called the Principle of Least Action.

In the same way, if the system move in the varied course not with the same energy, but in the same time, from the one given position to the other, then $\int L dt$ is a minimum.

Maupertuis conceived that he could establish à priori by theological arguments that all mechanical changes must take place in the world so as to occasion the least possible quantity of *action*. In asserting this it was proposed to measure the action by the product of the velocity and space; and this measure being adopted, mathematicians, though they did not generally assent to Maupertuis' reasonings, found that his principle expressed a remarkable and useful truth, which might be established on known mechanical grounds. Whewell's *History of the Inductive Sciences*, Vol. II. p. 119.

Euler, at the end of his *Traité des Isopérimètres*, 1744, established the truth of the principle for isolated particles describing orbits about centres of force. This was afterwards extended by Lagrange to the motion of any system of bodies acting in any manner on each other. In deducing conversely the equations of motion from the principle of Least Action, Lagrange seems to have fallen into some errors which were pointed out by Ostrogradsky in his *Mémoire sur les équations différentielles relatives au problème des Isopérimètres* published in the *Memoirs of the Academy of Sciences* at St Petersburgh in 1850. The theorem $\int L dt$ is a minimum when the time is constant was first given in this treatise 1877.

450.   If some of the co-ordinates appear in the Lagrangian function L only through their velocities (i.e. their differential coefficients with regard to t) their corresponding momenta are constant throughout the motion. As explained in Vol. I. Art. 422, it is then sometimes convenient to eliminate these velocities by modifying the Lagrangian function and using it thus changed in the ordinary Lagrangian

equations. Supposing that the co-ordinates q_1, q_2 appear only through q_1', q_2' we write $L_1 = L - \Sigma pq'$ where Σ implies summation for the co-ordinates q_1, q_2, then L_1 is the modified L. The general expression for L_1 after the elimination of q_1', q_2' is given in Vol. I., Art. 421. In the same way $2T_1 = 2T - \Sigma pq'$; where T_1 is the modified T.

If as supposed above the momenta p_1 and p_2 are constant throughout the motion, we have $\qquad \delta \int pq' dt = p \delta \int q' dt = p (\delta q_1 - \delta q_0),$
provided the variations are limited to those in which p retains its constant value. Since the initial and final positions are supposed to be fixed in the principle of least action it follows that $\delta \int pq' dt = 0$. We therefore infer that $\int L_1 dt$ and $\int T_1 dt$, retain *the max-min property under the same conditions as before provided the variations are restricted to be such as do not disturb the constancy of the momenta.* This theorem is given by Larmor, *Math. Soc.* 1884.

451. Motion deduced from the Calculus of Variations.

By making the first variation of either V or S equal to zero (under the given conditions) according to the rules of the Calculus of Variations, we may conversely find the co-ordinates q_1, q_2, &c. as functions of t. Amongst these functions of the time we shall certainly find the motions given by Lagrange's equations, because we have just proved that these make the first variations equal to zero. But it is possible that there may exist other courses or modes of conducting the system from the initial to the terminal position which (though contrary to mechanical laws) may make V or S a minimum. It is easy to see that some other courses must exist, for the two positions may be so placed that it is impossible to project the system from the initial position with a given energy so as to pass through the terminal position. Thus suppose it is required to project a particle under the action of gravity from an initial position with a given velocity so as to pass through a position B on the horizontal line through A, but beyond the maximum range. We know that this cannot be done with real conditions of projection in a real time. Yet some course of minimum action from A to B must exist. We shall now show, (1) that the ordinary processes of the Calculus of Variations, which are founded on the supposition that the variations of the independent co-ordinates may have any sign, lead only to Lagrange's equations; (2) that there are certain other modes of motion which are so situated that the co-ordinates (along some part at least of the course) cannot be made to vary on one side without introducing imaginary quantities, and that when these impossible variations are omitted such courses may give a maximum or minimum.

452. Continuous Motions.

Beginning with the first of these two propositions, let us make δS and δV equal to zero according to the rules of the Calculus of Variations.

Taking $\delta \int L dt = 0$ *where the time of transit is given*, we immediately have, Art. 442,
$$\int_{t_0}^{t_1} \Sigma \left(\frac{dL}{dq} - \frac{d}{dt} \frac{dL}{dq'} \right) \delta q \, dt = 0.$$

for all variations. Since the δq's are all arbitrary and independent, it follows that each coefficient under the integral sign must vanish separately. In this manner we are led directly to Lagrange's equations of motion.

453. *If the action is to be a minimum* some further considerations are necessary because the condition that the energy $T - U$ should be constant may act as a limit to the variations which can be given to the co-ordinates. Let h be this constant, then following Lagrange's rule in the Calculus of Variations we put

$$W = T + \lambda (T - U - h), \text{ and make } \delta \int W dt = 0,$$

without regard to the given condition. Afterwards we choose the arbitrary quantity λ so that the given condition is satisfied. Then $\delta \int W dt$ being zero for all variations of the co-ordinates, it immediately follows that $\delta \int T dt$ is also zero for all variations which do not violate the given condition. With the same notation as before we have, Art. 442,

$$\delta \int W \delta t = [W \delta t] + \Sigma \int \left(\frac{dW}{dq} - \frac{d}{dt} \frac{dW}{dq'} \right) (\delta q - q' \delta t) + \left[\Sigma \frac{dW}{dq'} (\delta q - q' \delta t) \right] = 0,$$

where the integrals and the quantities in square brackets are to be taken between the given limits, which are omitted for the sake of brevity.

First, let us consider the part outside the integral sign. The initial and final positions being given, each $\delta q = 0$. We therefore have

$$\{W - \Sigma (dW/dq') q'\} \delta t = 0.$$

This equation is satisfied by $\delta t = 0$, but since the time of transit is not to be the same in the actual and varied motions this factor is to be rejected. Also T is a homogeneous quadratic function of the q''s, hence $\Sigma (dT/dq') q' = 2T$. Substituting for W its value and using this equation we find $(1 + \lambda) T + \lambda (U + h) = 0$. But λ is such that $T - U = h$. Hence $(1 + 2\lambda) T = 0$, and therefore $\lambda = -\frac{1}{2}$.

Next, consider the part under the integral sign. By the rules of the Calculus of Variations we have (since the δq's are all arbitrary) the typical equation

$$\frac{dW}{dq} - \frac{d}{dt} \frac{dW}{dq'} = 0.$$

Substituting for W and giving λ its value just found, we have the typical Lagrange's equation.

454. **Ex.** If we add to the conditions used in the principle of Least Action the condition that the time of transit as well as the energy is to be the same in all the varied motions, show that the minimum does not in general lead to Lagrange's equations. Following the same notation as in the last article, show that the minimum for a given time (not necessarily equal to the time of free transit), leads to $\lambda = -\frac{1}{2} + A/T$, where A is a constant to be so chosen that the energy has its given value. Show also that when the time of transit is given so that $A = 0$, the minimum thus found is the least.

455. **Discontinuous Motions.** Turning now to the second proposition mentioned in Art. 451, let us investigate if there can be any other modes of motion besides those just found, which make the first variation of the action equal to zero. In obtaining these equations it is assumed that all the δq's are independent; but, if the conditions of the question imply any boundary, this may not be true for any actual motion which takes the system in the immediate neighbourhood of that boundary. Thus, in our case, since T cannot be negative, all positions of the system outside the boundary $U + h = 0$ are excluded. In the immediate neighbourhood of this boundary the variations of the co-ordinates may not be susceptible of

all signs*. It follows that a motion along the boundary may be a course of mini-mum action though not given by the *ordinary* equations of the Calculus of Variations.

It is evident that we cannot make the system travel along the boundary whose equation is $U + h = 0$ because this requires all the velocities to be zero. But the system may travel as near as we please to this boundary with a total "action" as small as we please. The following discontinuous motion may therefore be a course of minimum action. First project the system from its given initial position A with such velocities and directions of motion, but with the given energy, that every particle may come simultaneously to rest. Assuming the equations to give real conditions of projection, the system, when it comes to rest, is situated on the boundary. Let this position be called B. Next move the system close to the boundary until it reaches such a position C that on being set free without velocity it passes through the given terminal position D under the action of the forces represented by U. The motions from A to B and C to D are courses of minimum action, while the action from B to C may be made as small as we please.

456. We may show that the action along this discontinuous course is really a minimum. To prove this, let us take any neighbouring motion beginning at A and ending at D. Let B', C' be any positions of the system on the neighbouring course near B and C respectively. Since $\delta h = 0$, the action (Art. 443) along AB' exceeds that along AB by $\delta V = \left[\Sigma (dT/dq') \delta q \right]_{t_0}^{t_1}$. This vanishes at the lower limit since both courses begin at A. Since T is a quadratic function of the velocities, dT/dq' contains a velocity in every term and all these velocities vanish in the position B, i.e. at the upper limit. We therefore have $\delta V = 0$. We infer that the difference of the actions along AB and AB' is of the order of the quantities neglected in investigating this expression for δV. Thus the difference of these two actions is of the order of the squares and products of δq and $\delta q'$.

Next let M' be any position on the neighbouring motion $B'C'$ so that the change of place $B'M'$ is finite. The velocities in every position of the system between B' and M' are of the order $\delta q'$, and hence the semi vis viva T is of the order $(\delta q')^2$. But the time of transit from B' to M' varies inversely as the mean velocity, hence the $\int T dt$, i.e. the action from B' to M', is of the first order of small quantities, viz. $\delta q'$. This action is essentially positive, and we have just proved that it is infinitely greater than the difference of actions along AB and AB'. Hence the action along AM' is greater than that along AB.

In the same way if N' be a position of the system properly chosen on the neigh-bouring course nearer C', we may show that the action along $N'D$ is greater than that along CD. The action along $M'N'$ is also greater than that along BC. It

---

* Exceptional cases, similar to these, occur in the theory of maxima and minima in the differential calculus. When the independent variable is not capable of unlimited increase, but is bounded in one or both directions, its value at either boundary sometimes corresponds to a maximum or minimum value of the dependent variable, though this is not found by making the differential coefficient equal to zero.

In the calculus of variations some instances in which the variations at the boundaries are not susceptible of every sign are given in De Morgan's *Differential Calculus*, 1842, page 460, &c. These appear to have been rediscovered by Dr Tod-hunter in his *Researches in the Calculus of Variations*, 1871, Art. 18. See also Chap. VIII. of his *Researches &c.*

follows therefore that so long as the separation in space between the positions B and C is finite, the action along $ABCD$ is less than that along any neighbouring course.

457. Ex. If we use the principle of least action in the manner explained in Art. 453 we virtually remove the restriction on the variation of the co-ordinates. Show that in the discontinuous course the first variation of $\int W dt$ is zero if we regard λ as a discontinuous function which is equal to $-\frac{1}{2}$ along the courses AB, CD and equal to zero along the course BC.

458. **Is the Action an actual minimum?** To determine whether an integral is a maximum or a minimum or neither, we must examine the terms of the second order in the variation of the integral to ascertain if their sum keeps one sign or not for all variations of the independent variables. This is a very troublesome process, but it is unnecessary to discuss it. It will be sufficient to remind the reader of some remarks of Jacobi, given in the seventeenth volume of *Crelle's Journal*, 1837, and translated in Dr Todhunter's *History of the Calculus of Variations*, page 250.

Suppose a dynamical system to start from any given position which we shall call A, and to arrive at some position B. If the time be given, the motion is found by making $\delta \int L dt = 0$; if the energy be given, by making $\delta \int T dt = 0$. The constants which occur in integrating the differential equations supplied by the calculus of variations are to be determined by means of the given limiting values; but as this involves the solution of equations there will in general be several systems of values for the arbitrary constants, so that several possible modes of motion from A to B may be found which satisfy the same differential equation and the same limiting conditions. Let us suppose that when B and A are near each other there is but one mode of motion from A to B, then by Art. 448 that mode makes $\int T dt$ a minimum. Now let the position B recede from A so as always to be on this one mode of motion. Suppose that when B reaches the position C another possible mode of motion from A to B is indefinitely near to the former motion. We deduce from Jacobi's criterion that C determines the boundary up to which or beyond which the integration must not extend if the integral is to be a minimum.

Jacobi illustrates his rule by considering the principle of least action in the elliptic motion of a planet. Let S be the sun, and let the particle start from A towards aphelion to arrive at a point B. The path is known to be an ellipse with S for focus. Since we use the principle of least action, the energy of the motion is given: hence the major axis of the ellipse is known, let this be $2a$. The other focus H of the ellipse is the intersection of two circles described with centres A and B and radii $2a - SA$, $2a - SB$ respectively. The two intersections give two solutions which only coincide when the circles touch, that is when the line AB passes through the focus H. Thus if we draw a chord AC through H to cut the ellipse described by the particle in C, then the terminal position B must fall between A and C if the integral which occurs in the principle of least action is really to be a minimum for this ellipse. If B coincide with C, then the second variation cannot become negative, but it can become zero, so that the variation of the integral is then of the third order, and may therefore be either positive or negative. If B be beyond C the second variation itself can become negative.

If the particle start from A towards perihelion, then the extreme point C is determined by drawing a chord AC through the focus S to cut the ellipse in C. For if A and C are the limits we can obtain an infinite number of solutions by the revolution of the ellipse round AC. If in the last case the second limit B fall beyond C, Jacobi considered that there would be a curve of double curvature between the two given points for which the action is less than it is for the ellipse. But this supposition is unnecessary, for the discontinuous course spoken of in Art. 456 supplies the minimum for this case.

Examples. **Ex. 1.** A particle, under the action of a centre of force at O whose attraction varies as the distance, is projected from a given point A with a given velocity in such a direction as to reach another given point B. If C be the first point on the elliptic path at which the tangent is perpendicular to the direction of projection at A, prove that the "action" from A to B is or is not a minimum according as B is between A and C or beyond C.

If B lie within a certain ellipse having its centre at O and one focus at A, prove that there are two directions in which the particle can be projected from A to reach B and that the action is a minimum for one of these and not for the other. If B lie outside this bounding ellipse, the particle cannot reach B. If OA be produced to D, where D is such that the velocity of projection at A is equal to that acquired by a particle starting from rest at D and moving to A under the action of the central force, prove that the major axis of the bounding ellipse is equal to twice the distance OD.

If the point B be without the bounding ellipse, the particle can reach B only if properly conducted thither by some curve of constraint. The curve of minimum action can be found by the following construction. Produce OA, OB to meet the auxiliary circle of the bounding ellipse in E and F. The required path is indefinitely near to $AEFB$.

To prove these results, let us find the direction of projection from A that the particle may pass through B. We notice that if $OD = k$, the sum of the squares of any two semi-conjugate diameters is k^2. Bisect AB in N and let $ON = x$, $NA = NB = y$. Let the required direction of projection from A cut ON produced in T. Then from the equation to the ellipse we have a quadratic to find OT, showing that there are in general two elliptic paths which may be described in passing from A to B. Let the tangents at A to these intersect ON produced in T and U; we deduce from the quadratic that $OT \cdot OU = k^2$ and $NT \cdot NU = y^2$. These equations determine T and U.

We see at once that the two directions of projection coincide when $OT = k$, i.e. when the tangents at A and B, viz. AT and BT, are at right angles.

Describe two circles with centres O and N and radii equal to k and y respectively. Describe a third circle on TU as diameter. Since $OT \cdot OU = k^2$ this third circle cuts the circle with centre O at right angles. Similarly it cuts the circle with centre N at right angles. The tangents from the centre R of this third circle are therefore equal. The centre R is therefore on the radical axis of the circles whose centres are O and N. This gives an easy geometrical construction to find T and U.

The points T and U will be imaginary unless the radical axis lie outside the circles. The circles must therefore not intersect. Hence $ON + NA$ must be less than k. Produce AO to A' so that $OA' = OA$. Then we see that $AB + BA'$ must be less than $2k$. Hence unless B lies within an ellipse whose foci are A and A' and major axis $2k$, the particle cannot be projected from A to pass through B.

Ex. 2. A particle is projected from a given point A under the action of gravity and AC is a focal chord of the parabola described. Prove that the action from A to B is not a minimum unless B lies on the parabola between A and C. If B lies beyond C, find the path which makes the action a minimum.

The first result follows at once from Jacobi's example. To answer both these questions, we notice that there are *two* directions (if any) in which a particle may be projected from one given point A to pass through a second given point B. These have their foci S, S' one above and the other below the chord AB, so that SS' and AB bisect each other at right angles. These paths coincide when B is at C, and wherever B may be one of these has its focus below AB. This parabola is the path required.

Ex. 3. A particle, projected from a given point A with a given velocity, describes a circle about a centre of force on the circumference whose attraction varies inversely as the fifth power of the distance. If B be any other position on this circle through which the particle will pass before arriving at the centre of force, prove that the action from A to B is a minimum according to Jacobi's condition.

459. **The inversion of dynamical problems***. Since the equations of motion can be deduced from the principle of least action, it is clear that, if in applying the principle to two different problems we have to make the same expression a minimum under the same conditions, the general integrals of these two problems can be inferred the one from the other.

Consider the case of a single particle moving with a force function $U + C$ along a path APB beginning at one given point A and ending at another B. If $s = AP$, and v is the velocity of the particle, the path is such that $\int v \, ds$ is a minimum. If we invert the curve with regard to any point O, it follows that $k^2 \int v \, \dfrac{ds'}{r'^2}$ is a minimum for the inverse curve from A' to B', where accented letters refer to the inverse curve and k is the constant of inversion. It follows from the principle of least action that this curve will be the path of a free particle moving with such a force function $U' + C'$ that $v' = k^2 v / r'^2$. We have therefore from the principles of dynamics

$$r'v' = rv \quad \text{and} \quad \therefore \ r'^2 (U' + C') = r^2 (U + C) \dots \dots \dots \dots \dots \dots (1).$$

Since the radial angles are equal in a curve and its inverse, the first of these equations shows that the angular momenta about any axis through the centre of inversion at corresponding points in the two motions are equal.

---

* The substance of this article is taken from a paper by Larmor in Vol. xv. of the *Proceedings of the London Mathematical Society*, 1884.

We have therefore the following theorem :—*if a particle describe a path APB with a force function U + C, then a particle can describe the inverse path A'P'B' with a force function U' + C' given by* (1), *provided that at one set of corresponding points the velocities are related to each other by the equation r'v' = rv.*

Ex. 1. A particle, constrained to move on a smooth sphere and acted on by no forces, is known to describe a great circle. By inverting this theorem show that a particle, constrained to move on a smooth given sphere and acted on by a central force varying inversely as the fifth power of the distance from a point O, describes a circular path. Show also that this circle is the intersection of the given sphere with another sphere passing through O and a point O' which is the foot of the perpendicular from O on its polar plane.

Ex. 2. Prove that in a plane field of force of which the potential referred to polar co-ordinates is $\dfrac{a}{r^4} + \dfrac{\beta(1 + 3\sin^2\theta)}{r^6}$, a particle if projected in the proper direction with the velocity from infinity will describe a curve of the form

$$(r - a\sin\theta)(r - b\sin\theta) = ab,$$

provided $\dfrac{2}{ab} + \dfrac{4}{(a+b)^2} + \dfrac{a}{\beta} = 0.$ [Math. Tripos, 1886.

Ex. 3. A particle, constrained to move on an anchor ring of evanescent aperture, is acted on by a central force varying inversely as the fifth power of the distance from the aperture, prove that the path cuts all the meridians at the same angle.

We may also transform dynamical theorems by the help of conjugate functions. This method is analogous to that used in Chap. XIV. of this treatise to deduce the motion of a heterogeneous membrane from that of a homogeneous one. A list of the theorems required on these functions is given in that chapter.

Let (x, y), (ξ, η) be the co-ordinates of two points P, Π, moving in corresponding or conjugate planes, and so related that $\xi + \eta\sqrt{-1} = f(x + y\sqrt{-1})$. If μ be the modulus of transformation, then

$$\mu^2 = \left(\frac{d\xi}{dx}\right)^2 + \left(\frac{d\xi}{dy}\right)^2 = \left(\frac{d\eta}{dx}\right)^2 + \left(\frac{d\eta}{dy}\right)^2 \dots\dots\dots\dots\dots(\text{I}).$$

Let ds, $d\sigma$ be corresponding arcs of the paths described by the two points P, Π, then $d\sigma = \mu ds$. The motion of the particle Π in the plane (ξ, η) being given by $\delta\int v'd\sigma = 0$, that of P in the plane xy is given by $\delta\int v'\mu ds = 0$. The particles P and Π therefore move freely with velocities v and v' under force functions $U + C$ and $U' + C'$, provided

$$v = v'\mu \quad \text{and} \quad \therefore \ U + C = \mu^2(U' + C') \dots\dots\dots\dots\dots(\text{II}).$$

Ex. 4. A particle Π describes a central orbit whose polar equation is $f(\rho, \phi) = 0$ with a velocity v' such that $v' = F(\rho)$. Prove that a particle P can describe the central orbit $f(r^n, n\theta) = 0$ with a velocity $v = nr^{n-1}F(r^n)$ under a central force equal to $\frac{1}{2}dv^2/dr$. Show also that the ratio of the angular momenta of P and Π about the centres of force is equal to n and that the times of describing corresponding elementary arcs are in the ratio $1 : n^2 r^{2(n-1)}$.

460. **Lagrange's transformation.** Lagrange has given a general view of his transformation from Cartesian co-ordinates which seems worthy of notice. Let L be any function of x, x', &c., y, y', &c. and of t, not restricting ourselves to differential coefficients of the first order. Let the variables x, y, &c. be transformed to others q_1, q_2, &c. by writing for x, y, &c. any functions of q_1, q_2, &c. and of t. The function L is thus expressed in two ways. By comparing the two values of

$\delta\int L dt$, given by the Calculus of Variations when the time is not varied, we see that

$$\int_{t_0}^{t_1} \Sigma \left(\frac{dL}{dx} - \frac{d}{dt}\frac{dL}{dx'} + \&\mathrm{c}.\right) \delta x dt - \int_{t_0}^{t_1} \Sigma \left(\frac{dL}{dq} - \frac{d}{dt}\frac{dL}{dq'} + \&\mathrm{c}.\right) \delta q dt$$

is equal to the difference of the integrated portions of the two variations. Hence the expression under the integral sign must be a perfect differential with regard to t, quite independently of the operation δ. But this cannot be unless the expression is zero, because it contains only the variations δx, δq, &c. and not the differential coefficients of these variations. We have therefore the general equation of trans-

formation $\Sigma \left(\dfrac{dL}{dx} - \dfrac{d}{dt}\dfrac{dL}{dx'} + \&\mathrm{c}.\right) \delta x = \Sigma \left(\dfrac{dL}{dq} - \dfrac{d}{dt}\dfrac{dL}{dq'} + \&\mathrm{c}.\right) \delta q,$

where the Σ implies summation for all the variables x, y, &c., q_1, q_2, &c.

If x, y, &c. be Cartesian co-ordinates and if L be of the usual form $\Sigma mx'^2 + U$, the left-hand side of this equality vanishes by virtual velocities. Hence the right-hand side must also vanish. The q's being all independent, we are led to Lagrange's equations.

Ex. Supposing the Lagrangian function L to be a function of the typical variables q, q', q'' and that the differential equations have the type

$$\frac{dL}{dq} - \frac{d}{dt}\frac{dL}{dq'} + \frac{d^2}{dt^2}\frac{dL}{dq''} = 0,$$

show that the corresponding Hamiltonian forms are

$$r'' = \frac{dH}{dq} - \frac{d}{dt}\frac{dH}{dq'}, \text{ and } q'' = \frac{dH}{dr},$$

where $r = dL/dq''$ and H is a function of q, q', r.

Let H be the reciprocal function of L with regard to q'', then $L + H = \Sigma rq''$. By Vol. I., Art. 410, $dL/dq = -dH/dq$, $dL/dq' = -dH/dq'$. The first result follows by substitution in the Lagrangian equation of motion and the second follows from the definition of a reciprocal function.

461. **Cyclical Motions.** When the geometrical equations do not contain the time explicitly the symbol H or h may be used to express the energy of the system. If we represent the energy by E, Sir W. R. Hamilton's fundamental equation may

be written $2\delta \displaystyle\int_0^t T dt = t\delta E + \left[\Sigma \dfrac{dT}{dq'}\delta q\right]_0^t$ ....(1).

This equation has been applied to the motion of a system of bodies oscillating in such a manner that the motion repeats itself in all respects at some constant interval. Let this interval be i. Suppose that some disturbance is given to the system by the addition of a quantity of energy δE. Let the system be such that the motion still recurs after a constant interval, and let this interval be now $i + \delta i$. The symbols of variation in Hamilton's equation may be used to imply a change from one kind of motion to the other. If the time t is taken equal to the period i of complete recurrence, the initial and terminal states of motion are the same and therefore the last term vanishes when taken between the limits. The equation reduces to $2\delta \int_0^i T dt = i\delta E$. Let T_m be the mean vis viva of the system during a period of complete recurrence of the motion, then $\int_0^i T dt = iT_m$. We therefore have $\dfrac{\delta E}{T_m} = 2\dfrac{\delta(iT_m)}{iT_m}$.

This equation may be put into another form. Let P_m be the mean potential energy of the system during a period of complete recurrence ; then we have

$$\delta P_m + \delta T_m = \delta E, \qquad \delta P_m - \delta T_m = 2T_m \frac{\delta i}{i} \quad..................(2),$$

which serve to determine the change in the mean potential and kinetic energies when any additional energy δE is added to the system.

If the system is not performing a principal oscillation the motion does not recur at a constant interval i. Let us suppose that the motion is compounded of several principal oscillations or more generally let the motion be of the kind called *stationary motion* in the chapter on vis viva in Vol. I. If the means are now taken for any very long time i, the equations just arrived at are still true. To show this we recur to Hamilton's equation (1). Dividing by $t = i$, the last term on the right-hand side becomes very small because the motion is such that the δq's in that term do not continually increase with the time. We therefore have $2\delta (iT_m)/i = \delta E$, and the rest of the proof is the same as before.

These or equivalent equations have been applied by Bolzman, Clausius and Szily to the Dynamical Theory of Heat. The papers of the two latter are in various numbers of the *Philosophical Magazine* extending from 1870 onwards. The second of the equations (2) may be called Clausius' equation. The reader may also refer to a work by Prof. J. J. Thomson on the *Applications of Dynamics to Physics and Chemistry*, 1888.

462. **Ex. 1.** If the period of complete recurrence of a dynamical system is not altered by the addition of energy, prove that this additional energy is equally distributed into potential and kinetic energy. See Art. 73.

Ex. 2. A quantity of energy dE is communicated to a system whose mean semi vis viva during a period of complete recurrence is T_m. This is repeated continually, so that at last the mean vis viva and the period of complete recurrence are the same as at first. Prove that $\int \dfrac{dE}{T_m} = 0$. This example is due to M. Szily, and is important in the Dynamical Theory of Heat.

On the Solution of the General Equations of Motion.

463. **Hamilton's Solution.** Sir W. R. Hamilton has applied his fundamental theorem expressing the variation of the Principal and Characteristic functions to obtain a new method of solving dynamical problems.

Let $(\beta_1, \beta_1', \beta_2, \beta_2', \&c.)$ be the values of $(q_1, q_1', q_2, q_2', \&c.)$ when $t = t_0$, and let T_0 be the same function of $(\beta_1, \beta_1', \&c.)$ that T is of $(q_1, q_1', \&c.)$. We have then by Art. 442 when t is written for the upper limit

$$\delta S = \Sigma \frac{dT}{dq'} \delta q - \Sigma \frac{dT_0}{d\beta'} \delta\beta - H\delta t + H_0 \delta t_0,$$

$$\delta V = \Sigma \frac{dT}{dq'} \delta q - \Sigma \frac{dT_0}{d\beta'} \delta\beta + t\delta H - t_0 \delta H_0.$$

It is clear that both S and V may be regarded as functions of the time and the initial conditions of the system of bodies, i.e. we may regard either of these quantities as a function of t_0, t, β_1, β_2, &c., β_1', β_2', &c. Also the co-ordinates q_1, q_2, &c. are functions of t_0, t and the same initial conditions. Though these functions are in general unknown, yet we can conceive the initial velocities β_1', β_2', &c. eliminated, so that S and V are now functions of t_0, t, and β_1, β_2, &c., q_1, q_2, &c. the co-ordinates of the system at the times t_0 and t.

Let S be thus expressed, then, by the equation for δS, we have the typical equations

$$\frac{dS}{dq} = \frac{dT}{dq'}, \quad \frac{dS}{d\beta} = -\frac{dT_0}{d\beta'} \dots\dots\dots\dots(1).$$

Since T is not a function of q'', the first of these equations contains no differential coefficient of a co-ordinate higher than the first. *This equation, therefore, represents typically all the first integrals of the equations of motion.*

Since T_0 contains only the initial co-ordinates and the initial velocities, the second equation has no differential coefficient of any co-ordinate in it. *This equation, therefore, represents typically all the second integrals of the motion.*

Besides these we have the two equations

$$\frac{dS}{dt} = -H, \quad \frac{dS}{dt_0} = H_0 \dots\dots\dots\dots (2),$$

where, if the geometrical equations do not contain the time explicitly, we may put h for H, h being a constant. In this case these integrals may be used to connect the constant of vis viva with the constants (β, β', &c.).

Comparing Art. 447 with these results we see that S is such a function, that all the equations of motion and their integrals are included in the statement that δS is a known function of the variation of the limits. If we keep the limits fixed, we get Lagrange's equations; if we vary the limits we get the integrals.

464. In just the same way, if we regard q_1', q_2', &c. as functions of t, the initial co-ordinates and their initial velocities, we may eliminate t also by means of the equation

$$H = -U - T + \Sigma \frac{dT}{dq'} q'.$$

We may eliminate t_0 also by means of a similar equation giving H_0 in terms of the initial conditions. Both these reduce to $H = H_0 = T - U$ when the geometrical equations do not contain the time explicitly.

Let us suppose V to be expressed in this manner as a function of the initial co-ordinates, the co-ordinates at the time t, and of H and H_0. Then, by the equation for δV,

$$\frac{dV}{dq} = \frac{dT}{dq'}, \quad \frac{dV}{d\beta} = -\frac{dT_0}{d\beta'}, \quad \frac{dV}{dH} = t. \quad \frac{dV}{dH_0} = -t_0.$$

Supposing V to be known, the first of these equations gives in a typical form all the first integrals of the equations of motion. The second supplies as many equations as there are co-ordinates

19—2

(q_1, q_2, &c.). When the geometrical equations do not contain the time explicitly these do not contain t, but they all contain h. One of them, therefore, reduces to the relation between this constant and the constants (β, β', &c.). *The two last equations become $dV/dh = t - t_0$. This will give another second integral of the equations of motion containing the time.*

465. The typical expression dT/dq' has been called in Vol. I. the momentum corresponding to the co-ordinate q or, more briefly, the q component of the momentum. We may therefore say that the q component of the momentum is given by dS/dq or dV/dq according as we are using S or V.

The momenta corresponding to the co-ordinates q_1, q_2, &c. will be represented by the symbols p_1, p_2, &c., or typically by the single letter p.

By Lagrange's equations $dp/dt = dL/dq$, we may therefore also say that the rate of change of each momentum is equal to the differential coefficient of a single function, viz. L with regard to the corresponding co-ordinate.

466. If $Q = \int_{t_0}^{t} (\Sigma qp' + H) dt$, where $p = \dfrac{dT}{dq'}$, prove that $\delta Q = \left[H\delta t + \Sigma q\delta p \right]_{t_0}^{t}$.

Thence show that if Q be expressed as a function of the initial and terminal components of momentum, viz. (a_1, a_2, &c.) and (p_1, p_2, &c.) and of the times t_0 and t, then $\dfrac{dQ}{dp} = q$, $\dfrac{dQ}{da} = -\beta$, $\dfrac{dQ}{dt} = H$. This result is due to Sir W. R. Hamilton.

467. **Examples.** Ex. 1. A homogeneous sphere of unit mass rolls down a perfectly rough fixed inclined plane. If the position of the sphere is defined by the distance q of the point of contact from a fixed point on the inclined plane, show that

$$S = \frac{7}{10}\frac{(q-\beta)^2}{t} + \frac{1}{2}(q+\beta) gt - \frac{5}{168} g^2 t^3,$$

where g is the resolved part of gravity down the plane and $t_0 = 0$.

Thence obtain by substitution the Hamiltonian first and second integrals of the equation of motion.

We easily find, as in Vol. I., that $q = \beta + \beta't + \frac{5}{14} gt^2$. Also $T = \frac{7}{10} q'^2$, $U = gq$. To find S, we substitute in $S = \int_0^t (T + U) dt$. After integration we must eliminate β' by means of the equation for q.

Ex. 2. Taking the same circumstances of motion as in the last example, show that $V = \dfrac{2}{3g} \sqrt{\tfrac{14}{5}} \{ (gq + h)^{\frac{3}{2}} - (g\beta + h)^{\frac{3}{2}} \}$. Thence also deduce the Hamiltonian first and second integrals.

Ex. 3. Show how to deduce the equation of vis viva from the Hamiltonian integrals.

We have V a function of q_1, q_2, &c. and H. Hence $\dfrac{dV}{dt} = \Sigma \dfrac{dV}{dq} q' + \dfrac{dV}{dH}\dfrac{dH}{dt}$, which becomes by Hamilton's integrals $2T = \Sigma (dT/dq') q' + t (dH/dt)$. When T is a homogeneous quadratic function of (q_1', q_2', &c.) this gives $dH/dt = 0$, or $H = $ constant. The equation of vis viva may also be deduced from Hamilton's principal function.

Ex. 4. When the geometrical equations do not contain the time explicitly, show that no two of the Hamiltonian integrals can be the same and that no one can be deduced from two others.

If it were possible that two could be the same, the ratio of dT/dq_1' to dT/dq_2' must be some constant m. Integrating this partial differential equation, we find T to be a homogeneous quadratic function of $q_1' + mq_2'$, q_3', &c. It would, therefore, be possible to set the system in motion, with values of q_1' and q_2' which are not zero, and yet so that the system is without vis viva.

Ex. 5. In any dynamical system, if the co-ordinates q_1, q_2, q_3 and their corresponding momenta p_1, p_2, p_3 are expressed in terms of their initial values and the time elapsed, prove that the Jacobian of p_1, p_2, p_3, q_1, q_2, q_3 with regard to their initial values is equal to unity.

Ex. 6. A system whose co-ordinates are q_1, q_2, &c. is making small oscillations about a state of steady motion determined by $q_1 = 0$, $q_2 = 0$, &c. The Lagrangian function, as in Art. 111, is given by $L = L_0 + \Sigma Aq' + L_2$, where L_2 is a homogeneous function of the second order of the co-ordinates and their velocities. Prove that

$$S = L_0 (t - t_0) + \Sigma A (q - \beta) + \tfrac{1}{2} [\Sigma qdL_2/dq'],$$

where the last term is to be taken between the limits t_0 and t. Here the integrations have been effected, but in order to express S (Art. 463) as a function of the co-ordinates we must finally substitute for q' and β' in terms of these quantities.

Ex. 7. The position of a system making small oscillations as in Ex. 6 is defined by one co-ordinate q, so that

$$L = L_0 + A_1 q' + \tfrac{1}{2} A_{11} q'^2 + \tfrac{1}{2} C_{11} q^2 + G_{11} qq',$$

where the coefficients are all constants. Prove that when $t_0 = 0$

$$S = L_0 t + A_1 (q - \beta) + \tfrac{1}{2} G_{11} (q^2 - \beta^2) + \tfrac{1}{2} mA_{11} \frac{(q^2 + \beta^2)(e^{mt} + e^{-mt}) - 4q\beta}{e^{mt} - e^{-mt}},$$

where $m^2 = C_{11}/A_{11}$.

Ex. 8. A particle oscillates in a straight line about a centre of force which varies as the distance, show that the Hamiltonian function

$$S = \frac{\sqrt{\mu}}{2} \frac{(x_0{}^2 + x^2) \cos \sqrt{(\mu)} (t - t_0) - 2xx_0}{\sin \sqrt{(\mu)} (t - t_0)}.$$

Verify this by deducing the Hamiltonian Integrals.

468. Hamilton's Differential Equations.

By the preceding reasoning all the integrals of a dynamical system of equations can be expressed in terms of the differential coefficients of a single function. But the method supplies no means of discovering this function *à priori*. *We shall now show that this function must always satisfy a certain differential equation, so that the solution of all dynamical problems may be reduced to the integration of one differential equation.*

To construct this differential equation we first form the reciprocal of the Lagrangian function $L = T + U$ according to the rule given in the first volume of this treatise, Arts. 410 and 414. Briefly the rule is as follows, we put $dT/dq_1' = p_1$, $dT/dq_2' = p_2$, &c. as in Art. 465 of this volume; also putting

$$T + T_2 = \Sigma pq',$$

we eliminate the velocities $q_1{}'$, $q_2{}'$, &c. and express T_2 as a function of the co-ordinates q_1, q_2, &c. and the momenta p_1, p_2, &c. The reciprocal function of $L = T + U$ is then $H = T_2 - U$.

If the geometrical equations do not contain the time explicitly, the vis viva $2T$ is a homogeneous quadratic function of the velocities. If this function be

$$2T = A_{11}q_1{}'^2 + 2A_{12}q_1{}'q_2{}' + \&c.,$$

we have $\quad T_2 = -\dfrac{1}{2\Delta} \begin{vmatrix} 0 & p_1 & p_2 \cdots \\ p_1 & A_{11} & A_{12} \cdots \\ \multicolumn{3}{c}{\cdots\cdots\cdots\cdots} \end{vmatrix}$,

where Δ is the discriminant of T; see Vol. I. Art. 413. Thus H is a quadratic function of the momenta p_1, p_2, &c. We may shortly write this in the form

$$H = \tfrac{1}{2}B_{11}p_1{}^2 + B_{12}p_1p_2 + \ldots - U.$$

But $p_1 = dV/dq_1$, $p_2 = dV/dq_2$, &c. and the equation of vis viva gives $H = h$. Hence V must satisfy the equation

$$\tfrac{1}{2}B_{11}\left(\frac{dV}{dq_1}\right)^2 + B_{12}\frac{dV}{dq_1}\frac{dV}{dq_2} + \&c. - U = h \ldots\ldots (I).$$

In just the same way $p_1 = dS/dq_1$, $p_2 = dS/dq_2$, &c. and $H = -dS/dt$. Hence S must satisfy the equation

$$\tfrac{1}{2}B_{11}\left(\frac{dS}{dq_1}\right)^2 + B_{12}\frac{dS}{dq_1}\frac{dS}{dq_2} + \&c. - U = -\frac{dS}{dt} \ldots\ldots(II).$$

Here the coefficients B_{11}, B_{12}, &c. are all known functions of the co-ordinates q_1, q_2, &c.

We have supposed V to be expressed as a function of the co-ordinates at the time t, the initial co-ordinates and the energy h. *But in this equation we may also regard V to be a function of the co-ordinates at the time t, the energy h, and as many arbitrary constants as there are co-ordinates.* In this case these constants are really functions of the initial co-ordinates which we do not care to determine. The equations giving the momenta p_1, p_2, &c. at the time t as the differential coefficients of V with regard to q_1, q_2, &c. will still be true; but the equations expressing the initial momenta are supposed not to be wanted.

If we take as these constants the actual co-ordinates at any epoch $t = t_0$ we may form another equation of a form similar to (I) with β_1, β_2, &c. written for q_1, q_2, &c. and t_0 for t. It is then necessary that V should satisfy both these equations.

Summing up, we may form the Hamiltonian equation (I) by the following process. *We first form the Lagrangian function $L = T + U$ and thence its reciprocal function $H = T - U$ by the rule given in Vol. I. Art. 410. Equating this to a constant h, we have the equation of vis viva expressed in terms of the momenta. Lastly, we* *write for the momenta the differential coefficients of V with regard to the corresponding co-ordinates.*

469. When the equations contain the time explicitly, the vis viva $2T$ contains both first and second powers of the velocities. Let this be

$$2T = A_{11}q_{11}'^2 + 2A_{12}q_1'q_2' + \ldots\ldots + 2A_1q_1' + 2A_2q_2' + \ldots$$

Then the reciprocal function T_2 contains both first and second powers of p_1, p_2 &c., and may be written

$$T_2 = -\frac{1}{2\Delta} \begin{vmatrix} 0 & p_1 - A_1 & p_2 - A_2 \cdots \\ p - A_1 & A_{11} & A_{12} \cdots \\ \cdots\cdots\cdots & \cdots\cdots\cdots\cdots \end{vmatrix}$$

where Δ is the minor of the leading constituent in this determinant. Thus $H = T_2 - U$ when expanded takes the form

$$H = \tfrac{1}{2}B_{11}p_1^2 + B_{12}p_1p_2 + \ldots + B_1p_1 + B_2p_2 + \ldots - U.$$

We then substitute $p_1 = dV/dq_1$ &c. and obtain the equation

$$\tfrac{1}{2}B_{11}\left(\frac{dV}{dq_1}\right)^2 + \ldots + B_1\frac{dV}{dq_1} + \ldots = H + U \quad\ldots\ldots\ldots\ldots(\mathrm{III}).$$

Since the time t here occurs explicitly we suppose its value dV/dH to have been written for it. We have thus a partial differential equation to find V as a function of q_1, q_2 &c. and H. Supposing V to be properly found from this partial differential equation, the formulæ given in Art. 463 would determine all the integrals of the dynamical equations.

Ex. If the expression for T were of the form

$$T = T_0 + T_1 + T_2 + \ldots = \Sigma T_n$$

where T_m is a homogeneous function of q_1', q_2' &c. of the mth degree, show that the reciprocal function of T is $\Sigma(n-1)T_n$, which must of course be expressed in terms of the momenta. It may be noticed that T_1 is absent from the formula.

470. **Jacobi's Complete Integral.** We thus have, in general, a partial differential equation to find V or S. This equation admits of many forms of solution, but Sir W. R. Hamilton gave no rule to determine *which integral is to be taken.* This defect has been supplied by Jacobi in the following proposition.

Let there be n co-ordinates in the system. Suppose a complete solution to have been found containing $n-1$ constants (besides h) and the constants which may be introduced by simple addition to the function V. These constants need not be the initial values of q_1, q_2, &c., but may be any constants whatever. Let them be denoted by b_1, $b_2 \ldots b_{n-1}$, so that

$$V = f(q_1, q_2 \ldots q_n, b_1, b_2 \ldots b_{n-1}) + b_n \ldots\ldots\ldots\ldots(1).$$

Then the integrals of the dynamical equations will be

$$\frac{df}{db_1} = -a_1, \text{ &c. } \frac{df}{db_{n-1}} = -a_{n-1}, \frac{df}{dh} = t + \epsilon \ldots\ldots\ldots\ldots(2),$$

where a_1, $a_2 \ldots a_{n-1}$ and ϵ are n new arbitrary constants, and the first integrals of the equations may be written in the form

$$\frac{df}{dq_1} = \frac{dT}{dq_1'}, \quad \frac{df}{dq_2} = \frac{dT}{dq_2'}, \quad \text{&c.} = \text{&c.} \ldots\ldots\ldots\ldots(3).$$

It appears from Jacobi's proposition that *any integral, provided it is complete, will supply a solution to the dynamical problem.* We have also a sufficient number of constants, viz. $b_1 \dots b_{n-1}$, h, ϵ and $a_1 \dots a_{n-1}$ to satisfy any initial conditions.

An integral of a partial differential equation has been called by Lagrange "complete," when it contains as many arbitrary constants as there are independent variables. It is implied that the constants enter in such a manner into the integral that they cannot by any algebraic process be reduced to a smaller number. For instance, if two of the constants enter in the form $b_1 + b_2$, they amount on the whole to only one.

471. To prove these results we must show that, if the form of V given by (1) satisfies *identically* the equation

$$H = \tfrac{1}{2}B_{11}p_1^2 + B_{12}p_1p_2 + \dots - U = h \dots\dots\dots\dots(\text{I}),$$

where p stands for dV/dq, then the relations (2) will satisfy identically the two typical Hamiltonian equations

$$\frac{dH}{dp} = q', \qquad -\frac{dH}{dq} = p' \dots\dots\dots\dots(\text{II}).$$

It will immediately follow, since H and $T + U$ are reciprocal functions, that the relations (2) will also make

$$p = \frac{dT}{dq'} \dots\dots\dots\dots\dots\dots(\text{III}).$$

Since (I) is identically satisfied, we may differentiate it partially with regard to each of the n constants $b_1 \dots b_{n-1}$ and h. We thus obtain, after substitution from (1), $n - 1$ equations of the form

$$\frac{dH}{dp_1}\frac{dp_1}{db} + \frac{dH}{dp_2}\frac{dp_2}{db} + \dots = 0,$$

and an nth equation derived from this by writing h for b and unity for the zero on the right-hand side. We shall use these n equations to find dH/dp_1, dH/dp_2, &c.

But if we differentiate Jacobi's integrals (2) with regard to t we have $n - 1$ equations of the form

$$\frac{dq_1}{dt}\frac{d^2f}{dbdq_1} + \frac{dq_2}{dt}\frac{d^2f}{dbdq_2} + \dots = 0,$$

and an nth equation derived from this by writing h for b and putting unity on the right-hand side. We shall use these n equations to find dq/dt_1, dq_2/dt, &c.

Comparing these two sets of equations, we see that, when we substitute for the typical p its value derived from $p = df/dq$, the equations become identical. Hence,

$$\frac{dH}{dp_1} = \frac{dq_1}{dt}, \qquad \frac{dH}{dp_2} = \frac{dq_2}{dt}, \quad \&c.$$

Again, if we differentiate the identical equation (I) with regard to each of the co-ordinates $q_1 \ldots q_n$ in turn, we obtain after substitution from (1) the typical equation

$$\frac{dH}{dq} + \frac{dH}{dp_1}\frac{dp_1}{dq} + \frac{dH}{dp_2}\frac{dp_2}{dq} + \ldots = 0,$$

$$\therefore -\frac{dH}{dq} = \frac{dq_1}{dt}\frac{d^2f}{dq_1 dq} + \frac{dq_2}{dt}\frac{d^2f}{dq_2 dq} + \ldots.$$

But since $p = df/dq$, the right-hand side is the same as dp/dt, we therefore have

$$-\frac{dH}{dq_1} = \frac{dp_1}{dt}, \qquad -\frac{dH}{dq_2} = \frac{dp_2}{dt}, \text{ &c.}$$

472. When the geometrical equations contain the time explicitly the enunciation is slightly altered. Since the partial equation, as explained in Art. 469, has now $n+1$ variables, viz. $q_1,\ldots q_n$ and H, the complete integral has $n+1$ constants and may be written in the form

$$V = f(q_1 \ldots q_n,\ H,\ b_1 \ldots b_n) + b_{n+1} \quad\ldots\ldots\ldots\ldots\ldots\ldots(1).$$

Then the n integrals of the dynamical equations are

$$\frac{df}{db_1} = -a_1, \quad \text{&c.,} \quad \frac{df}{db_n} = -a_n \ldots\ldots\ldots\ldots\ldots\ldots(2),$$

where $a_1 \ldots a_n$ are n new arbitrary constants. These integrals contain $q_1 \ldots q_n$ and H, but H may be eliminated and t introduced by using the equation $dV/dH = t$. The n first integrals are

$$\frac{df}{dq_1} = \frac{dT}{dq_1'}, \quad \text{&c.} \quad \frac{df}{dq_n} = \frac{dT}{dq_n'}, \quad\ldots\ldots\ldots\ldots\ldots\ldots (3),$$

and H may be eliminated as before.

When the geometrical equations do not contain the time explicitly the partial differential equation (III) of Art. 469 contains H but not dV/dH, this last having been introduced merely to eliminate t. The complete integral has therefore n constants instead of $n+1$. We now write h for H and by Art. 464 we have $dV/dh = t - t_0$. Putting ϵ for $-t_0$, we see that the place of the missing constant in (2), viz. a_n, is filled by the constant ϵ.

473. **Geometrical Remarks.** To simplify the argument let us suppose that the dynamical system depends only on two co-ordinates q_1, q_2. The Hamiltonian equation (I) therefore takes the form

$$\tfrac{1}{2}B_{11}\left(\frac{dV}{dq_1}\right)^2 + B_{12}\frac{dV}{dq_1}\frac{dV}{dq_2} + \tfrac{1}{2}B_{22}\left(\frac{dV}{dq_2}\right)^2 = U + h \ldots\ldots\ldots\ldots(1).$$

Let us suppose that a complete integral has been found, viz.,

$$V = f(q_1,\ q_2,\ b_1) + b_2 \ldots\ldots\ldots\ldots\ldots\ldots (2).$$

Regarding q_1, q_2 and V as the Cartesian co-ordinates of a point P, this is the equation to a double system or family of surfaces. Let us select any family we please, so that the constants b_1, b_2 are now related by some equation $b_2 = \psi(b_1)$. The characteristics of this chosen family are given by

$$\left.\begin{array}{l} V = f(q_1,\ q_2,\ b_1) + \psi(b_1) \\ 0 = df/db_1 + d\psi/db_1 \end{array}\right\} \ldots\ldots\ldots\ldots\ldots\ldots (3).$$

where b_1 is regarded as a constant.

The general integral is obtained by eliminating b_1 between the two equations (3).

Here b_1 in the first equation is to be regarded as a function of q_1, q_2, determined by the second equation. This of course is merely following Lagrange's rule to find the general integral when any complete integral is known.

In the same way we find that Lagrange's singular solution is at infinity.

It appears from this that all the characteristics of all the families of surfaces included in the complete integral (1) are used to build up the general integral. We choose any set of characteristics we please so that a surface can be made to pass through every member of the set. This surface is a particular case of the general solution.

474. According to Jacobi's theorem the path of the dynamical system is defined by $df/db_1 = -a_1$. Looking at the second of equations (3) we see that this is equivalent to asserting that $d\psi/db_1$ and therefore b_1 is constant. It follows that the possible *paths of the dynamical system are the characteristics of the families which may be chosen out of the complete integral.*

475. Since Lagrange's method of finding the general integral will give a solution whatever the form of $\psi(b_1)$ may be, we may use that process to obtain other complete integrals. If we write $\phi(m, b_1) + n$ for $\psi(b_1)$ and proceed to eliminate b_1 we obtain a solution which contains two constants, viz. m and n, and which is therefore a complete integral. Here ϕ may be any function we please, and b_1 is to be regarded as a function of q_1, q_2 determined by the second of the equations (3).

The paths derived from this new complete integral by Jacobi's method are given by $(df/db_1 + d\psi/db_1)\, db_1/dm + d\psi/dm = -a_1$.

By the second of equations (3) the term in brackets is zero. The path therefore is defined by equating to a constant a function of b_1 and m. The paths are therefore given by equating b_1 to a constant. It follows that *the two complete integrals lead to the same set of dynamical paths.*

476. If the Hamiltonian equation

$$\tfrac{1}{2}B_{11}(dV/dq_1)^2 + B_{12}(dV/dq_1)(dV/dq_2) + B_{22}(dV/dq_2)^2 = U + h$$

is such that all the coefficients on the left side and also U are functions of one coordinate only, say q_2, then a complete integral can be found by writing $V = W + b_1 q_1$, where W is a function of q_2 only. Substituting this in the Hamiltonian equation we have a differential equation with one independent variable, viz. q_2. The solution of this can be effected by the ordinary method of separating the variables. Thus we easily find by solving a quadratic that dV/dq_2 is a known function of q_2 and b_1. Integrating this we have a value for V with one additional constant. This therefore is a complete integral.

477. **Examples.** Ex. 1. Taking the same problem as that in Ex. 1 of Art. 467, show that Hamilton's differential equation for V is $\tfrac{5}{14}(dV/dq)^2 - gq = h$. Integrate this equation and thence find the motion.

Ex. 2. Let us next consider a more complicated case in which there are two coordinates. The simplest example we can take is that of the motion of a projectile under the action of gravity.

If q_1, q_2 be its co-ordinates the equation of vis viva may be written $\tfrac{1}{2}(q_1'^2 + q_2'^2) = -gq_2 + h$. Following the rule of Art. 468 we see that the Hamiltonian equation is $\tfrac{1}{2}(dV/dq_1)^2 + \tfrac{1}{2}(dV/dq_2)^2 = -gq_2 + h$. To solve this we notice that all the coefficients on the left side are constants and that U is a function of q_2 only. By Art. 476 we therefore assume $V = W + b_1 q_1$. Substituting and integrating we find W, so that finally $V = b_1 q_1 - \dfrac{1}{3g}(2h - b_1^2 - 2gq_2)^{\frac{3}{2}} + b_2$.

Following Jacobi's rule (Art. 470), the motion is given by

$$dV/db_1 = q_1 + \frac{b_1}{g}(2h - b_1{}^2 - 2gq_2)^{\frac{1}{2}} = -a_1 \Big\}$$
$$dV/dh = \quad -\frac{1}{g}(2h - b_1{}^2 - 2gq_2)^{\frac{1}{2}} = t + \epsilon \Big\}$$

These easily reduce to the ordinary formulæ for the motion of a projectile.

Ex. 3. A particle describes an orbit about a centre of force which attracts according to the law of nature. If r, θ be its polar co-ordinates referred to the centre of force as origin, show that the Hamiltonian equation is

$$(dV/dr)^2 + (dV/rd\theta)^2 = 2\mu/r + 2h.$$

Show also that a complete integral may be found (as in the last example) by putting $V = W + b\theta$.

Variation of the Elements.

478. Lagrange's Theorem. Let the co-ordinates of a system be $q_1, q_2, \ldots q_n$ and let the corresponding momenta be $p_1, p_2, \ldots p_n$. If the Hamiltonian function be

$$H = f(q_1 \ldots q_n, p_1 \ldots p_n, t)\ldots\ldots\ldots\ldots\ldots(1),$$

the equations of motion may be written in the typical form

$$p' = -\frac{dH}{dq}, \quad q' = \frac{dH}{dp}\ldots\ldots\ldots\ldots\ldots\ldots(2),$$

where accents denote differentiations with regard to the time.

Let two independent variations be given to these letters, which we shall represent by the symbols δ and Δ. We may imagine these to be produced by varying in two different ways the initial conditions.

$$\therefore \quad \delta H = \Sigma\left(\frac{dH}{dp}\delta p + \frac{dH}{dq}\delta q\right) = \Sigma(q'\delta p - p'\delta q)\ldots\ldots(3),$$

the time t not being varied. Performing the operation Δ on both sides of the equation, we have

$$\Delta\delta H = \Sigma(\Delta q'\delta p - \Delta p'\delta q + q'\Delta\delta p - p'\Delta\delta q)\ldots\ldots\ldots(4).$$

But reversing the order of the operations, we find

$$\delta\Delta H = \Sigma(\delta q'\Delta p - \delta p'\Delta q + q'\delta\Delta p - p'\delta\Delta q)\ldots\ldots\ldots(5).$$

Subtracting and remembering that $\delta\Delta = \Delta\delta$ we have

$$\Sigma(\Delta q'\delta p - \delta q'\Delta p - \Delta p'\delta q + \delta p'\Delta q) = 0.$$

Since both the operations Δ and δ are independent of d/dt, this gives

$$\frac{d}{dt}\Sigma(\Delta q\delta p - \Delta p\delta q) = 0 \ldots\ldots\ldots\ldots\ldots(6).$$

Thus the total differential with regard to t of the quantity summed is zero throughout the motion; that quantity is therefore constant.

Let us suppose that the co-ordinates q_1, &c. and their momenta

p_1, &c. have been found by solving the equations of motion and that each is expressed as a function of t and the constants of integration, say a, b, c &c. Let these constants receive any two independent variations, represented by δa, Δa, &c., the time not being varied, then the corresponding variations δq, Δq, &c. may be found by simple differentiation in terms of t, the constants a &c. and their variations. *The theorem asserts that, on substituting these in the expression*

$$\Sigma\,(\Delta q \delta p - \Delta p \delta q)\dots\dots\dots\dots\dots(7),$$

the time t will disappear from the result, so that the result is a function only of the constants and their variations.

Let t_0 be any time other than t and let $\alpha_1 \dots \alpha_n$, $\beta_1 \dots \beta_n$ be the values of p_1, &c. q_1, &c. at that time. For example we may let t_0 denote the time of the initial motion, and $\alpha_1 \dots \alpha_n$, $\beta_1 \dots \beta_n$ the initial values of the variables p_1 &c., q_1 &c. We then have

$$\Sigma\,(\Delta q \delta p - \Delta p \delta q) = \Sigma\,(\Delta \beta \delta \alpha - \Delta \alpha \delta \beta)\dots\dots\dots(8).$$

Lagrange deduces the theorem from his own general equations of motion, see page 304, Vol. 1, of the *Mécanique Analytique*. The proof just given is due to Boole; see *Cambridge Mathematical Journal*, Vol. 11, p. 100.

479. *Extension of Lagrange's Theorem.* In Lagrange's theorem the quantities q, $q + \Delta q$, $q + \delta q$ are *contemporary* values of the co-ordinate q. It is however sometimes convenient to vary the time also, just as in the calculus of variations we ascribe a variation to the abscissa as well as to the ordinate. Let then q, $q + \Delta q$, $q + \delta q$ represent the values of any co-ordinate in the undisturbed and varied motions at the times t, $t + \Delta t$, $t + \delta t$ respectively, where Δt and δt are any small arbitrary functions of the time. On this supposition we must alter Lagrange's theorem by writing $\Delta q - q'\Delta t$ and $\delta q - q'\delta t$ &c., for Δq and δq &c., see Art. 445. In the same way, if Δt_0 and δt_0 be the arbitrary changes in the initial time, we write $\Delta a - a'\Delta t$ &c. for Δa &c.

Let also H_0 represent the same function of t_0, $a_1 \dots a_n$, $\beta_1 \dots \beta_n$ that H is of t, $p_1 \dots p_n$, $q_1 \dots q_n$. Then, making these substitutions in (8) and remembering that

$$\Delta H = \Sigma\,(q'\Delta p - p'\Delta q) + H'\Delta t\dots\dots\dots\dots\dots(9),$$

with similar expressions for δH, ΔH_0 and δH_0, we find

$$\Sigma(\Delta q \delta p - \Delta p \delta q) + \Delta H \delta t - \Delta t \delta H = \Sigma\,(\Delta \beta \delta \alpha - \Delta \alpha \delta \beta) + \Delta H_0 \delta t_0 - \Delta t_0 \delta H_0 \dots(10).$$

If the geometrical equations do not contain the time explicitly, H is not a function of t and therefore $H = H_0 = h$. The equation (10) then becomes

$$\Sigma\,(\Delta q \delta p - \Delta p \delta q) + \Delta h \delta\,(t - t_0) - \Delta\,(t - t_0)\,\delta h = \Sigma\,(\Delta \beta \delta \alpha - \Delta \alpha \delta \beta)\dots\dots(11).$$

480. As an example of this theorem, let the symbol Δ represent simply d/dt. Then Δq is the difference between the values of the co-ordinate q in the undisturbed motion at the times t and $t + \Delta t$, no change being made in the initial conditions. It follows that $\Delta a = 0$, $\Delta \beta = 0$, $\Delta t_0 = 0$, $\Delta H_0 = 0$. Dividing equation (10) by Δt, we have therefore $\qquad \delta H = \Sigma\,(q'\delta p - p'\delta q) + H'\delta t,$
which is a symbolical method of writing the Hamiltonian equations.

In the same way we may let Δ represent differentiations with regard to some other letter. For example, we may regard H as the independent variable, and express p_1, &c., q_1, &c. and t in terms of H and the constants of integration; then taking Δ to represent d/dH, the constants not being varied, we obtain the Hamiltonian equations with t and H, p and q interchanged.

481. **Ex. 1.** Assuming $H = \frac{1}{2}p^2 - qt$, $H_0 = \frac{1}{2}a^2 - \beta t_0$, solve the Hamiltonian equations of motion and express p, q and H in terms of t and the initial values of p and q. Thence verify by substitution both Lagrange's variation theorem and the extension of that theorem.

Ex. 2. Let q_1, $q_2 ... q_n$ be the co-ordinates of a dynamical system and let the corresponding momenta be p_1, $p_2 ... p_n$. Taking these in pairs, let $(p_1 q_1)$, $(p_2 q_2)$,... be the Cartesian rectangular co-ordinates of n moving points P_1, P_2,..., P_n whose positions in space at the time t therefore determine the position of the system. Suppose that, when any two small arbitrary changes are given to the initial values of the p's and q's, these points take the positions Q_1, Q_2,...; R_1, R_2,... at the same time t. Prove that the sum of the area of the triangles $P_1 Q_1 R_1$, $P_2 Q_2 R_2$, &c. is constant throughout the motion.

Prove also that, if the Hamiltonian function H be expressed as a function of the Cartesian or polar co-ordinates of the points P_1, P_2,..., then H acts as a stream function, so that its partial differential coefficients give the resolved velocities of the points P_1, P_2,... in any directions.

Ex. 3. *Brassinne's extension of Lagrange's variation formula.* Supposing the Lagrangian function L to be a function of the typical variables q, q', q'' and the differential equations of motion to have the form

$$\frac{dL}{dq} - \frac{d}{dt}\frac{dL}{dq'} + \frac{d^2}{dt^2}\frac{dL}{dq''} = 0,$$

show that, when the time is not varied, Lagrange's variation formula becomes

$$(\Delta q \delta p - \Delta p \delta q) + (\Delta q' \delta r - \Delta r \delta q') + (\Delta r' \delta q - \Delta q \delta r') = \text{constant},$$

where $p = dL/dq'$, $r = dL/dq''$. *Liouville's Journal*, Tome XVI. 1851.

Brassinne deduces the result from Lagrange's equations, but it follows more easily from the corresponding Hamiltonian forms. Following Boole's method the result is arrived at by equating $\delta \Delta H$ and $\Delta \delta H$.

482. **Normal Transformations.** We have supposed that the constants $a_1 ... a_n$, $\beta_1 ... \beta_n$ are the values of the variables $p_1 ... p_n$, $q_1 ... q_n$ at some time $t = t_0$. But this restriction is not necessary. Let the $2n$ independent integrals of the equations of motion be

$$f_1(p_1 ... p_n, q_1 ... q_n, t) = f_1(a_1 ... a_n, \beta_1 ... \beta_n, t_0), \quad f_2(\&c.) = f_2(\&c.), \quad \&c. = \&c....(A).$$

It is evident that we may combine these together in an arbitrary manner so as to arrive at $2n$ other independent equations, which may equally serve as integrals. Thus supposing, we write

$$\left.\begin{array}{l} \phi_1(a_1 ... a_n, \beta_1 ... \beta_n) = a_1, \quad \phi_2(\&c.) = a_2, \quad \&c. \\ \psi_1(a_1 ... a_n, \beta_1 ... \beta_n) = b_1, \quad \psi_2(\&c.) = b_2, \quad \&c. \end{array}\right\} ....(B),$$

where a_1, &c., b_1, &c. are $2n$ new constants, the new forms of the integrals are obtained by eliminating $a_1 ... a_n$, $\beta_1 ... \beta_n$, between (A) and (B). The resulting forms contain t_0, but, if desired, we may eliminate t_0 also, either by giving it some definite value or by properly introducing it into the functions ϕ_1, &c., ψ_1, &c. The former course is the simpler of the two.

The only restriction on the arbitrary functions ϕ_1, &c. which it is necessary to make for our present purpose is that the variations of the two sets of constants should obey Lagrange's variation formula, viz.

$$\Sigma (\Delta b \delta a - \Delta a \delta b) = \Sigma (\Delta \beta \delta a - \Delta a \delta \beta) ....(12).$$

Supposing this to be the case, let H_0 be expressed in terms of the new constants and t_0. The extended Lagrangian variation formula then takes the form

$$\Sigma \left(\Delta q \delta p - \Delta p \delta q \right) + \Delta H \delta t - \Delta t \delta H - \Sigma \left(\Delta b \delta a - \Delta a \delta b \right) - \Delta H_0 \delta t_0 + \Delta t_0 \delta H_0 = 0 \ldots (13),$$

where the letters a_1, &c., b_1, &c. are either the values of the elements at some arbitrary time t_0, or some constants derived from them by a normal transformation; the terms containing δt_0 and δH_0 being omitted if the arbitrary time t_0 is not varied.

There are many ways of so choosing the relations between the two sets of constants that the variation formula (12) may hold. It will be presently proved that, K being any arbitrary function of the quantities $a_1 \ldots a_n$, $\beta_1 \ldots \beta_n$, the equation (12) is satisfied if the two sets are so related that each $b = dK/da$ and each $a = dK/d\beta$.

When quantities $(a_1$ &c.), $(\beta_1$ &c.) are changed into others $(a_1$ &c.), $(b_1$ &c.) by relations such that each $b = dK/da$ and each $a = dK/d\beta$, the *transformation has been called normal* by Donkin, see *Phil. Trans.* 1855. We shall however extend the meaning of this term to include all transformations which satisfy equation (12).

483. Conjugate elements. We notice that the elements or letters used in equations (10) or (13) run in pairs, so that in using the theorem it will be convenient to write them in two rows, thus :—

$$q_1, q_2, \ldots q_n, \quad H_0, \quad b_1, \quad b_2, \ldots \quad b_n, H,$$
$$p_1, p_2, \ldots p_n, \quad -t_0, \quad -a_1, \quad -a_2, \ldots -a_n, \quad t,$$

where one or both of the columns containing H, t ; H_0, t_0 are omitted when we do not wish to vary t or t_0. The letters or elements here placed in any column are usually called *conjugates*. If x, y be any two conjugates the equation (13) may be shortly written $\quad \Sigma \left(\Delta x \delta y - \Delta y \delta x \right) = 0 \quad \ldots \ldots \ldots \ldots \ldots \ldots \ldots \ldots (14)$.

We further notice that *Lagrange's theorem is not altered by interchanging any two conjugates provided we change one of their signs.* For instance we may write the letters in the order $\quad q_1, \ldots q_n, \quad H_0, \quad a_1, \ldots a_n, \quad H,$

$$p_1, \ldots p_n, \quad -t_0, \quad b_1, \ldots b_n, \quad t.$$

It is evident that the effect of the change of order in (a, b) is exactly counteracted by the change of sign.

484. Two ways of expressing the solutions. Supposing H to be a given function of p_1 &c., q_1 &c. and t, we can form the Hamiltonian equations of motion ; let these be solved and let the constants of integration be expressed in terms of either the initial elements at the time t_0 or the functions of them represented by a_1 &c., b_1 &c. In this way we have $2n$ equations connecting the variables p_1 &c., q_1 &c. with the $2n$ constant elements and the two times t and t_0. If necessary we may join to these the two equations connecting H and H_0 with the same letters. These $2n+2$ equations may be combined together in a great variety of ways and (with some exceptions) we may express any $2n+2$ of the letters in terms of the remaining $2n+2$ as independent variables. Two combinations are generally used, though others may be imagined.

(1) Suppose the elements written in two rows having conjugate elements in the same column, as in Art. 483, then the elements in either row may be regarded as functions of those in the other.

(2) Omitting the columns which contain H, t and H_0, t_0 and arranging the remaining columns so that the p's and q's are on one side of the middle vertical line and the a's and b's on the other; the letters on either side of the middle line may be regarded as functions of those on the other side together with t and t_0.

485. **Various Potential functions.** Writing the letters in the order

$$q_1, \ldots q_n, \quad b_1, \ldots \quad b_n, \quad t, \quad t_0,$$
$$p_1, \ldots p_n, \quad -a_1, \ldots -a_n, \quad -H, \quad H_0,$$

let the elements in the upper row be regarded as the independent variables. Let us choose the operation Δ so that the variation of every element in the upper row except one is zero ; let this one be q_r. The variations of the elements in the lower row due to Δ are not zero, but taking any one of them say p, $\Delta p = \Delta q_r \cdot dp/dq_r$. In the same way, we shall so choose the operation δ that the variation of every element in the upper row, except one, say q_s, is zero, then as before $\delta p = \delta q_s \cdot dp/dq_s$. The theorem expressed by equation (14) then becomes

$$\Delta q_r \delta p_r - \Delta p_s \delta q_s = 0.$$

It immediately follows that $\dfrac{dp_r}{dq_s} = \dfrac{dp}{dq_r}$.

By interchanging conjugate elements and changing the sign of one of them we may obtain a number of similar equations. In whichever of these orders the rows are written, it follows that, *if the elements in either row are independent, the differential coefficients of any two dependent elements, each taken with regard to the conjugate of the other, are equal.*

486. The equality of these differential coefficients expresses the fact that

$$p_1 dq_1 + \ldots + p_n dq_n - a_1 db_1 - \&c. - H dt + H_0 dt_0$$

is a perfect differential of some function of the co-ordinates $q_1 \ldots q_n$, $b_1 \ldots b_n$, t and t_0. If S be this function we have the typical equations

$$p = \frac{dS}{dq}, \quad -a = \frac{dS}{db}, \quad -H = \frac{dS}{dt}, \quad H_0 = \frac{dS}{dt_0}.$$

In the same way, if we interchange the conjugate elements $(-H, t)$, $(H_0 t_0)$ and give the proper change of sign we see that

$$p_1 dq_1 + \&c. - a_1 db_1 - \&c. + t dH - t_0 dH_0$$

is a perfect differential of some function of the co-ordinates q_1 &c., b_1 &c., H and H_0. If V be this function we have

$$p = \frac{dV}{dq}, \quad -a = \frac{dV}{db}, \quad t = \frac{dV}{dH}, \quad -t_0 = \frac{dV}{dH_0}.$$

To discover the meanings of the functions here called S and V we recall the letters L and H as defined in Art. 442. Putting L for the Lagrangian function and remembering that H is its reciprocal, (Vol. I. Art. 410), we have $L + H = \Sigma pq'$. From the equation giving the total differential of S we have $dS/dt = \Sigma pq' - H = L$. If the constant elements are the initial values of q_1 &c., p_1 &c., we have in the same way, $dS/dt_0 = -L_0$, where L_0 is the initial value of L. We therefore have $S = \int L dt$ where the limits are $t = t_0$ and $t = t$.

Again, comparing the total differentials of S and V, we see that

$$d(S - V) = -d(Ht) + d(H_0 t_0),$$

whence $S = V - Ht + H_0 t_0$. This leads to the same value of V as that given in Art. 443.

487. It is evident that we may obtain a variety of functions besides S and V which possess analogous properties. We have only to interchange two conjugate elements, changing the sign of one of them, and a new function may be deduced at once from the new arrangement. The relations between these functions may be put more generally as follows.

Let any two series of variables be represented by the two rows

$$x_1, \quad x_2, \ldots x_n$$
$$\xi_1, \quad \xi_2, \ldots \xi_n.$$

Firstly, let each element ξ_r in the lower row be obtained by differentiating some function A_1 of the elements in the upper line with regard to the conjugate viz. x_r.

This series of equations we may write typically $\xi = \dfrac{dA_1}{dx}$.

Then $$\Delta A_1 = \Sigma \xi \Delta x,$$
$$\therefore \; \delta \Delta A_1 = \Sigma (\delta \xi \Delta x + \xi \delta \Delta x).$$
Similarly $$\Delta \delta A_1 = \Sigma (\Delta \xi \delta x + \xi \Delta \delta x).$$

Equating these results exactly as in Art. 478, we have $\Sigma (\Delta x \delta \xi - \Delta \xi \delta x) = 0$, which corresponds to Lagrange's theorem.

It follows by the same reasoning as in Art. 485 that each x is the differential coefficient of some function A_2 of the elements in the lower line with regard to its conjugate viz. ξ. Thus $x = \dfrac{dA_2}{d\xi}$.

Since $dA_1 = \Sigma \xi dx$ and $dA_2 = \Sigma x d\xi$ we have by addition and integration $A_1 + A_2 = \Sigma x\xi$.

Hence A_1 and A_2 are *reciprocal functions* according to the definition given in Vol. I. Art. 410.

Let us next reverse the order of one of the conjugate elements, writing the scheme in the form $x_1, x_2 \ldots x_{n-1}, \xi_n,$

$$\xi_1, \xi_2 \ldots \xi_{n-1}, -x_n,$$

we then have $\xi_r = dB_n/dx_r$ from $r = 1$ to $r = n - 1$, and $-x_n = dB_n/d\xi_n$ where B_n is some function of the elements $x_1 \ldots x_{n-1}$ and ξ_n. Since

$$dB_n = \xi_1 dx_1 + \ldots + \xi_{n-1} dx_{n-1} - x_n d\xi_n,$$

we have $B_n = A_1 - x_n \xi_n$. Referring to Vol. I. Art. 418 we see that B_n *is the modified function of A_1 for the conjugate set* (x_n, ξ_n).

488. We may now express in a convenient manner the relation between the constant elements $a_1 \ldots a_n$, $b_1 \ldots b_n$ and the initial values of $p_1 \ldots p_n$, $q_1 \ldots q_n$. Putting $a_1 \ldots a_n$, $\beta_1 \ldots \beta_n$ for these initial values, we have by Art. 482

$$\Sigma (\Delta b \delta a - \Delta a \delta b) - \Sigma (\Delta \beta \delta a - \delta \beta \Delta a) = 0.$$

If then we write the letters in the order

$$a_1, \ldots a_n, \quad a_1, \ldots \quad a_n,$$
$$b_1, \ldots b_n, \quad -\beta_1, \ldots -\beta_n,$$

each letter in either row is the differential coefficient with regard to its conjugate of some function. Thus, if K be any arbitrary function of the letters in the upper row, we have $b = dK/da$ and $-\beta = dK/da$. Other orders of the letters give other rules.

489. **Canonical elements.** We shall now return to Lagrange's equation and show how we may arrive at another set of relations by treating it in a different manner.

Writing the letters in the order

$$p_1, p_2 \ldots p_n \mid b_1, b_2 \ldots b_n$$
$$q_1, q_2 \ldots q_n \mid a_1, a_2 \ldots a_n$$

we shall regard the elements on one side of the vertical bar as functions of those on the other together with t and t_0. As we are about to use Lagrange's variation theorem the constants must be either the initial values of the variables or those derived from them by a normal transformation. Since the time will not be varied in what immediately follows the presence of t or t_0 is not material.

We shall now prove that *the partial differential coefficient of an element in one row on one side of the bar with regard to any element in the other row on the other side of the bar is equal to the partial differential coefficient of the conjugate of the latter with regard to the conjugate of the former.*

To prove this we use Lagrange's theorem. Let the symbol Δ mean that the variation of every letter on the left-hand side except p_r is zero, so that Δ represents $\Delta p_r \cdot d/dp_r$. Let δ mean that the variation of every letter on the right-hand side except b_s is zero, so that δ represents $\delta b_s \cdot d/db_s$. We then have $\Delta p_r \delta q_r - \Delta a_s \delta b_s = 0$,

$$\therefore \frac{dp_r}{da_s} = \frac{db_s}{dq_r},$$

which proves the theorem.

If we interchange the conjugates on the right-hand side of the vertical bar, changing the signs of one of the rows, we deduce at once $\dfrac{dp_r}{db_s} = -\dfrac{da_s}{dq_r}$.

The method of deriving the equality of these differential coefficients from Lagrange's theorem is due to Donkin.

490. We shall now introduce a new symbol due to Poisson. Let u, v be any two functions of the variables $p_1 \ldots p_n$, $q_1 \ldots q_n$, then

$$(u, v) = \Sigma \left(\frac{du}{dp_i} \frac{dv}{dq_i} - \frac{du}{dq_i} \frac{dv}{dp_i} \right),$$

where the summation is to be taken for all values of i from $i=1$ to $i=n$. We may also include the conjugate elements (H, t) if u, v are functions of H or t, but this term is not to be included unless it is expressly mentioned. In using the abridged notation (u, v) the order of the letters is to be attended to. The first factor on the right-hand side is du/dp not du/dq.

There is another summation which Lagrange has represented by the same symbol. To prevent confusion we shall slightly alter its form. Let u and v be two quantities of which the variables p_1, &c., q_1, &c. are functions, then

$$[u, v] = \Sigma \left(\frac{dp_i}{du} \frac{dq_i}{dv} - \frac{dq_i}{du} \frac{dp_i}{dv} \right),$$

where the summation is to be taken for all values of i, the denominators u, v being the same in every term.

We shall again apply Lagrange's variation theorem with new meanings to the operators Δ and δ. Considering the letters on the right hand of the vertical bar as the independent variables, Art. 489, let Δ denote differentiation with regard to b_s and δ differentiation with regard to a_r. We then have (Art. 478)

$$[a_r, b_s] = \Sigma_i \left(\frac{dp_i}{da_r} \frac{dq_i}{db_s} - \frac{dp_i}{db_s} \frac{dq_i}{da_r} \right) = 0 \text{ or } 1,$$

the right-hand side being zero or unity according as a_r, b_s are not or are conjugate elements.

It has already been shown that

$$\frac{dp_i}{db_s} = -\frac{da_s}{dq_i}, \qquad \frac{dp_i}{da_r} = \frac{db_r}{dq_i}, \qquad \frac{dq_i}{da_r} = -\frac{db_r}{dp_i}, \qquad \frac{dq_i}{db_s} = \frac{da_s}{dp_i}.$$

Substituting these values $(a_s, b_r) = \Sigma_i \left(\dfrac{da_s}{dp_i} \dfrac{db_r}{dq_i} - \dfrac{db_r}{dp_i} \dfrac{da_s}{dq_i} \right) = 0$ or 1,

the right-hand side being zero or unity according as a_r, b_s are not or are conjugate elements. In the same way $(b_r, b_s) = 0$, $(a_r, a_s) = 0$.

When the dynamical equations have been solved we have $2n$ equations giving the values of $(p_1$ &c.$)$, $(q_1$ &c.$)$ in terms of t and the constants $(a_1$ &c.$)$, $(b_1$ &c.$)$ of integration. If these constants are so chosen as to be the initial values of $(p_1$ &c.$)$, $(q_1$ &c.$)$, or if they are any constants derived from them by a normal transformation, we have just proved　　　　　　　　　　$(a, b) = 0$ or 1 ....(I).

But if the constants are merely those introduced at each integration it may happen that they do not satisfy the above relations. To distinguish these cases, *the constants are called canonical when they are so arranged that they satisfy the relations* (I).

491. There is another way of proving the relations (I) which has the advantage of showing how closely they are connected with those already proved in Art. 487.

Let two sets of conjugate variables be represented in the two rows

$$x_1, \ x_2 \ldots x_n$$
$$\xi_1, \ \xi_2 \ldots \xi_n,$$

and let them be connected together by the n relations

$$F_1 (x_1 \ldots x_n, \ \xi_1 \ldots \xi_n) = 0, \qquad F_2 = 0, \qquad \&c. = 0.$$

Then the $\tfrac{1}{2}n(n-1)$ equations typically represented by $d\xi_r/dx_s = d\xi_s/dx_r$ for all values of r and s are equivalent to the $\tfrac{1}{2}n(n-1)$ equations represented by

$$(F_r, \ F_s) = \Sigma \left(\frac{dF_r}{dx_i} \frac{dF_s}{d\xi_i} - \frac{dF_s}{dx_i} \frac{dF_r}{d\xi_i} \right) = 0,$$

where the Σ implies summation for all values of i. If either of these sets of equations is given, the other follows from it. A proof of this theorem is given in Forsyth's *Differential Equations* under the heading *Jacobi's general method*, Art. 211.

To apply this theorem to our present purpose we let the two sets of variables be

$$q_1 \ldots q_n, \ a_1 \ldots a_n,$$
$$p_1 \ldots p_n, \ b_1 \ldots b_n,$$

so that x corresponds to any letter in the upper row and ξ to any in the lower row. Let the $2n$ relations between these be

$$F_1 = \phi_1 - a_1 = 0, \qquad F_2 = \phi_2 - a_2 = 0, \ \&c.$$
$$F_{n+1} = \phi_{n+1} - b_1 = 0, \qquad F_{n+2} = \phi_{n+2} - b_2 = 0, \ \&c.$$

where each ϕ is some function of the variables $q_1 \ldots q_n$, $p_1 \ldots p_n$ and t. Then remembering that the sum $(F_r, \ F_s)$ extends over all the conjugate letters, while the sum $(\phi_r, \ \phi_s)$ extends only over the p's and q's, we have $(F_r, \ F_s) = (\phi_r, \ \phi_s) + 1$ or 0, where $r > s$ and 1 or 0 is taken according as the elements b_r, a_s are or are not conjugate. Substituting for ϕ_r, ϕ_s, the equation $(F_r, \ F_s) = 0$ may be written in the form $(b, a) = -1$ or 0.

It follows that the two following statements are equivalent to each other, viz.

(1)　　　　　$p_1 dq_1 + \ldots + b_1 da_1 + \ldots = $ a perfect differential ;

(2)　　the constants are such that (a, b) is unity or zero according as a and b are or are not conjugate.

It has also been shown in Art. 487 that the first statement is equivalent to the following

(3)　　　　　$\Sigma (\Delta p \delta q - \Delta q \delta p) - \Sigma (\Delta a \delta b - \Delta b \delta a) = 0.$

By giving the p's and q's their initial values in the third statement we see that when the constants are canonical, i.e. when the second statement holds, they can be derived from the initial values by a normal transformation.

It appears also that, when the constants a, b are any two of a canonical set, the summations represented by (a, b) and $[a, b]$ are equal.

492. **Ex.** *Helmholtz's Theorem.* The natural motion of a conservative system would carry it from a position A to a position B in a time t; the system would also describe the reversed motion from B to A in the same time. Let its co-ordinates and momenta at A and B be respectively $b_1 \ldots b_n$, $a_1 \ldots a_n$ and $q_1 \ldots q_n$, $p_1 \ldots p_n$. Suppose that in passing through the position A the system receives some small impulse, so that the momentum a_r is increased by δa_r, all the other elements being unchanged, and that the co-ordinates after a time t are in consequence altered by $\delta q_1, \ldots \delta q_n$. Suppose again that, when passing through the position B in the reversed motion from B to A, a small impulse is given to the system by which the momentum p_s is increased by Δp_s, and let $\Delta b_1, \ldots \Delta b_n$ be the corresponding changes in the co-ordinates after a time t. Then $\delta q_s / \delta a_r = \Delta b_r / \Delta p_s$. *Crelle's Journal,* Vol. 100.

Prof. Lamb in commenting on this theorem gives a number of applications to Acoustics, Optics, &c. See *Reciprocal Theorems in Dynamics,* Vol. XIX. *of the Proceedings of the London Mathematical Society,* 1888.

493. **Poisson's Theorem.** *If any two integrals of the equations of motion are written in the forms* $c_1 = \phi_1 (p_1 \ \&c., \ q_1 \ \&c. \ t)$, $c_2 = \phi_2 (p_1 \ \&c., \ q_1 \ \&c. \ t)$, *then regarding c_1 and c_2 as functions of p_1, &c., q_1, &c., t being constant, the quantity (c_1, c_2) is constant throughout the motion.*

Since there cannot be more than the proper number of integrals of the equations of motion, it must be possible to derive these two from the $2n$ integrals with the initial values for the arbitrary constants. If $(a_1 \ \&c.)$, $(\beta_1 \ \&c.)$ be these initial values, we have therefore $c_1 = f(a_1 \ \&c., \ \beta_1 \ \&c.)$, $c_2 = F(a_1 \ \&c., \ \beta_1 \ \&c.)$, where $(a_1 \ \&c.)$, $(\beta_1 \ \&c.)$ are to be regarded as known functions of $(p_1 \ \&c.)$, $(q_1 \ \&c.)$.

Now $\dfrac{dc_1}{dp} = \dfrac{df}{da_1}\dfrac{da_1}{dp} + \dfrac{df}{da_2}\dfrac{da_2}{dp} + \&c.$, $\dfrac{dc_2}{dq} = \dfrac{dF}{da_1}\dfrac{da_1}{dq} + \dfrac{dF}{da_2}\dfrac{da_2}{dq} + \&c.$

$\therefore \dfrac{dc_1}{dp}\dfrac{dc_2}{dq} - \dfrac{dc_1}{dq}\dfrac{dc_2}{dp} = \Sigma \left(\dfrac{df}{da_1}\dfrac{dF}{da_2} - \dfrac{df}{da_2}\dfrac{dF}{da_1} \right) \left(\dfrac{da_1}{dp}\dfrac{da_2}{dq} - \dfrac{da_1}{dq}\dfrac{da_2}{dp} \right).$

$\therefore \ (c_1, c_2) = \Sigma \left(\dfrac{df}{da_1}\dfrac{dF}{da_2} - \dfrac{df}{da_2}\dfrac{dF}{da_1} \right)(a_1, a_2).$

Since the integrals a_1, a_2 &c. are canonical, $(a_1, a_2) = 0$ or ± 1. Also their coefficients in this series are all functions of a_1, a_2 &c. and are therefore constants. It follows that (c_1, c_2) is constant throughout the motion.

It follows from Poisson's theorem that *whenever two integrals, say $c_1 = \phi$, $c_2 = \psi$, of the differential equations are known, the relation $c_3 = (\phi, \psi)$ must be a third integral of the equations of motion, or an identity, or deducible from the two integrals already known.*

494. *Another proof.* Since the integral $c_1 = \phi_1 (p_1 \ \&c. \ q_1 \ \&c. \ t)$ satisfies the Hamiltonian equations, we shall obtain an identical result if we differentiate it totally and substitute for p' and q' their values given by the Hamiltonian equations.

We thus obtain $0 = \Sigma \left(- \dfrac{dc_1}{dp}\dfrac{dH}{dq} + \dfrac{dc_1}{dq}\dfrac{dH}{dp} \right) + \dfrac{dc_1}{dt}$ .... (1).

This equation may also be written in the compact form $0 = (H, c_1) + dc_1/dt$ and expresses the condition that $c_1 = \phi$ (&c.) is an integral of the equations of motion.

Let $A = \Sigma \left[\dfrac{dc_1}{dq_s}\dfrac{dc_2}{dp_s} - \dfrac{dc_1}{dp_s}\dfrac{dc_2}{dq_s} \right]$, we have to prove that A, being regarded as a function of p_1, q_1, &c. and t, the total differential coefficient $d.A/dt$ is zero. Now

$$\frac{d.A}{dt} = \frac{dA}{dt} + \Sigma \left\{ \frac{dA}{dp_r}\,p_r' + \frac{dA}{dq_s}\,q_r' \right\} .$$

The letters p_1, q_1, &c. enter into the expression for A only through c_1 and c_2. Let us consider only the part of $d \cdot A/dt$ due to the variation of c_1, then the part due to the variation of c_2 may be found by interchanging c_1 and c_2, and changing the sign of the whole. The complete value of $d \cdot A/dt$ is the sum of these two parts.

The part of $d \cdot A/dt$ due to the variation of c_1 is

$$\Sigma \left[\frac{dc_2}{dp_s} \left\{ \frac{d}{dq_s} \frac{dc_1}{dt} - \frac{d^2c_1}{dp_r dq_s} \frac{dH}{dq_r} + \frac{d^2c_1}{dq_r dq_s} \frac{dH}{dp_r} \right\} - \frac{dc_2}{dq_s} \left\{ \frac{d}{dp_s} \frac{dc_1}{dt} - \frac{d^2c_1}{dp_r dq_s} \frac{dH}{dq_r} + \frac{d^2c_1}{dq_r dp_s} \frac{dH}{dp_r} \right\} \right].$$

If we substitute for dc_1/dt its value given by the identity (1), we get

$$\Sigma \left[\frac{dc_2}{dp_s} \left\{ \frac{dc_1}{dp_r} \frac{d^2H}{dq_s dq_r} - \frac{dc_1}{dq_r} \frac{d^2H}{dp_r dq_s} \right\} - \frac{dc_2}{dq_s} \left\{ \frac{dc_1}{dp_r} \frac{d^2H}{dp_s dq_r} - \frac{dc_1}{dq_r} \frac{d^2H}{dp_r dp_s} \right\} \right].$$

If we now interchange c_1 and c_2 we get the same result. Hence when the two parts of $d \cdot A/dt$ are added together, the signs being opposite, the sum is zero.

495. Examples. Ex. 1. If $c_1 = H$ is the equation of vis viva and $c_2 = \phi_2$ (&c.), is any other integral not containing t, prove that (c_1, c_2) is identically zero. But if the integral c_2 contain t and is written in the form $c_2 = \phi_2$ (&c.) $- t$, then (c_1, c_2) is identically unity. [Bertrand.]

The results follow from $(H, c) + dc/dt = 0$.

Ex. 2. If $c_1 = \phi_1$ (&c.) be any integral not containing t, there must be at least one other integral $c_2 = \phi_2$ (&c.) such that (c_1, c_2) is not zero.

For if possible let $(c_1, c_i) = 0$ for all integrals $c_1 ... c_{2n}$. This equality may be regarded as a differential equation to find c_i, and it must comprehend all the solutions of $(H, c_i) = 0$, since this last equation expresses the fact that c_i is an integral of the equations of motion not containing t explicitly. But two linear equations having the same number of variables cannot have the same integrals unless they are identical. Hence c_1 or ϕ_1 is a function of H and the given integral is the equation of vis viva. But if c_1 is the equation of vis viva there is an integral which, combined with it, gives the result unity, viz. that one in which the constant is joined to the time. [Bertrand.]

Ex. 3. If $c_1 = \phi_1$ (&c.), $c_2 = \phi_2$ (&c.) are two *different* integrals and are each derived from the same two integrals, a_1, a_2, taken from a canonical set, then (c_1, c_2) is finite or zero according as a_1, a_2 are conjugates or not. See Art. 493.

496. We shall now prove that *the constants introduced in Jacobi's complete integral form a canonical set*. Referring to Art. 470, we see that if the elements are written according to the scheme

$$q_1 ... q_n, \quad h \quad , \quad b_1 ... b_{n-1},$$
$$p_1 ... p_n, \quad t + \epsilon, \quad - a_1 ... - a_{n-1},$$

each element in the lower row is the partial differential coefficient with regard to its conjugate of a function f. It follows therefore that Lagrange's theorem applies to this scheme of elements when we treat $t + \epsilon$ as one of them, Art. 487. But, when the elements on the right-hand side are regarded as functions of those on the left, Lagrange's theorem supplies all that is necessary to obtain the relations $(a, b) = 0$ or 1. Since t and ϵ enter in the form of the sum $t + \epsilon$, these relations reduce to

$$(a, b) = 0 \text{ or } 1, \quad (b, \epsilon) = 0, \quad (h, \epsilon) = 1.$$

The constants are therefore canonical. This theorem is given by Donkin.

Ex. Taking the example of the motion of a projectile given in Art. 477, show that the four integrals deduced from Jacobi's complete integral are

$$- a_1 = q_1 + p_1 p_2/g, \qquad b_1 = p_1,$$
$$2h = p_1^2 + p_2^2 + 2gq_2, \qquad t + \epsilon = - p_2/g.$$

Verify that these constants are canonical.

497. Ex. *Bertrand's Theorem.* Let $a = \phi\,(p_1$ &c., q_1 &c., $t)$ be an integral of the equations of motion and let β, γ, δ be three others of the same kind. Form the determinant in which the first row is da/dp_r, da/dq_r, da/dp_s, da/dq_s and the three other rows are deduced from the first by writing β, γ, δ for a. Let $(a, \beta, \gamma, \delta)$ represent the sum of these determinants for all values of r and s. Prove that $(a, \beta, \gamma, \delta)$ is constant throughout the motion. [*Comptes Rendus*, 1852.

Brioschi gives a short proof of this by expanding $(a, \beta, \gamma, \delta)$ in a series of determinants each of two rows. The expansion is

$$2\,(a, \beta)\,(\gamma, \delta) + 2\,(a, \gamma)\,(\delta, \beta) + 2\,(a, \delta)\,(\beta, \gamma),$$

which is constant by Poisson's theorem. If the constants are canonical this reduces to 2 or 0, according as there are or are not two pairs of conjugate elements. He also shows that

$$(a, \beta, \gamma, \delta, \eta, \xi) = 3\,(a, \beta)\,(\gamma\;\delta\;\eta\;\xi) + 3\,(a, \gamma)\,(\beta\;\delta\;\xi\;\eta)$$
$$+ 3\,(a, \delta)\,(\beta\;\gamma\;\eta\;\xi) + 3\,(a, \eta)\,(\beta\;\gamma\;\xi\;\delta) + 3\,(a, \xi)\,(\beta\;\gamma\;\delta\;\eta).$$

Tortolini, *Annali di Scienze matematiche e fisiche*, Vol. IV., 1853.

498. **Properties of** (u, v). As the symbol (u, v) has considerable importance in theoretical dynamics, it will be found useful to notice the following properties.

(1) $(u, v) = -(v, u)$.

(2) $(u, u) = 0$.

(3) $(p_i, q_i) = 1$ and $(p_i, q_j) = 0$.

(4)   Let $U = f(u_1, u_2 \ldots u_n)$, $V = F(u_1, u_2 \ldots u_n)$, where u_1, &c. are functions of the elements $(p_1,$ &c.$)$, $(q_1,$ &c.$)$.   Then

$$(U, V) = \Sigma\left(\frac{dU}{du_r}\frac{dV}{du_s} - \frac{dU}{du_s}\frac{dV}{du_r}\right)(u_r, u_s),$$

where Σ implies summation for all values of r and s. Bertrand, see notes to the *Mécanique Analytique* of Lagrange, 1853.

(5)   The following is a more general theorem.

Let
$$U = f\,(p_1 \ldots p_n,\; q_1 \ldots q_n,\; u_1 \ldots u_n),$$
$$V = F\,(p_1 \ldots p_n,\; q_1 \ldots q_n,\; u_1 \ldots u_n).$$

Then
$$(U, V) = (U;\; V) + \Sigma\left\{\frac{dU}{du_r}(u_r,\; V) + \frac{dV}{du_r}(U,\; u_r)\right\} + R,$$

where $(U;\; V)$ is partial with regard to p and q, and R stands for the result given in Theorem 4. This theorem is given by Imschenetsky, see the translation from Russian into French in *Grunert's Archiv*, 1869.

(6)   If u and v are functions of $(p_1$ &c.$)$, $(q_1$ &c.$)$ and any letter x, it follows by the rule for differentiating determinants that

$$\frac{d}{dx}\,(u, v) = \left(\frac{du}{dx},\; v\right) + \left(u,\; \frac{dv}{dx}\right).$$

Proceeding as in Leibnitz's theorem, we have

$$\frac{d^n}{dx^n}\,(u, v) = \left(\frac{d^n u}{dx^n},\; v\right) + n\left(\frac{d^{n-1}u}{dx^{n-1}},\; \frac{dv}{dx}\right) + n\,\frac{n-1}{2}\left(\frac{d^{n-2}u}{dx^{n-2}},\; \frac{d^2 v}{dx^2}\right) + \&c.$$

Imschenetsky.

(7)   If u, v, w are three functions of the variables, then

$$(u, (v, w)) + (v, (w, u)) + (w, (u, v)) = 0.$$

Jacobi. *Crelle's Journal*, LX. p. 42. A proof is given in Forsyth's *Differential Equations*.

499. Transformation of Co-ordinates. The Hamiltonian equations of motion may be written in the typical form

$$p' = - dH/dq, \qquad q' = dH/dp \dots\dots\dots\dots\dots\dots(1).$$

If we now change the co-ordinates $q_1 \dots q_n$ to others $Q_1 \dots Q_n$ connected with the former by equations of the form $q = f(Q_1 \dots Q_n)$, we know from dynamical considerations that the transformed equations take the typical form

$$P' = - dH/dQ, \qquad Q' = dH/dP \dots\dots\dots\dots\dots\dots(2),$$

where $P_1 \dots P_n$ are the momenta corresponding to $Q_1 \dots Q_n$ respectively and may be derived from the transformed value of the vis viva by the same rules as before.

In order to generalize this, let us enquire whether we can find any more general transformation, such as

$$q_1 = f_1(Q_1 \dots Q_n, P_1 \dots P_n), \&c \dots\dots(3), \qquad p_1 = F_1(Q_1 \dots Q_n, P_1 \dots P_n), \&c \dots\dots(4),$$

so that the Hamiltonian equations (1) when transformed will take the form (2). We suppose that H is any given function of $(p_1, \&c.)$, $(q_1, \&c.)$ and of t, but that the formulæ of transformation (3) and (4) do not contain t explicitly.

Since the Hamiltonian equations (Art. 479) may be written in the form

$$\Sigma (\Delta q \delta p - \Delta p \delta q) + \Delta H \delta t - \Delta t \delta H = \text{constant} \dots\dots\dots\dots(5),$$

it is clear that the transformation can be effected if we take

$$\Sigma (\Delta q \delta p - \Delta p \delta q) = \Sigma (\Delta Q \delta P - \Delta P \delta Q) \dots\dots\dots\dots(6),$$

where Δ and δ have the meanings given to them in Art. 478.

If we write the letters according to the scheme

$$p_1 \dots p_n, \qquad - P_1 \dots - P_n,$$
$$q_1 \dots q_n, \qquad Q_1 \dots \quad Q_n,$$

we can infer from Art. 487 the following rule, originally due to Jacobi, (see his *Dynamik*) : *assume any arbitrary function,* $\psi (q_1 \dots q_n, Q_1 \dots Q_n)$, *of the given co-ordinates and of the new set to be introduced, then the required relations* (3) *and* (4) *are equivalent to the typical relations* $p = d\psi/dq$ *and* $- P = d\psi/dQ$.

Other rules may be obtained by interchanging the conjugate elements with the necessary change of sign. Thus taking the order

$$q_1 \dots q_n, \qquad P_1 \dots P_n,$$
$$p_1 \dots p_n, \qquad Q_1 \dots Q_n,$$

we may obtain transformation formulæ equivalent to (3) and (4) by *putting* $q = d\psi/dp$ *and* $P = d\psi/dQ$ *where* ψ *is an arbitrary function of* $p_1 \dots p_n$, $Q_1 \dots Q_n$. This rule is also given by Jacobi, see the *Comptes Rendus*, 1837, Tome v. p. 66.

500. Examples. Ex. 1. Let us choose the arbitrary function ψ to be

$$\psi = p_1 f_1 (Q_1 \dots Q_n) + p_2 f_2 (Q_1 \dots Q_n) + \dots \quad\dots\dots\dots(1).$$

We then find by Jacobi's second rule that the required formulæ of transformation are $\qquad q_i = f_i (Q_1 \dots Q_n) \dots\dots(2), \qquad P_i = p_1 df_1/dQ_i + p_2 df_2/dQ_i + \dots \quad \dots\dots\dots(3).$

These are the ordinary formulæ of transformation when we change from one set of co-ordinates $q_1 \dots q_n$ to another $Q_1 \dots Q_n$.

By remembering the definition of p_1, p_2, &c. and noticing that Q_1', Q_2', &c., do not enter into (2) we easily find that

$$dT/dQ_i' = p_1 dq_1'/dQ_i' + p_2 dq_2'/dQ_i' + \&c.$$

This by differentiating (2) is seen to lead to the right-hand side of (3). It therefore follows that in this case P_i is the momentum corresponding to the co-ordinate Q_i.

Ex. 2. A system depends on two pairs of elements, viz. (p_1, q_1) (p_2, q_2); taking

Jacobi's arbitrary function to be $2\beta\psi = (q_1 - Q_1)^2 + (q_2 - Q_2)^2$ find the formulæ of transformation and examine what they become when $\beta = 0$.

Ex. 3. *Donkin's rule.* In Jacobi's rule the arbitrary function ψ is not to contain t explicitly. If we suppose ψ to be an arbitrary function of p_1, &c., Q_1, &c. and t, prove that the transformation formulæ typically written $q = d\psi/dp$, $P = d\psi/dQ$ will change the differential equations into others still of the Hamiltonian form but with $H - d\psi/dt$ written for H. [*Phil. Trans.* 1885.

Let x be such a function of the variables and t that the equation (6) Art. 499 is true after the addition of $\Sigma (\Delta t \delta x - \Delta x \delta t)$ to its right-hand side. The possibility of this assumption is proved by finding the proper form for x. The second scheme is then altered by the addition of another set of elements, viz. x to the upper and t to the lower line. It then follows by the same reasoning as before that $x = d\psi/dt$ and conversely. The equation (5) then shows that $H - x$ must be written for H in the Hamiltonian equations.

Ex. 4. *Mathieu's rule.* If the variables $(p_1$, &c.$)$, $(q_1$, &c.$)$ are changed into $(P_1$, &c.$)$, $(Q_1$, &c.$)$ by relations such that $\Sigma p \delta q = \Sigma P \delta Q$, prove that the Hamiltonian equations when so transformed retain the Hamiltonian form. Thence deduce the following rule to obtain a set of transformation formulæ. Assume any arbitrary function of the old and new co-ordinates say $\psi (q_1$, &c., Q_1, &c.$)$ and equate it to zero. The required relations may be typically written $p = \mu d\psi/dq$ and $- P = \mu d\psi/dQ$. We thus have $2n + 1$ equations to find $(P_1$, &c$)$, $(Q_1$, &c.$)$ and μ. *Liouville, XIX., 1874.*

To prove the first part of this theorem Mathieu remarks that the Hamiltonian equations may be written in the form

$$\delta H = \Sigma \{\delta (pq') - d/dt (p\delta q)\}.$$

Hence if we choose $\Sigma p \delta q = \Sigma P \delta Q$ for all variations the Hamiltonian form is unchanged.

To prove the second part, Mathieu notices that the equation $\Sigma p \delta q = \Sigma P \delta Q$ leads to $2n$ equations which may be typically written

$$p_1 \frac{dq_1}{dQ_i} + p_2 \frac{dq_2}{dQ_i} + \&c. = P_i \ldots\ldots (\text{I.}), \qquad p_1 \frac{dq_1}{dP_i} + p_2 \frac{dq_2}{dP_i} + \&c. = 0 \ldots\ldots (\text{II.}),$$

where i has any value from 1 to n. The set (II.) shows, by elimination, that the Jacobian of $q_1 \ldots q_n$ with regard to $P_1 \ldots P_n$ is zero. Hence the n equations (3) of Art. 499 are such that, if we eliminate $n - 1$ of the P's, the nth will also disappear, and leave an equation containing only $q_1 \ldots q_n$ and $Q_1 \ldots Q_n$. This is the equation he calls $\psi = 0$. Differentiating $\psi = 0$ with regard to $P_1 \ldots P_n$ in turn, the equations (II.) show that $p_i = \mu d\psi/dq_i$. Then, substituting in (I.) it follows that $P_i = - \mu d\psi/dQ_i$. It may be noticed that the unknown quantity μ is not restricted to be a function of $q_1 \ldots q_n$, $Q_1 \ldots Q_n$ only.

501. The use of changing the variables p_1 &c., q_1 &c. into others P_1 &c., Q_1 &c. is that if the arbitrary function ψ is properly chosen the expression for H can be simplified, while the Hamiltonian form of the differential equations is still retained. The letters (P_1, Q_1), (P_2, Q_2) &c. retain their dual character so far as the differential equations are concerned, but it does not follow that P represents the momentum corresponding to Q. Other conditions must be satisfied that this may be true.

Ex. 1. Supposing H not to be an explicit function of t and the formulæ of transformation to contain only the old and new variables without t, show that the semi vis viva T and the force function U may be expressed in terms of Q_1, Q_2 &c.

and Q_1', Q_2' &c. Show also that if each P is to be the momentum corresponding to each Q, i.e. if $P=dL/dQ'$ is to be true, it is necessary and sufficient that

$$\Sigma p\delta q = \Sigma P\delta Q.$$

We notice that, if L and M are the reciprocal functions of H with regard to $(p_1, \&c.)$ and $(P_1, \&c.)$ respectively, the typical equations $p=dL/dq'$, $P=dM/dQ'$ follow. To prove the second part of the example it is sufficient to make $M=L$. By the definition of a reciprocal this requires $\Sigma pq' = \Sigma PQ'$, and, since t does not anywhere enter explicitly, this leads to $\Sigma p\delta q = \Sigma P\delta Q$.

Ex. 2. Show that in Jacobi's first rule, Art. 499, the P's do not in general represent the momenta corresponding to the Q's.

If they did there would be a relation between the q's and Q's alone, Art. 500, Ex. 4. But Jacobi's formulæ do not admit of this.

502. **Hamilton's equations with indeterminate multipliers.** Let $q_1...q_n$, $p_1...p_n$ be the co-ordinates and momenta of the system, L the Lagrangian function and H its reciprocal. By the principle of virtual moments we have as in Vol. I. Art. 397

$$\Sigma \left(\frac{d}{dt}\frac{dL}{dq'} - \frac{dL}{dq}\right)\delta q = 0 \quad \dots\dots\dots\dots\dots\dots(1),$$

for all variations consistent with the geometrical relations. Again by the definition of a reciprocal function

$$H+L=\Sigma pq' \dots\dots\dots\dots\dots\dots\dots(2).$$

Taking the total variation of this as in Vol. I. Art. 410 we have

$$\delta H = -\Sigma \frac{dL}{dq}\delta q + \Sigma\left(-\frac{dL}{dq'} + p\right)\delta q' + \Sigma q'\delta p \quad \dots\dots\dots\dots\dots(3).$$

Remembering that $p=dL/dq'$ by definition and eliminating $\Sigma (dL/dq)\,\delta q$ by (1), we have

$$\delta H = -\Sigma p'\delta q + \Sigma q'\delta p \dots\dots\dots\dots \dots\dots\dots(4).$$

If all the p's and q's were independent, we could deduce at once from this the Hamiltonian equations. If however there are equations of condition between the variables we may use the method of indeterminate multipliers. Let there be r equations of condition and let these be expressed by

$$f_i(p_1...p_n, \quad q_1...q_n)=0 \dots\dots\dots\dots\dots\dots(5),$$

where i has any value from $i=1$ to r. Differentiating these, and subtracting them from (4) after multiplication by $\lambda_1, \lambda_2...\lambda_r$, we have

$$\delta H=\Sigma\left\{-p'-\lambda_1\frac{df_1}{dq}-\lambda_2\frac{df_2}{dq}-\&c.\right\}\delta q + \Sigma\left\{q'-\lambda_1\frac{df}{dp}-\&c.\right\}\delta p \dots\dots(6),$$

where the Σ implies summation for all the co-ordinates. From this we deduce the following n equations: which are typically written

$$\left.\begin{aligned}-p' &= \frac{dH}{dq}+\lambda_1\frac{df_1}{dq}+\lambda_2\frac{df_2}{dq}+\&c.\\ q' &= \frac{dH}{dp}+\lambda_1\frac{df_1}{dp}+\lambda_2\frac{df_2}{dq}+\&c.\end{aligned}\right\} \dots\dots\dots\dots\dots(7).$$

If we put

$$K=H+\lambda_1 f_1+\lambda_2 f_2+\&c. \dots\dots\dots\dots\dots(8),$$

we see that the equations (7), by virtue of (5) take the form

$$p' = -dK/dq, \qquad q'=dK/dp \dots\dots\dots\dots\dots\dots(9).$$

The r equations represented by (5) and the $2n$ equations represented by (7) or (9) are sufficient to determine the r multipliers and the $2n$ co-ordinates and momenta.

503.   The values of the r multipliers $\lambda_1...\lambda_r$ may be found as follows.   Differentiating (5) we have $\Sigma q' df/dq + \Sigma p' df/dp = 0$.

Substituting from (7) the values of p' and q', we find

$$(H, \ f) + \lambda_1 (f_1, \ f) + \lambda_2 (f_2, \ f) + ... = 0 ....(10),$$

where the symbol (u, v) has the meaning given to it in Art. 490.   Writing $f_1...f_r$ successively for f in this typical equation, we have r linear equations to find the multipliers.   Substituting their values in (7), we have $2n$ equations to find the co-ordinates.   The equations (10) are given by Mathieu in *Liouville's Journal*, 1874.

504.   The equations of condition (5) have been taken to contain the momenta as well as the co-ordinates, as this supposition made the investigation more symmetrical, but in most cases the momenta are absent and the results are accordingly simplified.

505.   **Variation of the elements.**   Let there be two dynamical problems in one of which the Hamiltonian function is H and in the other $H + K$.   Their differential equations are therefore respectively

$$p' = -\frac{dH}{dq}, \quad q' = \frac{dH}{dp} ....(1), \qquad p' = -\frac{dH}{dq} - \frac{dK}{dq}, \quad q' = \frac{dH}{dp} + \frac{dK}{dp} ....(2).$$

Let the integrals of the first problem be

$$c_1 = f_1 (p_1, \&c., \ q_1, \&c., \ t), \qquad c_2 = f_2 (p_1, \&c., \ q_1, \&c., \ t), \&c. ....(3).$$

If we consider c_1, c_2, &c., the constants of the solution of the first problem, to be functions of p_1, &c., q_1, &c. and t, we may suppose the solution of the second problem to be represented by integrals of the same form (3) as those of the first problem.   It is our object to discover what functions c_1, c_2, &c. are of p_1, &c., q_1, &c. and t.   The function K is called the disturbing function and is usually small compared with H.

Since the equations (3) are the integrals of the differential equation (1) when c_1, &c. are regarded as constants, we shall obtain identical equations by substituting from (3) in (1).   Hence, differentiating (3) and substituting for p' and q' their values given by (1), we have the typical equation

$$0 = -\frac{dc}{dp}\frac{dH}{dq} + \frac{dc}{dq}\frac{dH}{dp} + ... + \frac{dc}{dt} ....(4),$$

where c stands for any one of the constants c_1, c_2, &c.   See Art. 494.

But, when c_1, c_2,... are considered as variables, the equations (3) are the integrals of the differential equations (2).   Hence, repeating the same process, we have

$$c' = -\frac{dc}{dp}\frac{dH}{dq} + \frac{dc}{dq}\frac{dH}{dp} + ... + \frac{dc}{dt} - \frac{dc}{dp}\frac{dK}{dq} + \frac{dc}{dq}\frac{dK}{dq} + ...$$

where the differential coefficients on the left-hand side are total, and those on the right-hand side partial.

Hence, using the identities (4), we get $c_1' = -\dfrac{dc_1}{dp}\dfrac{dK}{dq} + \dfrac{dc_1}{dq}\dfrac{dK}{dp}$ .... (5),

with similar expressions for c_2', &c.

If K be given as a function of p, q, &c. and t, we have dc_1/dt, &c. expressed as functions of p, q, &c. and t.   Joining these equations to those marked (3) we find c_1, c_2... as functions of t.

If K be given as a function of c_1, c_2,... and t we may continue thus,

$$\frac{dK}{dp} = \frac{dK}{dc_1}\frac{dc_1}{dp} + \frac{dK}{dc_2}\frac{dc_2}{dp} + ..., \qquad \frac{dK}{dq} = \frac{dK}{dc_1}\frac{dc_1}{dq} + \frac{dK}{dc_2}\frac{dc_2}{dq} + ...$$

Substituting in the expression for c_1', we get

$$c_1' = \Sigma \left[\frac{dc_1}{dq} \frac{dc_2}{dp} - \frac{dc_1}{dp} \frac{dc_2}{dq} \right] \frac{dK}{dc_2} + \Sigma \left[\frac{dc_1}{dq} \frac{dc_3}{dp} - \frac{dc_1}{dp} \frac{dc_3}{dq} \right] \frac{dK}{dc_3} + \dots \quad \dots \dots (6),$$

where the Σ means summation for all values of p, q, viz. p_1, q_1, p_2, q_2, &c.

By using the abbreviated notation explained in Art. 490 this equation may be written in the compact form

$$c_1' = (c_2, c_1) \frac{dK}{dc_2} + (c_3, c_1) \frac{dK}{dc_3} + \dots \dots \dots \dots \dots (7).$$

506. The formulæ giving the variations of the constants are greatly simplified when the elements chosen are canonical. When this is the case the constants run in pairs, let these pairs be c_1, c_2; c_3, c_4; &c., then $(c_2, c_1) = 1$, $(c_1, c_3) = 0$ and so on. The formulæ then take the form

$$\left. \begin{array}{l} c_1' = dK/dc_2 \\ c_2' = -dK/dc_1 \end{array} \right\}, \qquad \left. \begin{array}{l} c_3' = dK/dc_4 \\ c_4' = -dK/dc_3 \end{array} \right\}, \&c. \dots \dots \dots (8).$$

507. Returning to the general equation (7) where the constants are unrestricted we notice that, when c_1, c_2 &c. are expressed as functions of (p_1, q_1) &c. and t, as in (3), the coefficients (c_2, c_1) &c. may be found by simple differentiation. It will usually be found more convenient to express them in terms of the constants c_1, c_2 &c. and t, by substituting for (p_1, q_1) &c. their values given by the integrals (3).

On effecting this substitution it is found that t disappears from the expressions. This follows at once from Poisson's theorem given in Art. 494. Thus *when the disturbing function is given in terms of the time and the constants of the undisturbed motion, the variations of those constants produced by the disturbing forces can be expressed in terms of the differential coefficients of the disturbing function without t appearing explicitly in any coefficient.*

508. As an example consider the case of a particle or planet describing an ellipse about a centre of force. The constants of the elliptic motion are usually taken to be the major axis $2a$, the eccentricity e, the longitude of one apse ω, &c. Supposing the motion of the particle to be disturbed by the attraction of some other particle, the object of Lagrange's method of treating the planetary theory is to find how these constants are altered by the disturbing forces. To effect this, the disturbing function K is first expressed in terms of the time and the constants a, e, ω &c., and secondly formulæ are found giving a', e', ω' &c. in terms of dK/da, dK/de &c. These formulæ do not contain t except implicitly through the disturbing function, and this remarkable characteristic is not restricted to these particular constants, but holds true whatever constants are chosen to fix the elliptic motion. We may also notice that this property holds when K is a function, not merely of the co-ordinates q_1, q_2 &c. but of both the co-ordinates and their corresponding momenta.

509. The equations (6) given above, expressing c_1', c_2' &c. in terms of the differential coefficients of K, are due to Poisson; the corresponding formulæ of Lagrange are differently arranged. Regarding K as a function of the co-ordinates and the momenta, we have

$$\frac{dK}{dc_1} = \frac{dK}{dq_1} \frac{dq_1}{dc_1} + \frac{dK}{dq_2} \frac{dq_2}{dc_1} + \&c. + \frac{dK}{dp_1} \frac{dp_1}{dc_1} + \&c. \dots \dots \dots \dots (9).$$

Taking the differential equations of the undisturbed motion in the Hamiltonian form (1), let their solutions be

$$q_1 = F_1(t, c_1, c_2 \&c.), \qquad q_2 = \&c. \dots \dots \dots \dots (10).$$

These if substituted in (1), treating c_1, c_2 &c. as constants, satisfy (1) identically. Hence, when they are substituted in (2), treating c_1, c_2 &c. as functions of t, all terms will cancel each other identically except those which contain c_1', c_2' &c. and the terms dK/dq, dK/dp. The excepted terms which contain c_1', c_2' &c. can enter only through p' and q', we therefore have

$$-\frac{dK}{dq} = \frac{dp}{dc_1}c_1' + \frac{dp}{dc_2}c_2' + \&c., \qquad \frac{dK}{dp} = \frac{dq}{dc_1}c_1' + \&c. \quad\ldots\ldots\ldots(11).$$

Substituting these in (9), we find that c_1' disappears from the result, and that

$$\frac{dK}{dc_1} = [c_1,\, c_2]\, c_2' + [c_1,\, c_3]\, c_3' + \ldots\ldots\ldots\ldots\ldots (12),$$

where $[c_1,\, c_2]$ has the meaning given to it in Art. 490. Similar relations hold for each of the differential coefficients dK/dc_2 &c., so that we have as many equations as there are constants.

Comparing the equations (7) and (12), we see that in both the disturbing function K is supposed to be known as a function of the constants of the undisturbed motion and t. To find the coefficients in (7), the integrals of the undisturbed motion must be expressed in the form (3), i.e. each constant must be given as a function of the variables and the time. To find the coefficients in (12), the integrals must be expressed in the form (10), i.e. each variable must be given as a function of the time and the constants. Again in (7) c_1', c_2' &c. are found directly in terms of dK/dc, &c., but in (12) a system of linear equations must be solved to find c_1', c_2' &c. In both (7) and (12) the coefficients $(c_1,\, c_2)$, $[c_1,\, c_2]$ &c. do not contain the time explicitly.

510. Lagrange shows that, when the constants are the initial values of the variables $(p_1,\ q_1)$ &c., these equations reduce to simpler forms like those in (8). Regarding any constants which may be introduced in the integrations as functions of these, he proceeds in the *Mécanique Analytique* to express their variations in terms of the differential coefficients of K in a form resembling (7).

511. One peculiarity of the method of the variation of constants is that the co-ordinates $q_1,\ldots q_n$ and the momenta $p_1,\ldots p_n$ are expressed by the same functions of c_1, c_2 &c. and t, whether the motion considered is the undisturbed or the varied motion. It immediately follows that the velocities $q_1',\ldots q_n'$ are also expressed by the same functions of c_1, c_2 &c. and t in both motions. To prove this it is sufficient to notice that, since $p_1 = dT/dq_1'$, $p_2 = dT/dq_2'$ &c., we can express q_1', q_2' &c. in terms of p_1, p_2 &c., q_1, q_2 &c.

512. The subject of Theoretical dynamics is so large that it is impossible to discuss it fully in a treatise which contains so many applications of dynamics. We can therefore only allude to Donkin's theorem that a knowledge of half the integrals of the Hamiltonian system will in certain cases lead to a determination of the rest, (*Phil. Trans.* 1854, 1855), or to Bour's method of reducing the number of variables when some of the integrals are known (*Liouville's Journal*, Vol. xx., 1855).

CHAPTER XI.

On the Potential.

513. *To find the potential of a body of any form at any external distant point.*

Let the centre of gravity G of the body be taken as the origin of co-ordinates, and let the axis of x pass through S the external point. Let the distance $GS = \rho$. Let (x, y, z) be the co-ordinates of any element dm of the body situated at any point P and let $GP = r$, then $PS^2 = \rho^2 + r^2 - 2\rho x$. The potential of the body is

$$V = \Sigma \frac{dm}{PS}; \quad \therefore V = \Sigma \frac{dm}{\rho} \left\{ 1 - \frac{2\rho x - r^2}{\rho^2} \right\}^{-\frac{1}{2}}$$

$$= \Sigma \frac{dm}{\rho} \left\{ 1 + \frac{1}{2} \frac{2\rho x - r^2}{\rho^2} + \frac{3}{8} \left(\frac{2\rho x - r^2}{\rho^2} \right)^2 + \frac{5}{16} \left(\frac{2\rho x - r^2}{\rho^2} \right)^3 + \frac{35}{128} \left(\frac{2\rho x - r^2}{\rho^2} \right)^4 + \ldots \right\};$$

arranging these terms in descending powers of ρ, we get

$$V = \Sigma \frac{dm}{\rho} \left\{ 1 + \frac{x}{\rho} + \frac{3x^2 - r^2}{2\rho^2} + \frac{5x^3 - 3xr^2}{2\rho^3} + \frac{35x^4 - 30x^2r^2 + 3r^4}{8\rho^4} + \ldots \right\}.$$

Let M be the mass of the body, then $\Sigma dm = M$. Also since the origin is at the centre of gravity, we have $\Sigma x \, dm = 0$.

Let A, B, C be the principal moments of inertia at the centre of gravity, I the moment of inertia about the axis of x, which in our case is the line joining the centre of gravity of the body to the attracted point. Then

$$\Sigma dm r^2 = \tfrac{1}{2} (A + B + C),$$

$$\Sigma dm x^2 = \Sigma dm \, (r^2 - y^2 - z^2) = \tfrac{1}{2} (A + B + C) - I.$$

Let l be any linear dimension of the body, then, if ρ be so great compared with l that we may neglect the fraction $(l/\rho)^3$ of the potential, we have

$$V = \frac{M}{\rho} + \frac{A + B + C - 3I}{2\rho^3}.$$

If we wish to make a nearer approximation to the value of V, we must take account of the next terms, viz. $\dfrac{5\Sigma mx^3 - 3\Sigma mxr^2}{2\rho^4}$.

Let (ξ, η, ζ) be the co-ordinates of m referred to any fixed rectangular axes having the origin at G, and let (α, β, γ) be the angles GS makes with these axes. Then

$$x = \xi \cos\alpha + \eta \cos\beta + \zeta \cos\gamma;$$
$$\therefore \Sigma mx^3 = \cos^3\alpha \Sigma m\xi^3 + 3\cos^2\alpha \cos\beta \Sigma m\xi^2\eta + \ldots\ldots$$

If the body is symmetrical about any set of rectangular axes meeting at G, we have $\Sigma m\xi^3 = 0$, $\Sigma m\xi^2\eta = 0$, &c. $= 0$, so that the next term in the expression for the potential vanishes altogether. Thus the error of the preceding expression for V is comparable to only the fraction $(l/\rho)^4$ of the potential. This is the case with the earth, the form and structure of which are very nearly symmetrical about the principal axes at its centre of gravity.

514. In this investigation S has been supposed to be at a very great distance. *But the expression for the potential is also very nearly correct wherever the point S is situated, provided the body is an ellipsoid whose strata of equal density are concentric ellipsoids of small ellipticity.*

To prove this, we may use a theorem in attractions due to Maclaurin, viz., the potentials of confocal ellipsoids at any external point are proportional to their masses. Let us first consider the case of a solid homogeneous ellipsoid. Describe an internal confocal ellipsoid of very small dimensions and let a', b', c' be its semi-axes. Then, because the ellipticity is very small, we can take a', b', c' so small that S may be regarded as a distant point with regard to the internal ellipsoid. Hence the potential due to the internal ellipsoid is

$$V' = \frac{M'}{\rho} + \frac{A'+B'+C'-3I'}{2\rho^3},$$

where accented letters have the same meaning relatively to the internal ellipsoid that unaccented letters have with regard to the given ellipsoid. The error made in this expression is of the order $(a'/\rho)^4 V'$. Hence, by Maclaurin's theorem, the potential V of the given ellipsoid is

$$V = \frac{M}{\rho} + \frac{M}{M'}\frac{A'+B'+C'-3I'}{2\rho^3},$$

and the error is of the order $(a'/\rho)^4 V$.

If a, b, c be the semi-axes of the given ellipsoid, we have

$$a^2 - a'^2 = b^2 - b'^2 = c^2 - c'^2 = \lambda^2;$$
$$\therefore A = M\frac{b^2+c^2}{5} = M\left(\frac{b'^2+c'^2}{5} + \frac{2}{5}\lambda^2\right) = \frac{M}{M'}A' + \frac{2}{5}M\lambda^2.$$

Similarly, $B = \dfrac{M}{M'} B' + \dfrac{2}{5} M\lambda^2,$ $C = \dfrac{M}{M'} C' + \dfrac{2}{5} M\lambda^2.$

Also if (α, β, γ) be the direction-angles of the line GS with reference to the principal axes at G, we have

$$I = A \cos^2\alpha + B \cos^2\beta + C \cos^2\gamma = \dfrac{M}{M'} I' + \dfrac{2}{5} M\lambda^2.$$

Hence, substituting, we have $V = \dfrac{M}{\rho} + \dfrac{A + B + C - 3I}{2\rho^3}.$

If a, b, c are arranged in descending order of magnitude, we can by diminishing the size of the internal ellipsoid make c' as small as we please, though in the limit the ellipticities of both the sections containing $c'a'$ and $c'b'$ become equal to unity. In this case we have ultimately $a' = \sqrt{a^2 - c^2}$. Let ϵ be the ellipticity of the section containing a and c the greatest and least semi-axes. Then $a' = a\sqrt{2\epsilon}$, and the error of the above expression for V is of the order $4(a/\rho)^4\epsilon^2 V$.

The theorem being true for any solid homogeneous ellipsoid is also true for any homogeneous shell bounded by concentric ellipsoids of small ellipticity. For the potential of such a shell may be found by subtracting the potentials of the bounding ellipsoids, $A + B + C$ (see Vol. I.) being independent of the directions of the axes.

Lastly, suppose the body to be an ellipsoid whose strata of equal density are concentric ellipsoids of small ellipticity, the external boundary being homogeneous. Then the proposition, being true for each stratum, is also true for the whole body.

Ex. Verify this theorem by showing that when the attracting body is a homogeneous ellipsoid the terms of the fourth order given in Art. 513 are of the order

$$(a/\rho)^4 \epsilon^2 V.$$

We first show by integration that the terms of the fourth order are

$\dfrac{3}{8 \cdot 35} \dfrac{M}{\rho^5} [35 \, (\lambda^2 a^2 + \mu^2 b^2 + \nu^2 c^2)^2 - 20 \, (\lambda^2 a^4 + \mu^2 b^4 + \nu^2 c^4)$

$\qquad\qquad - 10 \, (\lambda^2 a^2 + \mu^2 b^2 + \nu^2 c^2) \, (a^2 + b^2 + c^2) + (a^2 + b^2 + c^2)^2 + 2 \, (a^4 + b^4 + c^4)],$

where (λ, μ, ν) are the direction-cosines of GS. If the ellipsoid is nearly spherical we put $b/a = 1 - \epsilon$ and $c/a = 1 - \epsilon'$. It is easily seen on substitution that not only are all the terms independent of ϵ, ϵ' zero but that the terms containing the first powers of ϵ and ϵ' disappear.

The theorem of Art. 513 is due to Poisson, but it was put into the convenient form given in that article by MacCullagh. The fact that this theorem is very nearly true even when the attracting body is close to the earth provided that the earth is ellipsoidal is given by Laplace, *Mécanique Céleste*, Book v. The proof in Art. 515 is nearly the same as that of MacCullagh. *Transactions of the Royal Irish Academy*, Vol. XXII. Parts I. and II. Science.

515. The following geometrical interpretation of the formula of Art. 513 is also due to Prof. MacCullagh. His demonstration and another by the Rev. R. Townsend may be found in the *Irish Transactions* for 1855.

A system of material points attracts a point S whose distance from the centre of gravity G of the attracting mass is very great compared with the mutual distances of the particles. If a tangent plane be drawn to the ellipsoid of gyration perpendicular to GS, touching the ellipsoid in T and cutting GS in U, then the resultant attraction on S lies in the plane SGT. The component P of the attraction on S in the direction $TU = -\dfrac{3M}{\rho^4} GU \cdot UT$. *The component of the attraction on S in the direction* $UG = \dfrac{M}{\rho^2} + \dfrac{3}{2} \dfrac{A+B+C-3I}{\rho^4}$.

These theorems are also true if we replace the ellipsoid of gyration by any confocal ellipsoid. Let a, b, c be the semi-axes of this confocal, and let p be the perpendicular GU on the tangent plane. Since (see Vol. I.) $A = Ma^2 + \lambda$, $B = Mb^2 + \lambda$, &c. where λ is some constant, we have $V = \dfrac{M}{\rho} + \dfrac{M(a^2 + b^2 + c^2 - 3p^2)}{2\rho^3}$.

To prove that the resultant force on S lies in the plane SGT, let us displace S to S' where SS' is perpendicular to this plane and is equal to $\rho d\psi$. Because V is a potential, the force on S in the direction SS' is $dV/\rho d\psi$. But after this displacement the tangent plane perpendicular to GS' intersects along TU the former tangent plane, hence $dp/d\psi = 0$, and $\therefore dV/d\psi = 0$.

To find the force P acting at S in the direction TU, let us displace S to S'', where

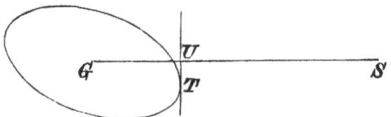

SS'' is parallel to TU and is equal to $\rho d\psi$. Since GU is perpendicular to UT we have $TU = dp/d\psi$. Hence $P = \dfrac{1}{\rho}\dfrac{dV}{d\psi} = -\dfrac{3M}{\rho^4} p \cdot TU$.

Lastly, $R = -\dfrac{dV}{d\rho} = \dfrac{M}{\rho^2} + \dfrac{3}{2}\dfrac{A+B+C-3I}{\rho^4}$.

Ex. Show that the product $GU \cdot TU$ is the same for all confocals.

516. **Examples on attractions.** Ex. 1. Let GP be a straight line through the centre of gravity such that the moment of inertia about it is equal to the mean of the three principal moments of inertia at G, then the resolved attraction of the body on any point S in the direction SG is, when S lies in GP, more nearly the same as if the body were collected into its centre of gravity than when S lies in any other straight line through G.

Show also that the moment of inertia about GP is equal to the mean of the moments of inertia about all straight lines passing through G.

If two of the principal moments of inertia are equal, prove that GP makes with the axis of unequal moment an angle equal to $\cos^{-1}(1/\sqrt{3})$. In the case of the earth this line is in latitude $54° . 45'$.

517. **Other laws of attraction.** Ex. 2. If the law of attraction had been $-\phi$ (dist.) instead of the inverse square, the potential of a body on any external point S would have been represented by $\Sigma m\phi_1(PS)$, where $\phi(\rho)$ is the differential coefficient of $\phi_1(\rho)$. In this case, by reasoning in the same way as in Art. 513, we get

$$V = M\phi_1(\rho) + \phi'(\rho)\dfrac{A+B+C}{4} - \dfrac{\rho}{2}\dfrac{d}{d\rho}\left(\dfrac{\phi(\rho)}{\rho}\right)I,$$

where A, B, C and I have the same meanings as before.

If (x', y', z') be the co-ordinates of S referred to the principal axes at G, the moment of the attraction of S about the axis of y is $= \dfrac{1}{\rho} \dfrac{d}{d\rho} \dfrac{\phi(\rho)}{\rho} \cdot (C - A) x'z'.$

518. *To find the Force-function due to the attraction of any body on any other distant body.*

Let G, G' be the centres of gravity of the two bodies, and let $GG' = R$. Let A, B, C; A', B', C' be the principal moments of inertia of the two bodies at G and G' respectively; I, I' the moments of inertia about GG', and let M, M' be the masses of the two bodies.

Let m' be any element of the body M' situated at a point S, and let $GS = \rho$. Then the potential of the body M at m' is

$$m' \left\{ \frac{M}{\rho} + \frac{A + B + C - 3I_1}{2\rho^3} \right\},$$ where I_1 is the moment of inertia of

the body M about GS. We have now to sum this expression for all values of m'. This gives $\quad M\Sigma \dfrac{m'}{\rho} + \Sigma m' \dfrac{A + B + C - 3I_1}{2\rho^3}.$

The first term by the same reasoning as before gives

$$\frac{MM'}{R} + M \frac{A' + B' + C' - 3I'}{2R^3}.$$

In the second term, let x', y', z' be the co-ordinates of m' referred to G' as origin. Then

$$\rho = R \left(1 + \frac{x'}{R} + \text{squares of } x', y', z' \right),$$

$$I_1 = I (1 + \alpha x' + \beta y' + \gamma z' + \text{squares}),$$

where α, β, γ are some constants. Substituting these, and remembering that $\Sigma m'x' = 0$, $\Sigma m'y' = 0$, $\Sigma m'z' = 0$, we get

$$M' \cdot \frac{A + B + C - 3I}{2R^3} \left\{ 1 + \left(\begin{matrix} \text{terms depending on the} \\ \text{squares of } x', y', z' \end{matrix} \right) \right\}.$$

Hence the required force-function is

$$V = \frac{MM'}{R} + M \frac{A' + B' + C' - 3I'}{2R^3} + M' \frac{A + B + C - 3I}{2R^3}.$$

The error of this expression is of the order $(ll'/R^2)^2 V$, where l, l' are any linear dimensions of the two bodies respectively.

519. **Moment of the Sun's force.** *To find the moment of the attraction of the sun or moon about one of the principal axes of the earth at its centre of gravity.*

Let the principal axes of the earth at its centre of gravity be taken as the axes of reference, and let α, β, γ be the direction-angles of the centre of gravity G' of the sun. Then, if V be the potential of the sun or moon on the earth, we have

$$V = \frac{MM'}{R} + M \frac{A' + B' + C' - 3I'}{2R^3} + M' \frac{A + B + C - 3I}{2R^3},$$

where unaccented letters refer to the earth, and accented letters to the sun or moon. Let θ be the angle which the plane through the sun and the axis of y makes with the plane of xy, then $dV/d\theta$ is the required moment in the direction in which we must turn the body to increase θ. From the above expression, since θ enters only through I, we have
$$\frac{dV}{d\theta} = -\frac{3}{2}\frac{M'}{R^3}\frac{dI}{d\theta}.$$

Now $I = A\cos^2\alpha + B\cos^2\beta + C\cos^2\gamma$, and by Spherical Trigonometry, we have $\cos\gamma = \sin\beta\sin\theta$, $\cos\alpha = \sin\beta\cos\theta$;
$$\therefore \frac{dI}{d\theta} = -2(A-C)\sin^2\beta\sin\theta\cos\theta;$$

$\therefore$ the moment required about the axis of y $\Big\} = -3\frac{M'}{R^3}(C-A)\cos\alpha\cos\gamma.$

In this expression the mass of the attracting body is measured in astronomical units. We may eliminate this unit in the following manner. Let n' be the mean angular velocity of the sun about the earth, R_0 its mean distance, so that if M be the mass of the earth, we have $(M'+M)/R_0^3 = n'^2$. Now M is very small compared with M', so small that M/M' is of the order of terms already neglected. Hence we may in the same terms put $M'/R_0^3 = n'^2$, and therefore

the moment of the sun's attraction about the axis of y $\Big\} = -3n'^2(C-A)\cos\alpha\cos\gamma\left(\frac{R_0}{R}\right)^3.$

Let n'' be the mean angular velocity of the moon about the earth, so that, if M'' be the mass of the moon, R'_0 the mean distance, we have $(M''+M)/R'^3_0 = n''^2$. Let ν be the ratio of the mass of the earth to that of the moon, then $M''(1+\nu)/R'^3_0 = n''^2$, and therefore, if R' be the distance of the moon,

the moment of the moon's attraction about the axis of y $\Big\} = -\frac{3n''^2}{1+\nu}(C-A)\cos\alpha\cos\gamma\left(\frac{R'_0}{R'}\right)^3.$

In the same way the moments about the other axes may be found. Putting κ for the coefficient, we have

moment about axis of $x = -3\kappa(B-C)\cos\beta\cos\gamma,$

moment about axis of $z = -3\kappa(A-B)\cos\alpha\cos\beta.$

520. Examples. Ex. 1. The force-function between a body of any form and a uniform circular ring whose centre is at the centre of gravity of the body and whose mass is M' is $V = \frac{MM'}{\rho} - M'\frac{A+B+C-3J}{4\rho^3}$,

where J is the moment of inertia of the body about an axis through its centre of gravity perpendicular to the plane of the ring, and A, B, C are the principal moments of inertia at the centre of gravity.

Thence show that Saturn's ring supposed uniform will have the same moments to turn Saturn about its centre of gravity as if half the whole mass were collected into a particle and placed in the axis of the ring at the same distance from Saturn, provided that the particle repelled instead of attracted Saturn.

Ex. 2. If the earth be formed of concentric spheroidal strata of small but different ellipticities and of different densities, show that the ratio of C to A may be found from the equation $C\int\rho d\,(a^5\epsilon)=(C-A)\int\rho d\,(a^5)$, where ϵ is the ellipticity and ρ the density of a stratum, the major-axis of which is a; the square of ϵ being neglected. It follows that if ϵ be constant, the ratio of C to A is independent of the law of density.

If we assume the law of density and the law of ellipticity usually taken for the Figure of the Earth, this formula gives $(C-A)/C=\cdot00313593$. See Pratt's *Figure of the Earth*.

Ex. 3. A body free to turn about a fixed straight line passing through the centre of gravity is in equilibrium under the attraction of a distant fixed particle. Show that the time of a small oscillation is $2\pi\left\{\dfrac{B\rho^5}{3M'\xi\left\{(C-A)\,\xi+F\eta\right\}}\right\}^{\frac{1}{2}}$, where the fixed straight line is the axis of y, the plane of xy in equilibrium passes through the attracting particle, and ξ,η are the co-ordinates of the particle. Also A,B,C,D,E,F are the moments and products of inertia of the body about the axes. If the straight line did not pass through the centre of gravity show that the time would be proportional to ρ.

Motion of the Earth about its Centre of Gravity.

521. *To find the motion of the pole of the earth about its centre of gravity when disturbed by the attraction of the sun and moon, the figure of the earth being taken to be one of revolution.*

Let us consider the effect of these two bodies separately. Then, provided we neglect terms depending on the square of the disturbing force, we can by addition determine their joint effect.

The sun attracts the parts of the earth nearer to it with a force slightly greater than that with which it attracts the parts more remote, and thus produces a small couple, which tends to turn the earth about an axis lying in the plane of the equator and perpendicular to the line joining the centre of the earth to the centre of the sun. It is the effect of this couple which we have now to determine. It clearly produces small angular velocities about axes perpendicular to the axis of figure. We shall suppose that the initial axis of rotation so nearly coincides with the axis of figure, that we may regard the angular velocities about axes lying in the plane of the equator to be small compared with the angular velocity about the axis of figure.

Let us take as axes of reference in the earth, GC the axis of figure, GA and GB moving in the earth with an angular velocity θ_3 round GC. Then, following the notation of Art. 10, we have $h_1 = A\omega_1, \quad h_2 = A\omega_2, \quad h_3 = C\omega_3,$

$\qquad\qquad\qquad\theta_1 = \quad\omega_1, \quad \theta_2 = \quad\omega_2.$

The equations of motion are therefore

$$
\left.
\begin{aligned}
A\,\frac{d\omega_1}{dt} - A\omega_2\theta_3 + C\omega_3\omega_2 &= L \\[2mm]
A\,\frac{d\omega_2}{dt} - C\omega_3\omega_1 + A\omega_1\theta_3 &= M \\[2mm]
C\,\frac{d\omega_3}{dt} &= 0
\end{aligned}
\right\}
\;\ldots\ldots\ldots\ldots\;(1).
$$

The last of these equations shows that ω_3 is constant. Let this constant be denoted by n.

The angular velocities ω_1 and ω_2 are to be found by solving the other two equations. The solution must be conducted by the method of continued approximation, ω_1 and ω_2 being regarded as small compared with n.

In the first instance let us suppose the orbit of the disturbing body to be fixed in space. This is very nearly true in the case of the sun, less nearly so for the moon. This limitation of the problem proposed will be found greatly to simplify the solution. We can now choose as our axes of reference in space two straight lines GX, GY at right angles to each other in the plane of the orbit and a third axis GZ normal to the plane.

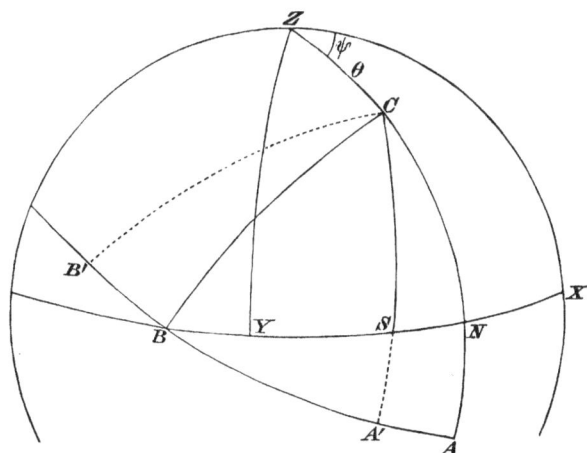

522. In these equations of motion the quantity θ_3 is at our choice, let it be so chosen* that the plane containing the

---

* We might also very conveniently have chosen as axes of reference, GC the axis of figure and axes GA', GB' moving on the earth so that GB' is the axis of the resultant couple produced by the action of the disturbing body on the earth. In this case the plane GA' moves so as always to contain the disturbing body S,

axes GC, GA also contains GZ. Then θ_3 is the angular velocity of
the plane ZGC round GC. The velocity of A in the direction AB
is therefore represented both by θ_3 and by $\sin ZA \cdot d\psi/dt$, where ψ
is the angle the plane ZGC makes with some fixed plane ZGX.
Equating these, as we do in forming the third Eulerian geometrical

equation, we have $\qquad \theta_3 = \cos\theta \dfrac{d\psi}{dt}$ ....(2).

If, as usual, θ represent the angle ZC, we have also the two
geometrical equations

$$\omega_1 = -\sin\theta \frac{d\psi}{dt} \qquad \omega_2 = \frac{d\theta}{dt} \dotfill (3).$$

These follow at once from a mere inspection of the figure, or we
may deduce them from Euler's geometrical equations (see Vol. I.)
by putting $\phi = 0$.

The terms $\theta_3\omega_1$ and $\theta_3\omega_2$ in the differential equations (1) contain
the squares of the small quantities to be found. As it will appear
that both $d\theta/dt$ and $d\psi/dt$ are of the same order as the disturbing
moments L and M, we shall *presently neglect these two terms*. Re-

thus θ_3 is the angular velocity of CS round C and is therefore a small quantity of
the order n'. We shall therefore reject the small terms $\omega_2\theta_3$ and $\omega_1\theta_3$ in equations
(1). The equations now become

$$A\frac{d\omega_1}{dt} + Cn\omega_2 = 0, \qquad A\frac{d\omega_2}{dt} - Cn\omega_1 = M = -3\kappa(C-A)\cos a \cos\gamma,$$

where the value of M is at once obtained from Art. 519, and in our case $a = \frac{1}{2}\pi - \gamma$.

Eliminating ω_2 we have $\qquad \dfrac{d^2\omega_1}{dt^2} + \left(\dfrac{Cn}{A}\right)^2 \omega_1 = -\dfrac{Cn}{A^2}M.$

Since the angular distance γ of the disturbing body from the pole of the earth
varies very slowly, the term on the right-hand side is very nearly constant. If
this be regarded as a sufficient approximation we have

$$\omega_1 = \frac{3\kappa}{2n}\frac{C-A}{C}\sin 2\gamma, \qquad \omega_2 = 0.$$

But in fact these are nearly true when we take account of the periodical term
provided that S moves slowly. For suppose $\qquad M = M_0 + \Sigma P \sin(pt+Q)$,

where p is small; we have in that case $\qquad \omega_1 = -\dfrac{M_0}{Cn} - \Sigma \dfrac{CnP}{C^2n^2 - A^2p^2}\sin(pt+Q)$,

neglecting the small term p^2 in the denominator we have as before $\omega_1 = -\dfrac{M}{Cn}$.

The motion of the axis C in space is therefore simply that due to an angular
velocity ω_1 about the axis A'. Since the plane $A'C$ moves so as always to contain
the disturbing body S, the axis of figure GC is at any instant moving perpendicular
to the plane containing it and the disturbing body (*i.e.* in the figure C is always
moving perpendicular to SC) with an angular velocity equal to $\dfrac{3\kappa}{2n}\dfrac{C-A}{C}\sin 2\gamma$. If
we resolve this in the directions along and perpendicular to ZC, we easily deduce the
equations (7) in the text, and the solution may be continued as above.

taining them for the present, to show how they affect the steady motion, the equations of motion take the form

$$\left.\begin{aligned} -\sin\theta\,\frac{d^2\psi}{dt^2} - 2\cos\theta\,\frac{d\theta}{dt}\,\frac{d\psi}{dt} + \frac{Cn}{A}\,\frac{d\theta}{dt} &= \frac{L}{A} \\ \frac{d^2\theta}{dt^2} - \sin\theta\cos\theta\left(\frac{d\psi}{dt}\right)^2 + \frac{Cn}{A}\sin\theta\,\frac{d\psi}{dt} &= \frac{M}{A} \end{aligned}\right\}\dots\dots\dots(4).$$

523. We have now to find the magnitudes of L and M. Let S be the disturbing body and let it move in the direction X to Y. According to the usual rule in Astronomy, we shall suppose the longitude l of S to be measured in the direction of motion from some fixed line in the plane of XY, say the axis of X. Then $SN = l - \psi$ and $BS = \frac{1}{2}\pi - (l - \psi)$. Also $\psi - \frac{1}{2}\pi$ is the longitude of the ascending node in which the plane of the orbit of S cuts the equator. When S represents the sun, this node is called the first point of Aries. By Art. 519 we have

$$L = -3\kappa\,(B - C)\cos\beta\cos\gamma = -3\kappa\,(A - C)\sin SN\cos SN\sin\theta$$
$$= \tfrac{3}{2}\kappa\,(C - A)\sin\theta\sin 2\,(l - \psi)\dots\dots(5).$$

$$M = -3\kappa\,(C - A)\cos\alpha\cos\gamma = -3\kappa\,(C - A)\cos^2 SN\sin\theta\cos\theta$$
$$= -\tfrac{3}{2}\kappa\,(C - A)\sin\theta\cos\theta\,\{1 + \cos 2\,(l - \psi)\}\dots\dots(6).$$

These values of L and M contain the two small multipliers κ and $(C - A)$. They are not the complete values of L and M, but only the principal terms (Art. 514). We shall therefore suppose that the square of $\kappa\,(C - A)$ is to be neglected. The mean value of κ is n'^2 where n' is the angular velocity of the disturbing body. The ratio n'/n is very small, being about $\frac{1}{27}$ for the moon and $\frac{1}{365}$ for the sun.

By referring to Art. 519 we notice that κ contains the factor $(R_0/R)^3$. If the eccentricity e' of the orbit of the disturbing body is not rejected, even when multiplied by $n'^2\,(C - A)$, we must substitute in (5) and (6) for R and also for l their values given by the theory of elliptic motion. The value of l is known to be of the form $\qquad l = n't + \epsilon' + 2e'\sin(n't + \epsilon' - \omega') + \&c\dots\dots\dots(7)$, and there is a similar expression for the reciprocal of R. In these series the coefficients of the trigonometrical terms and the coefficients of t in the arguments are all small compared with n.

524. **To find the steady motion or precession.** We notice that the quantities L and M contain only one term which is not an explicit function of the longitude of the disturbing body. We find the steady motion by taking this one term alone. We have therefore $L = 0$, $M = -\tfrac{3}{2}\kappa\,(C - A)\sin\theta\cos\theta$. The differential equations are satisfied if we put

$$\theta = \alpha, \quad d\psi/dt = \mu$$

where α and μ are two constants which satisfy

$$\sin \alpha \left\{ A \cos \alpha \mu^2 - Cn\mu - \tfrac{3}{2}\kappa \left(C - A \right) \cos \alpha \right\} = 0.$$

Since α in the case of the earth is about $23\tfrac{1}{2}°$, we must have the quadratic factor equal to zero. Since n is not small, this gives two values of μ, one nearly equal to $-\dfrac{3\kappa}{2n}\dfrac{C-A}{C}\cos\alpha$, and the other nearly equal to $\dfrac{C}{A} n \sec \alpha$.

As in the analogous case of the top, considered in Art. 207, either of these values of μ might be the true one if the proper initial conditions were given to the earth.

The latter value of μ gives, by (3), $\omega_1 = -\left(C/A \right) n \tan \alpha$, and in this case the axis of rotation can not closely coincide with the axis of figure. The initial conditions must therefore have been such as to give μ the smaller value.

The actual steady motion is therefore such that the pole C of the earth describes a small circle of radius α about the pole Z of the orbit of the disturbing body with a retrograde angular velocity equal to $\dfrac{3\kappa}{2n}\dfrac{C-A}{C}\cos\alpha$.

We notice that, if the angular velocity n of the earth about its axis were very small or zero, the roots of the quadratic to find μ would take a different form, so that the expression just found for the retrograde motion of the pole of the earth would cease to be even approximately true.

We may also notice that, if the pole of the equator were very close to the pole of the ecliptic or very nearly 90° from it, we should have a different state of steady motion. As in the case of the top already referred to, the oscillations or nutations about this state of motion would have to be treated by a different analysis.

525. To find the Nutation. We must next consider the terms in L and M which contain the longitude l of the disturbing body explicitly. At the same time to make the differential equations linear we might write $\theta = \alpha + \theta_1$, $d\psi/dt = \mu + d\psi_1/dt$, where the additional terms θ_1 and $d\psi_1/dt$ are so small that their squares can be neglected. This substitution is however unnecessary, for having now ascertained that the constant part μ of $d\psi/dt$ is of the order $\kappa (C - A)$ we may at once neglect the terms $\theta_3\omega_1$ and $\theta_3\omega_2$ in the differential equations (1). They now take the linear form

$$\left. \begin{aligned} A\frac{d\omega_1}{dt} + Cn\omega_2 &= L \\ A\frac{d\omega_2}{dt} - Cn\omega_1 &= M \end{aligned} \right\} \dots\dots\dots\dots\dots(8).$$

Since the motion of the disturbing body is very slow compared with the angular velocity of the earth about its axis, l is, and therefore L and M are, very nearly constant. If this be regarded as a sufficiently near approximation we have at once

$$\omega_1 = -\frac{M}{Cn}, \quad \omega_2 = \frac{L}{Cn}.$$

These give by (3), (5) and (6)

$$\left.\begin{aligned}
\omega_2 &= \frac{d\theta}{dt} = \frac{3\kappa}{2n}\frac{C-A}{C}\sin\alpha\sin 2\,(l-\psi) \\
-\frac{\omega_1}{\sin\alpha} &= \frac{d\psi}{dt} = -\frac{3\kappa}{2n}\frac{C-A}{C}\cos\alpha\,\{1+\cos 2\,(l-\psi)\}
\end{aligned}\right\}\dots(9).$$

526. To find the motion of the pole of the earth in space referred to the pole of the orbit of the disturbing body as origin, we integrate the equations (9). If we write for l its approximate value $l = n't + \epsilon'$ we find

$$\left.\begin{aligned}
\theta &= \alpha - \frac{3\kappa}{4nn'}\frac{C-A}{C}\sin\alpha\cos 2\,(l-\psi) \\
\psi &= \text{const.} -\frac{3\kappa}{2nn'}\frac{C-A}{C}\cos\alpha\,\{l+\tfrac{1}{2}\sin 2\,(l-\psi)\}
\end{aligned}\right\}\dots(12).$$

In these equations $l - \psi + \tfrac{1}{2}\pi$ is the longitude of the disturbing body measured from the ascending node of the orbit. This, as before mentioned, is the first point of Aries when the body is the sun.

If the origin of measurement of l and ψ is such that they vanish together, the constant of integration in the second equation is zero.

527. We may measure the degree of approximation of equations (9) in the following manner. If we eliminate ω_2 between the equations (8) we have

$$\frac{d^2\omega_1}{dt^2} + \frac{C^2n^2}{A^2}\,\omega_1 = \frac{1}{A}\frac{dL}{dt} - \frac{Cn}{A^2}M.$$

Since we reject the squares of $\kappa\,(C-A)$ we may, in calculating the value of the right-hand side from the expressions (5) and (6), put $\theta = \alpha$ and $\psi = \mu t + \nu$. Substituting the values of l and R given in (7), suppose we find

$$M - \frac{A}{Cn}\frac{dL}{dt} = \Sigma F\cos\,(\lambda t + f),$$

where the constant part of M is given by $\lambda = 0$ and all the other values of λ are small.

Then solving, we find $\quad \omega_1 = -\Sigma\dfrac{FCn}{C^2n^2 - A^2\lambda^2}\cos\,(\lambda t + f).$

Since F and λ^2 are both small we may reject the small term λ^2 in the denominator, we then have $\quad \omega_1 = -\dfrac{1}{Cn}\Sigma F\cos\,(\lambda t + f) = -\dfrac{M}{Cn} + \dfrac{A}{C^2n^2}\dfrac{dL}{dt}\dots\dots\dots\dots\dots(10).$

In the same way we find $\quad \omega_2 = \dfrac{L}{Cn} + \dfrac{A}{C^2n^2}\dfrac{dM}{dt}\dots\dots\dots\dots\dots\dots\dots(11).$

In this approximation we have rejected terms of the order $\lambda^2 M$ or $\lambda^2 L$. We see by (7) that this is equivalent to rejecting terms of the order $(n'/n)^2\,M$ or $(n'/n)^2\,L$.

By referring to (5) and (6), we see that the terms dL/dt and dM/dt contain, besides the small factor $\kappa\,(C-A)$, another small factor n' which arises from the differentiation of l. These terms are therefore of the order $(n'/n)\,L$ or $(n'/n)\,M$. As the first terms on the right-hand side of (10) and (11) give rise to nutations which are very small, or only just perceptible, it is unnecessary to take account of the second terms. As these are of the same general forms, viz. $P\cos 2l$ and $Q\sin 2l$, as the first terms we notice that they will not be divided on integration by any small factors which do not also divide the first term (see Art. 337). Rejecting then these terms we have the same result as (9).

528. The integration in Art. 526 by itself is not altogether satisfactory. For, when we substitute for l its full elliptic value given in (7), each of the moments L and M assumes the form of a series of terms such as $F\cos(\lambda t + f)$, where the values of λ are small. After integration these terms get magnified by the divisor λ and if any constant term should occur it would get multiplied by t after integration.

By a slight modification of these equations (suggested by Laplace) we may evade this difficulty. Taking for l its value given by the theory of *elliptic motion* we have $R^2 \dfrac{dl}{dt} = $ constant. This constant is evidently $R_0^2 n' (1 - e'^2)^{\frac{1}{2}}$. Substituting for κ its value given in Art. 519 and *taking l as the independent variable*, the equations (9) assume the form

$$\frac{d\theta}{dl} = \frac{PR_0 \sin 2\,(l - \psi)}{R\,(1 - e'^2)^{\frac{1}{2}}}, \qquad \frac{d\psi}{dl} = Q + \frac{P'R_0 \cos 2\,(l - \psi)}{R\,(1 - e'^2)^{\frac{1}{2}}},$$

where P, P' and Q are small constant terms.

From the equation to the ellipse, we have $\quad \dfrac{R_0(1 - e'^2)}{R} = 1 + e'\cos(l - L).$

If this value of R be substituted, the integrations can be effected without difficulty. It is clear however that the combinations of the one term $\cos(l - L)$ with $\sin 2\,(l - \psi)$ and $\cos 2\,(l - \psi)$ can produce only periodic terms. These are of the form

$$\begin{matrix} \sin \\ \cos \end{matrix} \left\{ 2\,(l - \psi) \pm (l - L) \right\}$$

and after integration are divided only by the same small factor n' that divides the terms independent of e'.

Since e' is small, we see that the terms which depend on the eccentricity of the orbit of the disturbing body retain always their relative insignificance compared with the principal terms calculated in equations (12).

529. Let us now examine the geometrical meaning of the equations (12). For the sake of brevity, let us put $S = \dfrac{3\kappa}{2nn'}\dfrac{C - A}{C}$, so that, by Art. 519, $S = \dfrac{3}{2}\dfrac{C - A}{C}\dfrac{n'}{n}$ or $S = \dfrac{3}{2}\dfrac{C - A}{C}\dfrac{n''}{n}\dfrac{1}{1 + \nu}$ according as the sun or moon is the disturbing body, the orbit of the disturbing body being in both cases regarded as circular.

Let us consider first the term $-S\cos\theta l$ in the value of ψ. Let a point C_0 describe a small circle round Z the pole of the orbit of the disturbing planet, the distance CZ being constant and equal to the mean value of θ. Let the velocity be uniform and equal to $Sn'\cos\theta\sin\theta$, and let the direction of motion be *opposite* to that of the disturbing body. Then C_0 represents the motion of the pole of the earth so far as this term is concerned. This uniform motion is called Precession.

Next let us consider the two terms

$$\delta\theta = \tfrac{1}{2}\,S \sin\theta \cos 2l, \qquad \delta\psi = \tfrac{1}{2}\,S \cos\theta \sin 2l.$$

If we put $x = \sin\theta\,\delta\psi$, $y = \delta\theta$, we have

$$\frac{x^2}{(\tfrac{1}{2}S \cos\theta \sin\theta)^2} + \frac{y^2}{(\tfrac{1}{2}S \sin\theta)^2} = 1,$$

which is the equation to an ellipse.

Let us then describe round C_0 as centre an ellipse whose semi-axes are $\tfrac{1}{2}S \cos\theta \sin\theta$ and $\tfrac{1}{2}S \sin\theta$ respectively perpendicular to and along ZC; and let a point C_1 describe this ellipse in a period equal to half the periodic time of the disturbing body. Also let the velocity of C_1 be the same as if it were a material point attracted by a centre of force in the centre varying as the distance. Then C_1 represents the motion of the pole of the earth as affected both by Precession and the principal parts of Nutation.

If we had chosen to include in our approximate values of θ and ψ any small term of a higher order, we might have represented its effect by the motion of a point C_2 describing another small ellipse having C_1 for centre. And in a similar manner by drawing successive ellipses we can represent geometrically all the terms of θ and ψ.

530. **Numerical results.** The preceding investigations are of course approximations. In the first instance we neglected in the differential equations the squares of the ratios of ω_1 and ω_2 to n, and afterwards some periodical terms which are an (n'/n)th of those retained. We see by equations (3) and (12) that the second set of terms rejected is much greater than the first, and yet when the sun is the disturbing body these terms are only about $\frac{1}{365}$th part of those retained, and when the moon is the disturbing body they are only $\frac{1}{27}$th part of terms which themselves are imperceptible.

We have also regarded the earth as a solid of revolution so that $A - B$ may be taken zero, a supposition which cannot be strictly correct.

531. In the case of the sun we have $S = \dfrac{3}{2}\dfrac{C - A}{C}\dfrac{n'}{n}$, so that the precession in one year is $\dfrac{3}{2}\dfrac{C - A}{C}\dfrac{n'}{n} \cos\theta\, 2\pi$. It is shown in treatises on the Figure of the Earth that there is reason to suppose that $(C - A)/C$ lies between ·0031 and ·0033. Also we have $n'/n = \frac{1}{365}$, and $\theta = 23°.8'$. This gives a precession of about $15''\!\cdot\!42$ per annum. Similarly the coefficients of Solar Nutation in ψ and θ are respectively found to be $1''\!\cdot\!23$ and $0''\!\cdot\!53$. If we sup-

posed the moon's orbit to be fixed, we could find in a similar manner the motion produced by the moon referred to the pole of the moon's orbit. In this case $S = \dfrac{3}{2} \dfrac{C - A}{C} \dfrac{n''}{n} \dfrac{1}{1+\nu}$. The value of θ varies between the limits $23° \pm 5°$. Putting $n'/n = \frac{1}{27}$, $\nu = 80$, $\theta = 23°$, we find a precession in one year a little more than double that produced by the sun. But the coefficients of what would be the nutations are about one-sixth of those produced by the sun.

532. The complementary functions. In this solution we have not yet considered the complementary functions. If we abstract all the disturbing forces and regard the earth as simply set in rotation and left to itself, the equations of Art. 525 take the form $A\omega_1' + Cn\omega_2 = 0$, $A\omega_2' - Cn\omega_1 = 0$.
We easily find
$$\omega_1 = H \sin(qt + K), \qquad \omega_2 = -H \cos(qt + K),$$
where $q = Cn/A$, and H, K are two arbitrary constants. The effect of these terms, if of sensible magnitude, would be to produce a small oscillation in the earth's axis. *This is sometimes called the Eulerian nutation.*

As the initial values of ω_1 and ω_2 are unknown, the magnitude of H must be determined by observing the changes produced in the position of the pole of the earth. Since the latitudes of places on the earth are very nearly constant we conclude that the magnitude of H is nearly insensible.

533. The effect of these complementary functions on the motion of the pole of the earth has been already considered in Arts. 180—182. Let i and γ be the inclinations of the instantaneous axis GI and the invariable line GL to the axis of figure GC. Then $\tan i = H/n$ and $\tan \gamma = \tan i \cdot A/C$. In the case of the earth A and C are very nearly equal and $1 - A/C$ has been variously estimated to lie between ·0031 and ·0033. Thus γ and i differ at most by $\frac{1}{300}$th part of either and must therefore be regarded as very nearly equal.

As explained in the articles just referred to, the instantaneous axis GI describes a right cone in space whose axis is GL and whose angular radius is equal to $i - \gamma$, the time of a complete revolution being nearly equal to a sidereal day. The instantaneous axis is therefore nearly fixed in space and coincident with GL.

The instantaneous axis and the invariable line describe right cones in the body whose common axis is the axis of figure, the time of a revolution being $\sin \gamma / \sin (i - \gamma)$th part of a day. The period is therefore 306 to 325 days according to the value taken for A/C. This is often called *Euler's ten monthly period.*

534. The common method of finding the latitude of a place P depends on observations made on a star at an interval of half a day. The latitude found is therefore the angle between GP and the invariable line. As the invariable line travels in the body round the axis of figure, the latitude should have a ten-monthly period whose magnitude is H. For the purpose of detecting the possible changes of latitude special methods have been used, but they cannot be described here.

A series of observations made at Berlin in 1884—86, to determine the coefficient of aberration, led to the result that the latitude had decreased $0''\cdot2$ in one year. Afterwards, at Berlin, Potsdam and Prague, observations made in 1889—90 showed that small periodic changes of latitude do occur amounting to half a second. As the changes at these three places have all the same sign and follow very nearly the same law it is impossible that they could be due to purely local causes. They appear to indicate a yearly inequality. We learn from No. 3055 of *Ast. Nach.* that the observations have been continued in 1891 and that these confirm the previous results. The latitude of a place can be observed, by using the best instruments and taking the utmost care, to within a tenth of a second. This corresponds roughly to three yards on the surface of the earth, so that a change of place of the instrument in the same room can be detected (Flammarion, *Astronomie*, April 1891). When so much can be done we may expect that before long the uncertainties remaining in this problem will be removed. In a paper, read to the Geographical Society at Berlin 1891, Prof. Forster stated that simultaneous observations were to be made for this purpose at Berlin and Honolulu continuously for a year.

These places being nearly on opposite meridians, their latitudes should be altered by equal but opposite quantities if the changes are caused by movements of the instantaneous axis. We now learn from the presidential address of Sir W. Thomson that the results of the first three months of observation at Honolulu show that movements of the instantaneous axis do occur sufficiently great to cause sensible changes of latitude at that place in the direction expected.

535. These changes of latitude may be due to other causes acting jointly with the Eulerian nutation, and amongst these we must include the yearly meteorological changes of the earth. The consequence is that the change of latitude appears to have a double oscillation, the period of one being ten months and of the other a year. The least common multiple of these is five years, so that the changes should repeat themselves in this time. Again, when the two oscillations are compounded together, the rate at which the latitude changes is not uniform. At one time the magnitudes of the two oscillations are both increasing and their rates of change are added together, at another time they are subtracted from each other ; see Art. 89 on the transference of oscillations. It follows, as Prof. Forster remarks, that one series of observations may be favourable to exhibit the change of latitude, while another series made at a different time may show but faint traces of change.

In connection with this double oscillation the problem of Helmert given at the end of Art. 23 is interesting. It should also be noticed that the displacement of the instantaneous axis has been magnified by the nearness of the period of Euler's nutation to that of the meteorological disturbance.

536. We should notice that the complementary functions are not strictly represented by the Eulerian nutation. Taking only the forces which produce precession, the equations of motion are, by Art. 522,

$$- A \sin \theta \psi'' - 2 A \cos \theta \theta' \psi' + C n \theta' = 0,$$

$$A \theta'' - A \sin \theta \psi'^2 + C n \sin \theta \psi' = - \tfrac{3}{2} \kappa (C - A) \sin \theta \cos \theta,$$

where accents denote differentiations with regard to the time. The precession being determined by writing $\theta = a$ and $\psi' = \mu$, we substitute $\theta = a + x$ and $\psi = \mu t + y$ to find the nutations. We shall evidently find on the right-hand side terms which contain the first power of x, and, though these are of the second order of small quantities, they should be examined into for the reason given in Art. 356. Like the terms in the Lunar theory of the form $c\theta - a$, they modify the first approximation, Art. 359.

Making these substitutions we find that one effect of these terms is to alter the Eulerian period. If $2\pi/q_1$ is the altered period, we have

$$q_1 = \frac{Cn}{A} + \frac{C-A}{A}\frac{3\kappa}{n}(\cos^2 a - \tfrac{1}{4}\sin^2 a).$$

The value of q_1 differs very slightly from that defined by q in Art. 530. Since the Eulerian period is not yet known with sufficient accuracy to make this difference of importance (Art. 532), it is unnecessary to discuss at length the effects of these small terms.

537. Examples. Ex. 1. If the earth were a homogeneous shell bounded by similar ellipsoids, the interior being empty, the precession would be the same as if the earth were solid throughout.

Ex. 2. If the earth were a homogeneous shell bounded externally by a spheroid and internally by a concentric sphere, the interior being filled with a perfect fluid of the same density as the earth, show that the precession would be greater than if the earth were solid throughout.

Let (a, a, c) be the semi-axes of the spheroid, r the radius of the sphere. Then, since the precession varies as $(C-A)/C$ by Art. 529, the precession is increased in the ratio $a^4c : a^4c - r^5$.

Ex. 3. If the sun were removed to twice its present distance, show that the solar precession per unit of time would be reduced to one-eighth of its present value; and the precession per year to about one-third of its present value.

Ex. 4. A body turning about a fixed point is acted on by forces which tend to *produce* rotation about an axis at right angles to the instantaneous axis, show that the angular velocity cannot be uniform unless the momental ellipsoid at the fixed point is a spheroid.

The axis about which the forces tend to produce rotation is that axis about which it would begin to turn if the body were placed at rest.

Ex. 5. A body free to turn about its centre of gravity, which is fixed, is in stable equilibrium under the attraction of a distant fixed particle. Show that the axis of least moment is turned toward the particle. Show also that the times of the principal oscillations are respectively $2\pi\left\{\dfrac{B\rho^3}{3M'(C-A)}\right\}^{\frac{1}{2}}$ and $2\pi\left\{\dfrac{C\rho^3}{3M'(B-A)}\right\}^{\frac{1}{2}}$.

If the body be the earth and M' be the sun, show that the smaller of these two periods is about ten years.

538. Unequal moments of inertia. The method used in Art. 521 is well adapted to find the precession and nutation of the earth both to a first and to higher degrees of approximation when we regard the earth as a uniaxal body. Though the method may be used when A is not equal to B, yet it loses much of its brevity. We shall therefore adopt a different method to determine how the precession and nutation are altered when we regard all the three principal moments of inertia as unequal, though of course their ratios are supposed not to differ much from unity.

Referring the motion to the principal axes, the Eulerian equations become

$$\left.\begin{aligned}A\omega_1' - (B-C)\omega_2\omega_3 &= L = -3\kappa(B-C)\cos\beta\cos\gamma\\ B\omega_2' - (C-A)\omega_1\omega_3 &= M = -3\kappa(C-A)\cos a\cos\gamma\\ C\omega_3' - (A-B)\omega_1\omega_2 &= N = -3\kappa(A-B)\cos a\cos\beta\end{aligned}\right\}\dots\dots\dots\dots(1),$$

$$\left.\begin{aligned}\theta' &= \omega_1\sin\phi + \omega_2\cos\phi\\ -\sin\theta\,\psi' &= \omega_1\cos\phi - \omega_2\sin\phi\\ \cos\theta\,\psi' + \phi' &= \omega_3\end{aligned}\right\}\dots\dots\dots\dots\dots(2).$$

We see by (1) that ω_1, ω_2 are of the order of the constant of precession, i.e. $\kappa(C-A)/C$; hence, as we reject the square of this quantity, we shall reject the second term on the left-hand side of each of the equations (1). To find $A\omega_1$, $B\omega_2$, $C\omega_3$ we have therefore merely to integrate the right-hand sides of these equations.

Proceeding as in Art. 523 we find

$$\left.\begin{array}{l} \cos\alpha = \sin(l-\psi)\sin\phi + \cos(l-\psi)\cos\theta\cos\phi \\ \cos\beta = \sin(l-\psi)\cos\phi - \cos(l-\psi)\cos\theta\sin\phi \\ \cos\gamma = \cos(l-\psi)\sin\theta \end{array}\right\} \quad \ldots\ldots\ldots\ldots (3).$$

The third of equations (2) shows that ϕ' differs from a constant, viz. n, by quantities of the order of the precession, hence, if we reject the product of this quantity by n'/n, we may on the right-hand sides of (1) put $\phi = nt + \epsilon$, and in the integration treat l, θ, and ψ as constants. We therefore see that $n\int\cos\beta dt$ and $-n\int\cos\alpha dt$ differ from $\cos\alpha$ and $\cos\beta$ only by constant terms. These constant terms may be omitted, as they represent the complementary functions which are considered separately; it is also evident from (2) that small constant additions to ω_1 and ω_2 only give rise to small daily periodic terms in θ' and ψ'.

We therefore have

$$\omega_1 = \frac{3\kappa}{n}\frac{C-B}{A}\cos\alpha\cos\gamma, \qquad \omega_2 = \frac{3\kappa}{n}\frac{C-A}{B}\cos\beta\cos\gamma.$$

Since we have rejected the squares of $(C-B)/A$ and $(C-A)/B$, we may to the same degree of approximation write C for A and B in the denominators of these expressions.

Substituting these values of ω_1, ω_2 in the equations (2) we find

$$\theta' = \frac{3\kappa}{2n}\frac{2C-A-B}{2C}\sin\theta\sin 2(l-\psi) + R_1,$$

$$\psi' = -\frac{3\kappa}{2n}\frac{2C-A-B}{2C}\cos\theta\{1+\cos 2(l-\psi)\} + R_2,$$

where R_1 and R_2 contain only terms whose period is about half a day and whose coefficients contain the small factor $\kappa(A-B)/C$. The value of $(A-B)/C$ has not been determined but it is known to be very much less than $(C-A)/C$. As only terms of long periods can rise into importance after integration with regard to t, R_1 and R_2 may be altogether rejected.

Omitting the terms R_1 and R_2 as being quite insensible, we see that *both the precession and nutation of the earth, with unequal principal moments of inertia A, B, C, are the same as those for a uniaxal earth with principal moments of inertia* $\frac{1}{2}(A+B)$, $\frac{1}{2}(A+B)$ *and C*.

539. Let us now consider the third of equations (1). As the constancy of the angular velocity ω_3 is a matter of great importance, it may be proper to examine it to a higher degree of approximation. This equation as it appears in (1) is accurate except that on the right-hand side we have rejected some small terms depending on the higher inverse powers of the distance of the disturbing body; see Art. 513. The values of ω_1 and ω_2 have been found rejecting terms of the order $\kappa(C-A)/C$ when multiplied by n'/n. Substituting these in the small terms we have

$$C\omega_3' - (A-B)\left(\frac{3\kappa}{n}\right)^2\frac{(C-A)(C-B)}{AB}\cos^2\gamma\cos\alpha\cos\beta = -3\kappa(A-B)\cos\alpha\cos\beta.$$

If we now write for $\cos\alpha$, $\cos\beta$, $\cos\gamma$ their values given by (3) we obtain only a long series of trigonometrical terms whose periods are about half a day and whose coefficients are very small. It is unnecessary to calculate these at length, it being sufficient to notice that the periods are not such that the coefficients are magnified by integration. We infer that *the attractions of the sun or moon cannot produce sensible changes in the period of rotation of the earth.*

It is possible that the angular velocity of the earth might be altered by other causes such as its gradual refrigeration, tidal friction, &c., for these have not been included in the above discussion.

540. **Ex.** If the principal axes of the earth remain fixed in position but the magnitudes A, B, C alter slowly and become equal to $A + at$, $B + bt$, $C + ct$ after a time t, show that the *secular* inequality in the obliquity is given by

$$\frac{d\theta}{dt} = \frac{P}{2n} \cdot \frac{a + b - 2c}{C - A} \sin \theta,$$

where P is the precession of the equinoxes, i.e. $50''$. Darwin, *Phil. Trans.*, 1876.

To prove this, we may begin with the equations used in Art. 24, Ex. 2 and proceed as in Art. 538.

The existing difference $C - A$ between the moments of inertia of the earth corresponds to an excess of the equatorial over the polar radius of thirteen miles. Unless we can suppose that geological changes could produce alterations of level comparable with this, it is clear that the coefficient of $\sin \theta$ in the expression given above will be a small fraction of P.

541. *To give a general explanation of the manner in which the attraction of the Sun causes Precession and Nutation.*

In order to explain the effect of the sun's attraction on the earth it will be convenient to refer to Poinsot's construction for the motion of a body, described in 140 and the following articles.

If a body be set in rotation about a fixed point O under the action of no forces, we know that the momenta of all the particles are together equivalent to a couple which we shall represent by G about an axis OL called the invariable line. Let T be the vis viva of the body. If a plane be drawn perpendicular to the axis of G at a distance $\epsilon^2 \sqrt{MT}/G$ from the fixed point, then the whole motion is represented by making the momental ellipsoid whose parameter is ϵ roll on this plane. In the case of the earth, the axis OI of instantaneous rotation so nearly coincides with OC, the axis of figure, that the fixed plane on which the ellipsoid rolls is very nearly a tangent plane at the extremity of the axis of figure. This is so very nearly the case that we shall neglect the *squares* of all small terms depending on the resolved part of the angular velocity about any axis of the earth perpendicular to the axis of figure.

Let us now consider how this motion is disturbed by the action of the sun. The sun attracts the parts of the earth nearer to it with a slightly greater force than it attracts those more remote. Hence, when the sun is either north or south of the equator, its attraction will produce a couple tending to turn the earth about that axis in the plane of the equator which is perpendicular to the line joining the centre of the earth to the centre of the sun. Let the magnitude of this couple be represented by α, and let us suppose that it acts impulsively at intervals of time dt.

At any one instant this couple will generate a new momentum αdt about the axis of the couple α. This has to be compounded with the existing momentum G to form a resultant couple G'. If the axis of α were exactly perpendicular to that of G we should have $G' = \sqrt{G^2 + (\alpha dt)^2} = G$ ultimately.

Let θ be the angle that the axis of G makes with OC, then θ is a quantity of that order of small quantities whose square is to be neglected. Taking the case when OC, the axis of G, and the axis of α are in one plane, for this is the case in which G' will most differ from G, we have

$$G'^2 = (G \cos \theta)^2 + (G \sin \theta + \alpha dt)^2$$
$$= (G^2 + 2G\alpha \sin \theta dt \dots\dots\dots\dots\dots\dots\dots\dots\dots(1).$$

Then α and θ being of the same order of small quantities, the term $\alpha \sin \theta$ is to be neglected. Hence we have $G' = G$. But the axis of G is altered in space by an angle $\alpha dt/G$ in a plane passing through it and the axis of α.

Next let us consider how the vis viva T is altered. If T' be the new vis viva, we have

$$T' - T = \text{twice the work done by the couple } \alpha,$$
$$= 2\alpha (\omega \cos \beta) \, dt \dots\dots \dots\dots\dots\dots\dots\dots\dots\dots(2),$$

where $\omega \cos \beta$ is the resolved part of the angular velocity about the axis of α. For the same reason as before the product of this angular velocity and α is to be neglected. Hence we have $T' = T$. It follows from these results that the distance $\epsilon^2 \sqrt{MT}/G$ of the fixed plane from the fixed point is unaltered by the action of α.

Thus the fixed plane on which the ellipsoid rolls keeps at the same distance from the fixed point, so that the three lines OC, OI, OL, being initially very near each other, will always remain very close to each other. But the normal OL to this plane has a motion in space, hence the others must accompany it. This motion is what we call Precession and Nutation.

Lastly the small terms which have been neglected will not continually accumulate so as to produce any sensible effect. As the earth turns round in one day, the axis OC will describe a cone of small angle θ round OL. The axis about which the sun generates the angular velocity α is always at right angles to the plane containing the sun and OC. Hence, regarding the sun as fixed for a day, the angle θ in equation (1) changes its sign every half day. Thus G' is alternately greater and less than G. Similarly, since the instantaneous axis describes a cone about OL, it may be shown that T' is alternately greater and less than T.

542. **Solar Precession and Nutation.** The three axes in the earth which are the most important in our theory are (1) the axis of figure OC, (2) the instantaneous axis of rotation OI, (3) the

invariable line OL. It has just been proved in the last article that, if these three be at any one instant very nearly coincident with each other, they will, notwithstanding the sun's attraction, always remain very close together. It will therefore be sufficient for our present purpose to find the motion in space of any one of the three.

Let OA, OB be two perpendicular axes in the earth's equator and let the earth turn round OC in the positive direction AB. Let the sun S at the time t be in the plane COA and on the positive or north side of the equator. The sun's attraction during the time dt generates a couple αdt about the axis OB, which acts in the negative direction AC. It follows from the last article that OL (which is very nearly coincident with OC) moves in space in the plane BOC with an angular velocity equal to α/G in the direction BC. Since the sun moves round O in the same direction as the earth turns round its axis OC, it follows that, when α is positive, the axes OL and OC move very nearly at right angles to the plane COS in a direction opposite to the sun's motion.

Knowing the motion produced in these axes by the sun in the time dt, we now proceed to sum up the whole effects produced by the sun in one year. For simplicity we shall speak only of the axis of figure, viz. OC.

Describe a sphere whose centre is at O, and let us refer the

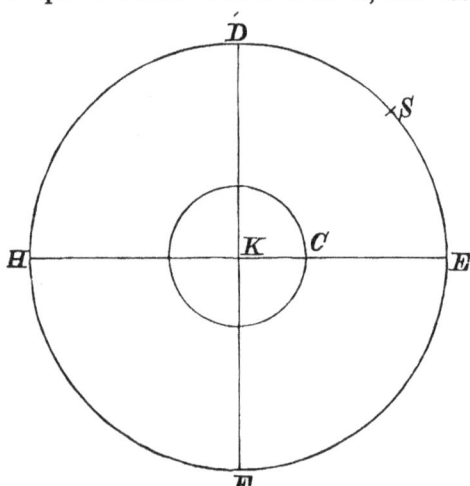

motion to the surface of this sphere. Let K be the pole of the ecliptic, and let the sun S describe the circle $DEFH$ of which K is the pole. Let DF be a great circle perpendicular to KC, then since OC and the axis of figure of the earth are so close that we may treat them as coincident, D and F will be the intersections of the equator and ecliptic. When the sun is north or south of the equator, its attraction generates the couple α, which will be positive or negative according as the sun is on one side or the other. This couple vanishes when the sun is passing through the equator at D or F. If the sun be anywhere in DEF, i.e. north of the equator, C is moved in a direction perpendicular to the arc CS towards D. If the sun be anywhere in FHD, α has the opposite sign, and hence C is again moved perpendicular to the instantaneous position of CS but still towards D. Considering the whole effect produced in one year while the sun describes the circle $DEFH$, we see that C will be moved a very small space towards D, i.e. in the direction opposite to the sun's motion. Resolving this along the tangent to the circle, centre K and radius KC, we see that the motion of C is made up of (1) a uniform motion of C along this circle backwards, which is called Precession and (2) an inequality in this uniform motion which is one part of Solar Nutation. Again, as the sun moves from D to E, C is moved inwards so that the distance KC is diminished, but, as the sun moves from E to F, KC is as much increased. So that on the whole the distance KC is unaltered, but it has an inequality which is the other part of Solar Nutation.

It is evident that each of these inequalities goes through its period in half a year.

543. Lunar Nutation. *To explain the cause of Lunar Nutation.*

The attraction of the sun on the protuberant parts at the earth's equator causes the pole C of the earth to describe a small circle with uniform velocity round K the pole of the ecliptic with two inequalities, one in latitude and one in longitude, whose period is half a year. These two inequalities are called Solar Nutations. In the same way the attraction of the moon causes the pole of the earth to describe a small circle round M, the pole of the lunar orbit, with two inequalities. These inequalities are very small and of short period, viz. a fortnight, and are therefore generally neglected. All that is taken account of is the uniform motion of C round M. Now K is the origin of reference, hence if M were fixed the motion of C round M would be represented by a slow uniform motion of C round K, together with two inequalities whose magnitude would be equal to the arc MK, or 5 degrees, and whose period would be very long, viz. equal to that of C round K produced by the uniform motion. But we know by Lunar Theory that M describes a circle round K as centre with a velocity much more rapid than that of C. Hence the motion of C will be represented by a slow uniform motion round K, together with two inequalities which will be the smaller as the velocity of M round K is greater, and whose period will be nearly equal to that of M round K. This period we know to be about 19 years. These two inequalities are called the Lunar Nutations.

It will be perceived that their origin is different from that of Solar Nutation.

544. **Motion of the plane of the disturbing body.** In the reasoning in Art. 521 the plane of the orbit of the disturbing body was treated as if it were fixed in space. In order to discuss the Lunar Nutations it will be necessary to determine how far its motion affects the precession. We shall continue to take the principal axis OA so that the plane OCA is perpendicular to the instantaneous position of the orbit at the moment under consideration. The quantity θ_3 will not be the same as before*, but, if the motion of the orbit in space be very slow, θ_3 will still be very small. We may therefore neglect the small terms $\theta_3\omega_1$ and $\theta_3\omega_2$ as before. The dynamical equations will not therefore be materially altered. With regard to the geometrical equations (3), it is clear that ω_2, ω_1 will continue to express the resolved parts of the velocity of C in space along and perpendicular to the instantaneous position of ZC. These velocities are therefore expressed by the values of $d\theta/dt$ and $-\sin\theta d\psi/dt$ given in equations (9). To this degree of approximation, therefore, all the change that will be necessary is to refer the velocities as given by equations (9) to axes fixed in space, and then by integration we shall find the motion of C. This is the course we shall pursue to find the lunar nutation.

545. *To calculate the Lunar Precession and Nutation.*

Let K be the pole of the ecliptic, M that of the lunar orbit, C the pole of the earth. Let KX be any fixed arc, $KC = \theta$, $XKC = \psi$, then we have to find θ and ψ in terms of t. In calculating the lunar precession and nutation we are, by Art. 543, to take account only of one part of the motion of C, viz. that called the uniform motion of C round M. By Art. 529 we know that this motion is represented by a velocity equal to $-Sn''\sin MC\cos MC$ in a direction perpendicular to the arc MC. Substituting for S its value given in that Article, it follows that the velocity of C in space is at any instant in a direction perpendicular to MC, and is equal to

$$-\frac{3n''^2}{2n}\frac{C-A}{C}\frac{1}{1+\nu}\cos MC\sin MC.$$

For the sake of brevity let the coefficient of $\cos MC\sin MC$

---

* The value of θ_3 may be found in the following manner. The orbit at any instant is turning about the radius vector of the planet as an instantaneous axis. Let u be this angular velocity, which we shall suppose known. Let Z, Z'; B, B' be two successive positions of the pole of the orbit and the extremity of the axis of B respectively. Then $ZB = $ a right angle $= Z'B'$. Hence the projections of ZZ', BB', on ZB are equal. This gives, since ZB is at right angles to both CZ and SB, $B\hat{S}B'\sin BS = Z\hat{C}Z'\sin ZC$. Now the angle $Z\hat{C}Z' = -\delta\theta_3$ and the angle $B\hat{S}B' = u$, hence $\delta\theta_3 . \sin\theta = -u\sin l$. The value of $\delta\theta_3$ must be added to the former value of θ_3.

be represented by P. Then resolving this velocity along and perpendicular to KC, we have.

$$d\theta/dt = -P \sin MC \cos MC \sin KCM \left. \right\}$$
$$\sin\theta \, d\psi/dt = -P \sin MC \cos MC \cos KCM \left. \right\} \cdot$$

By Lunar theory we know that M regredes round K uniformly, the distance KM remaining unaltered. Let then $KM = i$, and the angle $XKM = -mt + a$. Now, by spherical trigonometry,

$$\cos MC = \cos i \cos\theta + \sin i \sin\theta \cos MKC,$$

$$\sin MC \cos KCM = \frac{\cos i - \cos MC \cos\theta}{\sin\theta}$$

$$= \cos i \sin\theta - \sin i \cos\theta \cos MKC,$$

$$\sin MC \cdot \sin KCM = \sin i \sin MKC.$$

Substituting these values, we have

$$d\theta/dt = -P \{\sin i \cos i \cos\theta \sin MKC + \tfrac{1}{2} \sin^2 i \sin\theta \sin 2MKC\},$$
$$\sin\theta \, d\psi/dt = -P \{\sin\theta \cos\theta (\cos^2 i - \tfrac{1}{2} \sin^2 i)$$
$$- \sin i \cos i \cos 2\theta \cos MKC - \tfrac{1}{2} \sin^2 i \sin\theta \cos\theta \cos 2MKC\}.$$

For a first approximation we may neglect the variations of θ and ψ when multiplied by the small quantity P. Hence $d\theta/dt$ contains only periodic terms, and the inclination θ has no permanent alteration. But $d\psi/dt$ contains a term independent of MKC; considering only this term, we have

$$\psi = \text{constant} - P \cos\theta (\cos^2 i - \tfrac{1}{2} \sin^2 i)\, t.$$

This equation expresses the precessional motion of the pole due to the attraction of the moon. We may write this equation in the form $\psi = \psi_0 - pt$.

To find the nutations, we must substitute for MKC its approximate value $MKC = (-m + p)\, t + a - \psi_0$.

We then have, after integration,

$$\theta = \text{const.} - \frac{P \sin i \cos i \cos\theta}{m - p} \cos MKC - \frac{P \sin^2 i \sin\theta}{4\,(m - p)} \cos 2MKC.$$

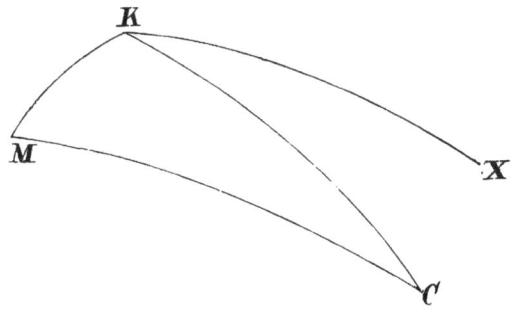

The second of these two periodic terms, being about one-fiftieth part of the first, which is itself very small, is usually neglected. Also p is very small compared with m, hence we have

$$\theta = \theta_0 - \frac{P \sin i \cos i \cos \theta}{m} \cos MKC.$$

This term expresses the Lunar Nutation in the obliquity. The coefficient of the periodical term $\cos MKC$ lies between $8''$ and $9''$.

In the same way by integrating the expression for ψ, and neglecting the very small terms, we have

$$\psi = \psi_0 - P \cos \theta \left(\cos^2 i - \tfrac{1}{2} \sin^2 i\right) t - P \frac{\sin 2i}{2m} \cdot \frac{\cos 2\theta}{\sin \theta} \sin MKC.$$

The angle MKC is the longitude of the moon's descending node, and the line of nodes is known to complete a revolution in about 18 years and 7 months. If we represent this period by T, we have $MKC = -2\pi t/T + \text{constant}$. The coefficient of $\sin MKC$ lies between $16''$ and $17''$.

The pole M of the lunar orbit moves round the point of reference K with an angular velocity which is rapid compared with p, but yet is sufficiently small to make the Lunar Nutations greater than the Solar. We may also notice that, if M had moved round K with an angular velocity more nearly equal to p, the Nutations would have been still larger.

546. We may also make some allowance by this method for the effect of the motion of the ecliptic. We now let M be the pole of the moving ecliptic at any time t, K that of some fixed circle of reference. Assuming that the chief effect of the solar precession is to make the pole C of the earth move perpendicularly to the arc CM with a velocity equal to $P \cos MC . \sin MC$, we find the same values for $d\theta/dt$ and $\sin \theta d\psi/dt$ as before. The motion of the ecliptic is so slow that, if we take as the fixed point K the pole of the ecliptic at some not very remote date, we may neglect the squares of i. We thus have

$$d\theta/dt = -Pi \cos \theta \sin MKC,$$

$$\sin \theta d\psi/dt = -P (\sin \theta \cos \theta - i \cos 2\theta \cos MKC),$$

where $KC = \theta$ and the angle $CKX = \psi$.

Since the pole of the lunar orbit describes very nearly a small circle with a uniform motion we were able in Art. 545 to substitute for the angle MKC its value $(p-m) t + \&c.$ In the case of the ecliptic we proceed otherwise. Let $t = 0$ be the time at which the pole of the ecliptic is at K and let the arc KX join K to the pole C_0 of the equator at the same time. Let the resolved velocities of K along and perpendicular to KX be g' and g. Assuming that the time t is not so long that the direction and velocity of K has had time to change sensibly, we may regard $g't$ and gt as the co-ordinates of M referred to KX as axis of x. Hence

$$i \sin MKC = gt \cos \psi - g't \sin \psi, \qquad i \cos MKC = g't \cos \psi + gt \sin \psi.$$

Now ψ is zero when $t = 0$ and increases at about $50''$ per year, so that in a hundred years ψ amounts to a little over one degree. Since P, g, g' and ψ are all small quantities, we shall write in the small terms $i \sin MKC = gt$ and $i \cos MKC = g't$.

Substituting these in the above expressions and integrating we have

$$\theta = \theta_0 - \tfrac{1}{2}P\cos\theta_0 g t^2, \quad \psi = -P(\cos\theta_0 t - \tfrac{1}{2}\cos 2\theta_0 \operatorname{cosec}\theta_0 g' t^2),$$

where θ_0 is the angular distance of the pole C_0 of the earth from the pole K of the ecliptic at some chosen epoch, and θ, ψ are the co-ordinates of C after a time t referred to the same point as origin.

547. It is sometimes more convenient to refer the motion of C to the pole M of the ecliptic at the time t. Putting $MC = \theta_1$ and the angle $CMC_0 = \psi_1$ we evidently have $\theta_1 = \theta - g't$. Remembering that KM is less than one degree while the four arcs CK, CM, C_0K, C_0M are each about $23°$, we have $\psi_1 \sin\theta_1$ and $\psi \sin\theta$ each nearly equal to C_0C. We therefore have $\psi_1 = \psi(1 + g't\cot\theta)$. Thus, when θ and ψ are known, the values of θ_1 and ψ_1 follow at once.

Ex. If the pole M of the ecliptic, starting from K, describe a great circle KX with a constant angular velocity v, prove that the motion of the pole C of the earth is given by $d\theta/dt = -v\cos\psi$, $d\psi/dt = v\cot\theta\sin\psi - P\cos\theta$, where $\theta = MC$, $\psi = CMX$ and P has the meaning given to it in Art. 546. Show also that, if the square of v/P is neglected, these equations are satisfied by

$$\theta = a - (v/P)\sec a \sin(P\cos at), \quad -\psi = P\cos at - (v/P)\sec^2 a \operatorname{cosec} a \cos(P\cos at).$$

If there were no precession, i.e. if P were zero, the changes in the obliquity due to the motion of the ecliptic would be nearly given by $\theta = a - vt$, but we see that here one effect of the precession is to bring the possible changes of the obliquity within narrow bounds.

The actual motion of the pole of the ecliptic is very different from that supposed in this example, but Laplace has shown that, when we take the co-ordinates of K supplied by the planetary theory, a similar theorem is still true. One effect of the precession is to cause the plane of the equator to move with the plane of the ecliptic so that the possible change of obliquity is less than it would be if there were no precession; *Mécanique Céleste*, Vol. II. p. 367.

548. **Numerical results.** Let BDE and DA be the positions of the ecliptic and equator at some fixed epoch, say Jan. 1, 1850; CAE and BCF their positions after a time t measured in Julian years i.e. years of 365·25 mean solar day each.

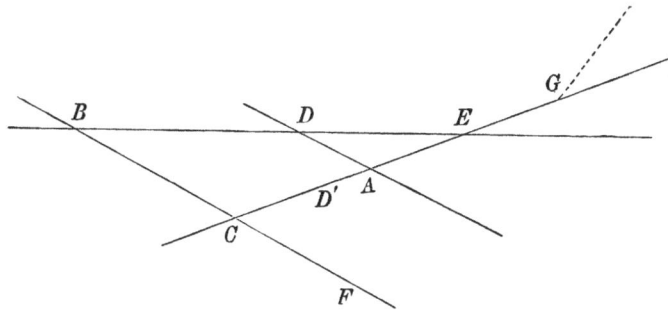

BDE is the fixed ecliptic, DA the fixed equator, CAE the moving ecliptic and BC the moving equator.

Consider first the precession. That part of the precession which is due to the action of the sun and moon on the earth is called *luni-solar precession*. This is referred to the fixed ecliptic and is represented in the figure by BD, we have

$$\psi = BD = 50''\,37140t - 0''\!\cdot\!000108806t^2.$$

The inclination of the equator to the ecliptic would be constant if the motion of the ecliptic did not modify the forces (Art. 524). The inclination CBD of the ecliptic to the equator is therefore given by

$$\theta = CBD = 23° 27' 32'' + 0''{\cdot}00000719 t^2.$$

To these values of ψ and θ we must add the geometrical effect of the motion of the ecliptic, or as it is usually called *planetary precession*. The resultant of luni-solar and planetary precession is called *general precession*. Taking a point D' on the moving ecliptic so that $ED' = ED$, the arc $D'C$ represents the general precession. We have $\psi_1 = D'C = 50''{\cdot}23572 t + 0''{\cdot}00011290 t^2,$

$$\theta_1 = ACF = 23° 27' 32'' - 0''{\cdot}47566 t - 0''{\cdot}00000149 t^2.$$

The planes of the moving ecliptic and equator determined by these angular co-ordinates are usually called the *mean ecliptic and mean equator at the time t.*

The coefficient of t in the expression for ψ_1 is usually called *the constant of precession*. It represents the sum of the precessions due to the sun and moon found in Arts. 524 and 545 together with the correction depending on $g' \cot \theta$ mentioned in Art. 546.

549. *Consider next the Nutations.* These are so small that the amounts to be added to ψ or ψ_1, θ or θ_1 are the same; let these be called respectively Ψ and Θ. Then $\Psi = -17''{\cdot}251 \sin \Omega + 0''{\cdot}207 \sin 2\Omega - 1''{\cdot}269 \sin 2\odot$

$$- 0''{\cdot}204 \sin 2\mathbb{D} + 0''{\cdot}069 \sin A_m + 0''{\cdot}128 \sin A_s,$$

$$\Theta = 9''{\cdot}223 \cos \Omega - 0''{\cdot}090 \cos 2\Omega + 0''{\cdot}551 \cos 2\odot + 0''{\cdot}089 \cos 2\mathbb{D}.$$

Let the dotted line in the figure represent the Lunar orbit, so that G is its ascending node, then $\Omega = CG$ is the longitude of G measured on the true ecliptic from the true spring equinox, but it is sufficient in these small terms to regard Ω as representing the longitude of the mean node measured from the mean equinox. Similarly in these terms $\odot$ and $\mathbb{D}$ are the longitudes of the sun and moon measured on the moving ecliptic from either the true or mean equinox. The symbols A_s and A_m represent the mean anomalies of the sun and moon in their elliptic orbits.

Several terms are here exhibited which have been rejected in the preceding theory in order that the relative magnitudes of the term may be more clearly understood.

The coefficient of $\sin \Omega$ in the expression for Ψ is called *the constant of nutation*. It represents the coefficient of $\sin MKC$ in the expression for ψ in Art. 545.

The terms in Ψ and Θ containing $\sin 2\Omega$ and $\cos 2\Omega$ are discussed in Art. 545, and then rejected. The terms with $\sin 2\odot$ and $\cos 2\odot$ are the solar nutations, see Art. 526. The terms containing $\sin 2\mathbb{D}$ and $\cos 2\mathbb{D}$ are discussed in Art. 531 and it is pointed out in Art. 543 that they are usually neglected. The terms depending on A_m and A_s are alluded to in Art. 528.

The numerical values of the several terms are variously given by different calculators, though the variations are not important. Those here followed are given by Serret, *Annales de l'Observatoire*, t. v. 1859. Another list differing from these is given in Main's *Astronomy* (1863), where Bessel's constants are used. In these the year 1750 is taken as the fixed epoch from which the time is measured.

550. **Nutation of the earth's axis when the mean obliquity is zero.** When the instantaneous obliquity is small, a very slight change in the position of the equator may greatly alter its line of intersection with the ecliptic. It is therefore not convenient to measure our angles from the first point of Aries. Let GZ be a normal to the ecliptic, GC the axis of figure, then we wish to find the small

oscillations of GC about GZ. Let GX, GY be axes fixed in the ecliptic and let the longitude of the sun be measured from GX. Let $(P, Q, 1)$ be the direction cosines of GC referred to the axes of X, Y, Z. It is unnecessary to go through all the steps of the investigation, it is enough to say that the equations of motion to find P and Q take the form given in Art. 15. Remembering that the disturbing couple due to the sun's attraction is equal to $-3k\,(C-A)\sin CS\,.\cos CS$, and that its axis makes an angle $l+\tfrac{1}{2}\pi$ with GX, we obtain the equations

$$AQ'' - CnP' + fQ = -f\sin 2l\,.\,P + f\cos 2l\,.\,Q$$
$$AP'' + CnQ' + fP = -f\cos 2l\,.\,P - f\sin 2l\,.\,Q$$

where $f=\tfrac{3}{2}k\,(C-A)$ and $l=n't$. The small terms fP and fQ must be retained in the first approximation, for the reason given in Art. 356. The first approximation is then found by omitting the right-hand side and assuming

$$P = H\cos(\rho t + \epsilon), \qquad Q = K\sin(\rho t + \epsilon).$$

We then find the quadratic $A\rho^2 - Cn\rho - f = 0$, so that the two values of ρ are nearly equal to Cn/A and $-f/Cn$. Also $K=H$. If ρ and ρ' be the roots of the quadratic, we have for a second approximation

$$P = H\cos(\rho t + \epsilon) + X\cos\{(2n' - \rho)\,t - \epsilon\} + H'\cos(\rho' t + \epsilon') + X'\cos\{(2n' - \rho')\,t - \epsilon'\},$$

$$Q = H\sin(\rho t + \epsilon) + X\sin\{(2n' - \rho)\,t - \epsilon\} + H'\sin(\rho' t + \epsilon') + X'\sin\{(2n' - \rho')\,t - \epsilon'\},$$

where $X\{A\,(2n' - \rho)^2 - Cn\,(2n' - \rho) - f\} = X'\{A\,(2n' - \rho')^2 - Cn\,(2n' - \rho') - f\} = Hf$.
It may be noticed that, when k is small, it has not been assumed that A and C are nearly equal. The method of approximation adopted requires that X and X' should be small compared with H, and this will be true if n'/n is small and C/A not small. It will also be true if $n=0$ and C is nearly equal to A.

Poisson attached so much importance to this problem that he wrote at least two memoirs on it. The first was published in the *Connaissance des Tems* for 1837, where he criticises a dynamical argument of Laplace on this subject in the *Exposition du système du monde*, livre IV. chap. xiii. Soon afterwards he returns to the subject, giving a new solution in the fourteenth volume of the *Mémoires de l'Académie des Sciences*, 1838. He refers the motion to a set of axes different from those used above, though the equations are afterwards reduced to a somewhat similar form. He then obtains an accurate solution of the equations, but the easy approximations here given are sufficient for our present purpose.

CHAPTER XII.

551. In the theory of precession ànd nutation the earth is generally regarded as a uniaxal body. This is a sufficient approximation in the case of the earth, for we have seen in Art. 538 that no important phenomenon of the motion is caused by the slight differences which really exist between the equatorial moments. But in the case of the moon the supposition would cause us to miss some of the most interesting peculiarities of the motion. Besides this there are other differences so great that the two theories are perfectly distinct.

As our object is to examine the mode in which the disturbing forces alter the several motions of the moon about its centre of gravity, rather than to obtain arithmetical results of the greatest possible accuracy, we shall separate the problem into two. In the first place we shall suppose the moon to describe an orbit which is very nearly circular, in a plane which is one of the principal planes at its centre of gravity. In the second case we shall remove the latter restriction, and examine the effects of the obliquity of the moon's orbit to the moon's equator.

552. *The moon describes an orbit about the centre of the earth which is very nearly circular. Supposing the plane of the orbit to be one of the principal planes of the moon at its centre of gravity, it is required to find the motion of the moon about its centre of gravity.*

Let GA, GB, GC be the principal axes at G the centre of gravity of the moon, and let GC be the axis perpendicular to the plane in which G moves. Let A, B, C be the moments of inertia about GA, GB, GC respectively, and let M be the mass of the moon, and let accented letters denote corresponding quantities for the earth.

Let O be the centre of the earth, and let Ox be the initial line. Let $OG = r$, $GOx = \theta$. Let us suppose that the moon turns round its axis GC in the same direction that the centre of gravity describes its orbit about O, and let the angle $OGA = \phi$.

The mutual potential of the earth and moon is, by Art. 518,

$$V = \frac{MM'}{r} + M\frac{A' + B' + C' - 3I'}{2r^3} + M'\frac{A + B + C - 3I}{2r^3}.$$

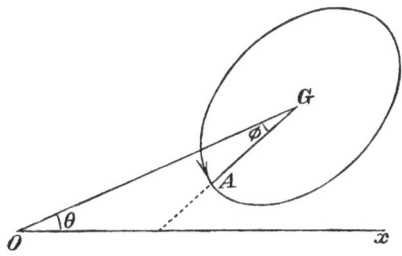

Here $I = A\cos^2\phi + B\sin^2\phi$, and therefore the moment of the forces tending to turn the moon round GC is

$$\frac{dV}{d\phi} = -\frac{3}{2}\frac{M'}{r^3}(B - A)\sin 2\phi \quad\dots\dots\dots\dots (1).$$

Since $\theta + \phi$ is the angle which GA, a line fixed in the body, makes with Ox, a line fixed in space, the equation of the motion of the moon round GC is

$$\frac{d^2\theta}{dt^2} + \frac{d^2\phi}{dt^2} = -\frac{3}{2}\frac{M'}{r^3}\frac{B - A}{C}\sin 2\phi \quad\dots\dots\dots (2).$$

The motion of the centre of gravity of the moon referred to the centre of the earth as a fixed point is found in the Lunar Theory. It is there shown that r and θ may be expressed in the form

$$r = c\{1 + L\cos(pt + \alpha) + \&c.\},$$
$$d\theta/dt = n + \beta t + Mp\cos(pt + \alpha) + \&c.,$$

where βt is a very small term which represents a secular change in the moon's angular velocity about the earth, and is really the first term of the expansion of a trigonometrical expression.

If we substitute the value of $d\theta/dt$ in equation (2), we have the following equation to determine ϕ,

$$\frac{d^2\phi}{dt^2} = -\frac{1}{2}q^2\sin 2\phi - \beta + Mp^2\sin(pt + \alpha) + \&c.\dots\dots(3),$$

where for the sake of brevity we have put $n^2\dfrac{3}{2}\dfrac{B - A}{C} = \dfrac{q^2}{2}$.

Now we know by observation that the moon always turns the same face towards the earth, so that amongst the various motions which may result from different initial conditions, the one which we wish to examine is characterized by ϕ being nearly constant. Let us then introduce into this equation the assumption that ϕ is nearly constant; we may then deduce from the integral how far this assumption is compatible with any given initial conditions which we may suppose to have been imposed on the moon. Putting $\phi = \phi_0 + \phi'$, where ϕ_0 is supposed to contain all the constant part of ϕ, we easily find

$$\left. \begin{aligned} \tfrac{1}{2} q^2 \sin 2\phi_0 &= -\beta \\ \frac{d^2\phi'}{dt^2} + q^2 \cos 2\phi_0 \phi' &= Mp^2 \sin(pt + \alpha) + \&c. \end{aligned} \right\} \dots\dots(4).$$

Solving the second equation, we find,

$$\phi = H \sin(qt + K) + \phi_0 + M \frac{p^2}{q^2 \cos 2\phi_0 - p^2} \sin(pt + \alpha) + \&c.\dots(5),$$

where H and K are two arbitrary constants whose values depend on the initial conditions. The angular velocity of the moon about its axis is therefore given by the formula

$$\frac{d\theta}{dt} + \frac{d\phi}{dt} = n + \beta t + Hq \cos(qt + K) + M \frac{pq^2 \cos 2\phi_0}{q^2 \cos 2\phi_0 - p^2} \cos(pt + \alpha) + \&c.\dots(6).$$

In this investigation the axis GA which makes the angle ϕ with the radius vector GO drawn to the earth may be either of the principal axes in the moon's equator. If we choose GA to be that axis whose mean position makes the lesser angle with the radius vector GO, the quantity $\cos 2\phi_0$ will be positive. The quantity q^2 will be positive or negative according as that axis GA has the least or greatest moment. In the solution just written down q^2 has been taken to be positive.

If q^2 were negative or zero, the character of the solution of (3) would be altered. In the former case the expression for ϕ would contain real exponentials. If the initial conditions were so nicely adjusted that the coefficient of the term containing the positive exponent were zero, the value of ϕ' would still be always small. But this motion would be unstable, the smallest disturbances would alter the values of the arbitrary constants, and then ϕ' would become large. If we also examine the solution when $q^2 = 0$, we easily see that ϕ' could not remain small. The complementary function would then take the form $Ht + K$, and as before some small disturbance might cause ϕ' to become great. We therefore infer that, of the axes GA, GB of the moon, the axis of least moment is turned more towards the earth than the other, and that these two principal moments are not equal.

In order that the expression (5) for ϕ may represent the actual motion it is necessary and sufficient that H when found from the initial conditions should be small. We see, by differentiation, that Hq is of the same order of small quantities as $d\phi/dt$. Hence H will be small if at any instant the angular velocity, viz. $d\theta/dt + d\phi/dt$, of the moon about GC is so nearly equal to the angular velocity, viz. $d\theta/dt$, of its centre of gravity round the earth, that the ratio of the difference to q is very small.

We see from the first of equations (4) that the magnitude of the constant part ϕ_0 of the angle which the axis of least moment in the moon's equator makes with the radius vector drawn to the earth depends on the ratio $2\beta/q^2$. The value of β is found in the Lunar Theory and is known to be extremely small. It represents an increase in each century of the angular velocity of the moon in her orbit round the earth of about 25 seconds per century. The numerical value of q^2 depends on the structure of the moon, and is not properly known. Its value can only be found by comparing the results of this or some other investigation with those of observation. It will presently be shown that according to Nicollet $3(B-A)/C = \cdot00167$. This would make ϕ_0 to be so small as to be inappreciable.

The first of equations (4) shows that 2β must be less than q^2; so that unless the moments of inertia A and B in the moon are sufficiently unequal to satisfy this condition, the moon could not move so as always to turn the same face to the earth.

If we enquire what can be the physical cause of the difference between the moments of inertia about the two principal axes in the moon's equator we naturally think of the attraction of the earth on that body. This attraction, either in the past or in the present time, would tend to lengthen that diameter which is directed to the earth. Taking the suppositions usually made in the theory of the Figure of the Earth, Laplace has attempted to deduce from this the value of q^2. The only result we are here concerned with is that the ratio $2\beta/q^2$ is so small that we may reject its square. Assuming this, we again see that ϕ_0 must also be very small. It follows also that we may write $-\beta/q^2$ for ϕ_0 and unity for $\cos 2\phi_0$ in equations (5) and (6).

If therefore we suppose the moon at any instant to be moving with its axis of least moment pointed towards the earth, and its angular velocity about its axis of rotation to be nearly equal to that of the moon round the earth, then the axis of least moment will continue always to point very nearly to the earth. The mean angular velocity of the moon about its axis will immediately become equal to that of the moon about the earth and will partake of all its secular changes. This is Laplace's theorem. It shows that the present state of motion of the moon is stable, rather than explains how the angular velocity about the axis came to be so nearly equal to the angular velocity about the earth.

553. The statement that the moon always presents the same face to the earth must be understood with some limitation. The angular velocity of the moon about its axis is very nearly uniform, but the angular velocity in its orbit about the earth is not constant, and hence there arises an inequality or libration in longitude which may amount to as much as six degrees. Again, the axis of rotation of the moon is not quite perpendicular to the plane of its orbit, so that there is a libration in latitude. Lastly, as the observer is not situated at the centre of the earth there is

a diurnal libration which arises from parallax and may amount to nearly one degree. These are called the apparent or geometrical librations. After all these have been allowed for, there remains a real libration in the angular velocity of the moon about its axis and it is this last inequality or libration which we are here considering.

554. If the longitude of the centre of the moon as seen from the centre of the earth be $\qquad \theta = \theta_1 + M \sin (pt + a) + \&c.$,

where $\theta_1 = nt + \frac{1}{2}\beta t^2 + \epsilon$, then the longitude of any spot on the moon as seen from the centre of the moon and measured from the first point of Aries is

$$L = l + \pi + \theta_1 + H \sin (qt + K) + \frac{Mq^2}{q^2 - p^2} \sin (pt + a) + \&c.,$$

where l is some constant. Any lunar meridian whose longitude is given by this expression is fixed on the moon and moves with it. That particular meridian whose longitude is defined by this expression when l is omitted is called the first meridian, and l is the longitude of the spot under consideration measured from the first meridian. If the periodic terms in the expression for L are omitted as being almost insensible, the first meridian will be defined by the longitude $L = \pi + \theta_1$, and this meridian will bisect the visible disc of the moon, supposing it to move in the ecliptic with an angular velocity $n + \beta t$ about the earth, to rotate with the same angular velocity about an axis perpendicular to the ecliptic, and to be seen from the centre of the earth.

555. To determine the numerical values of the coefficients of the periodic terms in the expression for L, the oscillations of some spot conveniently situated on the apparent disc of the moon must be observed. Bouvard measured the difference of the right ascension and declination of the spot Manilius from the bright rim or border of the moon. Subtracting these from the calculated semi-diameter of the moon, the co-ordinates of the spot referred to the centre of the visible disc are known. A great variety of astronomical corrections have to be made and the result has to be referred to the centre of the moon as origin. Finally the longitude of the spot measured on the ecliptic from Aries up to the descending node of the lunar equator and then along that equator is determined.

By equating the longitudes of a spot on the moon observed at different times to those deduced from theory we may form a sufficient number of equations to determine the values of any unknown constants in the theory. In this way we may attempt to discover the value of $(B - A)/C$. The observations of Bouvard and Nicollet however show that the amount of the true libration is so small as to be almost insensible. The extent of the oscillation in *lunar* longitude on each side of the mean position is about $4'. 45''$ or $285''$.

If the term $Hq \cos (qt + K)$ could be detected by observations we should deduce the value of $(B - A)/C$ from its period. Among the other terms of the expression for the angular velocity of the moon about its axis those will be best suited to discover the value of q which have the largest coefficients, that is, those in which either the numerator M is the greatest, or the denominator $q^2 - p^2$ the least, possible. The term with the largest M is the elliptic inequality, and if $(B - A)/C$ were as great as ·03, Laplace has shown that it could be recognized by observation. The term with the least value of p is the annual equation, and here $n/p = 13·36$, $M = -669''$. If we ascribe the variation of the spots wholly to this inequality we have $Mq^2/(p^2 - q^2) = 285$. We easily deduce $(B - A)/C = ·00057$.

The spot Manilius was selected as being both distinct and not far from the centre of the visible disc, and was observed by Bouvard at Paris at every opportunity during the four years 1806—10. The choice was afterwards objected to by Beer and Mædler because its aspect differs according to the mode of illumination. They suggested the crater Mæsting A, which is described by Webb (*Celestial objects*) as minute and very luminous. This spot was accordingly observed for two years and a half at Königsberg.

556. **Motion of the centre of gravity of the Moon.** We may also deduce from the potential given in Art. 552 the radial and transverse forces which act on the centre of gravity of the moon due to the mutual attractions of the earth and moon. Since the principal moments of the moon are nearly equal, and its linear size small compared with its distance from the earth, these forces are very nearly the same as if the moon were collected into its centre of gravity. The effect of the small forces neglected by this assumption will be insignificant compared with the other forces which act on the centre of gravity of the moon. The motion of the centre of gravity of the moon is therefore very nearly the same as if the whole mass were collected into its centre of gravity.

Ex. The centre of gravity G of a rigid body describes an orbit which is nearly circular about a very distant fixed centre of force O attracting according to the Newtonian law and situated in one of the principal planes through G. If $r = c(1 + \rho)$, $\theta = nt + n\psi$ be the polar co-ordinates of G referred to O, show that the equations of motion are

$$\left. \begin{array}{c} \dfrac{d^2\rho}{dt^2} - 3n^2\rho - 2n^2 \dfrac{d\psi}{dt} = -\dfrac{9}{4}n^2\gamma' - \dfrac{9}{4}n^2\gamma \cos 2\phi \\[2mm] 2\dfrac{d\rho}{dt} + \dfrac{d^2\psi}{dt^2} = \dfrac{3}{2}n\gamma \sin 2\phi \\[2mm] \dfrac{d^2\phi}{dt^2} + n\dfrac{d^2\psi}{dt^2} = -\dfrac{q^2}{2}\sin 2\phi \end{array} \right\} ,$$

where $\gamma = \dfrac{B-A}{Mc^2}$, $\gamma' = \dfrac{2C-A-B}{3Mc^2}$.

We may notice that the values of γ and γ' are much smaller than that of q^2 and might therefore be rejected in a first approximation.

If the body always turns the same face to the centre of force so that ϕ is nearly constant and is small, show that there will be two small inequalities in the value of ϕ of the form $L\sin(pt + a)$, where p is given by

$$(p^2 - n^2)(p^2 - q^2) - 3n^2\gamma(p^2 + 3n^2) = 0,$$

one of these periods being nearly the same as that of the body round the centre of force, and the other being very long.

If the body turns very nearly uniformly round its axis GC, so that $\phi = n't + \epsilon'$ nearly, show that there will be two small inequalities in the value of ϕ, one in which $p = n$ and another in which $p = 2n'$.

557. **Examples.** Ex. 1. Show that the moon always very nearly turns the same face to that focus of her orbit in which the earth is not situated. [Smith's Prize.]

Ex. 2. If the centre of gravity G of the moon is constrained to describe a circle with a uniform angular velocity n about a fixed centre of force O attracting according to the Newtonian law, show that the axis GA of the moon will oscillate on each side of GO, or will make complete revolutions relatively to GO, according as the angular velocity of the moon about its axis at the moment when GA and GO

coincide in direction is less or greater than $n+q$, where q has the meaning given to it in Art. 552. Find also the extent of the oscillations.

Ex. 3. A particle m moves without pressure along a smooth circular wire of mass M with uniform velocity under the action of a central force situated in the centre of the wire attracting according to the law of nature. Show that this system of motion is stable if $\dfrac{m}{M} > \dfrac{8+12\sqrt{6}}{25}$. The disturbance is supposed to be given to the particle or to the wire, the centre of force remaining fixed in space.

Ex. 4. A uniform ring of mass M and of very small section is loaded with a heavy particle of mass m at a point on its circumference, and the whole is in uniform motion about a centre of force attracting according to the law of nature. Show that the motion cannot be stable unless $m/(M+m)$ lies between ·815865 and ·8279.

This example shows (1) that if a ring, such as Saturn's ring, be in motion about a centre of force, its position cannot be stable, if the ring be uniform; and (2) that if, to render the motion stable, the ring be weighted, a most delicate adjustment of weights is necessary. A very small change in the distribution of the weights will change a stable combination to one that is unstable. This example is taken from *Prof. Maxwell's Essay on Saturn's Rings.*

Ex. 5. The centre of gravity of a body of mass M, symmetrical about the plane of xy, is G; and O is a point such that the resultant attraction of the body on O is along the line GO. Then, if the body be placed with O coinciding with a fixed centre of force S, and be set in rotation about an axis through O perpendicular to the plane of xy with an angular velocity ω, G will, if undisturbed, revolve uniformly in a circle, always turning the same face towards O, provided that $Ma\omega^2$ is equal to the resultant attraction along GO, where a is the distance GO. It is required to determine the conditions that this motion should be stable.

The motion being disturbed, O will no longer coincide with the centre of force S. Let two straight lines at right angles revolving uniformly round S as origin with an angular velocity ω be chosen as co-ordinate axes, and let x be initially parallel to OG. Let (x, y) be the co-ordinates of O, ϕ the angle OG makes with the axis of x, then x, y, ϕ are all small. Let V be the potential of the body at O, and let $d^2V/dx^2 = \alpha$, $d^2V/dxdy = \gamma$, $d^2V/dy^2 = \beta$. Let S be the amount of matter in the centre of force. The equations of motion of a particle referred to axes moving in one plane round a fixed origin are given in Vol. I. These equations may also be deduced from Arts. 4 and 5 of this volume by putting $\theta_1 = 0$ and $\theta_2 = 0$. In this way the equations of motion of G reduce to

$$\left(\frac{d^2}{dt^2} - \omega^2 - \frac{S}{M}\alpha\right)x - \left(2\omega\frac{d}{dt} + \frac{S}{M}\gamma\right)y - 2\omega a\frac{d}{dt}\phi = 0,$$

$$\left(2\omega\frac{d}{dt} - \frac{S}{M}\gamma\right)x + \left(\frac{d^2}{dt^2} - \omega^2 - \frac{S}{M}\beta\right)y + a\frac{d^2}{dt^2}\phi = 0,$$

and the equation of angular momentum about S will lead to

$$2\omega ax + a\frac{d}{dt}y + (a^2 + k^2)\frac{d}{dt}\phi = 0,$$

where k is the radius of gyration of the body about O. Combining these equations as a determinant, and reducing we find that the differential equation in ξ, η, or ϕ is of the form $\qquad A\dfrac{d^4}{dt^4} + B\dfrac{d^2}{dt^2} + C = 0.$

The condition of stability is that the roots of this equation should be real and negative. Hence A, B, C must be of the same sign and $B^2 > 4AC$. This proposition is due to Sir W. Thomson and is given in *Prof. Maxwell's Essay on Saturn's Rings*.

558. Cassini's theorem on the Moon's equator. Before we proceed to the theoretical discussion of this problem it will be convenient to mention the most striking of the results arrived at. There are three planes with which we are concerned, viz. (1) the plane of the moon's orbit round the earth or, which is the same thing, the plane of the earth's orbit as seen from the moon; (2) a plane drawn through the centre of the moon parallel to the ecliptic, i.e. parallel to the plane of the earth's orbit round the sun ; (3) the plane of the moon's equator. This last is a plane perpendicular to that axis of figure which most nearly coincides with the axis of rotation. Now Cassini discovered that these three planes all intersect in the same straight line, so that the plane of the moon's equator has to follow the plane of the moon's orbit as it regredes along the ecliptic. He also discovered that the plane parallel to the ecliptic always lies *between* the other two planes, *Mémoires de l'Académie des Sciences*, vol. VIII. These results were afterwards confirmed by T. Mayer, who undertook a series of observations on the spots of the moon during the years 1748 and 1749. He also corrected the inclinations of the three planes as given by Cassini. Subsequently Lalande confirmed Cassini's theorems a second time, see the *Mémoires de l'Académie des Sciences*, 1764. Lastly, Bouvard undertook a more complete set of observations which extend over the years 1806—1810. These were reduced and discussed by Nicollet, who published his results in the *Connaissance des Tems* for the year 1822 published in 1820. These observations, thus reduced, still remain the standard set of observations and are generally referred to as the proof of Cassini's theorem. According to Nicollet the inclination of the moon's equator to the ecliptic is constant and equal to $1° 28'$. He also found that a meridian drawn on the moon through any spot oscillates on each side of its mean position though an angle of only about $4'$ to $5'$.

These relations between the three planes are so interesting and extraordinary that a theoretical explanation was soon sought after. D'Alembert in 1754 was the first to attempt the solution. But his results were far from complete. The Academy of Sciences offered their prize of 1764 for a complete theory of the moon's libration. This was gained by Lagrange. In 1780 he proved that, if the three planes originally coincided, the attraction of the earth on the moon would maintain the coincidence. *Mémoires de Berlin*, 1780. Laplace showed further that these theorems are disturbed neither by the secular inequalities of the mean motion of the moon nor by the secular changes of the ecliptic. Poisson repeated and extended Lagrange's theory and discovered some new inequalities in the motion. These results may be found in the *Connaissance des Tems* for 1821. For a further account of the history the reader may consult *Grant's History of Physical Astronomy* and the *Connaissance des Tems* for 1822.

559. Theoretical investigation of Cassini's theorem. The motion of a rigid body about a distant centre of force has been investigated on the supposition that the motion takes place entirely in one plane. We see by equation (2) of Art. 552 that the case in which the centre of gravity describes a circular orbit, and the rigid body always turns a principal axis towards the centre of force, is one of *steady motion*. The preceding investigation also shows that this motion is *stable* for all disturbances which do not alter the plane of motion, provided that the moment of inertia about that principal axis which is directed towards the centre of force is less than the moment of inertia about the other principal axis in the plane of motion.

It remains now to determine the effect of these disturbances in the more general case when the motion takes place in three dimensions.

Statement of problem. The problem we have to consider may therefore be summed up thus. The moon turns about its centre of gravity G and is acted on by a centre of force E which moves in a given manner. The instantaneous axis is very nearly coincident with one principal axis GC, and is nearly perpendicular to the plane of the ecliptic. The mean angular velocity is equal to that of E round G, so that a principal axis GA is nearly pointed to E. The centre of force E moves in a nearly circular orbit in a plane which is very nearly perpendicular to GC. This plane is known to have a slow motion in space, so that the normal GM to its instantaneous position describes a cone of small angle round GZ the normal to the ecliptic. The two normals GM and GZ maintain a nearly constant inclination of about 5° 8′. The motion of the normal GM round GZ is nearly uniform, and a complete revolution is effected in about 18 years and 7 months. Thus the nodes of the orbit of E round G *regrede* on the ecliptic at a rate about 1/250th part of the angular velocity of E round G.

Before proceeding further it will be useful to state the numerical magnitudes of some of the small terms. The direction cosines of E are λ, μ, ν. Now the inclinations of the moon's equator and the moon's orbit to the ecliptic are respectively $1\frac{1}{2}°$ and 5°. Hence the greatest value of ν is $\sin 6\frac{1}{2}°$, which is about $\frac{1}{10}$. It appears from Art. 552 that the mean value of μ is zero, while the libration in longitude is about 4 or 5 minutes. This would make the greatest value of $\mu = \sin 5' = \frac{1}{700}$. Thus $\lambda = 1 - \frac{1}{200}$. Again $r = \cos CZ = \cos 1\frac{1}{2}° = 1 - \frac{1}{3000}$ nearly. Hence $p^2 + q^2 = \frac{1}{1500}$ so that the greatest value of either p or q is about $\frac{1}{40}$. We shall now be able to estimate the magnitudes of the small terms rejected in the following investigation.

The figure has been drawn so that the direction cosines (λ, μ, ν) and (p, q, r) are positive. The poles C, Z, M are actually on a great circle and Z lies between C and M.

560. It will clearly be convenient to refer the motion to axes GX, GY, GZ fixed in space such that GZ is normal to the ecliptic. Let GA, GB, GC be the principal axes of the moon at the centre of gravity G. Let (p, q, r) be the direction-cosines of GZ referred to the co-ordinate axes GA, GB, GC. Then we have by Art. 18, since GZ is fixed in space,

$$p' - \omega_3 q + \omega_2 r = 0, \qquad q' - \omega_1 r + \omega_3 p = 0, \qquad r' - \omega_2 p + \omega_1 q = 0 \ldots\ldots\ldots(I),$$

where accents denote differential coefficients with regard to the time.

Let GC be the axis of rotation of the moon, and as before let the moment of inertia about GA be less than that about GB.

Now our object is to find the small oscillations about the state of steady motion in which GZ, GC, GM coincide. We shall therefore have p, q, ω_1, ω_2 small, and r very nearly equal to unity. The equations (I) therefore become

$$p' - nq + \omega_2 = 0, \qquad q' - \omega_1 + np = 0,$$

where n is the mean value of ω_3.

Let λ, μ, ν be the direction-cosines of the centre of force E as seen from G. Then we have by Euler's equations and Art. 519,

$$\left.\begin{array}{l} A\omega_1' - (B - C)\,\omega_2\omega_3 = -3n'^2 (B - C)\,\mu\nu \\ B\omega_2' - (C - A)\,\omega_3\omega_1 = -3n'^2 (C - A)\,\nu\lambda \\ C\omega_3' - (A - B)\,\omega_1\omega_2 = -3n'^2 (A - B)\,\lambda\mu \end{array}\right\} \ldots\ldots\ldots\ldots\ldots(II).$$

In the case of steady motion, the rigid body always turns the axis (GA) of lesser moment towards the centre of force, and $\omega_3 = n$. We have then both μ and ν small quantities, so that in the first equation we may neglect their product $\mu\nu$, and in the second equation we may put $\nu\lambda = \nu$. Also, we may put $\omega_3 = n = n'$ in the small terms.

If l be the latitude of the earth as seen from the moon, we have

$$\sin l = \cos ZE = p\lambda + q\mu + r\nu = p + \nu \text{ nearly.}$$

Hence the two first of Euler's equations take the form

$$\left. \begin{array}{l} A\omega_1' - (B - C)\, n\omega_2 = 0 \\ B\omega_2' - (C - A)\, n\omega_1 = -3n^2\, (C - A)\, (-p + \sin l) \end{array} \right\} \quad \text{.............. (III).}$$

If the earth, as seen from the moon, be supposed to move in a circular orbit in a plane making a constant inclination k with the ecliptic, and the longitude of whose ascending node is $-gt + \beta$, we shall have $\quad \sin l = k \sin(nt + gt - \beta)$.

In this expression g measures the rate at which the node regredes, and is about the two hundred and fiftieth part of n. We shall therefore regard g/n as a small quantity.

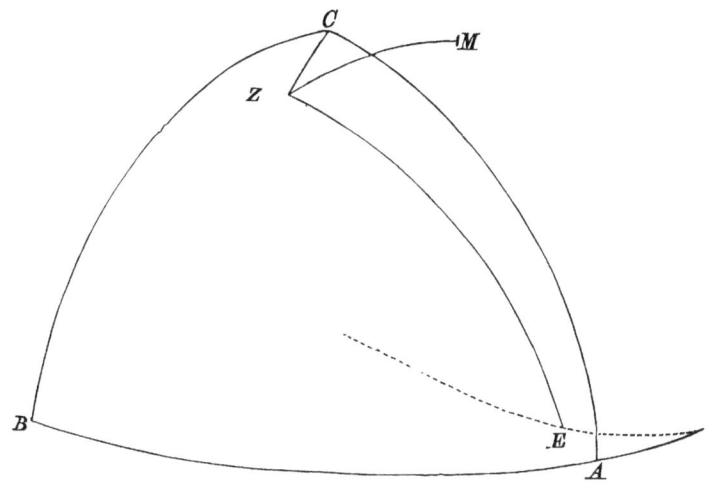

To solve these equations, it will be found convenient to substitute for ω_1, ω_2 their values in terms of p, q. We then have, as in Art. 15,

$$\left. \begin{array}{l} Aq'' + (A + B - C)\, np' - n^2\, (B - C)\, q = 0 \\ Bp'' - (A + B - C)\, nq' + 4n^2\, (C - A)\, p = 3n^2\, (C - A)\, \sin l \end{array} \right\} \quad \text{......... (IV).}$$

To find p, q, let us put $\quad p = P \sin\{(n + g)\, t - \beta\}, \quad q = Q \cos\{(n + g)\, t - \beta\}$, where P, Q are some constants to be determined by substitution in the equation.

We have $\quad \left. \begin{array}{l} Q\{A\,(n + g)^2 + (B - C)\, n^2\} = P\,(A + B - C)\, n\,(n + g) \\ P\{B\,(n + g)^2 - 4\,(C - A)\, n^2\} - Q\,(A + B - C)\, n\,(n + g) = -3n^2 k\,(C - A) \end{array} \right\}$.

We may solve these equations and find P and Q accurately. In the case of the moon the ratios $\dfrac{A - B}{C}$, $\dfrac{B - C}{A}$, $\dfrac{C - A}{B}$ and $\dfrac{g}{n}$ are all small; if we neglect the products of these small quantities, we have

$$\frac{Q}{P} = 1 - \frac{g}{n}, \qquad P = \frac{3nk\,(C - A)}{3n\,(C - A) - 2Bg}.$$

561.　The complementary functions. To find these we put

$$p = F \sin(st + H), \qquad q = G \cos(st + H).$$

R. D. II.　　　　　　　　　　　　　　　　　　　　　　　　　　　　　23

On substituting we have the quadratic

$$ABs^4 - \{(A+B-C)^2 - B(B-C) - 4A(A-C)\}n^2s^2 + 4(A-C)(B-C)n^4 = 0$$

to find s^2, and

$$\frac{G}{F} = \frac{(A+B-C)ns}{As^2 + (B-C)n^2},$$

to find the ratio of the coefficients of corresponding terms in p and q. If the roots of this equation were negative p and q would be represented by exponential values of t, and thus they would in time cease to be small. It is therefore necessary for stability that the coefficient of s^2 should be negative and the product $(A-C)(B-C)$ positive. Both these conditions are probably satisfied in the case of the moon. For since $B-C$ and $A-C$ are both small, the term $(A+B-C)^2$ is much greater than the two other terms in the coefficient of s^2. Also, since the moon is flattened at its poles, we shall probably have A and B both less than C.

We may approximate to the roots of this biquadratic in the following manner. Since the product of the roots (as indicated by the last term) is very small, and the sum of the roots (as indicated by the coefficient of s^2) is nearly equal to n^2; we see that one of the roots is very small and the other is nearly equal to n^2. To find the latter we put $s^2 = n^2 + x$, substitute in the equation, and neglect the squares of x. This gives $x = 3n^2(C-A)/C$ nearly. We thus find

$$s = n\left(1 + \frac{3}{2}\frac{C-A}{C}\right).$$

To find the former we reject s^4, writing s_1 for this root we have

$$s_1 = 2n\left(\frac{C-A}{C}\frac{C-B}{C}\right)^{\frac{1}{2}}.$$

Substituting these values in the expression for G/F, we find in these two cases

$$\frac{F}{G} = \frac{s}{n}, \qquad \frac{F_1}{G_1} = -\frac{1}{2}\left(\frac{C-B}{C-A}\right)^{\frac{1}{2}}.$$

It will be presently shown that $(C-A)/C = \cdot000597$ and $(C-B)/C = \cdot000033$. Taking these we see that the period of one of the complementary functions is very little less than a month, and the period of the second is about 3571 months or 274 years.

562. It appears therefore that each of the expressions for p and q contains three periodic terms but no constant terms. The periodic terms are the forced vibration due to the term $\sin l$ in equations (III), Art. 560, and the two complementary functions. We may approximately write these expressions in the form

$$p = -M\left(1 + \frac{g}{2n}\right)\sin\{(n+g)t - \beta\} + Ns\sin(st+H) + N_1s_1\sin(s_1t+H_1)$$
$$q = -M\left(1 - \frac{g}{2n}\right)\cos\{(n+g)t - \beta\} + Nn\cos(st+H) - 4N_1n\rho\cos(s_1t+H_1)$$

where $M = \dfrac{3(n - \frac{1}{2}g)\rho k}{2g - 3n\rho}$ and $\rho = \dfrac{C-A}{B}$. The numerical value of M is $1° 28'$, see Art. 559, so that M is about two-sevenths of k. It will presently appear that $(C-A)/B$ is $\cdot0006$ and, since g/n is about $\cdot0043$, it follows that M is positive.

563. *To find the motion of the principal axis GC in space and to deduce Cassini's theorem.* Let M be the pole of the orbit of E as seen from the centre of the moon, then M is the pole of the dotted line in the figure of Art. 560. If the longitude of E, viz. $\theta = (n+g)t - \beta$, is measured in the ecliptic from the ascending node of the orbit, the angle EZM measured positively in the direction of motion is $\frac{1}{2}\pi + \theta$.

Again, since p and q are the co-ordinates of Z referred to tangents at C to CA,

CB as axes and E never deviates far from A, we have $\cos EZC = -p/\surd(p^2+q^2)$, and $\sin EZC = q/\surd(p^2+q^2)$, where the radical has the positive sign. Hence

$$\sin CZM = \sin EZC \cos EZM - \cos EZC \sin EZM = \frac{-q \sin \theta + p \cos \theta}{\surd(p^2+q^2)},$$

$$\sin^2 CZ = p^2 + q^2.$$

Firstly, taking the forced vibration only, Art. 562, we write $p = -M(1+g/2n)\sin\theta$, and $q = -M(1-g/2n)\cos\theta$. We easily find that

$$\sin CZM = (-g/2n)\sin 2\theta, \qquad \sin CZ = M\{1-(g/2n)\cos 2\theta\}.$$

Thus the mean value of the angle CZM is zero. *The three points C, Z and M therefore make very small oscillations about a state of steady motion such that all three lie on the same great circle. At the same time the arc CZ is sensibly constant throughout the motion*.*

Next, if we include the complementary functions in the values of p and q, we find more complicated values for $\sin CZM$ and $\sin CZ$. Supposing however that N/M (Art. 562) is so small that we may reject all terms beyond its square, we again find that $\sin CZM$ is periodic, and that $\sin CZ$ differs from a constant only by periodic terms. Thus we again arrive at the result that *the three poles C, Z, M lie very nearly on the same great circle, at distances apart which are sensibly constant.*

We may show that the pole Z always lies between C and M by examining the relative positions when the longitude of E has any convenient value. When $\theta = \frac{1}{2}\pi$, the disturbing body E lies on the great circle MZ, so that the points M, Z and E lie on the circle AC very nearly. Also, since E is then in north latitude, EM is greater than EZ, i.e. AM is greater than AZ. But, when θ has the value $\frac{1}{2}\pi$, the expressions for p and q in Art. 562 show that p is negative, if we assume that the magnitude of the forced oscillation is greater than that of both the free oscillations. The arc AZ is therefore greater than AC. It follows, on these suppositions, that Z lies between C and M.

564. *Inclination of the moon's equator to the ecliptic and the numerical value of*

---

* If we represent by $d\psi/dt$ the angular velocity of GC round GZ we have by Art. 19 $\quad (p^2+q^2)\psi' = $(ang. vel. about GZ) $-$ (ang. vel. about GC) $\cos CZ$

$$= \omega_1 p + \omega_2 q + \omega_3 r - \omega_3 r.$$

Substituting for ω_1, ω_2 from equations I. of Art. 560 we have

$$\frac{d\psi}{dt} = \frac{\omega_3}{r} + \frac{pq' - qp'}{(p^2+q^2)r}. \qquad \text{Also } \sin^2 CZ = p^2 + q^2.$$

This expression for $d\psi/dt$ is accurate, and therefore when we substitute for p and q their approximate values we shall be able to estimate the effect of rejecting any small terms. This result may also be deduced from Euler's geometrical formula since $p = -\sin\theta\cos\phi$, $q = \sin\theta\sin\phi$.

Effecting the substitution and retaining the squares of N/M, we find that $d\psi/dt$ differs from $-g$, and $\sin CZ$ from M, only by small periodic terms.

It is known that the pole M moves backwards round the pole Z of the ecliptic with a mean angular velocity which we have called g. Thus M and C regrede round Z with the same mean angular velocity. It follows that the angle MZC remains very nearly constant throughout the motion.

By examining the value of the angle MZC when E is 90° from the node of its orbit and remembering that E is very close to the meridian CA we easily find that the angle MZC is very small.

23—2

$(C-A)/B$. It appears from what precedes that, when we neglect the free oscillations, the inclination CZ of the moon's equator to the ecliptic is given by

$$CZ = M - M\frac{g}{2n}\cos 2\{(n+g)t - \beta\}.$$

This is very nearly constant, the variations from its mean value being at most $\frac{1}{500}$th part of the inclination itself. The period of these variations is about half a month, strictly half a synodic month of the moon and node.

The mean inclination is M, a quantity not arbitrary but depending on the values of $(C-A)/B$ and g. Now g is well known, we may therefore use the expression for M given in Art. 562 to deduce an approximate value of $(C-A)/B$. The actual numerical value of the inclination has been found by Mayer and Nicollet to be $1°\ 28'$.

Neglecting all the periodical inequalities as being at most only a small fraction of M, Laplace found in this way $(C-A)/B = \cdot000599$, which is nearly equal to $\frac{1}{2}g/n$.

565. *Motion of the instantaneous axis in the body.* Taking the complete values of p and q with the complementary functions given in Art. 562 we easily find ω_1, ω_2 by the help of the formulæ $\omega_1 = np + dq/dt, \quad \omega_2 = nq - dp/dt,$ given in Art. 560. We thus find

$$\omega_1 = N_1 n s_1 \sin (s_1 t + H_1),$$
$$\omega_2 = 2gM\cos\{(n+g)t - \beta\} - 3Nn^2\frac{C-A}{C}\cos (st + H) - 4N_1 n^2\frac{C-A}{C}\cos (s_1 t + H_1).$$

If we disregard all but the forced vibration we have

$$\omega_1 = 0, \qquad \omega_2 = 2gM\cos\{(n+g)t - \beta\}.$$

Thus the *instantaneous axis moves in that principal plane which is at right angles to the axis pointed to the earth.* It oscillates about the axis of figure GC with a period which is about a month. The extent of the oscillation is however very small since the maximum value of ω_2/n is about $45''$.

566. *Ex.* Taking the same degree of approximation as before, deduce from the third of equations (II) in Art. 560 that the rotation of the moon, as found in Art. 552, is not affected by the obliquity of the ecliptic to the lunar equator.

567. *Effect of the motion of the ecliptic.* The dynamical equations (II) in Art. 560 are referred to axes fixed in the body and are therefore unaltered by making the pole Z move. In that article we substituted in these equations $\sin l - p$ for ν and $k \sin\{(n+g)t - \beta\}$ for $\sin l$ or $\tan l$. To make this correct it is sufficient to regard p, q, r as the direction cosines of the instantaneous position of GZ. We must therefore put on the right-hand sides of the geometrical equations (I) the resolved velocities of Z in space parallel to the axes GA, GB, GC. (See Art. 18.) Writing α and β for these additions we see that the equations (III) are altered only by such small terms as $Ad\beta/dt$ and $(B-C)na$.

Let GZ be referred to axes X_1, Y_1, Z_1 fixed in space and let each of the Eulerian co-ordinates of GZ, viz. θ_1, ψ_1 be expressed in a series of the form $\Sigma H \sin (ht + \epsilon)$, the values of the constants in the several terms being supplied by the planetary theory. The angular velocities in space of GZ resolved along and perpendicular to the plane $Z_1 GZ$ may therefore be represented by two series of the form $\Sigma Hh \cos (ht + \epsilon)$. To resolve these velocities parallel to the axes GA, GB, we multiply them by the cosines of the angles their directions make with those axes. Since the axes GA and GB are turning round GC with an angular velocity n, these cosines take the form $\cos (nt + \gamma)$. Thus on the whole we see that each of the quantities α and β may be represented by a series of the form $Kh \sin \{(n \pm h)t - l\}$.

Owing to the extreme smallness of all the values of H and h such terms as these may be regarded as insensible unless they rise into greater importance after solving the differential equations. Referring to Art. 560 we see that it is unnecessary to repeat the solution, for the expression for sin l in terms of t has the same form as the general term of the series for a or β. If we replace $-3n^2k$ by Knh and g by h, the expressions already found for P and Q will give the effect of the term added to the second of equations (III). In this way we see that amplitudes of p and q corresponding to the general terms of a and β are of the order P, where

$$P = \frac{Kh\,(C-B)}{3n\,(C-A)-2Bh}.$$

All the values of h are so small that the ratio of h/n to $(C-A)/C$ is insensible, hence the order of P is the same as that of $\frac{1}{3}Kh/n$. The general effect of the additional terms a and β is therefore to introduce periodical terms of the order Kh into the expressions for p and q.

The conclusion is that *the motion of the lunar equator relatively to the true ecliptic is independent of the motion of that ecliptic, so that the mean inclination of the lunar equator to the true ecliptic remains always the same notwithstanding the displacements of the latter.* This theorem is given by Laplace, *Mécanique Céleste,* Vol. II. p. 420.

568. *Second Approximation. Poisson's term of long period.* Having obtained a first approximation to the values of p and q in Art. 560 we may proceed to a second approximation by substituting the values thus found in the terms which were rejected in the first approximation. The terms of the first approximation being themselves very small, we can expect those of the second to become sensible only when they are magnified in the solution as explained in Art. 338. By referring to that article we see that those terms are magnified whose periods are nearly the same as those of the complementary functions. Hence, by Art. 561, those terms of the second order will be magnified whose periods are very long or nearly equal to that of the moon round the earth. We shall look for such terms, and if any be found we can then determine if they are sufficiently magnified to become sensible.

The only term which thus rises into importance is one of long period discovered by Poisson, see the *Connaissance des Tems* for 1821 published in 1819. The phase of this term is the difference between the longitudes of the apse of the Moon's orbit and its node on the ecliptic. The former of these advances slowly at the rate of 3° per month, while the latter regredes at the rate of $1\frac{1}{2}$° per month. Thus the period at which they separate by 360° is very long and equal to about 80 months or six years. Let h be the rate at which the apse advances, then the longitude of the moving apse is $a_1 = ht + a$. The longitude of the moving node is $\beta_1 = -gt + \beta$. For the sake of brevity we shall put

$$E = (g+h)\,t + a - \beta,$$

then E is the phase of Poisson's term, and $g + h = \frac{1}{80}n$.

569. To investigate the coefficient of Poisson's term we must recur to equations (II) of Art. 560. We must examine the terms $\mu\nu$ and $\nu\lambda$ to discover what combinations will give rise to terms of the form sin E or cos E.

Let us begin with the term $\mu\nu$. Since $\mu = \cos EB$ and is positive when E is in front of A, we see that $-\mu$ is the same as ϕ in Art. 552. Taking the elliptic inequality, we have

$$\mu = \frac{2e}{1 - 3\,(B-A)/C}\sin\,(nt - a_1).$$

Again $\nu = \sin l - p = (k + M) \sin \{(n + g) \, t - \beta\}$. Combining these two and rejecting all terms in the product except those of long period, we have

$$\mu\nu = e \, (k + M) \cos E.$$

We have also rejected the small term $3 \, (B - A)/C = \cdot 00167$ in the denominator, as this only alters the result by about one six hundredth part.

Let us next examine the term $\nu\lambda$ in the second of equations (II). If θ be the longitude of the moon we have, as in Art. 560, $\sin l = k \sin (\theta + gt - \beta)$.

But, by the theory of elliptic motion, $\theta = nt + 2e \sin (nt - a_1)$.

Substituting and retaining only that term of the second order whose phase is E, we have $\sin l = k \sin \{(n + g) \, t - \beta\} - ke \sin E$.

In the Lunar Theory * we find an additional term in the expression for $\sin l$, so that we should write $\sin l = k \sin \{(n + g) \, t - \beta\} - ke \, (1 - 3m^2) \sin E$,

where m is the ratio of the angular velocities of the sun and moon round the earth and is about equal to $\frac{1}{13}$. But this additional term is a very small fraction of those retained, and is of only slight importance. As in Art. 560 we have $\sin l = p\mu + q\mu + r\nu$. Now $q = - M \cos \{(n + g) \, t - \beta\}$ and $\mu = 2e \sin (nt - a_1)$. Also $\lambda = 1$ and $r = 1$. Hence, substituting and retaining only that term of the second order in which E is the phase, we find $\sin l = p + \nu + Me \sin E$.

Hence, substituting for $\sin l$, we have, since $\lambda = 1$,

$$\nu\lambda = (k + M) \left[\sin \{(n + g) \, t - \beta\} - e \sin E\right] + 3m^2 \, ke \sin E.$$

Again, referring to Art. 519, we see that the moment of the forces about the axis of y contains in the denominator the factor R^3. Hence we must multiply the term $- 3n^2 \, (C - A) \, \nu\lambda$ on the right-hand side of equation (II) in Art. 560 by $1 + 3e \cos (\theta - a_1)$. Effecting the multiplication, and retaining only those terms of the second order in which the phase is E, we have

$$(k + M) \left[\sin (nt + gt - \beta) - e \sin E + \frac{3e}{2} \sin E\right] + 3m^2 \, ke \sin E.$$

We thus find for the right-hand side of the second of equations (II)

$$- \tfrac{3}{2}n^2 \, (C - A) \, [2 \, (k + M) \sin (nt + gt - \beta) + (k + M + 6m^2 k) \, e \sin E].$$

Though the period of the term we are seeking is as long as 6 years, yet that of the long free vibration is over 200 years, see Art. 561. Thus though $d\omega_1/dt$ and $d\omega_2/dt$ contain the small factor dE/dt, yet this latter is about 35 times as great as the small coefficients $n \, (C - B)/A$ or $n \, (C - A)/B$ which occur in the second terms on the left-hand side of equations (II). If then we reject these terms, we only lose about $\frac{1}{35}$th part of a very small inequality, while we greatly simplify our result. Omitting then these two terms, the equations (III) of Art. 560 now take the form

$$\frac{d\omega_1}{dt} = 3n^2 \frac{C - B}{A} \, (k + M) \, e \cos E,$$

$$\frac{d\omega_2}{dt} = - 3n^2 \frac{C - A}{B} \, \{(k + M) \sin (nt + gt - \beta) + \tfrac{1}{2} \, (k + M + 6m^2 k) \, e \sin E\}.$$

---

* Referring to Godfray's Lunar Theory, Arts. 31 and 46, we find that the differential equation to determine the latitude is

$$\frac{d^2 s}{d\theta^2} + s = - \tfrac{3}{2} \frac{m' u'^3}{h^2 u^4} \, s + \ldots\ldots$$

$$= - \tfrac{3}{2} m^2 k \, \{1 - 4e \cos (c\theta - a)\} \sin (g\theta - \gamma) + \ldots$$

$$= - \tfrac{3}{2} m^2 k \, \{1 - 4e \cos (\theta - a_1)\} \sin (\theta - \gamma_1) + \ldots$$

where a_1 and γ_1 are the longitudes of the moving apse and node. Combining these and retaining only the term with the phase E, we find the above

$$= + \tfrac{3}{2} m^2 k \, . \, 2e \sin E.$$

Solving the differential equation as in Godfray, Art. 51, we find $s = 3m^2 ke \sin E$.

The first term on the right-hand side of the second equation may be added to the principal term already considered in the first approximation. We may omit it for the moment, as we merely want the term whose phase is E.

Integrating these we find

$$\left.\begin{aligned} np + \frac{dq}{dt} = \omega_1 &= 3n^2 \frac{C-B}{A} \frac{k+M}{g+h} e \sin E \\ nq - \frac{dp}{dt} = \omega_2 &= \frac{3n^2}{2} \frac{C-A}{B} \frac{k+M+6m^2k}{g+h} e \cos E \end{aligned}\right\}.$$

It is not difficult to solve these in the usual way. But it is evident that the terms dp/dt and dq/dt on the left-hand side contain the small factor $g+h$, and therefore are about $\frac{1}{30}$th of np and nq. Rejecting these we have

$$p = 3n \frac{C-B}{A} \frac{k+M}{g+h} e \sin E, \qquad q = \frac{3n}{2} \frac{C-A}{B} \frac{k+M+6m^2k}{g+h} e \cos E.$$

570. Representing Poisson's terms by $\omega_1 = Rn \sin E$ and $\omega_2 = Sn \cos E$, and including these in the first approximation we have, by Art. 565,

$$\omega_1 = Rn \sin E, \qquad \omega_2 = Sn \cos E + 2gM \cos D,$$

where $D = (n+g) t - \beta$, so that D is the mean angular distance of the moon from the ascending node. We have omitted the complementary functions as they appear to be almost insensible. Substituting these values of ω_1 and ω_2 in Euler's geometrical equations, and writing $\phi = D - \frac{1}{2}\pi$ we have

$$\left.\begin{aligned} \theta - \theta_0 &= - (gM/2n) \cos 2D - R \sin D \sin E - S \cos D \cos E \\ (\psi - \psi_0) \sin \theta_0 &= - (gM/2n) \sin 2D + R \cos D \sin E - S \sin D \cos E \end{aligned}\right\}.$$

571. The theory of the term of long period is given by Poisson in the *Connaissance des Tems for* 1821, and the numerical values of the coefficients are deduced from Nicollet's measures in the volume for the succeeding year. These coefficients have been improved by C. Simon in the third volume of the *Annales de l'école normal*, 1866. The coefficients as calculated from the Königsberg observations are very different from Poisson's. They may be found in Tisserand's *Mécanique Céleste*, 1891. As it cannot be considered that the ratios of the moments of inertia A, B, C have yet been determined with accuracy, it seems needless to examine into these differences. Merely to indicate the order of the several terms, we reproduce Simon's result

$$\left.\begin{aligned} \theta &= \theta_0 - (10''\cdot7) \cos 2D - (10'''\cdot5) \sin D \cos E - (94''\cdot15) \cos D \cos E \\ \psi &= \psi_0 - (414''\cdot7) \sin 2D + (405''\cdot5) \cot D \sin E - (3649''\cdot3) \sin D \cos E \end{aligned}\right\}.$$

These equations give the nutations of the polar axis of the moon. The precession of that axis is included in the term ψ_0 and has been determined in Art. 563. The real libration round the polar axis has been found in Art. 552. The visible oscillation of any spot is the resultant of all three.

572. Ex. 1. Let the motion of the moon be given by the equations

$$p = - M \sin D + R \sin E, \qquad q = - M \cos D + S \cos E,$$

see Arts. 562 and 569. Let l and λ be the lunar longitude and latitude of a spot referred to the moon's first meridian and equator; L and Λ the longitude and latitude of the same spot referred to the ecliptic and the first point of Aries. Prove that if λ be small

$$L = \theta_1 + \pi + l - M \tan \lambda \cos (D + l) - R \tan \lambda \sin l \sin E + S \tan \lambda \cos l \cos E,$$

$$\Lambda = \lambda + M \sin (D + l) - R \cos l \sin E - S \sin l \cos E,$$

where θ_1 is the mean longitude of the moon seen from the earth.

Poisson remarks that, in the case of the spot Manilius for which $l = 8° 46'$, $\lambda = 14° 26'$, the inequalities dependent on the angle E are very small and may be omitted. This circumstance however is peculiar to the spot chosen and would not be true for spots remote from the equator and from the apparent centre of the lunar disc.

Ex. 2. If the moon move so as always to turn the same face to the earth and if the instantaneous axes be nearly fixed in the body and nearly perpendicular to the axis pointed to the earth, prove that these two axes are principal axes.

This follows from equations (II) Art. 560.

573. **A difficulty in the figure of the moon.** It appears from Bouvard's and Nicollet's observations on the moon's true libration in longitude that $(B - A)/C = \cdot000564$, (Art. 555); and from Mayer's observations on the inclination of the moon's equator to the ecliptic Laplace found that $(C - A)/C = \cdot000599$, Art. 564. We therefore have $(C - B)/C = \cdot000035$. These values may appear very small, but they are much larger than could have been expected if the moon's surface had the form of equilibrium given by theory. Supposing the moon to be homogeneous and attracted by the earth, we may deduce from the principles of hydrostatics (as Laplace does) that $(B - A)/C = \cdot0000003618\lambda$ and $(C - A)/C = \cdot0000004824\lambda$, where λ is the ratio of the mass of the earth to that of the moon. Nicollet remarks that, even if we put $\lambda = 1000$ (instead of 80), these cannot be made as large as the values deduced from his observations on Manilius. Laplace observes that for a heterogeneous moon, if we suppose the density to increase from the surface to the centre, the hydrostatic theory would give values for $(B - A)/C$ &c. even less than for a homogeneous moon. He therefore concludes that *the moon has not the figure of equilibrium which it would have if originally fluid.* Laplace considers that the high mountains and other inequalities on the moon have a very sensible effect on the moments of inertia, and that this effect is the greater because the ellipticity of the moon's surface is small and its mass is inconsiderable. Poisson considers that the omission of the complementary functions by Nicollet may partly explain the difficulty; he thinks it doubtful that these functions should have entirely disappeared; *Connaissance des Tems* for the year 1822.

When it is remembered that the real libration of $4\frac{1}{2}'$ observed by Nicollet only subtends $1\frac{1}{8}''$ at the earth, it may be well believed that the errors of his observations may account for much of the discrepancy. This is rendered more probable when we learn from Tisserand that the more recent observations at Königsberg make the real libration in longitude about half that found by Nicollet. On the other hand these later observations make

$$(B - A)/C = \cdot000315, \quad (C - A)/C = \cdot000614, \quad (C - B)/C = \cdot000299,$$

and thus do not help to explain the discrepancy between the hydrostatic theory of the figure of the moon and the observations made on its surface.

CHAPTER XIII.

MOTION OF A STRING OR CHAIN.

The Equations of Motion.

574. **Cartesian equations.** *To determine the general equations of motion of an inextensible string under the action of any forces*.*

Let Ox, Oy, Oz be any axes fixed in space. Let $Xmds$, $Ymds$, $Zmds$ be the impressed forces that act on any element ds of the string whose mass is mds. Let u, v, w be the resolved parts of the velocities of this element parallel to the axes. Then, by D'Alembert's principle, the element ds of the string is in equilibrium under the action of the forces

$$mds\left(X - \frac{du}{dt}\right), \quad mds\left(Y - \frac{dv}{dt}\right), \quad mds\left(Z - \frac{dw}{dt}\right) \ldots\ldots (1),$$

and the tensions at its two ends.

Let T be the tension at the point (x, y, z), then $T\,dx/ds$, $T\,dy/ds$, $T\,dz/ds$ are its resolved parts parallel to the axes.

---

* The Cartesian equations of Art. 574 agree with those given by Poisson, *Journal de l'École Polytechnique*, 1820, and reproduced by him in his *Traité de Mécanique*. The geometrical equation is not there given, being replaced by Hooke's law. He thence deduces the differential equations of the motion of a tight string given in Art. 612. The proofs of the tangential and normal equations (1) to (4) for two dimensions in Art. 577 are very nearly the same as those given in Vol. IV. of the *Quarterly Journal*. Though the date of the volume is subsequent to that of the first edition of this treatise yet that of the paper itself must have been so nearly the same, that the solutions should be regarded as having been obtained independently. The author has not met with the two equations (5) and (6) of Art. 580 in any place with a date earlier than that of their publication in this treatise. Their application to initial motions is given further on. The two equations (1) and (2) for impulsive forces in Art. 583 appear to have been first given in College examination papers. The author believes the first to be due to the late Dr Todhunter.

The resolved parts of the tensions at the other end of the element

will be $\qquad T\dfrac{dx}{ds} + \dfrac{d}{ds}\left(T\dfrac{dx}{ds}\right)ds,$

and two similar quantities with y and z written for x.

Hence the equations of motion are

$$\left.\begin{aligned}
m\frac{du}{dt} &= \frac{d}{ds}\left(T\frac{dx}{ds}\right) + m\,X\\
m\frac{dv}{dt} &= \frac{d}{ds}\left(T\frac{dy}{ds}\right) + m\,Y\\
m\frac{dw}{dt} &= \frac{d}{ds}\left(T\frac{dz}{ds}\right) + m\,Z
\end{aligned}\right\} \dots\dots\dots\dots(2).$$

In these equations the variables s and t are independent. For any the same element of the string, s is always constant, and its path is traced out by variation of t. On the other hand, the curve in which the string hangs at any proposed time is given by variations of s, t being constant. In this investigation s is measured from any arbitrary point, fixed in the string, to the element under consideration.

To find the geometrical equations. We have

$$\left(\frac{dx}{ds}\right)^2 + \left(\frac{dy}{ds}\right)^2 + \left(\frac{dz}{ds}\right)^2 = 1 \dots\dots\dots\dots(3).$$

Differentiating this with respect to t, we get

$$\frac{dx}{ds}\frac{du}{ds} + \frac{dy}{ds}\frac{dv}{ds} + \frac{dz}{ds}\frac{dw}{ds} = 0 \dots\dots\dots\dots(4).$$

The equations (2) and (4) are sufficient to determine x, y, z, and T, in terms of s and t.

575. The equations of motion may be put under another form. Let ϕ, ψ, χ be the angles made by the tangent at x, y, z, with the axes of co-ordinates. Then the equations (2) become $\qquad m\dfrac{du}{dt} = \dfrac{d}{ds}(T\cos\phi) + mX \dots\dots\dots\dots (5),$

with similar equations for v and w.

To find the geometrical equations, differentiate $\cos\phi = dx/ds$ with respect to t;

$$\therefore -\sin\phi\,\frac{d\phi}{dt} = \frac{du}{ds} \dots\dots\dots\dots (6).$$

Similarly, by differentiating $\cos\psi = dy/ds$ and $\cos\chi = dz/ds$, we get two similar equations for ψ and χ. Taking these six equations in conjunction with the following, $\qquad \cos^2\phi + \cos^2\psi + \cos^2\chi = 1 \dots\dots\dots\dots (7),$ we have seven equations to determine u, v, w, ϕ, ψ, χ and T.

If the motion takes place in one plane, these reduce to the four following

equations : $\qquad\left.\begin{aligned} m\dfrac{du}{dt} &= \dfrac{d}{ds}(T\cos\phi) + mX\\ m\dfrac{dv}{dt} &= \dfrac{d}{ds}(T\sin\phi) + mY \end{aligned}\right\} \dots\dots\dots\dots(8),$

$$-\sin\phi\,\frac{d\phi}{dt}=\frac{du}{ds}, \qquad\qquad \cos\phi\,\frac{d\phi}{dt}=\frac{dv}{ds}\ldots\ldots\ldots\ldots\ldots(9).$$

The arbitrary constants and functions which enter into the solutions of these equations must be determined from the peculiar circumstances of each problem.

576. **Elastic Strings.** Let σ be the unstretched length of the arc s, and let $md\sigma$ be the mass of an element $d\sigma$ of unstretched length or ds of stretched length. Then, by the same reasoning as before, the equations of motion become

$$m\frac{du}{dt}=\frac{d}{d\sigma}\left(T\frac{dx}{ds}\right)+mX \ldots\ldots\ldots\ldots\ldots\ldots\text{(i)},$$

and two similar equations for v and w. To find the geometrical equations we must differentiate $\left(\dfrac{dx}{d\sigma}\right)^{2}+\left(\dfrac{dy}{d\sigma}\right)^{2}+\left(\dfrac{dz}{d\sigma}\right)^{2}=\left(\dfrac{ds}{d\sigma}\right)^{2},$

the independent variables being now σ and t. Differentiating with regard to t, we have

$$\text{have}\qquad \frac{dx}{d\sigma}\frac{du}{d\sigma}+\frac{dy}{d\sigma}\frac{dv}{d\sigma}+\frac{dz}{d\sigma}\frac{dw}{d\sigma}=\frac{ds}{d\sigma}\frac{d}{dt}\left(\frac{ds}{d\sigma}\right).$$

But, if λ be the modulus of elasticity of the string, we have $\dfrac{ds}{d\sigma}=1+\dfrac{T}{\lambda}\ \ldots\ldots$ (ii).

$$\text{Substituting we have}\qquad \frac{dx}{d\sigma}\frac{du}{d\sigma}+\frac{dy}{d\sigma}\frac{dv}{d\sigma}+\frac{dz}{d\sigma}\frac{dw}{d\sigma}=\left(1+\frac{T}{\lambda}\right)\frac{1}{\lambda}\frac{dT}{dt}\ \ldots\ldots\ldots\ldots\text{(iii)}.$$

The two equations (ii), and (iii) together with the three equations (i), will suffice for the determination of u, v, w, s and T in terms of σ and t.

If we wish to use the equations of motion in the forms corresponding to (5) or (8), the dynamical equations become

$$m\frac{du}{dt}=\frac{d}{d\sigma}\left(T\cos\phi\right)+mX,$$

with similar equations for v and w.

The geometrical equations corresponding to (6) or (9) may be found thus. We have

$$\frac{dx}{d\sigma}=\cos\phi\,\frac{ds}{d\sigma}=\cos\phi\left(1+\frac{T}{\lambda}\right).$$

Differentiating, we have $\dfrac{du}{d\sigma}=-\sin\phi\,\dfrac{d\phi}{dt}+\dfrac{1}{\lambda}\dfrac{d}{dt}\left(T\cos\phi\right),$

with similar expressions for v and w.

577. **Tangential and Normal Resolutions.** When the motion of the string takes place in one plane, it is often convenient to resolve the velocities along the tangent and normal to the curve.

Let u, v be the resolved parts of the velocity of the element ds along the tangent and normal to the curve at that element. Let ϕ be the angle which the tangent at the element makes with the axis of x. Let $Pmds$, $Qmds$ be the impressed forces on the element ds, resolved respectively in the directions of the tangent and normal. Then, by Chap. IV. of Vol. I., or by putting $\theta_3 = d\phi/dt$, $\theta_1 = 0$, $\theta_2 = 0$ in Art. (5) of this Volume, the equations of motion are

$$\frac{du}{dt}-v\frac{d\phi}{dt}=P+\frac{dT}{mds}\ldots\ldots\ldots\ldots\ldots\ldots(1).$$

$$\frac{dv}{dt}+u\frac{d\phi}{dt}=Q+\frac{T}{m\rho}\ \ldots\ldots\ldots\ldots\ldots(2).$$

The geometrical equations may be obtained as follows. If u_x be the resolved velocity parallel to Ox, we have

$$u_x = u \cos \phi - v \sin \phi.$$

Differentiating with respect to s, we have, by Art. 575,

$$-\frac{d\phi}{dt}\sin\phi = \left(\frac{du}{ds} - v\frac{d\phi}{ds}\right)\cos\phi - \left(\frac{dv}{ds} + u\frac{d\phi}{ds}\right)\sin\phi,$$

Since the axis of x is arbitrary in position, let us take it so that the tangent to the element during its motion is parallel to it at the instant under consideration; then $\phi = 0$, and we have

$$0 = \frac{du}{ds} - v\frac{d\phi}{ds} \quad\text{.........................} (3).$$

Similarly, by taking the axis of x parallel to the normal,

$$\frac{d\phi}{dt} = \frac{dv}{ds} + u\frac{d\phi}{ds} \quad\text{.........................} (4).$$

These four equations are sufficient to determine u, v, ϕ and T in terms of s and t.

If the string is extensible, the dynamical equations become

$$\left.\begin{aligned}\frac{du}{dt} - v\frac{d\phi}{dt} &= P + \frac{dT}{md\sigma} \\ \frac{dv}{dt} + u\frac{d\phi}{dt} &= Q + \frac{T}{m\rho}\frac{ds}{d\sigma}\end{aligned}\right\}.$$

To find the geometrical equations, we may differentiate $u_z = u\cos\phi - v\sin\phi$ with regard to σ. This gives by Art. 576

$$-\sin\phi\frac{d\phi}{dt} + \frac{1}{\lambda}\frac{d}{dt}(T\cos\phi) = \left(\frac{du}{d\sigma} - \frac{v}{\rho}\frac{ds}{d\sigma}\right)\cos\phi - \left(\frac{dv}{d\sigma} + \frac{u}{\rho}\frac{ds}{d\sigma}\right)\sin\phi.$$

By the same reasoning as before, this reduces to

$$\frac{1}{\lambda}\frac{dT}{dt} = \frac{du}{d\sigma} - \frac{v}{\rho}\left(1 + \frac{T}{\lambda}\right),$$

$$\frac{d\phi}{dt}\left(1 + \frac{T}{\lambda}\right) = \frac{dv}{d\sigma} + \frac{u}{\rho}\left(1 + \frac{T}{\lambda}\right).$$

578. The equations (2) and (3) may also be obtained in the following manner. The motion of the point P of the string being represented by velocities u and v along the tangent PA and the normal PC at P, the motion of a consecutive point Q will be represented by velocities $u + du$ and $v + dv$ along the tangent QB, and normal QC at Q. Let the arc $PQ = ds$, and let QN be a perpendicular on PA. Since the string is inextensible, the resultant velocity of Q resolved along the tangent at P must be ultimately the same as the resolved part of the velocity of P in the same direction. Hence

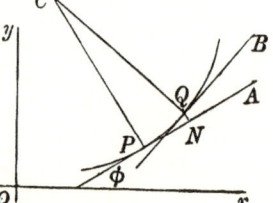

$$(u + du)\cos d\phi - (v + dv)\sin d\phi = u,$$

or, proceeding to the limit,

$$du - vd\phi = 0; \qquad \therefore \frac{du}{ds} - \frac{v}{\rho} = 0.$$

Again, $d\phi/dt$ is the angular velocity of PQ round P. Hence the difference of the velocities of P and Q resolved in any direction which is ultimately perpendicular to PQ must be equal to $PQ\, d\phi/dt$;

$$\therefore\ (u+du)\sin d\phi+(v+dv)\cos d\phi-v=ds\,\frac{d\phi}{dt},$$

or in the limit
$$\frac{d\phi}{dt}=\frac{dv}{ds}+\frac{u}{\rho}.$$

579. **Examples.** Ex. 1. If V be the vis viva of any arc AB of a chain; T_1, T_2 the tensions at the extremities of the arc; u_1', u_2' the velocities of the extremities resolved along the tangents at those extremities, u, v, w the Cartesian components of the velocity at any point, prove that

$$\tfrac{1}{2}dV/dt=T_2u_2'-T_1u_1'+\int(Xu+Yv+Zw)\,mds,$$

the integration extending over the whole arc.

Ex. 2. Investigate the polar equations of motion of a string in two dimensions. Let u, v be the resolved parts of the velocity of the element ds along and perpendicular to the radius vector, let $Pmds$, $Qmds$ be the resolved forces in the same directions, then

$$\frac{du}{dt}-\frac{v^2}{r}=\frac{dT}{ds}\cos\phi-\frac{T}{\rho}\sin\phi+P,\qquad \frac{dv}{dt}+\frac{uv}{r}=\frac{1}{r}\frac{d}{ds}(Tp)+Q,$$

$$-\sin\phi\,\frac{d\phi}{dt}=\frac{du}{ds},\qquad \cos\phi\,\frac{d\phi}{dt}=\frac{dv}{ds}+\frac{u\sin\phi}{r}-\frac{v\cos\phi}{r},$$

where ϕ is the angle the radius vector makes with the tangent and p is the perpendicular on the tangent.

Ex. 3. An elastic ring without weight, whose length when unstretched is given, is stretched round a circular cylinder. The cylinder is suddenly annihilated, show that the time which the ring will take to collapse to its natural length is $(Ma\pi/8\lambda)^{\frac{1}{2}}$, where M is the mass of the string, λ its modulus of elasticity, and a is the natural radius.

Ex. 4. A homogeneous light inextensible string is attached at its extremities to two fixed points, and turns about the straight line joining those points with uniform angular velocity. Let the straight line joining the fixed points be the axis of x. Show that the form of the string, supposing its figure permanent, is a plane curve whose equation is $1+(dy/dx)^2=b\,(a-y^2)^2$, where a and b are two constants.

580. The four equations of motion of Art. 577 may be reduced to two by the elimination of u and v. It will be found that we thus obtain two equations of convenient form which contain only the two unknown quantities T and ϕ. By eliminating T we may reduce these two equations to one and thus make the determination of the motion of the string depend on the solution of one differential equation. The elimination presents no difficulty but the result is not very simple.

Differentiating equation (1) with regard to s and (3) with regard to t we have

$$\frac{d\dot u}{ds}-\frac{dv}{ds}\,\dot\phi-v\,\frac{d\phi}{ds}=\frac{dP}{ds}+\frac{1}{m}\frac{d^2T}{ds^2}$$

$$\frac{d\dot u}{ds}-\dot v\,\frac{d\phi}{ds}-v\,\frac{d\dot\phi}{ds}=0,$$

where the dot represents differentiation with regard to t. Subtracting and substituting for $\dot{v}$ and dv/ds from (2) and (4), we have

$$\frac{d^2T}{ds^2} - T\left(\frac{d\phi}{ds}\right)^2 + m\left(\frac{dP}{ds} - Q\frac{d\phi}{ds}\right) = -m\left(\frac{d\phi}{dt}\right)^2 \quad \ldots\ldots\ldots(5).$$

In the same way, differentiating the equation (2) with regard to s, (4) with regard to t, and substituting, we have

$$\frac{1}{T}\frac{d}{ds}\left(T^2\frac{d\phi}{ds}\right) + m\left(P\frac{d\phi}{ds} + \frac{dQ}{ds}\right) = m\frac{d^2\phi}{dt^2} \quad \ldots\ldots\ldots\ldots(6).$$

If the string is heterogeneous m is a function of s. Putting $mds = d\sigma$, we find in the same way

$$\frac{d^2T}{d\sigma^2} - T\left(\frac{d\phi}{d\sigma}\right)^2 + \frac{dP}{d\sigma} - Q\frac{d\phi}{d\sigma} = -\frac{1}{m}\left(\frac{d\phi}{dt}\right)^2,$$

$$\frac{1}{T}\frac{d}{d\sigma}\left(T^2\frac{d\phi}{d\sigma}\right) + P\frac{d\phi}{d\sigma} + \frac{dQ}{d\sigma} = \frac{1}{m}\frac{d^2\phi}{dt^2}.$$

The equations (5) and (6) are of considerable utility. If the forces P, Q, the angular velocity $\dot{\phi}$, and the angular acceleration $\ddot{\phi}$ of each element are known in terms of s, we can deduce the tension of the string and the intrinsic equation of the curve in which it lies. Conversely if the distribution of tension, the curve of the string and the forces are known, the angular velocity and acceleration of every element are given at once.

581. Consider the position of the string at any instant. Let M be any point on the string, draw a straight line ON from the origin O parallel to the tangent at M and proportional in length to the tension of the string at M. The locus of N for all positions of M represents (as a kind of hodograph) the instantaneous distribution of tension along the string.

To simplify matters, let us suppose that the impressed forces P and Q are zero. The equations (5) and (6) show that the instantaneous values of T, ϕ, s, $-\dot{\phi}^2$, $\ddot{\phi}$ for a string are connected together just as the radius vector, longitude, time, radial and transversal forces are connected for a particle describing the hodograph.

By this analogy we may sometimes translate a question as to the instantaneous distribution of tension along a string into a more familiar problem on the motion of a single particle. If the string form a closed curve the allied curve is also closed. If the string have two ends, the terminal conditions must be made to correspond in the two curves.

582. **Examples.** Show how to deduce the analogy of Art. 581 from the Cartesian equations of motion of a string, Art. 574 ; and thence deduce equations (5) and (6) from the analogy. Show also that the analogy holds when the string moves in three dimensions.

Ex. 2. Determine the intrinsic equation to the form of a closed string and the distribution of tension when it is given that initially the square of the angular velocity of each element is proportional to the tension of that element, and that the angular velocity remains constant for a time dt. It is supposed that there are no impressed forces.

In this case, equations (5) and (6) become

$$\frac{d^2T}{ds^2} - T\left(\frac{d\phi}{ds}\right)^2 = -\mu T, \qquad \frac{1}{T}\frac{d}{dt}\left(T^2\frac{d\phi}{ds}\right) = 0.$$

If s represented the time these would be the equations of motion of a particle moving under a central force varying as the distance. This particle must describe an ellipse. Thus we have

$$\frac{1}{T^2} = \frac{\cos^2 \phi}{a^2} + \frac{\sin^2 \phi}{b^2}, \qquad \tan \phi = \frac{b}{a} \tan \sqrt{\mu s}.$$

These give the distribution of tension and the intrinsic equation. If l be the length of the string we see that $\sqrt{\mu l} = 2\pi$. If $a = b$ the curve is a circle.

Ex. 3. Show that the resultant acceleration of any point M of a string is represented in direction and magnitude by the tangent at N to the allied curve and in magnitude by the ratio of an elementary arc at N to the corresponding arc at M. Put $X = 0$, $Y = 0$ in the equations of Art. 574.

583. Impulsive forces. When the forces are impulsive the equations undergo some modifications. These may all be deduced in the usual manner from the corresponding equations for finite forces by integrating with regard to the time. But generally it will be found simpler to obtain them from first principles.

A string rests on a smooth horizontal table and is acted on at one extremity by an impulsive tension, to find the impulsive tension at any point and the initial motion.

Let T be the impulsive tension at any point P, $T + dT$ the tension at a consecutive point Q, then the element PQ is acted on by the tensions T and $T + dT$ at the extremities. Let ϕ be the angle made by the tangent at P to the string with any fixed line; u, v the initial velocities of the element resolved respectively along the tangent and normal at P to the string. Then, resolving along the tangent and normal, we have

$$\left. \begin{array}{l} m u \, ds = (T + dT) \cos d\phi - T \\ m v \, ds = (T + dT) \sin d\phi \end{array} \right\} ;$$

therefore, proceeding to the limit, $\qquad u = \dfrac{1}{m} \dfrac{dT}{ds}, \qquad v = \dfrac{1}{m} \dfrac{T}{\rho}.$

But, by Art. 577, we have $du/ds = v/\rho$. Hence the equation to find T becomes

$$\frac{d^2 T}{ds^2} - \frac{T}{\rho^2} = 0 \quad\dotfill (1).$$

If the chain be heterogeneous we easily find in the same way

$$\frac{d}{ds}\left(\frac{1}{m}\frac{dT}{ds}\right) = \frac{1}{m}\frac{T}{\rho^2} \dotfill (2).$$

If ω be the initial angular velocity of the element ds, we have

by Art. 577, $\qquad \omega = \dfrac{dv}{ds} + \dfrac{u}{\rho} = \dfrac{1}{T}\dfrac{d}{ds}\left(\dfrac{T^2}{m\rho}\right)\dotfill(3).$

584. If the string be in motion just before the action of the impulsive tension at one extremity, only a very slight modification of these equations is necessary. Let (u_1, v_1) (u_2, v_2) be the resolved velocities of the element PQ just before and just after the impulse.

We then simply modify the equations of the last article by writing

$$u = u_2 - u_1, \qquad v = v_2 - v_1.$$

Each of the resolutions $(u_1 v_1)$, $(u_2 v_2)$ must of course satisfy the geometrical equations obtained in Art. 577.

585.   Ex. If T_1, T_2 be the impulsive tensions at the extremities of any arc of the chain, u_1, u_2 the initial velocities at the extremities resolved along the tangents at the extremities, prove that the initial kinetic energy of the whole arc is

$$\tfrac{1}{2}(T_2 u_2 - T_1 u_1).$$

This readily follows by integrating $m(u^2 + v^2)\,ds$ along the whole length of the arc. But it also follows at once from the proposition proved in Vol. I. that the work due to an impulse is the product of the impulse into the mean of the resolved velocities of the point of application just before and just after the action of the impulse. Hence, since the string starts from rest the work done at either extremity is the product of the tension into half the initial tangential velocity.

586.   *To find the impulsive tension and the initial motion when the string forms a curve of double curvature.*

Let u, v, w be the resolved initial velocities of an element ds in the directions of the principal axes of the curve at that element ; the axis of x being the principal normal, that of y the tangent, and, z the binormal. Since the only forces on the element are the impulsive tensions at the extremities we have as in Art. 583,

$$u = \frac{1}{m}\frac{T}{\rho}, \qquad v = \frac{1}{m}\frac{dT}{ds}, \qquad w = 0 \dots\dots\dots\dots\dots(1).$$

To find the geometrical equations, we notice that while (u, v, w) represent the resolved velocities at one extremity A of the element ds along the principal axes at A, $(u + du, \&\text{c}.)$ represent the resolved velocities at the other extremity B of the same element along the principal axes at B. It follows that the relative velocities $(\delta u, \delta v, \delta w)$ of the extremities A and B resolved along the principal axes at A are given by Art. 21, where $d\phi_1$, $d\phi_2$, $d\phi_3$ are the angular displacements by which the principal axes at A are screwed into the positions of those at B. If $d\tau$ and $d\epsilon$ are the angles of torsion and contingence, we have $d\phi_1 = 0$, $d\phi_2 = -d\tau$, $d\phi_3 = -d\epsilon$. But, if ω_1, ω_2, ω_3 are the angular velocities of the element ds in space about the principal axes at A, we have $\delta u = -\omega_3 ds$, $\delta v = 0$, $\delta w = \omega_1 ds$. Equating these two sets of values of δu, δv, δw, we have

$$\omega_1 = \frac{u}{r}, \qquad \frac{dv}{ds} - \frac{u}{\rho} = 0, \qquad \frac{du}{ds} + \frac{v}{\rho} = -\omega_3 \dots\dots\dots\dots\dots(2),$$

where r and ρ are the radii of torsion and contingence.

Substituting from (1) in (2) we find

$$\omega_1 = \frac{T}{m\rho r}, \qquad \frac{d}{ds}\left(\frac{dT}{mds}\right) - \frac{T}{m\rho^2} = 0, \qquad -\omega_3 = \frac{1}{T}\frac{d}{ds}\left(\frac{T^2}{m\rho}\right)\dots\dots(3).$$

The second of these determines the initial tension when the form of the string is known, it is the same as the corresponding equation in two dimensions, so that the initial tension does not depend on the angle of torsion of the curve. The other two equations determine the initial angular velocities of the element, the angular velocity about the tangent not being required to find the initial motion.

We may verify these equations by a geometrical proof similar to that given in Art. 578 for a string in two dimensions.

587.   If the form of the string is given by its intrinsic equation $\rho = F(s)$, the

initial tension is to be found by solving the equation $\dfrac{d^2T}{ds^2} - \dfrac{T}{\rho^2} = 0$ ....(1).

The solution is known to be of the form $T = A\phi(s) + B\psi(s)$ ....(2),
where ϕ and ψ are some determinate functions of s and A and B are two undetermined constants. These constants must then be determined from the known values of the tension at the two extremities of the string.

The tension at any point of the string having been found, the velocity and direction of motion of any element may be deduced from the expressions given for the components u and v, Art. 583.

It is thus apparent that the determination of the motion depends on the solution of the differential equation (1). We have therefore thought it worth while to state in order a few solutions likely to be useful.

In some problems we have an additional term, say $f(s)$, on the right-hand side of the differential equation (1). The two first terms of the solution (1) constitute the complementary function, and when this has been found the particular integral due to $f(s)$ can be deduced by some one of the various rules given in the theory of differential equations. Perhaps the most convenient method is to substitute $T = z\phi(s)$ or $T = z\psi(s)$; the differential equation then takes a linear form from which z may be found. In what follows therefore it will be sufficient to suppose that the right-hand side of the differential equation is zero.

Case 1. Let ρ be constant, say $\rho = a$. *The form of the string is then a circle.* The solution is evidently $T = Ae^{s/a} + Be^{-s/a}$.

Case 2. Let ρ be a linear function of s, say $\rho = a + bs$. *The form of the string is then an equiangular spiral whose angle is* $\cot^{-1} b$. To solve the equation we put $a + bs = e^x$, the equation then takes the form considered in the last case. The complementary function reduces to $T = A(a + bs)^m + B(a + bs)^n$,
where m and n are the roots of the quadratic $b^2\kappa(\kappa - 1) = 1$.

Case 3. Let ρ be a quadratic function of s, say $\rho = a + bs + cs^2$. If the factors are real we may write this $\rho = c(s - a)(s - \beta)$. Assume as a trial solution

$$T = A(s - a)^m (s - \beta)^n.$$

Substituting in the differential equation and dividing by $(s - a)^{m-2}(s - \beta)^{n-2}$ we find

$$(m + n - 1)\left\{ \begin{array}{l} (m + n)s^2 - 2(an + \beta m)s \\ + a^2n(n - 1) + 2a\beta mn + \beta^2 m(m - 1) - c^{-2} \end{array} \right\} = 0.$$

The equation is satisfied if we choose m and n so that the coefficients of the several powers of s are zero. The two first powers lead to $m + n = 1$, and the last then gives $mn(a - \beta)^2 + c^{-2} = 0$. The required solution is therefore

$$T = A(s - a)^m (s - \beta)^n + B(s - a)^n (s - \beta)^m,$$

where m and n are the roots of the quadratic $x^2 - x = \{(a - \beta)c\}^{-2}$. This solution is given by Sir G. Stokes in the eighth volume of the *Cambridge Phil. Trans.*, 1849.

If the factors of the quadratic $\rho = a + bs + cs^2$ are imaginary, we may deduce the solution by rationalizing the value of T just found. But, putting $\rho = c\{(s + a)^2 + \beta^2\}$, it will be more convenient to proceed thus. If we put $s + a = \beta \tan \theta$, the differential equation takes the form

$$\frac{d^2}{d\theta^2}(T \cos \theta) + \left(1 - \frac{1}{\beta^2 c^2}\right)T \cos \theta = 0.$$

The solution of this equation is well known, and is trigonometrical or exponential according as βc is greater or less than unity.

If the factors of the quadratic $\rho = a + bs + cs^2$ are equal, we may solve the equation by writing $T = (s - a)\, z$ and $s - a = 1/x$.   The equation then reduces to $\dfrac{d^2 z}{dx^2} - \dfrac{z}{c^2} = 0$.

We therefore have $\qquad\qquad T = (s - a) \left\{ A e^{\frac{1}{c(s-a)}} + B e^{-\frac{1}{c(s-a)}} \right\}.$

If $c\rho = s^2 + c^2$ *the string has the form of a catenary.*   The solution is then

$$T = y\,(A\theta + B),$$

where y is the ordinate measured from the directrix, and θ is the angle the tangent makes with the horizon.   This result may be found, as just explained, by writing $s = c \tan \theta$.   But it may also be easily obtained by another process.   We notice that $T = y$ is one solution; putting $T = yz$ we have to find dz/ds a linear equation of the first order.   *See Cambridge Senate House Problems for* 1860 *with Solutions, page* 65.

Another solution is given in the ninth volume of *Liouville's Journal,* 1844, by Besge, who reduces the equation to one solved by Euler.

Let us write the equation in the form $\qquad \dfrac{d^2 T}{ds^2} = \dfrac{AT}{(a + 2bs + cs^2)^2}.$

Putting $\log T = \int U ds$, we find by substitution $\qquad \dfrac{dU}{ds} + U^2 = \dfrac{A}{(a + 2bs + cs^2)^2}.$

The denominator on the right-hand side suggests that a solution can be found of the form $\qquad\qquad U = \dfrac{V}{a + 2bs + cs^2}.$

Substituting in the differential equation we find

$$\frac{dV}{ds}\,(a + 2bs + cs^2) + (V - b - cs)^2 = (b + cs)^2 + A.$$

Now it is obvious that if we put $V - b - cs = k$, where k is some constant, the equation reduces to $\qquad\qquad ac - b^2 + k^2 = A.$

Thus we have two values for k.   Two particular integrals have therefore been found, viz. $\qquad\qquad \log T = \int \dfrac{b + cs \pm k}{a + 2bs + cs^2}\, ds.$

Each of these integrations can be effected in finite terms.   If the values of T thus found be $\phi\,(s)$ and $\psi\,(s)$, the general integral required is $T = M\phi\,(s) + N\psi\,(s)$, where M and N are two arbitrary constants.

Case 4.   If ρ^2 (not ρ) be a quadratic function of s, say $\rho^2 = a + bs + cs^2$, we may find a solution in finite terms of the form

$$T = A_0 + A_1 s + \ldots\ldots + A_n s^n,$$

provided the quadratic $cn\,(n-1) = 1$ gives a *positive* integral root.   This quadratic expresses the condition that the series for T has a highest term, it is therefore easily remembered by substituting only the highest power $A_n s^n$ of the series in the differential equation and rejecting all lower powers as they occur.   The relation between the successive coefficients may be easily found by substitution.   This relation will be much simplified by previously *clearing the quadratic for* ρ^2 *of either of the terms bs, or* a.   This is effected by writing $s = s' + m$ and choosing the constant m properly.

If n be an integral root of the quadratic $cn\,(n-1) = 1$, a solution may be written in either of the forms

$$T = A\rho^2 \left(\frac{d}{ds} \right)^n \rho^{2(n-1)}, \qquad\qquad T = A' \left(\frac{d}{ds} \right)^{n-2} \rho^{2(n-1)},$$

see a paper by the author in the *Proceedings of the Mathematical Society,* April 1885.

Case 5. If $1/\rho^2$ be a quadratic function of s, say $1/\rho^2 = a + bs + cs^2$, put $T = ze^{as + \beta s^2}$. Substituting, and choosing a and β properly, we reduce the equation to the form

$$\frac{d^2z}{ds^2} + (2a + 4\beta s)\frac{dz}{ds} + hz = 0.$$

This artifice is attributed to Liouville.

Putting $a + 2\beta s = \sigma$, a solution in the form of a finite series, viz.

$$z = A\left[\sigma^n + \tfrac{1}{2}n(n-1)\beta\sigma^{n-2} + \tfrac{1}{2}\cdot\tfrac{1}{4}n(n-1)(n-2)(n-3)\beta^2\sigma^{n-4} + \&c.\right]$$

may be found by substitution when $4\beta n + h$ gives a positive integral value of n. It is also shown in the paper already quoted from the Math. Soc. that

$$z = \frac{A}{R}\left(\frac{d}{d\sigma}\right)^n R, \quad \text{or} \quad A\left(\frac{d}{d\sigma}\right)^{-n-1}\frac{1}{R}, \quad \text{where} \quad \log R = \frac{\sigma^2}{2\beta};$$

one form or the other being used according as n is a positive or negative integer.

588. **Ex. 1.** If the curve in which the string is placed be such that $\rho^2 = \dfrac{s^2 - a^2}{i(i+1)}$, where i is any positive integer, show that one solution is $T = \int P_i dx$, where $x = s/a$ and P_i is a Legendre's function of x of the ith order.

Ex. 2. Trace the curve $\beta\rho = (s - a)(s - b)$.

The curve has three branches; the first extends from $s = a$ to b, the curvature is always in one direction and the branch terminates at each extremity with an infinite number of diminishing convolutions, being ultimately an equiangular spiral whose angle is $\tan^{-1}\beta/(a - b)$. The second branch extends from $s = b$ to ∞, it unwinds like an equiangular spiral with an infinite number of turns. The winding and unwinding branches have the same directions of curvature when the arc in each is measured from the infinitely small cusp. The unwinding branch finally proceeds to infinity, like one branch of the catenary $\beta\rho = s^2 + \beta^2$, the tangent being ultimately parallel to that at $s = \tfrac{1}{2}(a + b)$. The third branch extends from $s = -c$ to $-\infty$ and resembles the second branch.

Ex. 3. A string at rest on a table is jerked at one end, and begins to move so that the direction of motion of any element makes a constant angle with the tangent at that point. Prove that the curve in which the string rests is an equiangular spiral.

Ex. 4. An impulsive tension in the direction of the tangent is applied to one extremity of a uniform perfectly flexible heavy string lying on a smooth plane. If all the particles of the string start with equal velocities, prove that the string must lie in the form of a catenary or of a straight line.   [May Ex.

Ex. 5. An inelastic string, at rest in a circular tube which it just fills, is plucked at one end in the direction of the tangent at that end and begins to move with kinetic energy E. If the same string were unconfined and similarly plucked when at rest, show that it would move off with kinetic energy $2\pi E \coth(2\pi)$.

[Math. Tripos.

589. **Initial motions.** *A string in one plane is either at rest under the action of given forces or has its instantaneous motion known. Supposing a fracture or some other change to occur, it is required to find the initial changes of motion and the initial change of tension.*

Let $mPds$, $mQds$ be the resolved parts of the forces respectively along the tangent and radius of curvature at any element ds of the string. Let u, v be the resolved parts of the velocity in the same directions. Let mT be the tension. Let ψ be the angle which the tangent at the element ds makes with the axis of x, and let $\omega = d\phi/dt$ be the angular velocity of the element ds.

We have, by Art. 577, the equations

$$\dot{u} - v\omega = P + \frac{dT}{ds} \quad \ldots\ldots (1).$$

$$\frac{du}{ds} - \frac{v}{\rho} = 0 \quad \ldots\ldots (3).$$

$$\dot{v} + u\omega = Q + \frac{T}{\rho} \quad \ldots\ldots (2).$$

$$\frac{dv}{ds} + \frac{u}{\rho} = \omega \quad \ldots\ldots (4).$$

From these we deduce as in Art. 580 the two equations

$$\frac{d^2T}{ds^2} - \frac{T}{\rho^2} + \frac{dP}{ds} - \frac{Q}{\rho} = -\omega^2 \quad \ldots\ldots\ldots\ldots\ldots (5).$$

$$\frac{1}{T}\frac{d}{ds}\left(\frac{T^2}{\rho}\right) + \frac{P}{\rho} + \frac{dQ}{ds} = \frac{d\omega}{dt} \quad \ldots\ldots\ldots\ldots\ldots (6).$$

The instantaneous motion of the string being given and also the forces, ω, P and Q are all known functions of s. Thus (5) is the differential equation from which we have to find T. This differential equation is the same as the one already considered in Art. 587. We shall therefore suppose its solution to have been found. The constants of integration are to be determined by the given conditions at the extremities of the string. Thus the initial tension is found.

The initial values of u, v, ω, P, Q and T being known, the values of $\dot{u}$, $\dot{v}$ and $\dot{\omega}$ are found from (1), (2) and (6). Thus all the initial accelerations have been determined.

Differentiating (5) with regard to t, we have *another differential equation to find $\dot{T}$ of the same kind as before*. Having solved this, we may find $\ddot{u}$, $\ddot{v}$ and $\ddot{\omega}$ by differentiating (1), (2) and (6).

Proceeding in this way we may find the instantaneous values of all the differential coefficients of u, v, ω at the instant when the fracture occurs.

If u_t, v_t, ω_t be the values of these quantities after any time t, we have by Taylor's theorem (see Vol. I. Art. 199)

$$u_t = u + \dot{u}t + \tfrac{1}{2}\ddot{u}t^2 + \ldots$$

with similar expressions for v_t and ω_t. Thus the initial motion has been found to any degree of approximation.

590. To find the initial radius of curvature R of the path in space of any element of the string, we resolve the forces on that element in a direction perpendicular to the tangent to its path and equate the result to $(u^2 + v^2)/R$. The direction of motion of the element makes angles with the tangent and normal to the string whose *sines* are $v/(u^2 + v^2)^{\frac{1}{2}}$ and $u/(u^2 + v^2)^{\frac{1}{2}}$. The forces on the element are $P + dT/ds$ and $Q + T/\rho$. We therefore have

$$\frac{(u^2 + v^2)^{\frac{3}{2}}}{R} = u\left(Q + \frac{T}{\rho}\right) - v\left(P + \frac{dT}{ds}\right) \quad \ldots\ldots\ldots\ldots (7).$$

To find the rate at which the radius of curvature of the string begins to change, we notice that $\dfrac{1}{\rho} = \dfrac{d\psi}{ds}$. Hence $\dfrac{d}{dt}\dfrac{1}{\rho} = \dfrac{d\omega}{ds}$. Thus by differentiating (5) with regard to s, we find the rate at which the curvature of the string begins to change. By differentiating (6) with regard to s, we find the acceleration of the change of curvature.

591. If the string start from rest, u, v and ω are all zero. In this case the equations (5) and (6) of Art. 589 follow immediately from the corresponding equations for impulsive forces. Following Newton's argument in Prop. 1 of his second section we may treat the forces Pdt, Qdt as small impulses. The argument is then the same as that given in Art. 583.

The initial direction of motion of any element is found by compounding the velocities $\dot{u}dt$, $\dot{v}dt$ so that the direction of motion makes with the tangent to the string an angle equal to $\tan^{-1} \dot{v}/\dot{u}$. To find the initial radius of curvature of the path of any particle, we see by Vol. I. Art. 212, that we must find $\ddot{u}$, $\ddot{v}$ by differentiating twice the equations (1) and (2).

592. EXAMPLES. *A string is in equilibrium in the form of a circle about a centre of repulsive force in the centre. If the string be now cut at any point A, prove that the tension at any point P is instantaneously changed in the ratio*

$$e^{\pi} + e^{-\pi} - e^{\pi - \theta} - e^{-(\pi - \theta)} \;:\; e^{\pi} + e^{-\pi},$$

where θ is the angle subtended at the centre by the arc AP.

Let F be the central force, then $P = 0$, and $mQ = -F$. Let a be the radius of the circle. Then the equation of Art. 589 to determine T becomes $\dfrac{d^2T}{ds^2} - \dfrac{T}{a^2} = -\dfrac{F}{a}$.

Let s be measured from the point A towards P, then $s = a\theta$; also F is independent of s. Hence we have $\qquad T = Fa + A\epsilon^{\theta} + B\epsilon^{-\theta}$.

To determine the arbitrary constants A and B we have the condition $T = 0$ when $\theta = 0$ and $\theta = 2\pi$; also just before the string was cut $T = Fa$. Hence the result given in the enunciation follows.

Ex. 2. A string is wound round the under part of a vertical circle and is just supported in equilibrium at the ends of a horizontal diameter by two forces. The circle being suddenly removed, prove that the tension at the lowest point is instantly decreased in the ratio $4 : e^{\frac12 \pi} + e^{-\frac12 \pi}$

Ex. 3. The extreme links of a uniform chain can slide freely on two intersecting straight lines, which are at right angles and equally inclined to the vertical. The chain is in equilibrium under the action of gravity. If now the chain break at the lowest point, show that the tension at any point P is equal to the statical tension multiplied by $2\phi/(\pi + 2)$, where ϕ is the angle which the tangent at P makes with the horizon.

Ex. 4. A string rests on a smooth table in the form of an arc of an equiangular spiral, and begins to move from rest under the action of a central force F which tends from the pole and varies as the nth power of the distance, show that the initial tension is given by $T = -rF \dfrac{n\cos^2\alpha + \sin^2\alpha}{n(n+1)\cos^2\alpha - \sin^2\alpha} + Ar^p + Br^q$, where α is the angle of the spiral, and p, q are the roots of the quadratic $x(x-1) = \tan^2\alpha$. Show that the solution changes its form when α is such that the first term is infinite, and find the new form.

Ex. 5. A given heavy uniform inelastic chain is stretched nearly straight with the two ends at the same level; suddenly one end is released, prove that, to a first approximation, half the product of the tensions at the other end before and after release is equal to the square of the weight of the chain.

[Math. Tripos, 1888.

593. **Ex. 1.** An endless string in the form of a circle is rotating in its own plane with a uniform angular velocity ω. The string being cut at any point, find the initial tension, the initial radius of curvature of the path of any element, and the rate at which the tension is changing.

Let OCA be the diameter through the point of fracture A, and let the arc be measured from O. Let a be the radius and let $s = a\phi$. Since there are no impressed forces, $P = 0$, $Q = 0$. We have at once by (5), since $\rho = a$,

$$T = a^2\omega^2 + A \cosh \phi + B \sinh \phi,$$

where A and B are such that $T = 0$ when $\phi = \pm \pi$,

$$\therefore \ T = a^2\omega^2 (1 - \cosh \psi / \cosh \pi).$$

To find the radius of curvature of the path of any element, we notice that each element is moving with a velocity $u = a\omega$ along the tangent to the string. Resolving these along the normal to the string, we have $u^2/R = T/a$ whence $R = u^2 a/T$. This result follows at once from equation (7) since $v = 0$, $Q = 0$. To find $\dot{\omega}$, we have from (6), since $\rho = a$, $a^2\dot{\omega} = 2dT/d\psi$. By differentiating (5) with regard to t we obtain

$$\frac{d^2\dot{T}}{ds^2} - \frac{\dot{T}}{\rho^2} - \frac{2T}{\rho}\frac{d\omega}{ds} = -2\omega\dot{\omega}.$$

Since $d\omega/ds = 0$, we find by solving this differential equation,

$$\dot{T} = 2a^2\omega^3 \left(\frac{\phi \cosh \phi}{\cosh \pi} - \frac{\pi \sinh \phi}{\sinh \pi} \right).$$

By differentiating (5) and (6) with regard to s we may also show that the rate $\dot{\rho}$ at which the radius of curvature of the string is changing is initially zero and that the acceleration is initially equal to $2a\omega^2 \cosh \psi \cdot \operatorname{sech} \pi$.

Ex. 2. A string moves under the action of a central force $F(r)$ tending from the origin. The instantaneous motion being known, show that T may be found from

$$\frac{d^2T}{ds^2} - \frac{T}{\rho^2} + \frac{dF}{dr} \cos^2 \phi + \frac{F}{r} \sin^2 \phi = -\omega^2.$$

If the string start from rest and both its extremities are free, prove that dT/dt is initially zero throughout the string.

Ex. 3. A string of length $2a\alpha$ is at rest in the form of an arc of a circle of radius a and is acted on by a central force $F(r)$ tending from the centre of the circle. Show that the instantaneous tension at any point P is

$$T = aF(a)(1 - \cosh \theta / \cosh \alpha),$$

where θ is the angle subtended at the centre by the arc OP measured from the middle point O of the string.

Ex. 4. A heavy uniform string of given length is placed at rest on a rough table whose coefficient of friction is μ, and is acted on by a finite force at each end. If each element of the string begin to move in a direction making a given angle β with the tangent at the element, prove that the intrinsic equation to the string is

$$\frac{1}{\rho} = \frac{1}{a} e^{-2\phi \cot 2\beta} + \frac{1}{b} e^{-\phi \cot \beta},$$

where ϕ is the angle the tangent makes with a fixed straight line. Prove also that the force at either end must be $\mu b \sin \beta e^{\phi_1 \cot \beta}$ where ϕ_1 is the value of ϕ at that end. If a is infinite the curve is an equiangular spiral and the string is in equilibrium.

Ex. 5. If, in the last example, each element begin to move in a direction making an angle ϕ with the tangent, prove that the intrinsic equation is $a/\rho = 1 + b \sec^2 \phi$, where a and b are arbitrary constants and the force at either end is $\mu a \sin \phi$.

On Steady Motion.

594. DEF. When the motion of a string is such that the curve which it forms in space is always equal, similar, and similarly situated to that which it formed in its initial position, that motion may be called steady.

To investigate the steady motion of a homogeneous inextensible string.

It is obvious that every element of the string is animated with two velocities, one due to the motion of the curve in space, and the other to the motion of the string along the curve which it forms in space. Let a and b be the resolved parts along the axes of the velocity of the curve at the time t, and let c be the velocity of the string along its curve. Then, following the usual notation, we have

$$u = a + c \cos \phi, \quad v = b + c \sin \phi \quad \ldots\ldots\ldots\ldots (1).$$

Now a, b, c are functions of t only, hence $du/ds = - c \sin \phi \, d\phi/ds$. Therefore by equation (9) of Art. 575 we have

$$\frac{d\phi}{dt} = c \frac{d\phi}{ds} \ldots\ldots\ldots\ldots\ldots\ldots\ldots (2).$$

Substituting the values of u and v in the equations of motion, Art. 574, we get

$$\left. \begin{aligned} \frac{da}{dt} + \frac{dc}{dt} \cos \phi - c \sin \phi \frac{d\phi}{dt} &= X + \frac{d}{ds}\left(\frac{T}{m} \cos \phi\right) \\ \frac{db}{dt} + \frac{dc}{dt} \sin \phi + c \cos \phi \frac{d\phi}{dt} &= Y + \frac{d}{ds}\left(\frac{T}{m} \sin \phi\right) \end{aligned} \right\}.$$

Substituting for $d\phi/dt$, these equations reduce to

$$\left. \begin{aligned} \frac{da}{dt} &= \left(X - \frac{dc}{dt} \cos \phi\right) + \frac{d}{ds}\left\{\left(\frac{T}{m} - c^2\right) \cos \phi\right\} \\ \frac{db}{dt} &= \left(Y - \frac{dc}{dt} \sin \phi\right) + \frac{d}{ds}\left\{\left(\frac{T}{m} - c^2\right) \sin \phi\right\} \end{aligned} \right\} \quad \ldots\ldots (3).$$

The form of the curve is to be independent of t; hence, on eliminating T, the resulting equation must not contain t. This will not generally be the case unless da/dt, db/dt, dc/dt are constants. The motion is then called a *uniform steady motion*. In any case their values will be determined by the known circumstances of the problem. The above equations must then be solved,

s being supposed to be the only independent variable, and t being constant.

595.   If a, b, c are constants, these equations take a simpler form.   We then have

$$0 = mX + \frac{d}{ds}(T'\cos\psi), \qquad 0 = mY + \frac{d}{ds}(T'\sin\psi),\dots(4),$$

where $T' = T - mc^2$.   These are the equations of equilibrium of a string acted on by the same given forces, viz. mX and mY.   Thus we have a very convenient analogy between the steady motion and the equilibrium of a homogeneous string.

For example, if a string can move in a uniform steady motion under the action of gravity, we see that its form at any and every moment must be the same as that of a string in equilibrium under the action of gravity.   The form of the travelling curve must therefore be a catenary.   What catenary it is will depend on the terminal conditions, and if these are inconsistent with the properties of a catenary no uniform steady motion is possible.

Whatever catenary the string assumes, the tension T at any point of the moving string will exceed the tension at the corresponding point of the stationary catenary by mc^2.   We have therefore at any point $T = m(gy + c^2)$, where y is the ordinate of that point measured from the directrix.

More generally, we see from the equations (4) that a string cannot move in uniform steady motion unless every one of its positions is one in which a string could rest in equilibrium under the action of the instantaneous forces.   Supposing this condition to be satisfied, the conditions at the extremities (if the string form an unclosed curve) must also be consistent with this form of the string.   These are the necessary and sufficient conditions.

One important case of this theorem is when the string forms a closed curve which does not travel in space.   This case was first given in the *Solutions of Cambridge Problems*, 1854, *by Walton and Mackenzie*, who enunciated the theorem as follows.   If a uniform endless chain rest in any form subject to the action of forces depending only on the position of the particle acted on and to the reactions of smooth surfaces, it will continue to move in the same form if put in motion in such a manner that every point of the chain begins to move in the direction of the tangent at that point.

596.   **Examples.**   Ex. 1.   A horizontal cylinder revolves with uniform velocity about its axis and an endless chain passing round it revolves with it in such a manner that the form of the chain in space is always the same; show that the form of the curve is independent of the velocity.   [Math. Tripos, 1854.

Ex. 2.   A uniform string AB of any given length is placed in the form of an arc of an equiangular spiral, and is acted on by a centre of repulsive force situated in the pole O of the spiral whose accelerating force is equal to $\mu/(\text{distance})^2$.   Each element starts with a velocity u along the tangent to that element, and the extremities A, B are acted on by forces F_1, F_2.   If $F_1 = m(u^2 + \mu/OA)$ and $F_2 = m(u^2 + \mu/OB)$ where m is the mass of a unit of length, prove that the string will describe the spiral uniformly.

Ex. 3. A light flexible inextensible tube of small uniform section suspended from two points in the same horizontal line by its ends is full of water which flows through it with uniform velocity. Prove that it hangs in the form of the common catenary, and that the longitudinal tension is constant.   [Math. Tripos.

597. Ex. **Form of an electric cable.** *An electric cable is deposited at the bottom of a sea of uniform depth from a ship moving with uniform velocity in a straight line, and the cable is delivered with a velocity c equal to that of the ship. Determine the form of the string when the motion is steady.*

Consider the portion of the cable between the ship A and the ground B. If the friction of the water on the string is neglected, gravity diminished by the buoyancy of the water will be the only force acting on the string, let this be represented by g'. Then the form of the travelling curve is the common catenary, and the tension at any point exceeds the tension in the catenary (see Art. 595) by the weight of a length of string equal to c^2/g'.

To determine the particular catenary assumed by the string we consider the conditions at the extremities A and B. At the point B where the cable meets the ground the tangent to the catenary must be horizontal. For, if not, an element of string at B would have the tangents at its extremities inclined to each other at a *finite* angle. Then since T cannot be zero in a catenary, this elementary mass would be acted on by a finite resultant force. Hence the element would alter its position with an infinite velocity. The catenary therefore must be such that B is its vertex.

To fix the catenary one more condition is necessary. If l be the length of the portion of cable between the ship and the ground and h the depth of the sea, then the parameter γ of the catenary must satisfy the equation $(h+\gamma)^2 = l^2 + \gamma^2$.

The problem of the deposition of an electric cable appears first to have been considered by Longridge and Brooks (Institution of Civil Engineers, Feb. 1858). Another solution was given by Sir G. Airy in the *Phil. Mag.* for July, 1858. A further discussion by Mr Woolhouse may be found in the *Phil. Mag.* for May, 1860. All these include in their investigations the friction between the water and the cable.

598. We shall now consider how the solution is affected when the friction of the water on the cable is taken account of. We shall assume that *the friction on any element of the cable varies as the velocity in space of that element, and acts in a direction opposite to the direction of motion of the element.* Each element has motions both along the cable and transverse to it; and the coefficients of friction for these two motions are probably not strictly equal. In order however to simplify the formulae we here treat them as equal. Let μ be the coefficient of friction.

Let the axis of x be horizontal, and let x' be the abscissa of any point of the cable measured from the place where the cable touches the ground, in the direction of the ship's motion. Also let s' be the length of the curve measured from the same point. Then $x = x' + ct$, and $s = s' + ct$.

Following the same notation as before, we have

$$X = -\mu u, \qquad Y = -g' - \mu v.$$

But $\qquad\qquad u = c - c \cos\phi, \qquad v = -c \sin\phi.$

Hence the equations (3) of Art. 594 become

$$\left. \begin{aligned} 0 &= -\mu c + \mu c \cos\phi + \frac{d}{ds}\left\{\left(\frac{T}{m} - c^2\right)\cos\phi\right\} \\ 0 &= -g' + \mu c \sin\phi + \frac{d}{ds}\left\{\left(\frac{T}{m} - c^2\right)\sin\phi\right\} \end{aligned} \right\}.$$

To integrate these put $\sin \phi = dy/ds$, $\cos \phi = dx/ds$. Hence,

$$\left. \begin{aligned} g'A &= -\mu cs + \mu cx + \left(\frac{T}{m} - c^2\right)\cos\phi \\ g'B &= -g's + \mu cy + \left(\frac{T}{m} - c^2\right)\sin\phi \end{aligned} \right\} \quad\dots\dots\dots\dots\dots\dots(1),$$

where A and B are two arbitrary constants.

At the point where the cable meets the ground, we must have either $T=0$ or $\phi=0$. For if ϕ be not zero, the tangents at the extremities of an infinitely small portion of the string make a finite angle with each other. Then, if T be not zero, resolving the tensions at the two ends in any direction, we have an infinitely small mass acted on by a finite force. Hence the element will in that case alter its position with an infinite velocity. Firstly, let us suppose that $\phi=0$. Also, at the same point, $y=0$ and $z'=0$. Hence $B=-ct$.

Putting $\dfrac{\mu c}{g'}=e$, we get by division $\quad \dfrac{dy}{dx'} = \dfrac{s' - ey}{A - ex' + es'}$ ....(2).

This is the differential equation to the curve in which the cable hangs. To solve this equation we put p for dy/dx' and find s' in terms of the other quantities. Then differentiating, and writing $1+p^2$ for $(ds'/dx)^2$ and v for $A - ex' + e^2y$ we have

$$\frac{dv}{v} = \frac{-edp}{(1 - ep)\sqrt{1+p^2}}.$$

The variables are now separated, and the integrations can be effected. The equation can be integrated a second time, but the result is very long. The arbitrary constant A may have any value, depending on the length of the cable hanging from the ship at the time $t=0$.

The curve in its lowest part resembles a circular arc, or the lower part of a common catenary. But in its upper part the curve does not tend to become vertical, but tends to approach an asymptote making an angle $\cot^{-1} e$ with the horizon. The asymptote does not pass through the point where the cable touches the ground, but below it, the smallest distance being $A/e\,(e^2+1)^{\frac{1}{2}}$; the asymptote also passes below the ship.

If the conditions of the question are such that the tension at the lowest point of the cable is equal to zero, the tangent to the curve at that point is not necessarily horizontal. Let λ be the angle this tangent makes with the horizon. Referring to equations (1) of Art. 594 we have simultaneously

$$x'=0, \quad y=0, \quad s'=0, \quad T=0, \text{ and } \phi=\lambda.$$

Hence $Ag' = -c^2\cos\lambda, \qquad Bg' = -c^2\sin\lambda - g'ct.$

The differential equation to the curve now becomes

$$\frac{dy}{dx'} = \frac{-c^2\sin\lambda + g'\,(s' - ey)}{-c^2\cos\lambda + g'\,(es' - ex')} \quad\dots\dots\dots\dots\dots\dots(3).$$

which can be integrated in the same manner as before. One case deserves notice; viz. when $e=\cot\lambda$. The equation is then evidently satisfied by $y=x'/e$. The two constants in the integral of (3) are to be determined by the condition that, when $x'=0$, $y=0$, then $dy/dx'=\tan\lambda$. Both these conditions are satisfied by the relation $y=x'/e$. Hence this is the required integral. The form of the cable is therefore a straight line, inclined to the horizon at an angle $\lambda=\cot^{-1}e$; and the tension may be found from the formula $T=\dfrac{mg'y}{1+\cos\lambda}$.

Ex. 2. Let a cable be delivered with velocity c' from a ship moving with uniform velocity c in a straight line on the surface of a sea of uniform depth. If the resistance of the water to the cable be proportional to the square of the velocity, the coefficient B of resistance for longitudinal motion being different from the coefficient A for lateral motion, prove that the cable may take the form of a straight line making an angle λ with the horizon, such that $\cot^2 \lambda = \sqrt{e^4 + \tfrac{1}{4}} - \tfrac{1}{2}$, where e is the ratio of the speed of the ship to the terminal velocity of a length of cable falling laterally in water. Prove also that the tension will be found from the

equation $T = \left\{ y - \dfrac{B}{A} e^2 \left(\dfrac{c'}{c} - \cos \lambda \right)^2 \dfrac{y}{\sin \lambda} \right\} mg'.$ [*Phil. Mag.*

Small Oscillations of a Loose Chain.

599. **Chain suspended by one extremity.** *A heavy heterogeneous chain is suspended by one extremity, and hangs in a straight line under the action of gravity. A small disturbance being given to the chain in a vertical plane, it is required to find the equations of motion*.*

Let O be the point of support, let the axis Ox be measured vertically downwards, and Oy horizontally in the plane of disturbance. Let mds be the mass of any elementary arc whose length PQ is ds, and let T be the tension at P. Let l be the length of the string, and let us suppose that a weight Mg is attached to the lower extremity. The equations of motion, as in Art. 574, are

$$\frac{d^2x}{dt^2} = \frac{1}{m}\frac{d}{ds}\left(T\frac{dx}{ds}\right) + g, \qquad \frac{d^2y}{dt^2} = \frac{1}{m}\frac{d}{ds}\left(T\frac{dy}{ds}\right)\ldots\ldots\ldots(1).$$

Since the motion is very small, the point P will oscillate in a very small arc, the tangent at the middle point being horizontal. Hence we may put $dx/dt = 0$. For a similar reason we may put $dx = ds$. We therefore have by integrating the first equation

$$T = \text{constant} - g \int m dx.$$

But $T = Mg$ when $x = l$, hence we find

$$T = Mg + g \int_x^l m\, dx \ldots\ldots\ldots\ldots\ldots\ldots(2).$$

When the chain is homogeneous, this equation takes the simple form $T = Mg + mg\,(l - x)\ldots\ldots\ldots\ldots\ldots(3).$

It may be noticed (1) that this expression is independent of

---

* In the *Seventh Volume of the Journal Polytechnique*, Poisson discusses the oscillations of a heavy homogeneous chain suspended by one extremity. Putting $(l-x)^{\frac{1}{2}} \pm \tfrac{1}{2}g^{\frac{1}{2}}t$ equal to s or s' according as the upper or lower sign is taken, and $y' = y\,(l-x)^{\frac{1}{4}}$, he reduces the equation to the form $\dfrac{d^2y'}{ds\,ds'} = -\dfrac{1}{4}\dfrac{y'}{(s+s')^2}$. He obtains the integral by means of two definite integrals and two infinite series. After a rather long discussion of the forms of the arbitrary functions which occur in the integral, he finds that a solitary wave will travel up the chain with a uniform acceleration and down with a uniform retardation, each equal to half that of gravity.

the time ; (2) that the tension at any point of the chain is equal to the total weight of matter below that point.

The second equation may be written in either of the forms

$$\frac{d^2y}{dt^2} = \frac{1}{m}\frac{d}{dx}\left(T\frac{dy}{dx}\right) = \frac{1}{m}\,T\frac{d^2y}{dx^2} + \frac{1}{m}\frac{dT}{dx}\frac{dy}{dx}\ldots\ldots(4),$$

where T is a function of x given by the equation (2), or (3).

600. Let us suppose that the displacements of the particles forming any finite portion of the chain during a finite time are represented by $y = \phi(x, t)$, where ϕ is a continuous function of x and t. Let P be a geometrical point within this portion of the chain which moves so that the particle-velocity at P, i.e. dy/dt, is always equal to some constant quantity A. Let v be the velocity with which P moves, then, following in our mind the motion of P, we have by differentiating $dy/dt = A$ with regard to t

$$\frac{d^2y}{dt^2} + \frac{d^2y}{dxdt}\,v = 0\ldots\ldots\ldots(5).$$

Let Q be a point also within the portion, such that the tangent to the chain at Q makes with the vertical an angle whose tangent, i.e. dy/dx, is B/T, where B is some constant quantity. Let v' be the velocity with which Q moves, then

$$T\frac{d^2y}{dxdt} + \frac{d}{dx}\left(T\frac{dy}{dx}\right)v' = 0\ldots\ldots\ldots(6).$$

Eliminating the second differential coefficients of y from equations (4), (5) and (6), we easily deduce that, if P and Q coincide at any instant, $vv' = T/m\ldots\ldots\ldots(7).$

This reasoning requires that all the second differential coefficients should be finite, and that y should be a continuous function of x and t. It would not apply to any point P, if the discontinuous extremities of two waves were passing over P in opposite directions. But the consideration of these exceptions is unnecessary for our present purpose.

Let AB be a disturbed portion of the chain travelling in the direction AB on a chain otherwise in equilibrium. At the confines of the disturbance the two portions of the string must not make a finite angle with each other. If they did, an element of the string would be acted on by a finite moving force, namely, the resultant of the two finite tensions at its extremities. In such a case the disturbance would instantly extend itself further along the chain and assume some new form. Supposing we exclude any such case as this, we must have, as long as the motion is finite, both $dy/dt = 0$ and $dy/dx = 0$, at both the upper and lower extremity of the disturbance. If then P be a point at which $dy/dt = 0$, and Q a point at which $dy/dx = 0$, P and Q may be considered as taken just within the boundary of the wave; P and Q will therefore each travel with the velocity of that boundary. Hence, putting $v = v'$, we find for the velocity of either point

$$v^2 = T/m \ldots\ldots\ldots(8).$$

It appears therefore that if a solitary wave travel up the chain, the velocity increases as the wave approaches the upper extremity. The upper end of the wave will travel a little quicker than the lower end, because the tension at the upper end exceeds that at the lower; thus the length of the wave will gradually increase. When the wave travels down the chain, the velocity for the same reason decreases.

601. **Examples.** Ex. 1. If the chain be homogeneous, show that the boundaries of a solitary wave will travel up the chain with an acceleration equal to half that of gravity, and down the chain with a retardation of the same numerical amount.

Ex. 2. Let the law of density be $m = A\,(l + l' - x)^{-\frac{3}{2}}$ where l is the length of the chain and A, l' are two constants. Also let a weight equal to $2Ag\sqrt{l'}$ be fastened to the lower extremity, prove that

$$y = f\left\{(l + l' - x)^{\frac{1}{2}} - (\tfrac{1}{2}g)^{\frac{1}{2}}t\right\} + F\left\{(l + l' - x)^{\frac{1}{2}} + (\tfrac{1}{2}g)^{\frac{1}{2}}t\right\}.$$

This integration may be effected by writing $\theta = (l + l')^{\frac{1}{2}} - (l + l' - x)^{\frac{1}{2}}$. The equation of motion then takes the form $\dfrac{d^2y}{dt^2} = \dfrac{g}{2}\dfrac{d^2y}{d\theta^2}$, which can be solved in the usual manner.

Ex. 3. The chain is said to sound a harmonic note when its motion can be represented by an expression of the form $y = \phi\,(x)\sin(\kappa t + \alpha)$; so that the motion of every element repeats itself at the same constant interval. Show that the harmonic periods of the chain and weight are given by $\kappa l'^{\frac{1}{2}}\tan\kappa\left\{(l + l')^{\frac{1}{2}} - l'^{\frac{1}{2}}\right\} = 1$.

To prove this, we substitute $y = f(\theta)\sin(\kappa t + \alpha)$ in the differential equation obtained in the last Example; we thus find $f(\theta)$ to be trigonometrical. Since $y = 0$ when $x = 0$ for all values of t, the expression for y reduces to

$$y = \sin\kappa\theta\left\{A_\kappa\sin\kappa t\,(\tfrac{1}{2}g)^{\frac{1}{2}} + B_\kappa\cos\kappa t\,(\tfrac{1}{2}g)^{\frac{1}{2}}\right\}$$

where A_κ and B_κ are two arbitrary constants. But, when $x = l$, y must satisfy the equation of motion of the weight, viz. $d^2y/dt^2 = -gdy/dx$. Whence the result follows by substitution.

602. **Chain suspended by both extremities.** *An inelastic heterogeneous chain is suspended from two fixed points under the action of gravity. Any small disturbance being given in its own plane, it is required to find the small oscillations.*

Let the axis of x be horizontal and that of y vertical. Let C be any point on the chain when hanging in equilibrium, and let the arc s be measured from C. Let (x, y) be the co-ordinates of any point P determined by $CP = s$. Let T be the tension at P, $mgds$ the weight of an element ds situated at P. The equations of equilibrium are

$$\frac{d}{ds}\left(T\frac{dx}{ds}\right) = 0, \quad \frac{d}{ds}\left(T\frac{dy}{ds}\right) - mg = 0.$$

Let α be the angle which the tangent at P makes with the axis of x, then we easily find $T = \dfrac{wg}{\cos \alpha}$, $m = w \dfrac{d \tan \alpha}{ds}$ ....(1),

where w is an undetermined constant.

When the chain is in motion, let $(x + \xi,\ y + \eta)$ be the co-ordinates of the position of the particle P at the time t, and let the tension at that point be $T' = T + U$. The equations of motion will be

$$\frac{d^2\xi}{dt^2} = \frac{1}{m}\frac{d}{ds}\left\{T'\left(\frac{dx}{ds} + \frac{d\xi}{ds}\right)\right\}, \qquad \frac{d^2\eta}{dt^2} = \frac{1}{m}\frac{d}{ds}\left\{T'\left(\frac{dy}{ds} + \frac{d\eta}{ds}\right)\right\} - g,$$

which, by subtracting the equations of equilibrium, reduce to

$$\frac{d^2\xi}{dt^2} = \frac{1}{m}\frac{d}{ds}\left(T\frac{d\xi}{ds} + U\frac{dx}{ds}\right) \qquad \frac{d^2\eta}{dt^2} = \frac{1}{m}\frac{d}{ds}\left(T\frac{d\eta}{ds} + U\frac{dy}{ds}\right)\ldots(2),$$

when the squares of small quantities are neglected.

Since the string is inelastic, we have

$$(dx + d\xi)^2 + (dy + d\eta)^2 = (ds)^2.$$

Expanding, and rejecting the squares of small quantities, this becomes $\dfrac{dx}{ds}\dfrac{d\xi}{ds} + \dfrac{dy}{ds}\dfrac{d\eta}{ds} = 0$ ....(3).

We have thus three equations to find ξ, η and U as functions of s and t.

603.   **Velocity of a wave.**   *To find the velocity with which a solitary wave will travel along the chain.*

If we suppose a small disturbance to travel along this chain, so that there is no abrupt change of direction of the chain at the boundaries of the wave, we must have at those points $d\xi/ds = 0$, $d\eta/ds = 0$, $d\xi/dt = 0$, $d\eta/dt = 0$, and $U = 0$. Let v be the velocity with which one boundary of this wave travels along the chain, then, following that boundary in our mind, we have as in Art. 600

$$\frac{d^2\xi}{dt^2} + v\frac{d^2\xi}{ds\,dt} = 0, \qquad \frac{d^2\xi}{dt\,ds} + v\frac{d^2\xi}{ds^2} = 0,$$

and therefore $\dfrac{d^2\xi}{dt^2} = v^2 \dfrac{d^2\xi}{ds^2}$,

with a similar equation for η. Thus the dynamical equations become at the boundary

$$\left(v^2 - \frac{T}{m}\right)\frac{d^2\xi}{ds^2} = \frac{1}{m}\frac{dU}{ds}\frac{dx}{ds}, \qquad \left(v^2 - \frac{T}{m}\right)\frac{d^2\eta}{ds^2} = \frac{1}{m}\frac{dU}{ds}\frac{dy}{ds},$$

and the geometrical equation becomes $\dfrac{d^2\xi}{ds^2}\dfrac{dx}{ds} = -\dfrac{d^2\eta}{ds^2}\dfrac{dy}{ds}$.

From these we easily get $v^2 = T/m$. Substituting for T and m their values, we have, if ρ be the radius of curvature at P,

$$v = \sqrt{(g\rho \cos \alpha)} \ldots\ldots\ldots(4),$$

so that the velocity of either boundary of the wave is that due to one quarter of the vertical chord of curvature at that point.

Ex. A chain is in equilibrium under the action of any forces which are functions only of the position in space of the element acted on. Show that the velocity of either boundary of a solitary wave is that due to one quarter of the chord of curvature in the direction of the resultant force at that boundary.

604. **Intrinsic equation of motion.** *To solve as far as possible the equations of motion of a heavy slack heterogeneous chain.*

It will be convenient to express the unknown quantities ξ, η, U in terms of some *one* function ϕ.

Let $\alpha + \phi$ be the angle which the tangent at P makes with the horizon at the time t.

Then
$$\cos (\alpha + \phi) = \frac{dx + d\xi}{ds}, \qquad \sin (\alpha + \phi) = \frac{dy + d\eta}{ds} ;$$

$$\therefore - \phi \sin \alpha = \frac{d\xi}{ds}, \qquad \phi \cos \alpha = \frac{d\eta}{ds}\ldots\ldots\ldots(5) ;$$

$$\therefore \frac{d\xi}{d\alpha} = - \rho\phi \sin \alpha, \qquad \frac{d\eta}{d\alpha} = \rho\phi \cos \alpha \ldots\ldots\ldots(6),$$

$$\xi = - \int \rho\phi \sin \alpha \, d\alpha + A, \qquad \eta = \int \rho\phi \cos \alpha \, d\alpha + B\ldots\ldots\ldots(7),$$

where A and B are two undetermined functions of t.

The equations (2) now become by substitution from these and from (1)

$$\left.\begin{array}{l} \dfrac{d^2\xi}{dt^2} \dfrac{1}{\cos^2\alpha} = \dfrac{d}{d\alpha} \left(- g\phi \tan \alpha + \dfrac{U}{w}\cos \alpha \right) \\[3mm] \dfrac{d^2\eta}{dt^2} \dfrac{1}{\cos^2\alpha} = \dfrac{d}{d\alpha} \left(g\phi + \dfrac{U}{w}\sin \alpha \right) \end{array}\right\}\ldots\ldots\ldots(8).$$

For the sake of brevity let accents denote differentiations with regard to t. Expanding the differentiations on the right-hand side, these equations may be written in the form

$$\left.\begin{array}{l} - \xi'' \sin \alpha + \eta'' \cos \alpha - g \left(\phi \sin \alpha + \dfrac{d\phi}{d\alpha} \cos \alpha \right) = U\dfrac{\cos^2 \alpha}{w} \\[3mm] \xi'' \cos \alpha + \eta'' \sin \alpha + g\phi \cos \alpha = \dfrac{dU}{d\alpha}\dfrac{\cos^2\alpha}{w} \end{array}\right\}.$$

Differentiating the first with regard to α and adding the result to the second, we obtain

$$\frac{\rho\phi''}{\cos \alpha} - g \frac{d^2\phi}{d\alpha^2} = 2 \frac{d}{d\alpha}\left(\frac{U\cos\alpha}{w} \right).$$

Differentiating the second and subtracting the first from the result, we obtain

$$2g\frac{d\phi}{d\alpha} = \frac{d^2}{d\alpha^2}\left(\frac{U\cos\alpha}{w} \right).$$

These equations evidently give

$$U \cos \alpha = wg \left(2\textstyle\int\phi d\alpha + C\alpha + D \right)\ldots\ldots\ldots(9),$$

$$\frac{d^2\phi}{dt^2} = g\frac{\cos \alpha}{\rho}\left(\frac{d^2\phi}{d\alpha^2} + 4\phi + 2C \right)\ldots\ldots\ldots(10),$$

where C and D are two undetermined functions of t. These are the general equations to determine the small oscillations of a slack chain.

The undisturbed form of the curve being given, ρ is known as a function of a. We may then use the equation (10) to find ϕ as a function of a and t. The tension is then found from the equation (9), and the displacements ξ, η of any point of the chain by equations (7).

605.   The determination of the whole motion depends therefore on the solution of a single equation. Supposing the integration to have been effected, the expression for ϕ will contain two new arbitrary functions of a and t. These we may represent by $\psi(P)$ and $\chi(Q)$ where ψ and χ are arbitrary functions of two determinate combinations P and Q of the variables. The arbitrary functions A and B are not independent of C and D, and the relations between them may be found by substituting in equations (8).

We have thus four arbitrary functions whose values have to be determined from the conditions of the question. Let a_0, a_1, be the values of a which correspond to the two extremities of the string. Then the values of ϕ and $d\phi/dt$ are given by the question when $t=0$ for all values of a from $a=a_0$ to $a=a_1$; also the initial values of A and B are given. Thus the values of $\psi(P)$ and $\chi(Q)$ are determined for all values of P and Q between the two limits which correspond to $a=a_0$, $t=0$ and $a=a_1$, $t=0$. The forms of ψ and χ for values of P and Q exterior to these limits, and the values of A and B when t is not zero, are to be found from the conditions at the extremities of the chain. If the extremities be fixed, we have both ξ and η equal to zero for all values of t when $a=a_0$ and $a=a_1$. It may thus happen that the arbitrary functions A, B, ψ and χ are discontinuous.

In many cases the circumstances of the problem will enable us to determine at once the form of C. Thus, suppose the string when in equilibrium to be symmetrical about a vertical line, say the axis of y, and let the points of support be fixed in the same horizontal line. Then if the initial motion be also symmetrical about the axis of y, the whole subsequent motion will be symmetrical. Thus ϕ must be a function of a, containing when expanded only odd powers of a. Substituting such a series in equation (10) we see that C must be zero.

606.   **Oscillations of a cycloidal chain.** There are several cases in which the equation to find the small motions of a chain may be more or less completely integrated. One of the most interesting of these is that in which the chain hangs in equilibrium in the *form of a cycloid*. In this case we have, if b be the radius of the generating circle, $\rho = 4b \cos a$. The density of the chain at any point is given by $m = w/4b \cos^3 a$, so that all the lower part of the chain is of nearly uniform density, but the density increases rapidly higher up the chain and is infinite at the cusp.

The equation to find the oscillations now takes the simple form

$$\frac{d^2\phi}{dt^2} = \frac{g}{4b} \left\{ \frac{d^2\phi}{da^2} + 4\phi + 2C \right\} \quad\dots\dots\dots\dots\dots\dots\dots(11),$$

in which all the coefficients are constants.

There are two cases of motion to be discussed, (1) when the chain swings up and down, and (2) when it swings from side to side. The results are indicated in the two following examples.

Ex. 1.   *A heavy chain suspended from two points in the same horizontal line hangs under gravity in the form of a cycloid. Find the symmetrical oscillations of the chain, when the lowest point moves only up and down.*

In this case we have $C=0$. To find the nature and time of a small oscillation, we put $\phi = \Sigma R \sin \kappa t + \Sigma R' \cos \kappa t,$

where Σ implies summation for all values of κ, and R, R' are functions of α only.

Substituting, we have
$$\frac{d^2R}{d\alpha^2} + 4\left(1 + \frac{b\kappa^2}{g}\right)R = 0;$$

with a similar equation to find R'.   Therefore $R = L \sin 2\alpha \left(1 + \frac{b\kappa^2}{g}\right)^{\frac{1}{2}}$,

where L is an arbitrary constant, the other constant being determined by the consideration that the motion is symmetrical about the axis of y.   For the sake of brevity, put $\lambda = 2\sqrt{(1 + b\kappa^2/g)}$.   Substituting in (7), we find that the terms derived from R become
$$\xi = \Sigma L \frac{2b}{\lambda^2 - 4}\{\lambda \cos \lambda\alpha \sin 2\alpha - 2 \sin \lambda\alpha \cos 2\alpha\} \sin \kappa t,$$

$$\eta = \Sigma \left[-L\frac{2b}{\lambda^2 - 4}\{\lambda \cos \lambda\alpha \cos 2\alpha + 2 \sin \lambda\alpha \sin 2\alpha\} - L\frac{2b}{\lambda} \cos \lambda\alpha + H \right] \sin \kappa t,$$

where H is a constant depending on the position of the points of support.   The terms derived from R' must be added to these, but have been omitted for the sake of brevity.   They may be derived from those just written down by writing $\cos \kappa t$ for $\sin \kappa t$ and changing the constants L, H into two other constants L', H'.

Let the length of the chain be $2l$, then at either end $\sin \alpha_0 = l/4b$.   At both extremities we must have $\xi = 0$, $\eta = 0$.   All these four conditions can be satisfied if

$$\frac{\tan \lambda\alpha_0}{\lambda} = \frac{\tan 2\alpha_0}{2}.$$

This equation therefore determines the possible times of symmetrical vibration of a heterogeneous chain hanging in the form of a cycloid.

607.   *If α be not very large, the oscillations are nearly the same as those of a uniform chain* [*].   In this case, since α_0 is small but $\lambda\alpha_0$ is not necessarily small, the equation to determine λ is approximately

$$\tan \lambda\alpha_0 = \lambda\alpha_0.$$

The least value of $\lambda\alpha_0$ which can be taken is a little less than $\frac{3}{2}\pi$, viz. $\lambda\alpha_0 = 4\cdot4934$.   Hence λ is great, and therefore $\kappa = \lambda \left(g/4b\right)^{\frac{1}{2}}$ nearly.   The expressions for ξ and η now take the simple forms

$$\xi = \Sigma L \frac{4b}{\lambda^2}\{\lambda\alpha \cos \lambda\alpha - \sin \lambda\alpha\} \sin \left\{\left(\frac{g}{4b}\right)^{\frac{1}{2}} \lambda t + \epsilon\right\}$$
$$\eta = \Sigma L \frac{4b}{\lambda}\{\cos \lambda\alpha_0 - \cos \lambda\alpha\}\; \sin \left\{\left(\frac{g}{4b}\right)^{\frac{1}{2}} \lambda t + \epsilon\right\}$$

The terms depending on $\cos \kappa t$ have been included in these expressions for ξ and η by introducing ϵ into the trigonometrical factor.

The roots of the equation $\tan \lambda\alpha_0 = \lambda\alpha_0$ may be found by continued approximation.   The first is zero, but, since λ occurs in the denominator of some of the small terms, this value is inadmissible.   The others may be expressed by the formula $\lambda\alpha_0 = \frac{1}{2}(2i + 1)\pi - \theta$, where θ is not very large.   This makes the time of

---

* The reader who may wish to see another method of discussing the small oscillations of a suspension chain may consult a memoir by Mr Röhrs in the ninth volume of the *Cambridge Transactions*.   Mr Röhrs considers the chain to be homogeneous, symmetrical about a vertical line, and nearly horizontal from the beginning of the process.   In the second edition of this treatise the small oscillations were treated on the same hypotheses, but in a different manner.   That method, however, is not nearly so simple as the one here given in which the approximate oscillations for a catenary are deduced from the accurate ones for a cycloid.

vibration nearly equal to $\dfrac{4}{2i+1} \cdot \dfrac{l}{\sqrt{4gb}}$. Thus the times of vibration of the chain are all short.

This result will explain why the marching of troops in time along a suspension bridge may cause oscillations which are so great as to be dangerous to the bridge. It is clearly possible that the "marching time" may be equal to, or very nearly equal to, some one of the times of vibrations of the bridge. If this should occur it follows from Arts. 338 and 340 that the stability of the bridge may be severely strained.

It should be noticed that the terms in the expression for ξ have the square of λ in the denominator, while those in the expression for η have the first power of λ. Since λ is great we may as a first approximation reject the values of ξ altogether, and regard each element of the chain as simply moving up and down.

608. **Ex. 2.** *A heavy chain suspended from two points hangs under gravity in the form of a cycloid. If it swings from side to side in its own plane so that the middle point has only a lateral motion without any perceptible vertical motion, find the times of oscillation.*

As in the last example, we put $\phi = \Sigma R \sin \kappa t + \Sigma R' \cos \kappa t$,

where R and R' are functions of a only. Substituting in equation (11), we see that $2C = \Sigma h \sin \kappa t + \Sigma k \cos \kappa t$ where h and k are arbitrary constants. The equation to

find R becomes $\dfrac{d^2 R}{da^2} + 4\left(1 + \dfrac{b\kappa^2}{g}\right)R = -h$.

If we put $\lambda^2 = 4(1 + b\kappa^2/g)$ as before, we find $R = -h/\lambda^2 + L \sin(\lambda a + M)$.

Thence taking the term of ϕ which contains $\sin \kappa t$,

$$\frac{\xi}{\sin \kappa t} = \frac{h' - hb \cos 2a}{\lambda^2} + L\frac{2b}{\lambda^2 - 4}\{\lambda \cos(\lambda a + M)\sin 2a - 2 \sin(\lambda a + M)\cos 2a\},$$

where h' is an arbitrary constant introduced on integration. Substituting in equation (8), we find $h' = -h(b + g/\kappa^2)$. Also, we have in the same way

$$\frac{\eta}{\sin \kappa t} = -\frac{hb}{\lambda^2}(2a + \sin 2a)$$

$$-L\frac{2b}{\lambda^2 - 4}\{\lambda \cos(\lambda a + M)\cos 2a + 2 \sin(\lambda a + M)\sin 2a\} - L\frac{2b}{\lambda}\cos(\lambda a + M) + H.$$

If we suppose the two supports to be in the same horizontal line, we must have $\xi = 0$ and $\eta = 0$, when $a = \pm a_0$. These conditions may be satisfied if we take $M = \frac{1}{2}\pi$, $H = 0$, for then ξ becomes an even and η an odd function of a. In this case $\eta = 0$ at the lowest point of the chain. We have then two equations to find L/h; equating its values, we have

$$\frac{2\tan 2a_0 - \lambda \tan \lambda a_0 - \dfrac{\tan \lambda a_0}{\cos 2a_0}\dfrac{\lambda^2 - 4}{\lambda}}{2a_0 + \sin 2a_0} = \frac{\lambda \tan \lambda a_0 \tan 2a_0 + 2}{2\cos^2 a_0 + \dfrac{4}{\lambda^2 - 4}}.$$

609. If a_0 be small, this equation is very nearly satisfied by $\lambda a_0 = i\pi$, where i is any integer. In this case the complete expressions for ξ and η take the simple forms

$$\xi = \Sigma L \frac{4b}{\lambda^2}(\cos \lambda a_0 - \cos \lambda a - \lambda a \sin \lambda a)\sin\left\{\left(\frac{g}{4b}\right)^{\frac{1}{2}}\lambda t + \epsilon\right\}$$

$$\eta = \Sigma L \frac{4b}{\lambda}\sin \lambda a \sin\left\{\left(\frac{g}{4b}\right)^{\frac{1}{2}}\lambda t + \epsilon\right\}$$

610. **Examples.** Ex. 1. If we change the variables from α, t to p, q, where

$$p = t + \int \left(\frac{\rho}{g \cos \alpha} \right)^{\frac{1}{2}} d\alpha, \qquad q = -t + \int \left(\frac{\rho}{g \cos \alpha} \right)^{\frac{1}{2}} d\alpha,$$

show that the general equation (10) of small oscillations takes the form

$$\frac{d^2\phi'}{dpdq} + \frac{\mu^3}{4} \left(\frac{d^2\mu}{d\alpha^2} + 4\mu \right) \phi' = -\frac{\mu^3}{2} C,$$

where $\mu^4 = g \cos \alpha / \rho$ and $\phi = \mu\phi'$.

Show also that the coefficient of ϕ' is a function of $p + q$, the form of the function depending on the law of density of the chain.

This transformation may be useful, because it follows from Art. 603 that p is constant for the boundaries of a solitary wave travelling in one direction, and q for a wave travelling in the other direction.

Ex. 2. A heavy string hangs in equilibrium under gravity in such a form that its intrinsic equation is $\dfrac{\cos \alpha}{\rho} = \dfrac{b^4}{g} \sin^4 (2\alpha + c)$, where b and c are any constants. Show that its law of density is given by $m = w \dfrac{b^4}{g} \dfrac{\sin^4 (2\alpha + c)}{\cos^3 \alpha}$ If such a chain be set in motion in any symmetrical manner, prove that its motion is given by

$$\phi = b \sin (2\alpha + c) \left\{ F \left(t - \frac{\cot (2\alpha + c)}{2b^2} \right) + f \left(t + \frac{\cot (2\alpha + c)}{2b^2} \right) \right\}.$$

Ex. 3. If, in addition to gravity, each element of the chain be acted on by a small normal force whose magnitude is Fg, prove that the equation of motion of the chain is $\dfrac{\rho}{g \cos \alpha} \dfrac{d^2\phi}{dt^2} - \dfrac{d^2\phi}{d\alpha^2} - 4\phi - 2C = \dfrac{1}{\cos \alpha} \dfrac{dF}{d\alpha} + 2 \int \dfrac{F}{\cos \alpha} d\alpha.$

If the chain is nearly horizontal, so that α is very small, and if $F = f \sin(\alpha t - c\alpha)$, prove that the denominator of the corresponding term in the expression for ϕ is $g (c^2 - 4) - \rho a^2$.

Ex. 4. A heavy chain of length $2l$ is suspended from two points A, B in the same horizontal line whose distance apart is not very different from $2l$. Each particle of the chain is slightly disturbed from its position of rest in a direction perpendicular to the vertical plane through AB. Find the small oscillations of the chain.

Ex. 5. A heavy string is suspended from two fixed points A and B, and rests in equilibrium in the form of a catenary whose parameter is c. Let the string be initially displaced, the points of support A, B being also moved, so that

$$\phi = \sigma (1 + \cos 2\alpha) + \sigma' \sin 2\alpha,$$

where σ and σ' are two small quantities and the other letters have the same meaning as in Art. 604. If the string be placed at rest in this new position, prove that it will always remain at rest.

611. Ex. 1. A uniform string in the form of a circle of radius a rests on a smooth plane under a central repulsion whose measure at a distance r is ga^n/r^n. Show that, if the string be slightly displaced so that initially it is at rest and in the form $r = a + \Sigma a_m \cos m\theta$, then at any subsequent time t its form will be determined by

$$r = a + \Sigma a_m \cos m\theta \cos m \left\{ \frac{g}{a} \frac{m^2 + n - 2}{m^2 + 1} \right\}^{\frac{1}{2}} t$$

where Σ implies summation from $m = 1$ to $m = \infty$. Discuss the result (1) when $m = 1$ and $n = 1$, and (2) when $n = 3$.   [*Math. Tripos*, 1884.

Ex. 2.   A string is in equilibrium in the form of an equiangular spiral, of angle a, under the action of a centre of repulsion in the pole, the force on an element ds at a distance r_1 being $f ds / r_1{}^2$.   In order that, when the string is slightly disturbed from the position of equilibrium, the equations of motion may take a linear form with constant coefficients, we shall suppose that the string is loaded with non-attracted matter, so that the mass of an element ds becomes $a ds / r^3$, where r is the equilibrium distance of the element from the pole.   Let the particle whose equilibrium co-ordinates are (r, θ) occupy at the time t the position (r_1, θ_1) where $r_1 = r (1 + \xi)$, $\theta_1 = \theta + \eta$, and let the tension be $T_1 = T (1 + V)$ where T is the equilibrium tension.   Show that the equations of motion are

$$\xi + \sin a \cos a \frac{d\xi}{d\theta} + \sin^2 a \frac{d\eta}{d\theta} = 0$$

$$\frac{a}{f} \frac{d^2\xi}{dt^2} = - \xi + 2 \sin a \cos a \frac{d\xi}{d\theta} + \sin^2 a \frac{d^2\xi}{d\theta^2} - V + \sin a \cos a \frac{dV}{d\theta}$$

$$\frac{a}{f} \frac{d^2\eta}{dt^2} = 2 \sin^2 a \frac{d\xi}{d\theta} + \sin^2 a \frac{d^2\eta}{d\theta^2} + \sin^2 a \frac{dV}{d\theta}.$$

Hence show that the motion is represented by

$$\xi = Am \sin^2 a \sin m (vt - \theta),$$

$$- \eta = A \{ \cos m (vt - \theta) + m \sin a \cos a \sin m (vt - \theta) \},$$

where
$$v^2 = \frac{\cos^2 a + m^2 \sin^2 a}{1 + m^2 \sin^2 a} \cdot \frac{f \sin^2 a}{a}.$$

If the string is finite in length and its extremities A and B are fixed in space on the spiral so that the angle $AOB = \beta$, and if the period of vibration is $2\pi/p$, prove that the angle β must be a root of the equation

$$\tfrac{1}{2} (e^{h\beta} - e^{-h\beta}) \sin k\beta \, \frac{k^2 - h^2}{hk} = 2 - (e^{h\beta} + e^{-h\beta}) \cos k\beta,$$

where h^2 and $-k^2$ are the roots of the quadratic

$$x^2 f \sin^4 a - x \sin^2 a (f \cos^2 a - ap^2) - ap^2 = 0.$$

Ex. 3.   A heterogeneous string OA of length l, whose line density at a point distant x from O is $Da/(b^2 - x^2)^{\frac{3}{2}}$, has a particle of mass M attached at the end A where $Ml = Da (b^2 - l^2)^{\frac{3}{2}}$, and a, b are two given constants such that l is less than b. It is placed on a smooth horizontal plane and set rotating with an angular velocity ω about the end O as a fixed point so that each point describes a circle whose centre is O.   If it is slightly disturbed, show that a possible transverse oscillation is given by

$$\eta = A \sin \omega \sqrt{(q^2 - 1)} \, t . \sin (q \sin^{-1} x/b + Q),$$

where
$$Q = \cot^{-1} Mq/Da - q \sin^{-1} l/b,$$

and η is the distance of an element from the uniformly revolving line OA.

Small Oscillations of a Tight String.

612.   *An elastic string whose weight may be neglected and whose unstretched length is l has its extremities fixed at two points whose distance apart is l'.   The string being disturbed so that each particle is moved along the length of the string, find the equations of motion.*

Let A be one of the fixed points, and let AB be the string when unstretched and placed in a straight line. Let the extremity B be pulled until it reaches the other fixed point B'. Let PQ be any element of the unstretched string, $P'Q'$ the same element at the time t. Let $AP = x$ and let the abscissa AP' be x'. Let T and $T + dT$ be the tensions at P' and Q'. Let M be the mass of the whole string, m the mass of a unit of length of *unstretched* string. Since the mass of an element is $m\,dx$, the effective force on it is $(m\,dx)(d^2x'/dt^2)$. The difference of the tensions at the two extremities of the element is dT. Equating these, we find that the equation of motion is

$$m\,\frac{d^2x'}{dt^2} = \frac{dT}{dx}\ldots\ldots(1).$$

If E be the modulus of elasticity, we have by Hooke's law

$$\frac{dx'}{dx} = 1 + \frac{T}{E}\ldots\ldots(2).$$

Eliminating T, we have $\dfrac{d^2x'}{dt^2} = \dfrac{E}{m}\,\dfrac{d^2x'}{dx^2}\ldots\ldots(3).$

It should also be noticed that, assuming as usual the truth of Hooke's law, these equations and results are not merely approximations, but are strictly accurate.

It is often more convenient to select some particular state of the string as a standard of reference, and to express the actual position of any particle at the time t by its displacement from its position in this standard. Thus, if the unstretched state AB of the string be chosen as the standard of reference, we put $x' = x + \xi$, so that ξ is the displacement of the particle whose abscissa in the unstretched state is x. The equation of motion now takes the form

$$\frac{d^2\xi}{dt^2} = \frac{E}{m}\,\frac{d^2\xi}{dx^2}\ldots\ldots(4).$$

613. If the equilibrium position of the string when stretched between the fixed points A, B' is taken as the standard of reference, the equation of motion is somewhat different. Let x_1 be the abscissa of the equilibrium position of that point of the string which at the time t is at P', then $x_1/l' = x/l$. Let $x' = x_1 + \xi_1$, then substituting for x and x' in (3), the differential equation becomes

$$\frac{d^2\xi_1}{dt^2} = \frac{E}{m}\left(\frac{l'}{l}\right)^2 \frac{d^2\xi_1}{dx_1^2}\ldots\ldots\ldots\ldots(5).$$

614. If we put $E = ma^2$, the integral of the equation (4) may be written in the form $\quad \xi = f(at - x) + F(at + x)\ldots\ldots(6).$ The most general motion of the string is therefore obtained by

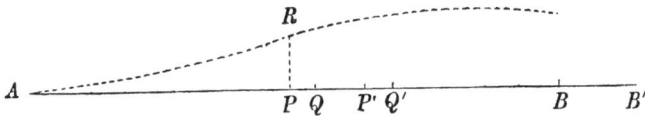

superimposing the motions determined by $X = f(at - x)$ and $X' = F(at + x)$, where $\xi = X + X'$. Let us consider these separately.

At each point P of the unstretched string draw an ordinate PR equal to the longitudinal displacement X of P at a given time t. The locus thus traced out by R exhibits to the eye the actual displacement at the time t of every point of the string. When t alters, this locus will change and adapt itself to the changing motion of the string. If the string vibrated transversely this construction would be unnecessary, for the displaced string would itself form the locus of R.

Let a point C starting from any position travel along the axis AB in such a manner that, if x be its abscissa, $at - x$ is constant and equal to c. The velocity of C is therefore uniform and equal to a. Since the displacement of the point of the string at any instant coincident with C is equal to $f(c)$, the displacement at C is always the same. If then C at starting coincide with the foot of an ordinate of given magnitude, it will always be at the foot of an ordinate of the same magnitude. This is the same thing as saying that every ordinate of the locus moves continually in the positive direction with a velocity equal to that of C without changing its magnitude. The locus travels along the axis as a wave travels on the surface of water.

The conclusion is that *the equation* $X = f(at - x)$ *represents a wave-like motion which travels in the positive direction with a uniform velocity equal to a*. In the same way the equation $X' = f(at + x)$ represents a wave motion which travels with a velocity equal to $- a$. Such a wave travels in the negative direction of the axis.

In the case of the string the velocity of either of these waves, when referred to the unstretched string as the standard, is $(E/m)^{\frac{1}{2}}$. If the equilibrium position of the string is taken as the standard, the velocity of either wave is $(E/m)^{\frac{1}{2}} . (l'/l)$. Shortly we may say that the velocity is such that *the time of traversing a length l of unstretched string or a length l' of stretched string is* $l(m/E)^{\frac{1}{2}}$. *It should be noticed that this time is independent both of the nature of the disturbance and of the tension of the string.*

615. Each of the waves into which the motion has been analysed may be further analysed by expanding the function into a series of sines and cosines. Let this expansion be

$$f(at - x) = A_1 \sin\{n_1(at - x) + \alpha_1\} + A_2 \sin\{n_2(at - x) + \alpha_2\} + \&c.$$

Taking any one term, say $X_n = A \sin\{n(at - x) + \alpha\}$,

the motion represented by X_n may be called a *simple wave* or a *harmonic wave*. The coefficient A expresses the maximum extent or *amplitude* of the oscillation; its square is usually called the *intensity* of the wave. The *period* of the oscillation of any particle is $2\pi/na$; the reciprocal of the period is called its *frequency*. This latter term is due to Lord Rayleigh. If we trace the curve whose abscissa is x and ordinate X_n, regarding t as constant, we see that the portions of the curve between the ordinates given by x, $x \pm 2\pi/n$, $x \pm 4\pi/n$ &c. are similar and equal to each other. In other words the values of the ordinates recur when x is increased by $2\pi/n$. *The quantity $2\pi/n$ is therefore called the length of the wave.* It follows that those waves in which n is least have their periods greatest and their lengths longest. Of two oscillations of unequal period, the one of shorter period is called the *sharper* of the two and the one of longer period is said to be the *flatter*.

616. *An elastic string, stretched as in the last proposition, is slightly disturbed in any manner, to find the equations of motion.*

Following the same notation as before, let (x', y', z') be the co-ordinates of P'. Proceeding exactly as in Art. 574, we may form the equations of motion. Since the mass of an element is $m\,dx$ instead of $m\,ds$, the equations will be

$$m\frac{d^2x'}{dt^2} = \frac{d}{dx}\left(T\frac{dx'}{ds'}\right)...(1), \quad m\frac{d^2y'}{dt^2} = \frac{d}{dx}\left(T\frac{dy'}{ds'}\right)...(2), \quad m\frac{d^2z'}{dt^2} = \frac{d}{dx}\left(T\frac{dz'}{ds'}\right)...(3),$$

where ds' is the length of the element $P'Q'$. If E be the modulus of elasticity we have by Hooke's law

$$\frac{ds'}{dx} = 1 + \frac{T}{E}.........(4).$$

Since the disturbance is very small, dy'/ds' and dz'/ds' are very small, and dx'/ds' is very nearly equal to unity. Hence the first equation takes the form $m\dfrac{d^2x'}{dt^2} = \dfrac{dT}{dx}.........(5)$,

and Hooke's equation takes the form $\dfrac{dx'}{dx} = 1 + \dfrac{T}{E}$,

which are the same as equations (1) and (2) of Art. 612, so that when the disturbance is small the longitudinal motion is independent of the motion transverse to the string.

In the second equation we may regard T as constant, its small variations being multiplied by the small quantity dy'/ds'. Hence we may put $T = T_0$, where $T_0 = E(l' - l)/l$.

This gives, by equation (4), $ds'/dx = l'/l$, and therefore $ds' = dx_1$, where as before $x_1/l' = x/l$. The equation of motion therefore becomes

$$\frac{d^2y'}{dt^2} = \frac{T_0 l}{ml'}\frac{d^2y'}{dx^2} \quad \text{or} \quad \frac{d^2y'}{dt^2} = \frac{T_0 l'}{ml}\frac{d^2y'}{dx_1^2}..............(6),$$

according as the unstretched or stretched string is the standard.

The third equation may be treated in the same way.

The velocity of a transverse vibration measured in units of length of unstretched string is therefore $(T_0 l/ml')^{\frac{1}{2}}$. *The time of traversing a length l of unstretched string or l' of stretched string is* $(mll'/T_0)^{\frac{1}{2}}$. *This velocity is independent of the nature of the disturbance, but depends on the tightness or tension of the string.*

If the string be very slightly elastic we may, in this last formula, put $l' = l$. We then obtain the results given in all treatises on Sound.

617. We may here notice one point of difference between the equations of motion of longitudinal and transverse vibrations. In the former, supposing that there are no transversal vibrations, no approximations are made, so that, as already pointed out, the equations (4) and (6) of Arts. 612 and 614 hold for large and small vibrations. In the latter, even if the longitudinal vibrations are insensible, we assume that ds'/dx and dx_1/dx are so nearly equal that we may write the one for the other. We have to the second order of small quantities

$$\frac{ds'}{dx} = \frac{dx_1}{dx} \left\{ 1 + \frac{d\xi_1}{dx_1} + \frac{1}{2} \left(\frac{dy'}{dx_1} \right)^2 \right\}.$$

If the string vibrate without sensible longitudinal vibrations ξ_1 is of the second order of small quantities, and as the substitution for ds'/dx is made on the right hand side of (2), which already contains the small quantity dy'/ds', the differential equation (6) is correct when we can neglect the cubes of small quantities. If however the string oscillates simultaneously with longitudinal and transversal vibrations, ξ_1 is of the first order of small quantities, so that the transversal and longitudinal vibrations are independent only when we can neglect the squares of small quantities.

618. There are two modes of applying the equations of motion to actual cases. We shall first illustrate these by solving a simple example by both methods, and we shall then make some remarks on the results.

An elastic string whose unstretched length is l rests on a perfectly smooth table and has its extremities fixed at two points A, B' whose distance apart is l', where l' is greater than l. The extremity B' is suddenly released, find the motion.

Solution by discontinuous functions. Following the same notation as in Art. 612, the motion is given by the equation

$$\xi = f(at - x) + F(at + x),$$

where ξ is the displacement of the particle whose abscissa in the unstretched string is x. The conditions to determine f and F are as follows :—

1. When $x = 0$, $\xi = 0$ for all values of t,
2. When $x = l$, $T = 0$ and $\therefore d\xi/dx = 0$ for all values of t,
3. When $t = 0$, $\xi = rx$ from $x = 0$ to $x = l$, where $l' = (r + 1) l$,
4. When $t = 0$, $d\xi/dt = 0$ from $x = 0$ to $x = l$.

From the first condition it follows that the functions F and f are the same with opposite signs. From the second condition we have $f'(at + l) = -f'(at - l)$, so that the values of the function f' recur with opposite signs when the variable is increased by $2l$. If then we knew the values of $f'(z)$ for all values of z from $z = z_0$ to $z = z_0 + 2l$ where z_0 has any value, then the form of the function is altogether known. Now the third condition gives $f(-x) - f(x) = rx$ and the fourth gives $f'(-x) = f'(x)$ from $x = 0$ to $x = l$. Hence $f'(x) = -\frac{1}{2}r$ from $x = -l$ to $x = l$. It follows that $f'(z) = -\frac{1}{2}r$ from $z = -l$ to l, $f'(z) = \frac{1}{2}r$ from $z = l$ to $3l$ and so on changing sign every time the variable passes the values l, $3l$, $5l$, &c. Let us consider the motion of any point P

of the string whose unstretched abscissa is x. Its velocity is given by the formula $v/a = f'(at - x) - f'(at + x)$. Since $x < l$ we have $v/a = -\frac{1}{2}r + \frac{1}{2}r = 0$; hence the particle does not move until $at + x = l$. The second function then changes sign, and we have $v/a = -\frac{1}{2}r - \frac{1}{2}r = -r$. The particle continues to move with this velocity until $at - x = l$, when the first function changes sign, and so on. Let AB be the unstretched string, and let a point R starting from B move continually along the string and back again with velocity a. Then it is easy to see that when R is on the same side of P as the loose end of the string, P will be at rest, and when R is on the same side of P as the fixed end, P will be moving with a velocity alternately equal to $\pm ra$. The general character of the motion is; the equilibrium of the string being disturbed at B, a wave of length $4l$ travels along the string, so that P does not begin to move until the wave reaches it. This wave is reflected at A and returns.

619. Solution by Trigonometrical series. The second method of conducting the solution is as follows. Taking as before the expression

$$\xi = f(at - x) + F(at + x),$$

let us expand each function in a series of sines and cosines, so that we have

$$\xi = \Sigma [A \sin\{n(at - x) + a\} + B \sin\{n(at + x) + \beta\}],$$

where Σ implies summation for all values of n, and A, B, a and β are constants which are different in every term, and may conveniently be regarded as functions of n.

Since the motion is oscillatory, we may suppose that all the values of n are real, and it is clear that without loss of generality we may restrict n to be positive. We do not propose to discuss the circumstances under which these suppositions may be correctly made. For these we must refer the reader to Fourier's theorem. We may here regard the assumptions as justified by the result, because we can thus satisfy all the data of the question.

The four conditions of the problem enable us to determine the constants. From the first condition we have $\beta = a + \kappa\pi$, $B = (-1)^{\kappa+1}A$, where κ is any integer. It easily follows, by expanding, that ξ may be written in the form

$$\xi = \Sigma(C \sin nat + D \cos nat) \sin nx,$$

where C and D are to be regarded as functions of n. From the second condition we have $\cos nl = 0$, hence $nl = \frac{1}{2}(2i + 1)\pi$, where i is any positive integer. The periods of the principal oscillations (Art. 53) of the string, with proper initial disturbances, one end being fixed and the other loose, are therefore included in the form $4l/(2i + 1)a$.

The initial disturbance is given by the third and fourth conditions. We have

$$\Sigma D \sin nx = rx, \qquad \Sigma Cn \sin nx = 0.$$

To find the value of D in any term, we multiply the first equation by the coefficient of D in that term, and integrate throughout the length of the string, i.e. from $x = 0$ to $x = l$. This gives

$$D\frac{l}{2} = r \int_0^l x \sin nx\, dx = r\, \frac{\sin nl}{n^2}.$$

The other terms all vanish, since $\int_0^l \sin nx \sin n'x\, dx = 0$, when n and n' are numerically unequal. This follows also from the rule given in Art. 398.

Treating the second equation in the same way, we find $C = 0$. Hence the motion is given by

$$\xi = \Sigma \frac{2r}{l}\, \frac{\sin nl}{n^2} \cos nat \sin nx.$$

Writing for i its values 1, 2, 3, &c. successively, this equation becomes when written at length

$$\xi = \frac{8rl}{\pi^2} \left\{ \cos\frac{\pi at}{2l} \sin\frac{\pi x}{2l} - \frac{1}{3^2}\cos\frac{3\pi at}{2l}\sin\frac{3\pi x}{2l} + \frac{1}{5^2}\cos\frac{5\pi at}{2l}\sin\frac{5\pi ax}{2l} - \&c. \right\}.$$

This is a convergent series for ξ, and it may be a sufficient approximation to the motion to take only the first few terms. For example, let us reject all beyond the first two terms, and, in order to compare the result with that obtained in the first solution, let us put $at = \frac{1}{2}l$. If we trace the curve whose ordinate is $-d\xi/dt$ and abscissa x, we find that it resembles $\xi = 0$ for small values of x, then rises with a point of contrary flexure, and becomes nearly horizontal as x approaches l. This agrees very well with the result found in Art. 618.

620. If we examine these solutions, we shall see that we have two kinds of conditions to determine the arbitrary functions. (1) There are the conditions at the two extremities of the string. The peculiarity of these is, that they hold for all values of t. (2) There are the initial conditions of motion. The peculiarity of these is, that they do *not* hold for all values of x, but only for all values within a certain range limited by the length of the string. The first set of conditions is used to determine the mode in which the values of the functions recur, so that, when their values are known through a certain limited range, they will become known for all those values of the variable which occur in the problem. The second set of conditions is used to determine their values during this limited range.

In the second form of the solution we replace the arbitrary functions by a convergent series of harmonic expressions. Taking a finite number of terms as an approximation, we have a perfectly continuous solution whose initial conditions differ but slightly from those of the proposed problem. This difference is less and less, the more terms of the series are included in the solution.

In comparing the two results, we see that each form has its advantages. The first determines the motion by a simple formula. The second is more convenient when the harmonic periods are required.

In both of these solutions the arbitrary functions were found to be discontinuous. The discontinuity is plainly exhibited in that of Art. 618, though in Art. 619 it is concealed in the series. It may be objected that no notice is taken of any possible discontinuity in forming the equations of motion, (Art. 612), and that therefore these equations cannot be applied, without further examination, to any cases which require the arbitrary functions introduced into the solution to be discontinuous. This question has been much discussed, but we have not space here to do more than make a very few remarks on it. We must refer the reader to De Morgan's *Differential Calculus*, Chap. XXI. page 727, where a short history of the dispute between D'Alembert and Lagrange, and a discussion of the difficulty, may be found. In the *Mécanique Analytique, Seconde Partie*, page 385, Lagrange shows that we may avoid the use of discontinuous functions by regarding the string as the limit of a light string loaded with masses in the manner described in Art. 402. Poisson gives other reasons in his *Traité de Mécanique*. It is now generally admitted that the functions may be discontinuous.

The discontinuity in the solution of Art. 618 has its origin in the contradiction between the condition (2), viz. that $T = 0$ when $x = l$, and the condition (3) that $T = Er$ from $x = 0$ to l. But this contradiction is only apparent, for we may replace the given initial conditions by others which are without ambiguity, and which differ as slightly as we please from those given above. Let a be some finite quantity however small such that a tension less than Ea may be neglected; then the condition (3) may be replaced by a continuous function $\xi = \phi(x)$, where $\phi'(x)$ differs from r by less than α for all values of x between $x = 0$ and $x = l - \beta$, and then decreases to zero while x increases from $x = l - \beta$ to l. Since β can be taken as small as we please, it is evident that the solution given above is substantially unaltered by this change of the initial conditions. The difference is that the tension and the velocity, instead of changing suddenly, change only very quickly in the small time β/a. It is true that the mode in which this rapid change is effected is unknown, but that is because there is nothing in the initial conditions to determine it. By going to the limit when α and β are small we can make the new set of conditions represent the former as nearly as we please. Some examples of such changes may be found in De Morgan's *Differential Calculus*, pages 605—630.

621. *An elastic string, whose unstretched length is l, has its two extremities fixed at two points whose distance apart is l', and vibrates transversely. It is required to find the notes which can be sounded.*

Taking the equilibrium position of the string as the standard, let y be the transversal displacement of any particle. Let m be the mass per unit of length of unstretched string. The differential equation is then

$$\frac{d^2y}{dt^2} = a^2 \frac{d^2y}{dx_1^2},$$

where $a^2 = T_0 l'/ml$, as shown in Art. 616. Since the *notes* which can be sounded are asked for, we adopt the solution in trigonometrical series. We therefore put

$$y = \Sigma \left[A \sin \{n(at - x_1) + \alpha\} + B \sin \{n(at + x_1) + \beta\} \right].$$

When $x_1 = 0$, y is zero for all values of t, hence as in Art. 619

$$y = \Sigma (C \sin nat + D \cos nat) \sin nx_1.$$

When $x_1 = l'$, we have again $y = 0$; hence $nl' = i\pi$, where i is some integer. We therefore find

$$y = \Sigma \left(C \sin i\pi t \sqrt{\frac{T_0}{mll'}} + D \cos i\pi t \sqrt{\frac{T_0}{mll'}} \right) \sin \frac{i\pi x_1}{l'},$$

where the Σ implies summation for all integer values of i.

The motion given by taking only the terms which have any one period and neglecting the others is called *a note*. The notes which can be sounded from any instrument are called *the harmonics*. The note of longest period, i.e. that determined by $i = 1$, is called *the fundamental note*. The period of the fundamental note is $2 \sqrt{\dfrac{mll'}{T_0}}$. If this period be called τ, the periods of the harmonics in order are τ, $\frac{1}{2}\tau$, $\frac{1}{3}\tau$ and so on. The lengths of the corresponding waves are found by multiplying the periods by the velocity a. If the length of the wave of the fundamental note is λ, we have $\lambda = 2l'$, and the lengths of the harmonic waves are λ, $\frac{1}{2}\lambda$, $\frac{1}{3}\lambda$ and so on.

The points of intersection of the string with the straight line joining the extreme fixed points are called the *nodes*, and the points of the string most remote from this straight line are called the *loops*. Putting $y = 0$ we see that the nodes are given by $\sin i\pi x_1/l' = 0$; putting $dy/dx_1 = 0$, the loops are given by $\cos i\pi x_1/l' = 0$. Thus the fundamental note has one loop and no node intervening between the fixed extreme points. The next harmonic has two loops and one intervening node, and so on. It is important to notice (1) that the positions of the loops and nodes are fixed throughout the motion, (2) that the nodes and loops occur alternately, (3) that the distance between any node and the consecutive loop is one quarter of the length of either of the waves forming the note, the length being measured on the stretched string. See Art. 433.

In most cases in which strings are used as vibrating bodies, the stretched and unstretched lengths are so nearly equal that we may put $l = l'$. The results then become the same as those in ordinary use.

In order that the string may be made to sound any given harmonic, the initial conditions must be such that the amplitudes of all the other notes are zero. In practice this condition cannot be satisfied, and all that can be accomplished is that the amplitude of the proposed note shall be very much greater than those of the others. It follows that every note when sounded is accompanied by a number of subsidiary notes whose periods are different from that of the note intended to be sounded. When therefore notes of the given period are sounded by two instruments of different construction, they may be accompanied by different series of subsidiary notes. This is usually expressed by saying that the notes are of different *qualities*.

622. Examples. Ex. 1. A heavy elastic string AB, whose unstretched length is l, is suspended from a point A under the action of gravity. If ξ be the vertical displacement of any point whose distance from A is x when the string is unstretched, and if a be the velocity of a wave measured in units of unstretched length, prove that

$$\xi = -\frac{gx^2}{2a^2} + \frac{glx}{a^2} + f(at - x) - f(at + x),$$

where $f(z)$ recurs with an opposite sign when z is increased by $2l$. If the string is initially unstretched and at rest, prove that $f(z) = \pm \frac{gz^2}{4a^2} + \frac{glz}{2a^2}$,

the upper sign being taken when z lies between $-l$ and 0, and the lower when z lies between 0 and l. Thence show that the whole length oscillates between l and $l + gl^2/a^2$.

Taking the other form of solution, show that the harmonic periods are $p = \frac{4l}{(2i+1)a}$, where i is any integer. Show also that

$$\xi = -\frac{gx^2}{2a^2} + \frac{glx}{a^2} - \frac{16gl^2}{\pi^3a^2} \Sigma \frac{\sin\left(\frac{2i+1}{2}\frac{\pi x}{l}\right)\cos\left(\frac{2i+1}{2}\frac{\pi at}{l}\right)}{(2i+1)^3},$$

the summation extending from $i = 0$ to $i = \infty$.

Ex. 2. A string infinite in length in both directions has its initial state determined by $\xi = f(x)$ and $d\xi/dt = F(x)$. Show that the displacements at the time t are given by $\xi = \frac{1}{2}f(x + at) + \frac{1}{2}f(x - at) + \frac{1}{2a}\int_{x-at}^{x+at} F(\lambda)\,d\lambda.$

[Riemann's *Partial Differential Equations*.

Ex. 3. A string AB is stretched at a tension such that the velocity of a wave is equal to a. One extremity A is fixed, while the other B is agitated according to the law $y = C \sin pat$. If A be the origin, show that the forced vibration is $y = C\frac{\sin px}{\sin pl}\sin pat$. If the string start from rest, the additional free vibrations are $y = \Sigma M \sin mx \sin mat$, where $ml = i\pi$ and $M(p^2l^2 - i^2\pi^2) = -2Cpl(-1)^i$. The Σ implies summation for all integral positive values of i.

Ex. 4. If, as in the last example, the string start from rest and have the extremity A fixed, but the extremity B agitated according to the law $y = f(t)$, prove that $y = -\frac{2\pi a^2}{l^2}\Sigma i(-1)^i \sin\frac{i\pi x}{l}\cos\frac{i\pi at}{l}\int_0^t \left\{\sec^2\frac{i\pi at}{l}\int_0^t f(t)\cos\frac{i\pi at}{l}\,dt\right\}dt$

for all values of x between 0 and l, the latter being excluded. Show also by an application of Fourier's theorem that the result of the last example follows from this.

Ex. 5. A heavy string is suspended vertically by one extremity without any of its parts being stretched; if it be then left to the action of gravity, prove that the lower end will oscillate as if it were acted on by an acceleration equal to that of gravity tending to the middle point of its path.   [Smith's Prize.

Ex. 6. If a stretched string of length l be fastened to two equal masses M, controlled by springs of strength μ allowing transversal vibrations, and be plucked at its middle point, the period p of vibration will be given by

$$ma \tan \frac{\pi l}{pa} = \frac{p\mu}{2\pi} - \frac{2\pi M}{p},$$

where m is the line density and ma^2 the tension of the string. [Math. Tripos, 1881.

Ex. 7. An elastic rod of length l lies on a smooth plane, and is longitudinally compressed between two pegs at a distance l' apart. One peg is suddenly removed; show that the rod leaves the other peg just as it reaches its natural state, and then proceeds with a velocity equal to $(l - l')/l$ of the velocity of propagation of a longitudinal wave in the rod.   [Math. Tripos, 1883.

Ex. 8. A ring formed of elastic string, of mass M and natural length $2\pi l$, is stretched round a smooth circular cylinder; prove that the time in which a longitudinal pulsation will travel round the cylinder is independent of the size of the cylinder.

When the ring is in equilibrium, the ends of an arc subtending an angle $2a$ at the centre of the ring are drawn together until the arc attains its natural length, and these ends are then let go. Measuring θ from the diameter bisecting the angle $2a$ prove that at any subsequent time the displacement from the position of equilibrium of the end of the corresponding arc is equal to

$$-\frac{a - l}{a} \cdot \frac{2}{\pi - a} \cdot \sum_{1}^{\infty} \frac{\sin n\theta \sin na \cos n\omega t}{n^2},$$

where $Ml\omega^2 = 2\pi E$, a being the radius of the cylinder and E the modulus of elasticity.   [Math. Tripos, 1886.

The first part of this example follows from the theorem on the velocity of a wave given in Art. 614. In the second part the differential equation leads to $\xi = \Sigma M \sin n\theta \cos n\omega t$. The values of M are found by using Fourier's Theorem as in Art. 619.

623. **Several strings.** *Three elastic strings AB, BC, CD of different materials are attached to each other at B and C and stretched in a straight line between two fixed points A, D. If the particles of the string receive any longitudinal displacements and start from rest, find the subsequent motion.**

---

* The problem of finding the transversal vibrations of a tight string composed of two parts of different kinds appears to have been first solved by Poisson, *Journal de l'école Polytechnique*, tome XI. 1820. Poisson points out that Euler and Bernoulli, who had attempted the problem before him, had arrived at only incomplete results, *Mémoires de Pétersbourg*, 1771 and 1772. The latter had indeed obtained an equation giving the periods, but had not found the form assumed by the string at any time during the motion. The results of the latter were to a certain extent erroneous, as he had rejected the condition that the two parts of the string must have a

Let A be the origin, AD the direction in which x is measured. Let the unstretched lengths of AB, BC, CD be l_1, l_2, l_3. Let E_1, E_2, E_3 be their respective coefficients of elasticity, m_1, m_2, m_3 their masses per unit of length. For the sake of brevity let $E_1 = m_1 a_1^2$, $E_2 = m_2 a_2^2$, $E_3 = m_3 a_3^2$. Let the rest of the notation be the same as before.

When the string is stretched *in equilibrium* between the two fixed points A and D, let T_0 be the tension of the string. In this position the displacements of the elements of each string from their positions when unstretched may be written

$$\xi_1 = \frac{l_1' - l_1}{l_1} x = \frac{T_0}{E_1} x,$$

$$\xi_2 = l_1' - l_1 + \frac{l_2' - l_2}{l_2}(x - l_1) = \frac{T_0}{E_1} l_1 + \frac{T_0}{E_2}(x - l_1),$$

$$\xi_3 = \&c. = \frac{T_0}{E_1} l_1 + \frac{T_0}{E_2} l_2 + \frac{T_0}{E_3}(x - l_1 - l_2).$$

At the time t after the equilibrium has been disturbed, let these displacements be respectively $\xi_1 + \xi_1'$, $\xi_2 + \xi_2'$, $\xi_3 + \xi_3'$. We then have as in Art. 619

$$\xi_1' = \Sigma L_1 \sin(n_1 x + M_1) \cos n_1 a_1 t,$$
$$\xi_2' = \Sigma L_2 \sin\{n_2(x - l_1) + M_2\} \cos n_2 a_2 t,$$
$$\xi_3' = \Sigma L_3 \sin\{n_3(x - l_1 - l_2) + M_3\} \cos n_3 a_3 t,$$

where Σ implies summation for all the harmonics. The terms containing $\sin n_1 a_1 t$, $\sin n_2 a t_2$, &c. are omitted because the string starts from rest, and therefore $d\xi_1/dt$, $d\xi_2'/dt$, &c. must vanish with t.

In order to compare the coefficients of the same harmonic we must suppose $n_1 a_1 = n_2 a_2 = n_3 a_3 = 2\pi/p$, where p is the period of the harmonic. To find the constants we have the conditions

when $x = 0$, $x = l_1$, $x = l_1 + l_2$, $x = l_1 + l_2 + l_3$,

   $\xi_1' = 0$, $\xi_1' = \xi_2'$, $\xi_2' = \xi_3'$, $\xi_3' = 0$,

$$E_1 \frac{d\xi_1'}{dx} = E_2 \frac{d\xi_2'}{dx}, \quad E_2 \frac{d\xi_2'}{dx} = E_3 \frac{d\xi_3'}{dx}.$$

These give $M_1 = 0$,

$$\left. \begin{array}{l} L_2 \sin M_2 = L_1 \sin(n_1 l_1 + M_1) \\ E_2 n_2 L_2 \cos M_2 = E_1 n_1 L_1 \cos(n_1 l_1 + M_1) \end{array} \right\},$$

$$\left. \begin{array}{l} L_3 \sin M_3 = L_2 \sin(n_2 l_2 + M_2) \\ E_3 n_3 L_3 \cos M_3 = E_2 n_2 L_2 \cos(n_2 l_2 + M_2) \end{array} \right\},$$

$$0 = L_3 \sin(n_3 l_3 + M_3).$$

These give the following equations to find the M's;

$$0 = M_1, \quad \frac{\tan M_2}{E_2 n_2} = \frac{\tan(n_1 l_1 + M_1)}{E_1 n_1}, \quad \frac{\tan M_3}{E_3 n_3} = \frac{\tan(n_2 l_2 + M_2)}{E_2 n_2}, \quad 0 = \frac{\tan(n_3 l_3 + M_3)}{E_3 n_3}.$$

common tangent at the point of junction. The problem has been again considered by Bourget in the *Annales de l'école normale supérieure*, tome IV. 1867, where he corrects some of the results of Poisson. He also discusses the vibrations of a tight cord formed of three different parts, and gives a somewhat complicated rule to find the periods when the cord is composed of n different parts. Finally he describes ten different experiments showing the agreement between the theory and experience. These experiments are again discussed in tome IX. of the *Annales de l'observatoire de Paris*, 1868.

Solving these we find

$$\frac{\tan n_1 l_1}{E_1 n_1} + \frac{\tan n_2 l_2}{E_2 n_2} + \frac{\tan n_3 l_3}{E_3 n_3} = (E_2 n_2)^2 \frac{\tan n_1 l_1}{E_1 n_1} \cdot \frac{\tan n_2 l_2}{E_2 n_2} \cdot \frac{\tan n_3 l_3}{E_3 n_3}.$$

Substituting for n_1, n_2, n_3 in terms of p, we have an equation to find the period p of any principal oscillation.

624. The values of p being known, it is clear that the preceding equations determine all the constants except L_1. We have therefore one constant undetermined for each harmonic function of t. To find these we must have recourse to the initial conditions. The rule to effect this has been fully given in Art. 399.

The equations may be written in the forms

$$\xi_1' = \Sigma P_n \cos nat, \qquad \xi_2' = \Sigma Q_n \cos nat, \qquad \xi_3' = R_n \cos nat,$$

where P_n, Q_n and R_n stand for the coefficients as exhibited in the last article. The first of these three equations represents in a typical form the motion of any particle in the string AB, the second represents the motion of any particle in BC, and so on. Referring to Art. 399, the three sets of multipliers may be typically represented by

$$m_1 dx P_n, \qquad m_2 dx Q_n, \qquad m_3 dx R_n.$$

The summations spoken of in Art. 399 are here integrations and extend over the lengths of the three strings respectively.

Suppose now that we have initially $\xi_1' = f_1(x)$, $\xi_2' = f_2(x)$, $\xi_3' = f_3(x)$. We easily find

$$\int_0^{l_1} m_1 dx f_1(x) P_n + \int_{l_1}^{l_1+l_2} m_2 dx f_2(x) Q_n + \int_{l_1+l_2}^{l_1+l_2+l_3} m_3 dx f_3(x) R_n$$

$$= \int_0^{l_1} m_1 dx P_n^2 + \int_{l_1}^{l_1+l_2} m_2 dx Q_n^2 + \int_{l_1+l_2}^{l_1+l_2+l_3} m_3 dx R_n^2.$$

These integrations can be effected when the forms of $f_1(x)$, $f_2(x)$ and $f_3(x)$ are given. Thus we have an additional equation to find the L which corresponds to any value of p.

625. **Examples.** Ex. 1. If the three strings vibrate transversely, and a_1, a_2, a_3 be the velocities of a wave along them measured in units of length of unstretched string, prove that the periods of the notes are given by the equation

$$\frac{\tan n_1 l_1}{n_1} + \frac{\tan n_2 l_2}{n_2} + \frac{\tan n_3 l_3}{n_3} = n_2^2 \frac{\tan n_1 l_1}{n_1} \cdot \frac{\tan n_2 l_2}{n_2} \cdot \frac{\tan n_3 l_3}{n_3},$$

where $n_1 a_1 = n_2 a_2 = n_3 a_3 = 2\pi/p$. If the initial disturbance is given show how to find the subsequent motion.

Ex. 2. Two heavy strings AB, BC of different materials are attached together at B and suspended under gravity from a fixed point A. Prove that the periods of the vertical oscillations are given by the equation

$$\tan \frac{2\pi l_1}{a_1 p} \cdot \tan \frac{2\pi l_2}{a_2 p} = \frac{E_1 a_2}{E_2 a_1},$$

the notation being the same as before. If the two strings be initially unstretched, find their lengths at any time.

Ex. 3. Two strings AB, BC of different materials are attached at B to a particle of mass M, while their other extremities A and C are fixed in space. If the particles of the system vibrate along the length of the straight line AC, prove that the period p of any principal oscillation is a root of the equation

$$M \frac{2\pi}{p} = \frac{E_1}{a_1} \cot \frac{2\pi l_1}{a_1 p_1} + \frac{E_2}{a_2} \cot \frac{2\pi l_2}{a_2 p_2},$$

where l_1, l_2 are the unstretched lengths of the strings, E_1, E_2 their elasticities, and a_1, a_2 the velocities of a wave measured in units of unstretched length per unit of time. The values of p obtained by equating (when possible) the cotangents simultaneously to infinity are to be included.

If the system make small oscillations transverse to the straight line AC, the periods will be given by the same equation if we replace E_1 and E_2 by T_0 the tension of the string when in equilibrium.

Ex. 4. A particle is suspended from a fixed point by an elastic string and performs small oscillations in a vertical direction, supposing the string uniform in its natural state and of small finite mass, show that the time of a small oscillation will be approximately the same as if the string were without weight and the mass of the particle were increased by one third that of the string.   [Smith's Prize.

Ex. 5. Two uniform heavy elastic beams AB, CD equal in every respect are connected by a light inextensible string BC; the beam AB lies unstrained on a smooth horizontal table, while CD is suspended at rest under the action of gravity by a string which, being held at B, passes over a smooth pulley P at the edge of the table, PBA being a straight line. Investigate the motion of the string when set free; prove that its tension, after being instantaneously diminished by one half, remains constant, and that its velocity receives equal increments at equal intervals.
[Math. Tripos, 1876.

The problem is unaltered in its physical relations if we suppose the rods to be in one straight line on the table and CD only to be acted on by gravity; in this way the problem is simplified by eliminating the pulley. To keep the centre of gravity of the whole stationary let us next apply to every particle a force half that of gravity in the opposite direction. The result is that the rod AB is acted on by $\frac{1}{2}g$ in the direction BA, and the rod CD by $\frac{1}{2}g$ in the direction CD. The solution then follows the lines of Ex. 1, Art. 622.

Ex. 6. A particle is fixed to the middle point of a heavy string, which is stretched to double its length between two fixed points on a smooth horizontal table. The unstretched length of the string is $2l$, its modulus is n times, and the weight of the particle is r times, the weight of the string. The particle is then moved through a distance λl towards one of the fixed points, and when the string has been reduced to rest the particle is set free. Show that there are sufficient conditions to determine completely the four arbitrary functions, and indicate how they are to be employed. Prove that the velocity of the particle during the first interval $\frac{2l}{a}$ is $\lambda a\left(1 - e^{-\frac{at}{rl}}\right)$, where $a^2 = 2gnl$ and t is the time from rest.

[Caius Coll., 1871.

Ex. 7. Three strings OA, OB, OC of the same material but of different lengths are united at O and are kept tight by being fastened to fixed points A, B, C, the angles BOC, COA, AOB being denoted by α, β, γ. Show that the times of transversal vibration of the different notes sounded when O is free are determined by the equation for T

$$\sqrt{\sin \alpha} \cdot \cot \pi T_1/T + \sqrt{\sin \beta} \cdot \cot \pi T_2/T + \sqrt{\sin \gamma} \cdot \cot \pi T_3/T = 0,$$

where T_1, T_2, T_3 are the times of the gravest notes of OA, OB, OC when O is fixed.
[Math. Tripos, 1884.

Ex. 8. A uniform string of length $2l$ is stretched with tension T between two fixed points. Prove that, if the string is initially pulled aside by a force Y at a point distant b from one end, the motion of the string is given by

$$y = \frac{Y}{lm} \cdot \Sigma \sin \frac{i\pi b}{2l} \sin \frac{i\pi x}{2l} \frac{\cos nt}{n^2},$$

where m is the mass per unit of length, a the velocity of propagation of waves along the string, $2nl = i\pi a$, and the summation is from $i = 1$ to $i = \infty$.

The string has its ends fastened to two masses each equal to M which are kept in place by springs of strength μ and has a mass M' fastened at its middle point. Prove that if M' is plucked transversely the period of the vibrations is $2\pi/pa$ where

$$p \tan pl \left\{ M'a^2 (Mp^2a^2 - \mu) - 2T^2 \right\} = T \left\{ 2Mp^2a^2 - 2\mu + M'p^2a^2 \right\}.$$

[Math. Tripos, 1885.

In the second part of the question take the middle point as origin, then for the string on the positive side $y = (P \cos px + Q \sin px) \cos pat$. The conditions are (1) $M'd^2y/dt^2 = -2Tdy/dx$ when $x = 0$; (2) $Md^2y/dt^2 = -Tdy/dx - \mu y$ when $x = l$. Substituting for y and eliminating Q/P we obtain the result.

Ex. 9. The ends A, B of a string AHB are fastened to light rings which are free to move on smooth rods parallel to one another. At A, H, B forces act transversely to the string and parallel to the rods with intensities

$$X = F \cos \kappa t + G \sin \kappa t, \quad Y = L \cos \lambda t + M \sin \lambda t, \quad Z = R \cos \mu t + S \sin \mu t,$$

respectively. Show that at the time t the consequent displacement, in the direction opposite to that of the forces of any point P in AH is

$$\frac{aX}{T\kappa} \cdot \frac{\cos \kappa (l - x)/a}{\sin \kappa l/a} + \frac{aY}{T\lambda} \cdot \frac{\cos \lambda (l - h)/a \cdot \cos \lambda x/a}{\sin \lambda l/a} + \frac{aZ}{T\mu} \cdot \frac{\cos \mu x/a}{\sin \mu l/a},$$

where T is the tension of the string, a its wave velocity, and x, h, l the natural lengths of AP, AH, AB respectively.   [Math. Tripos, 1886.

Consider the forces separately. Taking Y first, let η, η_1 be the transversal displacements of two points one in each of the strings AH, HB distant x and x_1 from A and B respectively. The conditions are (1) $d\eta/dx = 0$, $d\eta_1/dx_1 = 0$ at A and B respectively; (2) $\eta = \eta_1$ and $T (d\eta/dx + d\eta_1/dx_1) = Y$ at H. The displacement due to Y having been found, that due to Z is deduced by writing $h = l$ and changing Y, λ into Z, μ. The displacement due to X may be deduced from that due to Z. Superimposing all three the result given is obtained.

Ex. 10. A metal rod fits freely in a tube of the same length but of different substance, and the extremities of each are united by equal perfectly rigid discs fitted symmetrically at the ends. Show that the periods of the notes emissible, which have a node at the centre of the system, are given by $2\pi l/x$, where $2l$ is the length of the rod or tube and x is a root of the equation

$$2Mx = ma \cot x/a + m'a' \cot x/a',$$

and where M, m, m' are the masses of a disc, the bar, and the tube, and a, a' are the velocities of propagation of sound along the bar and the tube.

Discuss the particular cases (1) when M is very large, and (2) when M is very small, especially when $ma = m'a'$.   [Math. Tripos, 1883.

Ex. 11. The extremities of a uniform bar of length l are attached to two fixed points, distant l apart, by springs of equal strength. If the longitudinal vibrations of the bar are represented by $\xi = \{P \sin mx/l + Q \cos mx/l\} \sin rt$,

prove that $(m^2q^2 - l^2\mu^2) \tan m + 2mql\mu = 0$, where μ is the strength of either of the springs and q the ratio of the tension to the extension of the bar.

[Math. Tripos, 1880.

Ex. 12. Supposing that the resistance of the air to a vibrating string may be represented by a force on each element which varies as the velocity of that element, the equation of motion takes the form $\dfrac{d^2\xi}{dt^2} = a^2 \dfrac{d^2\xi}{dx^2} - 2f \dfrac{d\xi}{dt}$.

Show that the motion is given by

$$\xi = e^{-ft} \Sigma (A \sin pt + B \cos qt) \sin qx,$$

where $p^2 = q^2a^2 - f^2$, and the summation extends to all the values of q which the conditions of the problem admit of.

If the string is unlimited in length, show that the velocity v of a series of waves of length λ is given by $v^2 = a^2 - (f\lambda/2\pi)^2$. Thus the velocities of all waves are diminished by the resistance, but if f is small the diminution depends on the square of f and is therefore insignificant. If the length of the wave is so great that v is imaginary, show that the motion is given by

$$\xi = e^{-ft} \Sigma (Ae^{\rho t} + Be^{-\rho t}) \sin qx,$$

where $\rho^2 = -q^2a^2 + f^2$. Thus the motion ceases to be oscillatory.

If the length of the string is finite and equal to l, and each end is fixed in space, show that the motion is given by writing $q = i\pi/l$, where i is any integer. Thence show (1) that, whatever note the string is sounding, the extent of the vibration is reduced by the resistance to the same fraction of its original value in any given time; (2) that the positions of the nodes and loops are the same as if the resistance were absent; (3) that the pitch of the note sounded is flattened by the resistance and the flattest notes are the most altered. If the initial displacement of the string is given by $\xi = A \sin \pi x/l$ and the string start from rest, determine and compare the subsequent motions in the two cases in which, (1) there is no resistance and (2) the resistance is such that f is greater than $\pi a/l$.

Ex. 13. One effect of viscosity is to resist the compression or extension of an element of string whose extremities are moving with slightly different velocities, see Art. 333, Ex. 2. To represent this analytically, let us suppose that the tension exerted by a stretched element of string, instead of being given simply by Hooke's law, has an additional term proportional to the relative velocity of the extremities of the element. Show that the equation of motion of longitudinal vibrations is

$$m \frac{d^2\xi}{dt^2} = E \frac{d^2\xi}{dx^2} + 2Fm \frac{d^3\xi}{dx^2 dt}.$$

Hence show that the motion is given by

$$\xi = e^{-Fq^2t} \Sigma (A \sin pt + B \cos pt) \sin qx,$$

where $mp^2 = Eq^2 - F^2mq^4$ and the summation extends to all existing values of q. If p is imaginary we replace the trigonometrical functions of t by real exponentials.

If the string is unlimited in length, show that the short waves are sensibly extinguished by the resistance more quickly than the long ones.

If the string has its extremities fixed, show that the sharp notes disappear quicker than the flat ones.

This differential equation follows from Art. 612. The relative velocity of the two extremities of an element is $\frac{d}{dx}\left(\frac{d\xi}{dt}\right)dx$; hence $T = E\frac{d\xi}{dx} + 2Fm\frac{d^3\xi}{dx^2 dt}$. Substituting this value of T in the differential equation of the article referred to, the result follows at once.

626. Impact of Rods. Ex. 1. Two perfectly elastic rods AB, CD of the same form and material but of lengths l_1, l_2 are placed in the same straight line. AB is projected with a velocity V to hit CD placed at rest, both rods being without initial compression. Supposing l_1 to be less than l_2 find when the rods separate.

We regard the rods as being in contact when the distance between the extremities B, C of the rods becomes equal to the distance of molecular action. The two rods may then be treated as if they formed portions of a single rod, with the condition that the two portions remain in contact as long as they push against each other, i.e. as long as the tension at the point of contact is negative. They separate when the common tension at B and C becomes positive. As soon as this occurs the rods begin to move as separate bodies, but their mutual action may recommence if this motion bring the extremities B and C again within the distance of molecular action.

The problem of the impulse of rods has been considered by Cauchy, *Académie des Sciences*, 1827 and *Bulletin des Sciences de la Société Philomathique*, 1826, and by Poisson, *Traité de Mécanique*, 1833, Tome II. In *Liouville's Journal*, Vol. XII., 1867 there is a long memoir of 140 pages by Saint-Venant in which he enters fully into the conditions of separation. These great authorities differ considerably in the interpretation of their results, and especially in the conditions of separation.

Let P be any point of either rod, v its velocity. Let s be the dilatation, or extension of an element at P per unit of length, then by Art. 612, $s = d\xi/dx$ and also $s = T/E$. We have if x is measured from A towards D

$$v = \phi\,(at - x) + \psi\,(at + x), \qquad as = -\phi\,(at - x) + \psi\,(at + x).$$

To find ϕ and ψ we use the following conditions: (1) when $t = 0$, $v = V$ from $x = 0$ to l_1; $v = 0$ from $x = l_1$ to $l_1 + l_2$; $s = 0$ from $x = 0$ to $l_1 + l_2$, (2) when $x = 0$, $s = 0$ always, and when $x = l_1 + l_2$, $s = 0$ always.

We easily find that the functions ϕ and ψ are the same, and that the curve $y = \phi\,(x)$ consists of a series of finite straight lines whose lengths are alternately $2l_1$ and $2l_2$, the ordinates being $\frac{1}{2}V$ and zero respectively. These are represented in the diagram. The axis of y divides the system symmetrically.

The figure having been drawn, the following easy rule enables us to find the state of motion at any time t of a point P distant x from A. Measure AP' equal to AP in the negative direction, and let two points R, R' starting from P and P' respectively travel each with velocity a in the positive direction. The equations show that at the time t after the commencement of the impact

$$v \text{ at } P = \text{ ordinate of } R' + \text{ordinate of } R,$$
$$as \text{ at } P = -\text{ ordinate of } R' + \text{ordinate of } R.$$

To determine if the rods separate, we must find when the common tension at B and C vanishes and becomes positive. Let therefore R and R' start from B and B'; at first the ordinates at R and R' are equal to zero and $\frac{1}{2}V$ respectively. After a time given by $at = 2l_1$ the point R' reaches B, and its ordinate falls to zero. Since l_1 is less than l_2 the point R has not reached E where $DE = BD$, and its ordinate is still zero. Both v and s therefore at this instant become zero at the point of contact B.

If R_1', R_1 are the guiding points for any particle of the rod AB, it is clear that R_1' and R_1 will always lie between R' and R, so that when R' reaches B the two guiding points of every particle of the rod lie on BE. It is therefore easy to see that at this instant both v and s are zero at every point of the rod AB, and must remain equal to zero until the point R which started from B or C arrives at E.

At the time given by $at = 2l_2$ the point R starting from C arrives at E, and its ordinate becomes equal to $\frac{1}{2}V$. The tension at C then becomes positive and the velocity becomes $\frac{1}{2}V$, so that from both these causes the end C begins to move away from the end B. The tension and velocity at B would immediately begin to undergo similar changes if the rods were to remain in contact, but this is not the case. Since the whole of AB is at that instant without velocity or tension, the end B remains at rest.

The results are (1) the rods push against each other for a time $2l_1/a$; (2) they remain in contact but without reaction for an additional time $2(l_2-l_1)/a$; (3) the rod CD then separates from AB, leaving the latter at rest and without tension in any part.

When the rods have different initial velocities, say V_1 and V_2, we may reduce the latter to rest by superimposing on every particle of both rods a velocity equal and opposite to V_2. The general results are unaltered, except that AB instead of remaining at rest has a final velocity V_2.

In the impact of these two rods the whole momentum MV of one has been transferred to the other, whose centre of gravity therefore moves away with a velocity Vl_1/l_2. The vis viva has also been transferred to the second rod, a part MV^2l_1/l_2 being transformed into vis viva of translation and the remainder, viz. $MV^2(1-l_1/l_2)$ into the energy (kinetic and potential) of the internal vibration. This internal energy is zero if the rods are equal in length.

It is useful to compare the results obtained by theory with those given by Newton's experimental law of impact; Vol. I. Art. 179. Since vis viva is apparently lost in the impact, the rods (though stated to be elastic) are in Newton's formula to be regarded as imperfectly elastic. We easily see by putting $u'=0$ in his formula that the coefficient e is equal to l_1/l_2. We notice that this does not depend only on the nature of the materials.

Ex. 2.　Two elastic bars AB, CD of length l_1, l_2 masses M_1, M_2 and initial velocities V_1, V_2 but without initial strain impinge in the same straight line ABC on each other. If A be the origin show that the displacements for the two rods are

$$\xi_1 = \frac{M_1V_1+M_2V_2}{M_1+M_2}t + (V_1-V_2)\frac{M_2a_1}{l_1}\Sigma\frac{2}{p^2}\frac{\sin(pl_1/a_1)\sec^2(pl_1/a_1)\cos(px/a_1)\sin pt}{M_1\sec^2(pl_1/a_1)+M_2\sec^2(pl_2/a_2)},$$

$$\xi_2 = \frac{M_1V_1+M_2V_2}{M_1+M_2}t - (V_1-V_2)\frac{M_2a_2}{l_2}\Sigma\frac{2}{p^2}\frac{\sin(pl_2/a_2)\sec^2(pl_2/a_2)\cos(px/a_2)\sin pt}{M_1\sec^2(pl_1/a_1)+M_2\sec^2(pl_2/a_2)},$$

$$\frac{M_1a_1}{l_1}\tan\frac{pl_1}{a_1}+\frac{M_2a_2}{l_2}\tan\frac{pl_2}{a_2}=0,$$

where Σ implies summation for all values of p given by the third equation and a_1, a_2 are the wave velocities in the two bars.　　　　[Saint-Venant.

Poisson also gives the corresponding expressions in the case considered by him.

Ex. 3.　Two rods AB, CD, lengths l_1, l_2 and velocities V_1, V_2, impinge in the same straight line, and at the moment of contact the tensions of the rods are Es_1 and Es_2 respectively. Show that the two rods immediately separate or remain in contact for a time according as V_2+as_2 is greater or less than V_1-as_1. [Saint-Venant.

Ex. 4.　Two rods, lengths l_1, l_2, impinge, and at the moment of contact are moving with velocities V_1, V_2 and have dilatations s_1, s_2 uniformly distributed over their lengths. If $V_1\pm as_1$, V_2+as_2 are positive, and both the values of the former greater than that of the latter, prove that the rods will push against each other for a time $2l_1/a$, remain in contact without reaction for a time $(l_2-2l_1)/a$, if $l_2>2l_1$, and then, if s_2 is negative, separate. If s_2 is positive they again push against each other for a time $2l_1/a$, cease to react for a time $(l_2-2l_1)/a$ and then separate.

627. Energy of a string. *An elastic string is stretched between two fixed points A and B' and is set in vibration, it is required to find the energy.*

Let the notation be the same as that used in Arts. 612 and 616.

First, let the vibrations be longitudinal. The equation of motion is $\dfrac{d^2\xi}{dt^2} = a^2 \dfrac{d^2\xi}{dx^2}$.

Hence we have $\xi = \dfrac{l'-l}{l} x + \Sigma [A \sin \{n(at-x)+a\} + B \sin \{n(at+x)+\beta\}]$.

Since ξ must vanish when $x=0$ and be equal to $l'-l$ when $x=l$, we find, as in Art. 619, $\xi = \dfrac{l'-l}{l} x + \Sigma C \sin nx \sin (nat+\gamma)$,

where $nl = i\pi$ and Σ implies summation for all positive integer values of i. The letters C and γ are constants which may be different in every term, and which depend on the initial disturbance. The kinetic energy of the whole string is

$$= \int_0^l \frac{1}{2} m dx \left(\frac{d\xi}{dt}\right)^2 = \int_0^l \frac{1}{2} m dx \{\Sigma Cna \sin nx \cos (nat+\gamma)\}^2.$$

Now $\int_0^l \sin nx \sin n'x dx = 0$ when n and n' are numerically unequal, since nl and $n'l$ are both integer multiples of π. Hence, when the square of the series is expanded, the integral of the product of any two terms is zero.

Since $\int_0^l \sin^2 nx dx = \frac{1}{2} l$, the kinetic energy becomes $= \frac{1}{4} m l a^2 \Sigma C^2 n^2 \cos^2 (nat+\gamma)$.

To find the potential energy; we notice that the work done in stretching an element from its unstretched length dx to its length $dx + d\xi$ is (see Vol. I.) equal to $\dfrac{1}{2} E \left(\dfrac{d\xi}{dx}\right)^2 dx$. Hence the whole work done in stretching the string is

$$= \int_0^l \frac{1}{2} E dx \left(\frac{d\xi}{dx}\right)^2 = \int_0^l \frac{1}{2} E dx \left\{\frac{l'-l}{l} + \Sigma Cn \cos nx \sin (nat+\gamma)\right\}^2.$$

Now $\int_0^l \cos nx \cos n'x dx = 0$ or $\frac{1}{2} l$ according as n and n' are numerically unequal or equal to each other; also $\int_0^l \cos nx dx = 0$. Hence, as before, the integral becomes

$$= \frac{1}{2} E \frac{(l'-l)^2}{l} + \frac{1}{4} E l \Sigma C^2 n^2 \sin^2 (nat+\gamma).$$

The first term is the work done in stretching the string from the unstretched length l to the stretched length l'. If we refer the potential energy to the position of the string when stretched in equilibrium between the extreme points A and B' as the standard position, we retain the latter term only.

The energy is the sum of the kinetic and potential energies. Since $E = ma^2$, this becomes   energy $= \frac{1}{4} m l a^2 \Sigma C^2 n^2$.

This result might have been deduced more simply from Art. 72, where it is shown that the energy of a compound vibration is the sum of the energies of the simple vibrations into which it may be resolved. The kinetic energy of any *single* harmonic is easily seen by integration to be $\frac{1}{4} m l a^2 C^2 n^2 \cos^2 (nat+\gamma)$. Hence the whole energy is $\frac{1}{4} m l a^2 \Sigma C^2 n^2$.

We may also notice that, as in Art. 73, the mean kinetic energy is equal to the mean potential energy, the means being taken for any very long period.

Next, let the vibrations be transversal.

Following the notation of Art. 616, the motion is given, as before, by

$$y' = \Sigma C \sin nx \sin (nat+\gamma),$$

where $nl = i\pi$, and Σ implies summation for all positive integer values of i.

The kinetic energy by the same reasoning as before is equal to

$$\frac{1}{4} m l a^2 \Sigma C^2 n^2 \cos^2 (nat+\gamma).$$

To find the potential energy, we notice that the work done in stretching an element from its unstretched length dx to its stretched length ds' is (see Vol. I.) equal to $\dfrac{1}{2} E \left(\dfrac{ds'}{dx} - 1\right)^2 dx$. Now $(ds')^2 = (dx')^2 + (dy')^2 = \left(\dfrac{l'}{l} dx\right)^2 + dy'^2$,

$$\therefore \frac{ds'}{dx} = \frac{l'}{l}\left\{1 + \frac{1}{2}\frac{l^2}{l'^2}\left(\frac{dy'}{dx}\right)^2\right\} \text{ nearly.}$$

Remembering that, by Art. 616, $ma^2 = E\,(l' - l)/l'$; we find that the whole work done in stretching the string is $\displaystyle\int_0^l \frac{1}{2}\,dx \left\{E\left(\frac{l'-l}{l}\right)^2 + ma^2\left(\frac{dy'}{dx}\right)^2\right\}$.

Substituting for y' and integrating, we find that the work is equal to

$$\frac{1}{2} E \frac{(l' - l)^2}{l} + \frac{1}{4} mla^2 \Sigma C^2 n^2 \sin^2(nat + \gamma).$$

If we take the position of equilibrium of the string when stretched between the extreme points A and B' as the position of reference, we find that the

$$\text{energy} = \tfrac{1}{4} mla^2 \Sigma C^2 n^2.$$

This we may call the energy of the disturbance.

Prof. Donkin in his treatise on Acoustics, page 128, has found the energy of a string vibrating transversely by an ingenious application of the method of subtractions.

Ex. An elastic rod AB has the end A fixed and B free. Being placed on a perfectly smooth table, it vibrates longitudinally. Show that the energy of a disturbance represented by $\xi = \Sigma C \sin nx \sin (nat + \gamma)$, where $nl = \frac{1}{2}(2i + 1)\pi$, is $\frac{1}{4}mla^2\Sigma C^2 n^2$.

628. Vibrations of rods. *A thin uniform straight rod is in equilibrium under the action of forces at its two extremities, and when disturbed it makes small oscillations in one plane. It is required to form the equations of motion.*

The line which passes through the centre of gravity of every perpendicular section of the rod is called the axis. Let the axis AB in the position of equilibrium be taken as the axis of x, and let the plane of vibration be the plane of xy. Let D be the density of the rod, ω the area of any perpendicular section, and ωk^2 the moment of inertia of that area about a straight line through its centre of gravity drawn perpendicular to the plane of vibration.

Let P be any point on the axis of the rod; the finite portion PB is in equilibrium under the action of the reversed effective forces and the forces at the extremities P and B. Let x be the abscissa of P in the position of equilibrium, $(x + \xi, \eta)$ its co-ordinates at the time t.

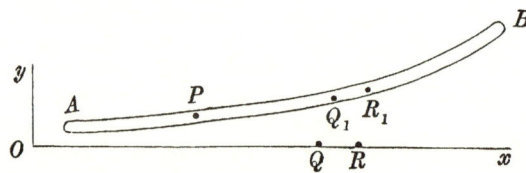

Let the action of the portion AP of the rod on PB be resolved into (1) two forces X, Y acting at P parallel to the axes, and (2) a couple L measured positively opposite to the direction in which the hands of a watch move. In the same way let the forces at the extremity B be resolved into the forces X_1, Y_1 and the couple L_1. In equilibrium both Y and Y_1 are zero, and $X = -T$, $X_1 = T$ where T is the given tension of the rod. Hence during the motion Y and Y_1 are small quantities and both X and X_1 differ from T in magnitude by small quantities.

Let QR be an element of the axis of the rod PB when in the position of equilibrium, Q_1R_1 its position at the time t. Let the co-ordinates of Q and Q_1 be respectively $(x_1, 0)$ and $(x_1+\xi_1, \eta_1)$; and let ψ be the small angle the tangent to Q_1R_1 makes with the axis of x. Consider the particles contained in an elementary slice of the rod bounded by two planes perpendicular to QR. The linear effective forces are respectively $D\omega dx\xi_1''$ and $D\omega dx\eta_1''$, where accents denote differential coefficients with regard to the time. It is also usually assumed that the angular momentum about an axis through the centre of gravity perpendicular to the plane of vibration is $Ddx \omega k^2\psi'$.

Taking moments about the instantaneous position of P, we have

$$L + \int[\xi_1'' (\eta_1-\eta) - \eta_1'' (x_1 + \xi_1 - x - \xi) - k^2\psi_1''] \omega Ddx_1 + L_1 + Y_1 (l - x - \xi) - X_1 (h - \eta) = 0,$$

where l and h are the co-ordinates of B at the time t, and the limits of the integral are $x_1 = x + \xi$ and $x_1 = l$. Rejecting some small quantities of the second order, this becomes

$$L - \int_x^l \omega\eta_1'' (x_1 - x) Ddx_1 - \int_x^l \omega k^2\psi_1'' Ddx_1 + L_1 + Y_1 (l - x) - T (h - \eta) = 0...(1).$$

By a theorem in statics we may write $L = \pm F/\rho$, where ρ is the radius of curvature at P, $F = k^2 (E\omega + T)$, and E is a constant which depends on the material of the rod and is usually called Young's Modulus. The moment L in the equation (1) has been taken in the positive direction, hence, since the rod tends to straighten itself, the constant F must have the negative sign.

Since we reject the squares of small quantities, we may write $\dfrac{1}{\rho} = \dfrac{d\psi}{dx} = \dfrac{d^2\eta}{dx^2}$.

Differentiating equation (1) with regard to x and remembering that L_1, Y_1, l, T and h are independent of x, we have

$$- F\frac{d^3\eta}{dx^3} + \int_x^l \omega \frac{d^2\eta_1}{dt^2} Ddx_1 + \omega k^2D \frac{d^2\psi}{dt^2} - Y_1 + T \frac{d\eta}{dx} = 0 \quad..........(2).$$

This differentiation is easily followed if we recollect the rule in the integral calculus, that

$$\frac{d}{dx} \int_x^l \phi (x, z) dz = - \phi (x, x) + \int_x^l \frac{d\phi (x, z)}{dx} dz.$$

Differentiating again with regard to x,

$$- F\frac{d^4\eta}{dx^4} - \omega D\frac{d^2\eta}{dt^2} + \omega k^2D \frac{d^4\eta}{dx^2dt^2} + T \frac{d^2\eta}{dx^2} = 0 \quad.................(3).$$

By resolving parallel to the axes of x and y we find in the same way

$$\frac{dX}{dx} + \omega D\frac{d^2\xi}{dt^2} = 0, \qquad \frac{dY}{dx} + \omega D\frac{d^2\eta}{dt^2} = 0 \quad....................(4).$$

Since ωk^2 is very small, the terms containing it may be neglected when it is not multiplied by E. It is therefore usual to omit the third term of equation (3). If the rod, when in the position of equilibrium, is unstretched, we have also $T = 0$. With these two simplifications the equation (3) takes the form

$$\frac{d^2\eta}{dt^2} + a^4\frac{d^4\eta}{dx^4} = 0.................................(5),$$

where $a^4 = k^2E/D$.

The theory of the transversal vibrations of rods was given by Poisson in his memoir on the equilibrium and the motion of elastic bodies, *Mémoires de l'Académie des Sciences*, vol. VIII., and also in his *Traité de Mécanique*, Vol. II. Art. 518. The term containing $\omega k^2\psi''$ is not found in Poisson's solution, but is given by Clebsch in his *Theory of Elasticity*; see also Donkin's *Acoustics*, 1870.

629. When the differential equation (5) has been solved, the arbitrary functions or constants which have been introduced must be determined by the conditions at the extremities and the initial motion.

At the extremity B we have $x = l$, and the integrals in both equations (1) and (2) vanish. Remembering also that the terms containing ψ'' and T are to be omitted, these equations become, when $x = l$,

$$- k^2 E \omega \frac{d^2\eta}{dx^2} + L_1 = 0, \qquad k^2 E \omega \frac{d^3\eta}{dx^3} + Y_1 = 0 \quad \dots\dots\dots(6).$$

If the extremity B is free, both L_1 and Y_1 are zero, the conditions (6) therefore become $\dfrac{d^2\eta}{dx^2} = 0$, $\dfrac{d^3\eta}{dx^3} = 0$ when $x = l$.

If the extremity B is fixed to a point on the axis of x, $L_1 = 0$ but Y_1 may have any value, the conditions are therefore $\eta = 0$, $\dfrac{d^2\eta}{dx^2} = 0$.

If the extremity B is clamped, both the point B and the tangent at B are fixed. The conditions are then $\eta = 0$, $\dfrac{d\eta}{dx} = 0$ when $x = l$. The reactions at B are then given by equations (6).

If the extremity B is free, except that a particle of mass M is attached to it, we put $L_1 = 0$ and replace Y_1 in equations (6) by $-M d^2 h / dt^2$, where h is the value of η when $x = l$. The terminal conditions are therefore $\dfrac{d^2\eta}{dx^2} = 0$, and $k^2 E \omega \dfrac{d^3\eta}{dx^3} - M \dfrac{d^2\eta}{dt^2} = 0$, when $x = l$.

630. Ex. 1. To find the oscillations of a rod with both ends free.

The equation of motion is $\qquad \dfrac{d^2\eta}{dt^2} + a^4 \dfrac{d^4\eta}{dx^4} = 0$.

Let $\qquad\qquad\qquad \eta = \Sigma \, (P \sin m^2 a^2 t + Q \cos m^2 a^2 t)$,

then P and Q are functions of x which satisfy

$$\frac{d^4 P}{dx^4} - m^4 P = 0, \qquad\qquad \frac{d^4 Q}{dx^4} - m^4 Q = 0 \, ;$$

$$\therefore \; P = A \sin mx + B \cos mx + \tfrac{1}{2} H \left(e^{mx} - e^{-mx} \right) + \tfrac{1}{2} K \left(e^{mx} + e^{-mx} \right).$$

At each extremity $d^2\eta/dx^2 = 0$ and $d^3\eta/dx^3 = 0$. When $x = 0$ these give $A = H$ and $B = K$, and when $x = l$

$$A \left(2 \sin ml - e^{ml} + e^{-ml} \right) = B \left(e^{ml} + e^{-ml} - 2 \cos ml \right),$$
$$B \left(2 \sin ml + e^{ml} - e^{-ml} \right) = A \left(2 \cos ml - e^{ml} - e^{-ml} \right).$$

Eliminating B/A, we have

$$\tfrac{1}{2} \left(e^{ml} + e^{-ml} \right) \cos ml - 1 = 0 \quad \dots\dots\dots\dots\dots\dots(1).$$

The equation for Q obviously leads to the same result.

If m_1, m_2, &c. are the roots of the equation (1), the periods of the possible oscillations of the rod are $2\pi / m_1^2 a^2$, $2\pi / m_2^2 a^2$, &c. This agrees with Poisson's result.

We easily see that the expression for η may be written in the form

$$\eta = \Sigma X_m \left(L \sin m^2 a^2 t + M \cos m^2 a^2 t \right)$$
$$X_m = \left(e^{ml} + e^{-ml} - 2 \cos ml \right) \left(\sin mx + \tfrac{1}{2} e^{mx} - \tfrac{1}{2} e^{-mx} \right)$$
$$+ \left(2 \sin ml - e^{ml} + e^{-ml} \right) \left(\cos mx + \tfrac{1}{2} e^{mx} + \tfrac{1}{2} e^{-mx} \right),$$

where the Σ implies summation for all values of m which satisfy equation (1), and L, M are two undetermined constants.

If the initial circumstances of the motion are given by $\eta = \phi \, (x)$ and $\eta' = \psi \, (x)$, we may find the values of L and M by the method of multipliers. Imagine the values of η for all the elements of the rod written down in successive rows, then by Art. 399 the proper multiplier to separate the column occupied by $\cos m^2 a^2 t$ is represented by the type $D dx X_m$. We therefore find by Art. 398

$$\int \phi \, (x) \, X_m dx = M \int X_m{}^2 dx, \qquad \int \psi \, (x) \, X_m dx = L a^2 m^2 \int X_m{}^2 dx,$$

where the limits of all the integrals are $x = 0$ and $x = l$.

Ex. 2. Show that the periods of oscillation of a straight rod clamped at one end and free at the other are given by $\frac{1}{2} \, (e^{ml} + e^{-ml}) \cos ml + 1 = 0$.   [Poisson.

Ex. 3. Two rods have equal sectional areas, and in one the section is a circle, in the other an equilateral triangle. Prove that the squares of the periods of their corresponding notes are as 2π to $3\sqrt{3}$.

Ex. 4. Having given the equation $d^2 u/dt^2 + d^4 u/dx^4 = 0$, and the values of u and of du/dt for all values of x when $t = 0$, find u in terms of t and x, from $x = -\infty$ to $x = \infty$.

An elastic wire indefinitely extended in one direction is firmly held in a clamp at one end. If a series of simple transverse waves travelling along the wire be reflected at the clamp, show that the reflected waves have the same amplitude as the incident waves, but that their phase is accelerated by one quarter of a wave length. What will be the result if the end be free instead of being clamped?

[Math. Tripos, 1879.

Ex. 5. Two equal and similar elastic rods AC, BC are hinged at C so as to form a right angle, while their other extremities are clamped. One vibrates transversely and the other longitudinally; prove that the periods are $2\pi l^2/f^2\theta^2$, where θ is given by $1 + \cos \theta \cosh \theta + \left(\dfrac{\sin \theta}{\theta} \cosh \theta - \cos \theta \, \dfrac{\sinh \theta}{\theta} \right) \dfrac{gl}{f^2} \cot \dfrac{\theta^2 f^2}{gl} = 0$, l is the length of either rod, and f, g are two constants depending on the material.

631. Ex. 1. The natural form of a thin inextensible rod when at rest is a circular arc, and the rod makes small oscillations about this form. If the arc is a complete circle prove that the periods, $2\pi/\rho$, are given by $\rho^2 \, (i^2 + 1) = ai^2 \, (i^2 - 1)^2$ where i is any integer and a is a constant depending on the flexibility of the rod. If the arc is not a complete circle, but has both ends free, show that it can be made to vibrate symmetrically about its middle point by suitable initial conditions in a period $2\pi/\rho$, provided that the angle 2θ which the arc subtends at its centre satisfies the equation

$$\frac{n \, (n^2 + 1) \, (n_1{}^2 - n_2{}^2)}{\tan n\theta} + \frac{n_1 \, (n_1{}^2 + 1) \, (n_2{}^2 - n^2)}{\tan n_1 \theta} + \frac{n_2 \, (n_2{}^2 + 1) \, (n^2 - n_1{}^2)}{\tan n_2 \theta} = 0,$$

where n^2, $n_1{}^2$, $n_2{}^2$ are the roots real or imaginary of the cubic $ax \, (x - 1)^2 = (x + 1) \, \rho^2$.

Deduce from this Poisson's expression for the periods of vibration of a straight rod with both ends free.

Let X, Y, L be the tension, shear and stress couple at any point P of the rod, then these are all small quantities of the order of the oscillation. Let the undisturbed and disturbed co-ordinates of P be a, θ, and $a \, (1 + u)$, $\theta + \phi$, respectively. The equations of motion, when the squares of small quantities are neglected, become

$$\frac{dX}{d\theta} + Y = ma^2 \, \frac{d^2\phi}{dt^2} \,, \qquad \frac{dY}{d\theta} - X = ma^2 \, \frac{d^2 u}{dt^2} \,, \qquad \frac{dL}{d\theta} + aY = 0,$$

where m is the mass per unit of length, and the couple L is measured as in Art. 628.

Let p and q be the proportional elongation and increase of curvature of an element of the rod at P, we find

$$p = \frac{d\phi}{d\theta} + u, \qquad\qquad q = -\left(u + \frac{d^2u}{d\theta^2}\right).$$

Since the rod is inextensible we have $p = 0$, and, by a theorem in statics, $L = -Eq$. Eliminating X, Y, L and u from these equations, we obtain the linear equation

$$(1 - \delta^2)\, d^2\phi/dt^2 = a\delta^2\,(\delta^2 + 1)^2\,\phi,$$

where δ stands for $d/d\theta$, and $a = E/ma^3$.

To solve this equation we put $\phi = \Sigma M \sin \rho t \sin (n\theta + \epsilon)$. Substituting, the equation reduces to $\rho^2 (n^2 + 1) = an^2 (n^2 - 1)^2$.

If the circle is complete, the values of ϕ must recur when θ is increased by 2π and therefore n must be an integer. If the circle is incomplete, the value of n is unrestricted, except that ρ and n must be connected by the above equation. It follows that each value of ρ has three corresponding values of n, so that ϕ takes the form

$$\phi = \Sigma \sin (\rho t + \zeta)\,\{M \sin (n\theta + \epsilon) + M_1 \sin (n_1\theta + \epsilon_1) + M_2 \sin (n_2\theta + \epsilon_2)\}.$$

The condition in the question is obtained by making X, Y and L vanish at each end of the rod.

The oscillations of a complete circle are discussed in a different way in Lord Rayleigh's Treatise on *Sound*, Vol. I. Art. 233. The equation giving ρ in terms of the integer i is ascribed to Hoppe who published it in Crelle, 1871.

Ex. 2. The natural form of a rod is a circle of radius a, and the rod is both extensible and flexible, it is required to find the small oscillations.

Consider the elementary portion of the undisturbed rod which is bounded by two planes normal to the axis at two consecutive points P, Q. Making the usual assumption that these planes continue to be normal to the axis after the curvature has been increased, we notice that the *unstretched lengths* of the fibres of the element which lie on either side of PQ are not equal to the unstretched length of PQ, but are longer on the convex side and shorter on the other. Let E be Young's modulus of elasticity, ω the area of the section at P, ωk^2 the moment of inertia about an axis through its centre of gravity perpendicular to the plane of bending, and a the undisturbed radius of the axis of the rod. We then find by integration that the resultant tension X of all the fibres which cross the section ω, and their bending moment L, are given by

$$X = E\omega p - E\,\frac{\omega k^2}{a^2}\,q = Ap - Bq,$$

$$-\frac{L}{a} = E\,\frac{\omega k^2}{a^2}\,q \quad = Bq,$$

where p and q have the same meanings as in the last example. In the same way we find that the potential energy of the fibres of an elementary length ds of the rod is

$$dV = \tfrac{1}{2}ds\,(Ap^2 + Bq^2),$$

this last result is however not wanted in the following solution.

Substituting these values of X and L in the dynamical equations of the last article, and eliminating Y, we find

$$ma^2\ddot{\phi} = A\delta p, \qquad ma^2\ddot{u} = B\,(\delta^2 + 1)\,q - Ap,$$

where δ stands for $d/d\theta$. Since the values of p and q are given in the last example, we thus have two equations from which ϕ and u may be found. To solve these we put $\phi = \Sigma M \sin (\rho t + \zeta) \sin (n\theta + \epsilon)$, $u = \Sigma N \sin (\rho t + \zeta) \cos (n\theta + \epsilon)$.

Substituting, and eliminating M/N in the usual way, we obtain

$$m^3 a^4 \rho^4 - m a^2 \rho^2 \left\{ A \left(n^2 + 1 \right) + B \left(n^2 - 1 \right)^2 \right\} + A B n^2 \left(n^2 - 1 \right)^2 = 0.$$

If the undisturbed form of the rod is a complete circle, n may be any integer, and the two periods, viz. $2\pi/\rho$, corresponding to each integer are given by this equation. If the undisturbed form is an arc, n is unrestricted, except that ρ and n must be connected by the equation. Each term in the expressions for ϕ and u defined by any value of ρ has, as in the last example, three corresponding values of n, and therefore contains three terms of the form $M \sin \left(n\theta + \epsilon \right)$.

The conditions that X, Y and L are zero at each extremity of the rod show that p, q and δq must vanish at the same points. These, as in the last example, determine ϵ, ϵ_1, ϵ_2, the ratios M_1/M, M_2/M, and give an equation connecting ρ with the length of the arc. The existing values of ρ and n being now known the series for ϕ and u have the two constants M and ζ in each term undetermined. Since each value of n has two corresponding values of ρ given by the quadratic, we may write each of these series in the form

$$\Sigma \left\{ M \sin \left(pt + \zeta \right) \sin \left(n\theta + \epsilon \right) + M' \sin \left(\rho' t + \zeta' \right) \sin \left(n\theta + \epsilon' \right) \right\},$$

where n, ϵ, ϵ' have been already determined. The relations between the constants in the two series have also been found, so that only four constants in each compound term, viz. M, M', ζ, ζ' remain undetermined. These are found by the use of Fourier's Theorem when the initial values of ϕ, $\dot{\phi}$, u and $\dot{u}$ are given for all values of θ.

Another solution. We may also find the oscillations by using Lagrange's equations. Let us regard the rod as made up of elements of equal mass, each separated from the next by a short angular distance, viz. $d\theta = l$. Let the co-ordinates of these in succession be (r_1, θ_1), (r_2, θ_2) &c. We then have

$$2T = \Sigma m a^2 \left(\dot{u}_n^2 + \dot{\phi}_n^2 \right),$$

$$2V = \Sigma \left(A p^2 + B q^2 \right)$$

$$= \Sigma \left[A \left\{ (\phi_{n+1} - \phi_n)/l + u_n \right\}^2 + B \left\{ (u_{n+2} - 2u_{n+1} + u_n)/l^2 + u_n \right\}^2 \right],$$

since $d\phi_n/d\theta = (\phi_{n+1} - \phi_n)/l$, &c. To obtain the two equations of motion we substitute these functions in

$$\frac{d}{dt} \frac{dT}{d\dot{\phi}_n} + \frac{dV}{d\phi_n} = 0, \qquad \frac{d}{dt} \frac{dT}{d\dot{u}_n} + \frac{dV}{du_n} = 0.$$

The resulting equations are easily seen to be the same as those already arrived at. If the particle represented by (u_n, ϕ_n) is close to either extremity, the Lagrangian equations give the conditions at that extremity of the rod.

See a note at the end of the volume.

CHAPTER XIV.

MOTION OF A MEMBRANE.

The transverse Oscillations of a plane Membrane.

632. LET us take as the subject of consideration a plane membrane equally stretched throughout, whose boundaries are either fixed or subject to given conditions. Let this plane be called the plane of xy. Suppose this membrane to be disturbed so that its particles are slightly displaced parallel to the axis of z. The membrane will now make small oscillations about the plane of xy. It is the laws of these oscillations which we wish to discover.

Let w be the displacement at the time t of a particle P whose co-ordinates when undisturbed are x, y. Taking an elementary area $dxdy$ at the point P, let $\rho dxdy$ be its mass; thus, if the membrane be homogeneous, ρ is the mass of a unit of area. The oscillations being transversal the effective force on the element will be

$$\rho dxdy \; d^2w/dt^2.$$

Let us now consider the action across any side, as dy, of the elementary area. In the general case of a lamina this might consist of a force and a couple. But since a membrane, like a string, can be folded in any manner, and can only exert a force along its length, it is implied that the couple is zero and that the force acts in the tangent plane. Further, the membrane being equally stretched in all directions, this force acts perpendicularly to the side across which it acts. Let us represent this force by Tdy, then T is called the *tension referred to a unit of length* and sometimes briefly the tension.

The actions across the two sides of the rectangular element which are parallel to the axis of y have to be resolved parallel to the axis of z. These resolved parts are clearly

$$- Tdy\frac{dw}{dx}, \qquad Tdy\left(\frac{dw}{dx} + \frac{d^2w}{dx^2}\,dx\right).$$

The resultant of the two is $T\dfrac{d^2w}{dx^2}\,dxdy$. In the same way the resultant of the two actions across the sides parallel to x is $T\dfrac{d^2w}{dy^2}\,dxdy$. Taking both these resultants, and equating them to the effective forces, we have the equation of motion *

$$\rho\frac{d^2w}{dt^2} = T\left(\frac{d^2w}{dx^2} + \frac{d^2w}{dy^2}\right).$$

---

* The reader will find a more complete discussion of those principles of the theory of elasticity on which this equation is founded in the *Leçons sur la théorie*

633. Since the axes of co-ordinates may be any whatever provided that they are rectangular, this equation must be the same whatever be the directions of the axes. If the membrane be referred to oblique axes inclined at an angle ϵ, we may show that the equation of motion is

$$\rho \frac{d^2w}{dt} = \frac{T}{\sin^2 \epsilon} \left(\frac{d^2w}{dx^2} - 2 \frac{d^2w}{dxdy} \cos \epsilon + \frac{d^2w}{dy^2} \right).$$

634. To obtain a solution of this equation of motion we notice that, if we disregard the boundaries, it must be possible for the membrane to vibrate as if it were constructed of a series of strings laid side by side whose lengths were all parallel to any fixed direction we pleased. Let a be the angle this fixed direction makes with the axis of x. Then, putting $T = m^2\rho$, one solution of the equation is certainly

$$w = f (x \cos a + y \sin a - mt) + F (x \cos a + y \sin a + mt),$$

where a is any arbitrary constant, and f, F are two arbitrary functions which may be continuous or discontinuous as explained in Art. 620. Either of these functions with a given value of a represents a wave travelling in the direction defined by a with a front which is always parallel to the straight line $x \cos a + y \sin a = 0$. A more complete solution may then be found by summing these for all values of a.

Since the motions under consideration are oscillatory, it will be more convenient to expand the functions f and F in sines and cosines. Taking only a principal oscillation, we write $w = P \sin pmt + Q \cos pmt,$

where P and Q may be written in either of the following equivalent forms, but with different constants,

$$\Sigma \{A \sin p (x \cos a + y \sin a) + B \cos p (x \cos a + y \sin a)\}$$
$$+ \Sigma \{C \sin p (x \cos a - y \sin a) + D \cos p (x \cos a - y \sin a)\}$$
$$= \Sigma L \frac{\sin}{\cos} (px \cos a) \frac{\sin}{\cos} (py \sin a).$$

The positive values of a are included in the first line and the negative values in the second line. It follows that the Σ here implies summation for all positive values of a.

635. **Rectangular Membrane.** *To find the oscillations of a homogeneous rectangular membrane whose four boundaries are fixed.*

Let $OACB$ be the membrane, and let the sides OA, OB, be taken as the axes of x and y. Let $OA = a$, $OB = b$. Then we have to find a solution which (1) makes $w = 0$ when $x = 0$ and when $x = a$, independently of any particular values of y and (2) makes $w = 0$ when $y = 0$ and when $y = b$, independently of any particular values of x. Such a solution can be at once selected from the general form given in Art. 634, viz. $w = \Sigma L \sin (px \cos a) \sin (py \sin a) \cos pmt,$

with a similar expression to contain $\sin pmt$. Here we must have

$$pa \cos a = i\pi, \qquad pb \sin a = i'\pi,$$

where i and i' are any two integers. The periods (viz. $2\pi/pm$) are therefore given by

$$\left(\frac{p}{\pi} \right)^2 = \left(\frac{i}{a} \right)^2 + \left(\frac{i'}{b} \right)^2.$$

Mathématique de l'élasticité des corps solides par M. G. Lamé. The equation itself was first given by Poisson in his *Mémoire sur l'équilibre et le mouvement des corps élastiques* in the eighth volume of the *Mémoires de l'Institut,* 1828. The oscillations of a rectangular membrane (Art. 635) were also first discussed by him.

The question arises whether this solution is perfectly general or not. The solution satisfies the equation of motion and all the boundary conditions. If then it can be made to satisfy the initial conditions of the membrane it will certainly include every case. Let the initial displacement be $w = \phi(x, y)$; then putting $t = 0$

we have
$$\phi(x, y) = \Sigma L \sin \frac{\pi i x}{a} \sin \frac{\pi i' y}{b},$$

for all values of x and y respectively less than a and b. But by an extension of Fourier's theorem such an expansion as this is always possible. The solution is therefore perfectly general.

Ex. The weight W of a rectangular membrane and its tension T referred to a unit of length are both given. Show that the gravest note is given when the membrane is square, and that in this case the period of the note is $(2W/gT)^{\frac{1}{2}}$. Thus the period is independent of the area.   [Poisson's Theorem.

636. When the period of vibration of a rectangular membrane is given by some value of p, all the possible modes of vibration are included in the form

$$w = \left[\Sigma L \sin \frac{i\pi x}{a} \sin \frac{i'\pi y}{b} \right] \cos pmt,$$

with a similar term containing $\sin pmt$. In this form i and i' represent any integers

which satisfy
$$\left(\frac{i}{a} \right)^2 + \left(\frac{i'}{b} \right)^2 = \left(\frac{p}{\pi} \right)^2.$$

If two sets of values of i and i' can satisfy the last equation, it easily follows that the squares of the sides are in the ratio of two integers. Supposing this condition not to be satisfied, each oscillation will be of the form

$$w = \sin \frac{i\pi x}{a} \sin \frac{i'\pi y}{b} (L \cos pmt + L' \sin pmt),$$

and will contain just two constants, viz. L and L'. In this case it will be seen that each of these oscillations will be a principal oscillation and that all the periods will be different.

But if several sets of values of i and i' accompany the same period there will be more than two constants in the expression for each oscillation. In this case it appears that there are several ways in which a membrane may be set in vibration so that the periods of oscillation may be the same. It follows therefore that the Lagrangian equation (Art. 57) giving the periods of the principal oscillations has a number of equal roots.

637. The *nodal lines* are those lines on the membrane which remain in their positions of equilibrium during the whole motion. If the period be such that the oscillation is accompanied by only one set of values of i and i', the nodal lines for that oscillation are of course given by

$$\sin \frac{i\pi x}{a} \sin \frac{i'\pi y}{b} = 0.$$

These values of x or y make the coefficients of both $\cos pmt$ and $\sin pmt$ equal to zero. The nodal lines are therefore straight lines parallel to the sides. But, if there are several sets of values of i and i' which give the same p, and if the initial conditions are such that the corresponding coefficients in the coefficients of $\cos pmt$ and $\sin pmt$ have the same ratio, the nodal lines will be given by the equation

$$\Sigma L \sin \frac{i\pi x}{a} \sin \frac{i'\pi y}{b} = 0.$$

They may assume a great variety of forms depending on the number of terms in the series and on the arbitrary values given to the coefficients represented by the letter L. Lamé in his *Theory of Elasticity* gives a brief sketch of these. Another analysis is given in Riemann's *Partial Differential Equations*. They both remark that if we take only two terms in the series, of the form

$$L \sin \frac{i\pi x}{a} \sin \frac{i'\pi y}{b} - L \sin \frac{i'\pi x}{a} \sin \frac{i\pi y}{b} = 0,$$

one nodal line will be the diagonal $x/a = y/b$. Here the integers i and i' have been interchanged in the two terms. But, since the equation connecting these integers with the given value of p must also be satisfied, we have

$$(i/a)^2 + (i'/b)^2 = (i'/a^2) + (i/b)^2,$$

which requires that $a = b$. The rectangle must therefore be a square.

From this we may deduce that the oscillations of a membrane bounded by an isosceles right-angled triangle are given by

$$w = \Sigma L \left[\sin \frac{i\pi x}{a} \sin \frac{i'\pi y}{a} - \sin \frac{i'\pi x}{a} \sin \frac{i\pi y}{a} \right] \cos pmt,$$

with a similar term containing $\sin pmt$, where i and i' are integers connected by the equation
$$i^2 + i'^2 = (ap/\pi)^2,$$
and a is a side of the square. See Lord Rayleigh's *Sound*.

Ex. 1. If the squares of the sides of a rectangular membrane do not bear to each other the ratio of any two integers, prove that the nodal lines of a rectangular membrane must be straight lines parallel to the sides. [Poisson's Theorem.

Ex. 2. If the sides of a rectangular membrane are such that two sets of values of i and i' give the same period of vibration, then by proper initial conditions a nodal line may be made to pass through any given point on the membrane.

638. Ex. Membrane bounded by an equilateral triangle. A membrane is bounded by an equilateral triangle and its boundaries are fixed. If ξ, η, ζ be the trilinear co-ordinates of any point within the triangle, show by actual substitution that the equation of motion is satisfied by

$$w = \Sigma L \sin \frac{i\pi\xi}{h} \sin \frac{i\pi\eta}{h} \sin \frac{i\pi\zeta}{h} \cos pmt,$$

where $p = 2i\pi/h$. Here h is the altitude of the triangle and i is any integer.

This result follows at once from the trigonometrical theorem that, if the sum of three angles is equal to $i\pi$, the sum of the products of their cotangents taken two and two is equal to unity.

This is not however the most general form of solution, because we have only one independent arbitrary integer, viz. i. We cannot therefore satisfy all the possible initial values of w.

It is shown in Lamé's Theory of Elasticity that a more general expression for the period is given by $p = (2\pi/h)(i^2 + i'^2 + ii')^{\frac{1}{2}}$,
which contains the two arbitrary integers i and i'.

639. Ex. 1. Loaded Membrane. A uniform rectangular membrane, whose sides are a and b and mass M, has a finite mass equal to μ attached to it at the point whose co-ordinates are h, k when referred to the sides as axes. Show that the periods $(2\pi/pm)$ of the small transversal vibrations are given by

$$\frac{M}{\mu} \frac{1}{4p^2} = \Sigma \frac{\sin^2 \dfrac{i\pi h}{a} \sin^2 \dfrac{i'\pi k}{b}}{\pi^2 \left(\dfrac{i^2}{a^2} + \dfrac{i'^2}{b^2} \right) - p^2},$$

where the Σ implies summation for all values of the integers i and i', and m (as before) is the ratio of the tension to the density of the membrane.

To prove this we shall suppose the mass μ to be distributed over a small area equal to $a\beta$. Let W be the displacement of this small area at the time t. The sum of the resolved tensional forces round the perimeter of the area is equal to $\mu \dfrac{d^2 W}{dt^2} = -R$. We have therefore to find the motion of a membrane acted on by a periodic force R at a given point h, k. Let us replace this single force by a continuous force $Z dx dy$ which acts at every point of the membrane, such that

$$Z = \Sigma C \sin (i\pi x/a) \sin (i'\pi y/b).$$

Since Z vanishes all over the membrane except in the immediate neighbourhood of the point h, k, and at this point $Za\beta = -\mu d^2 W/dt^2$, we have by Fourier's theorem

$$-\mu \frac{d^2 W}{dt^2} \sin \frac{i\pi h}{a} \sin \frac{i'\pi k}{b} = \tfrac{1}{4} Cab.$$

The equation of motion of the membrane is now

$$\rho \frac{d^2 w}{dt^2} = \rho m^2 \left(\frac{d^2 w}{dx^2} + \frac{d^2 w}{dy^2} \right) + Z.$$

To solve this we put $\quad w = f(x, y) \cos pmt$.

Substituting, we find by Theorem III. of Art. 265

$$\frac{M}{4\mu p^2} \frac{f(xy)}{f(hk)} = \Sigma \frac{\sin \dfrac{i\pi x}{a} \sin \dfrac{i'\pi y}{b} \sin \dfrac{i\pi h}{a} \sin \dfrac{i'\pi k}{b}}{\pi^2 \left(\dfrac{i^2}{a^2} + \dfrac{i'^2}{b^2} \right) - p^2}.$$

The form of the function f corresponding to any value of p has now been found. Putting $x = h$, $y = k$, we have an equation to find p.

Another solution is added in a note at the end of the volume.

Ex. 2. A rectangular membrane of mass M is oscillating with a period $(2\pi/pm)$ such that only one set of values of i, i' accompany this value of p. A small load of mass μ is placed at any point (h, k), prove that the new period of vibration, viz. $(2\pi/qm)$, is given by

$$q^2 = p^2 \left(1 - \frac{4\mu}{M} \sin^2 \frac{i\pi h}{a} \sin^2 \frac{i'\pi k}{b} \right).$$

This follows from the result given in the last example, for only one denominator on the right-hand side will be small. Rejecting all the terms except this one, we have the result.

Ex. 3. A membrane of mass M is bounded by two concentric circles whose radii are a and b, and the density varies inversely as the square of the distance from the centre. The period P of any symmetrical oscillation is given by $P = \dfrac{1}{q} \left(\dfrac{2\pi M}{T} \log \dfrac{a}{b} \right)^{\frac{1}{2}}$, where $q = i\pi$ if both the boundaries are fixed in space. But if the outer boundary only is fixed in space, while the inner is attached to a ring of mass μ, q is given by $q \tan q = M/\mu$.

If the ratio a/b is not very great, this membrane may be regarded as nearly homogeneous, with the inner parts slightly denser than the outer.

Ex. 4. Show that the equation to find the periods of vibration of a loaded membrane may be written in the form

$$\frac{M}{\mu} \frac{1}{4p^2} = \Sigma \frac{a}{2\phi} \frac{\sin \phi h \sin \phi (a - h)}{\sin \phi a} \sin^2 \frac{i'\pi k}{b},$$

where Σ implies summation for all values of the integer i' and $\phi^2 = p^2 - \pi^2 i'^2 / b^2$. This result may be obtained by expanding $\cos q\,(\pi - x)$ in a series of cosines, q not being an integer. We find

$$\frac{\cos x}{1 - q^2} + \frac{\cos 2x}{2^2 - q^2} + \dots = \frac{1}{2q^2} - \frac{\pi \cos q\,(\pi - x)}{2q \sin q\pi}.$$

The expansion holds from $x = 0$ to $x = \pi$, both inclusive. Putting $x = 0$, subtracting, and writing $2y$ for x, we have

$$\frac{\sin^2 y}{1 - q^2} + \frac{\sin^2 2y}{2^2 - q^2} + \dots = \frac{\pi \sin qy \cdot \sin q\,(\pi - y)}{2q \sin q\pi}.$$

The result given above easily follows.

640. Ex. **Membrane acted on by a given periodical force.** A rectangular membrane is bounded by the co-ordinate axes and the straight lines $x = a$, $y = b$. A finite accelerating force acts at the point (h, k) and is represented by $A \sin rt$. Show that the forced vibration is represented by

$$w = \frac{4A}{M} \Sigma \frac{\sin \dfrac{i\pi h}{a} \sin \dfrac{i'\pi k}{b} \sin \dfrac{i\pi x}{a} \sin \dfrac{i'\pi y}{b} \sin rt}{m^2 \pi^2 \left(\dfrac{i^2}{a^2} + \dfrac{i'^2}{b^2} \right) - r^2},$$

where Σ implies summation for all values of the positive integers i and i'.

The free vibrations have been found in Art. 636. Joining these to the forced vibration and supposing the membrane to start from rest in its position of equilibrium, we have $\quad w = \Sigma P \left(\sin rt - \dfrac{r}{pm} \sin pmt \right)$,

where P is the coefficient of $\sin rt$ in the forced vibration.

We may deduce from this expression *the effect of a force acting, like an impulse, for a very short time.* Let r be very great, and let the force $A \sin rt$ act only for the short time π/r. If F be the momentum communicated to the membrane, we have $F = \int A \sin rt\, dt$ where the limits are $t = 0$ and $t = \pi/r$. We thus have $F = 2A/r$. Substituting we find, when r is very great

$$w = \Sigma \sin \frac{i\pi h}{a} \sin \frac{i'\pi k}{b} \sin \frac{i\pi x}{a} \sin \frac{i'\pi y}{b} \left\{ - \frac{\sin rt}{r} + \frac{\sin pmt}{pm} \right\} \frac{2F}{M}.$$

The motion at the time $t = \pi/r$ is therefore given by

$$w = 0, \qquad \frac{dw}{dt} = \Sigma \sin \frac{i\pi h}{a} \sin \frac{i'\pi k}{b} \sin \frac{i\pi x}{a} \sin \frac{i'\pi y}{b} \cdot \frac{4F}{M}.$$

Motion of a heterogeneous membrane.

641. We propose to show in this section how by the use of the theory of conjugate functions we may deduce the motion of certain heterogeneous membranes from the corresponding motions of homogeneous membranes. The corresponding theorems for a network of particles are briefly given in Art. 421.

We shall begin by giving a list of those theorems on conjugate functions which we shall afterwards require, and in the next article we shall consider their application to the motion of membranes.

If we have two variables ξ, η connected with x, y by the relation

$$\xi + \eta \sqrt{-1} = f(x + y \sqrt{-1}),$$

where f is any real functional symbol, then ξ, η are called *conjugate functions*. By taking the first differential coefficients of this equation with regard to x and

y and equating the coefficients of the imaginary quantity we arrive at the well-known results
$$\frac{d\xi}{dx} = \frac{d\eta}{dy}, \quad \frac{d\xi}{dy} = -\frac{d\eta}{dx}.$$

Since we have also $x + y\sqrt{-1} = F(\xi + \eta\sqrt{-1})$ it follows in the same way that
$$\frac{dx}{d\xi} = \frac{dy}{d\eta} \text{ and } \frac{dy}{d\xi} = -\frac{dx}{d\eta}.$$

We may also show by a simple transformation of variables that
$$\frac{d^2w}{dx^2} + \frac{d^2w}{dy^2} = \left\{\frac{d^2w}{d\xi^2} + \frac{d^2w}{d\eta^2}\right\} \left\{\left(\frac{d\xi}{dx}\right)^2 + \left(\frac{d\xi}{dy}\right)^2\right\}.$$

Since we may interchange x, y and ξ, η in this formula, it easily follows that
$$\left\{\left(\frac{d\xi}{dx}\right)^2 + \left(\frac{d\xi}{dy}\right)^2\right\} \left\{\left(\frac{dx}{d\xi}\right)^2 + \left(\frac{dx}{d\eta}\right)^2\right\} = 1.$$

We shall also require a geometrical theorem. Let us draw two diagrams each referred to a set of rectangular axes. In one let ξ, η be the co-ordinates of a point which we shall call Π, in the other let x, y be the co-ordinates of a point which we shall call P. These points are said to correspond. In one diagram the loci defined by $\xi = a$, $\eta = b$, where a and b are constants, are straight lines parallel to the axes. In the other, where ξ and η are regarded as functions of x and y given above, the loci will in general be curved lines. In the same way the equation $\eta = \phi(\xi)$ will represent two corresponding curves, one on each diagram. Let the tangents to these curves at corresponding points Π and P make angles ϵ and e with the axis of x, then $\tan \epsilon = d\eta/d\xi$ and $\tan e = dy/dx$. Through P draw the curve $\eta = b$, where b has its proper constant value, and let the tangent to this curve make an angle A with the axis of x. Then denoting differential coefficients with regard to x and y by suffixes, we have $\eta_x + \eta_y \tan A = 0$. We also have, as proved above, $\xi_x = \eta_y$ and $\xi_y = -\eta_x$.

Since
$$\tan \epsilon = \frac{d\eta}{d\xi} = \frac{\eta_x dx + \eta_y dy}{\xi_x dx + \xi_y dy} = \frac{-\tan A + \tan e}{1 + \tan A \tan e},$$

we see that $\epsilon = e - A$. It immediately follows that *the angle made by any two curves which meet at P is equal to the angle between the corresponding curves which meet at Π.* In other words *corresponding angles are equal.*

If we draw two corresponding networks, one on each diagram, and if the meshes of each be infinitely small triangles, it follows from the equality of the angles that the *networks are similar to each other at corresponding points.* The scale or ratio of the networks is not however the same all over the diagrams.

It also follows from the equality of the angles that the curves defined by $\xi = a$, $\eta = b$ cut at the same angle in each diagram. *They therefore cut each other at right angles.*

642. Suppose that we know the motion of a homogeneous membrane with given bounding conditions vibrating transversely, say $w = \phi(\xi, \eta, t)$, where w represents the displacement of a point whose co-ordinates are (ξ, η). Then this value of w satisfies the equation
$$D_0 \frac{d^2w}{dt^2} = T\left(\frac{d^2w}{d\xi^2} + \frac{d^2w}{d\eta^2}\right),$$

where D_0 is the density and T is the tension of the membrane.

Let x, y be the co-ordinates of a point on another membrane which has sand strewed over it and fastened to it, so that the sand vibrates with the membrane. Let the density D of this heterogeneous medium be given by
$$\frac{D}{D_0} = \left(\frac{d\xi}{dx}\right)^2 + \left(\frac{d\xi}{dy}\right)^2.$$

Then the equation of motion of this new membrane is

$$D \frac{d^2w}{dt^2} = T \left(\frac{d^2w}{dx^2} + \frac{d^2w}{dy^2} \right).$$

But, since ξ, η are known functions of x, y, we obtain, by substitution in the equation $w = \phi\ (\xi,\ \eta,\ t)$, the new relation $w = \psi\ (x,\ y,\ t)$, which is the solution of the equation of motion of the new membrane.

Thus the motion of the new membrane is deduced from that of the first with corresponding bounding conditions.

643. Generally, we do not want the actual motion of the membrane, but only its possible periods of vibration and nodal lines. We may notice that the two membranes have the same periods of vibration and corresponding nodal lines.

644. In this transformation it is necessary that only one point of each membrane should correspond to any single point of the other membrane within the area considered. If this be not attended to, some difficulties in interpretation may occur.

645. The new membrane is of course heterogeneous, and it may be objected that the cases now considered are not such as occur in nature. If, however, the density is not very variable over the membrane, the results will nearly represent the motion of a homogeneous membrane. At the same time we must remember that the results to be obtained are not merely approximations, but are accurate solutions of the equations. Such a solution, if short, and obtained by some simple process, is sometimes preferable to one obtained by a long approximation, even though the latter may appear to be more directly applicable.

To take a simple example, the oscillations of a homogeneous loose heavy chain, suspended from two fixed points, can be found only by very troublesome algebraical approximations. But if we suppose the chain to be heterogeneous, we may obtain an accurate solution of the equations. This solution leads to nearly the same results as the approximate investigations for a homogeneous chain. See Art. 607.

To take another example, we may notice that the motion of a homogeneous membrane bounded by two radii vectores and two circular arcs, can be expressed by the help of Bessel's functions. But the motion of a membrane bounded in the same way and of the proper density, can be expressed by ordinary sines and cosines. This is much simpler than a solution in Bessel's functions, and helps us to understand the nature of the motion.

646. We may, if we please, express all this in geometrical language.

Consider first a heterogeneous membrane with any fixed boundary which vibrates according to the law $w = \psi\ (x,\ y,\ t)$,

where w is the displacement of the point P whose Cartesian co-ordinates are x, y. Trace on the membrane the two sets of curves whose equations are $f(x,\ y) = \xi$ and $F(x,\ y) = \eta$, where ξ and η are two parameters. These curves are to be such that, when the parameters ξ, η increase by a constant increment $d\xi = \alpha$ or $d\eta = \alpha$, the two sets of curves divide the membrane into elementary squares. That the corresponding increments of ξ and η should be equal when these curves form squares, follows from the proposition that the small corresponding figures formed on the two membranes by the method of conjugate functions are similar. It may, however, also be deduced from the relations mentioned in Art. 641. If $ABCD$ be one of the squares, draw a parallel to the axis of x through any corner A, and then draw perpendiculars BM and DN from the two adjacent corners on this parallel. We have thus two equal triangles ABM, ADN; the sides in each triangle being the dx and dy produced by varying first ξ only, and then η only. It follows from this that $\frac{dx}{d\xi}\,d\xi = \frac{dy}{d\eta}\,d\eta$ and $\frac{dx}{d\eta}\,d\eta = -\frac{dy}{d\xi}\,d\xi$. We therefore infer from Art. 641 that $d\xi = d\eta$.

The area of one of these squares is $\left(\dfrac{dx}{d\xi}\dfrac{dy}{d\eta} - \dfrac{dx}{d\eta}\dfrac{dy}{d\xi}\right)a^2.$

Thus, since the density D is given by $\dfrac{D_0}{D} = \left(\dfrac{dx}{d\xi}\right)^2 + \left(\dfrac{dx}{d\eta}\right)^2,$

it follows *that the mass of each elementary square is the same.*

Next, consider the corresponding homogeneous membrane. Draw on the membrane straight lines parallel to the axes of ξ, η at a distance a from each other, so that each straight line corresponds to one of the curves drawn on the heterogeneous membrane. Let a new boundary be drawn which cuts these straight lines at the same angles which the boundary of the heterogeneous membrane cuts the corresponding curves.

Then the motions of these two membranes are the same at corresponding points. We may consider each to be given by $\qquad w = \psi\,(x, y, t),$ according as we express w in terms of ξ, η or x, y.

647. We may notice that the two membranes are so related that the *masses of corresponding squares on the heterogeneous and homogeneous membranes are equal to each other. Thus the whole masses of the membranes are the same, but differently distributed.*

648. Similar theorems apply in changing from one heterogeneous medium to another, but as this case does not present any novelty, and is not so simple as the one just considered, we need not discuss it minutely.

649. Having traced on the membrane the two orthogonal sets of curves $f(x, y) = \xi$, $F(x, y) = \eta$, where ξ and η are constants, and the functions both satisfy Laplace's equation, we may trace a third set of curves given by

$$\left(\frac{d\xi}{dx}\right)^2 + \left(\frac{d\xi}{dy}\right)^2 = \left(\frac{d\eta}{dx}\right)^2 + \left(\frac{d\eta}{dy}\right)^2 = \text{constant}.$$

These are, of course, the curves of constant density.

A curve of constant density which passes through any point will cut the two members of the two orthogonal sets which pass through the same point at complementary angles. *Then we may show that the sines of these angles are as the radii of curvature of the two members at that point.*

To prove this, let us find $\tan\theta$, where θ is the angle that the curve of equal density makes with the curve $f(x, y) = \xi$. By simple differentiation, we find

$$\tan\theta = \frac{(f_y{}^2 - f_x{}^2)f_{xy} + 2f_x f_y f_{xx}}{2f_x f_y f_{xy} + (f_x{}^2 - f_y{}^2)f_{xx}},$$

where suffixes, as usual, imply differential coefficients. Since $f_x = F_y$ and $f_y = -F_x$, we see, by substituting in the numerator, that

$$\frac{\sin\theta}{\sin\theta'} = -\frac{(F_x{}^2 - F_y{}^2)F_{xx} + 2F_y F_x F_{xy}}{2f_x f_y f_{xy} + (f_x{}^2 - f_y{}^2)f_{xx}}.$$

But the radius of curvature ρ of the curve f is given by

$$(f_x{}^2 - f_y{}^2)f_{xx} + 2f_x f_y f_{xy} = \frac{(f_x{}^2 + f_y{}^2)^{\frac{3}{2}}}{\rho}.$$

Hence, we see that $\qquad \dfrac{\sin\theta}{\sin\theta'} = -\dfrac{\rho}{\rho'}.$

650. It is not every heterogeneous medium whose motion can be deduced from that of a *homogeneous one.* If we eliminate ξ between

$$\left(\frac{d\xi}{dx}\right)^2 + \left(\frac{d\xi}{dy}\right)^2 = \frac{D}{D_0}, \qquad \frac{d^2\xi}{dx^2} + \frac{d^2\xi}{dy^2} = 0,$$

we easily obtain

$$\frac{d^2 \log D}{dx^2} + \frac{d^2 \log D}{dy^2} = 0.$$

It immediately follows (from Art. 641) that

$$\frac{d^2 \log D}{d\xi^2} + \frac{d^2 \log D}{d\eta^2} = 0.$$

The density of the heterogeneous membrane must, therefore, be such that its logarithm satisfies Laplace's equation.

651. For convenience of reference, let (x, y) be the Cartesian co-ordinates, (r, θ) the polar co-ordinates of a point P on the heterogeneous membrane; (ξ, η) the Cartesian, (ρ, ω) the polar co-ordinates of the corresponding point Π on the homogeneous membrane. Suppose we take as our relation between the two points,

$$\xi + \eta\sqrt{-1} = c \log \frac{x + y\sqrt{-1}}{\beta}.$$

Then we find $\xi = c \log \dfrac{r}{\beta}, \qquad \eta = c\theta.$

Thus straight boundaries on the homogeneous membrane parallel to the axis of ξ correspond to straight boundaries on the heterogeneous membrane which pass through the origin. At the same time, straight boundaries parallel to the axis of η correspond to circles whose centre is at the origin.

The density D is given by $\dfrac{D}{D_0} = \left(\dfrac{d\xi}{dr}\right)^2 + \left(\dfrac{d\xi}{rd\theta}\right)^2 = \left(\dfrac{c}{r}\right)^2.$

If r vanish, we have D infinite ; it will therefore be necessary to exclude the origin from the area of the membrane.

If, then, we know the motion of a membrane bounded by a rectangle, the transformation immediately gives the motion of a heterogeneous membrane bounded by two circular arcs and any two radii vectores.

652. *Example.*—The motion of a rectilinear homogeneous membrane bounded by the straight lines $\xi = h_1$, $\xi = h_2$; $\eta = k_1$, $\eta = k_2$, is known to be given by the type

$$w = A \sin i\pi \frac{\xi - h_1}{h_2 - h_1} \sin i'\pi \frac{\eta - k_1}{k_2 - k_1} \cos pmt,$$

where the integers i, i' are any which satisfy $\dfrac{i^2}{(h_2 - h_1)^2} + \dfrac{i'^2}{(k_2 - k_1)^2} = \dfrac{p^2}{\pi^2}$,

and where $m^2 = T/D_0$.

It immediately follows that the motion of a heterogeneous membrane bounded by the arcs of concentric circles, whose radii are h'_1 and h'_2, and by two radii vectores $\theta = a_1$ and $\theta = a_2$, is given by

$$w = A \sin\left(i\pi \frac{\log r - \log h'_1}{\log h'_2 - \log h'_1}\right) \sin\left(i'\pi \frac{\theta - a_1}{a_2 - a_1}\right) \cos pmt,$$

where the integers i and i' satisfy $\dfrac{i^2}{(\log h'_2 - \log h'_1)^2} + \dfrac{i'^2}{(a_2 - a_1)^2} = \dfrac{c^2 p^2}{\pi^2}$,

and the density D of the membrane is given by $\dfrac{D}{D_0} = \left(\dfrac{c}{r}\right)^2.$

653. Another useful relation between the corresponding points P and Π is

$$\xi + \eta \sqrt{-1} = c \left(\frac{x + y \sqrt{-1}}{c} \right)^n.$$

This gives
$$\xi = c \left(\frac{r}{c} \right)^n \cos n\theta, \qquad \eta = c \left(\frac{r}{c} \right)^n \sin n\theta;$$

and therefore, in polar co-ordinates,
$$\rho = c \left(\frac{r}{c} \right)^n, \qquad \omega = n\theta.$$

By this transformation all radii vectores are turned round the origin and altered in a known manner.

Also, the density D of the heterogeneous membrane is given by $\dfrac{D}{D_0} = n^3 \left(\dfrac{r}{c} \right)^{2(n-1)}$.

Since $\theta = $ constant makes $\omega = $ constant, we see that straight lines through the origin correspond to straight lines through the origin. Also circles whose centres are at the origin correspond to circles whose centres are at the origin.

If we choose $n = -1$, we have the ordinary case of inversion; thus

$$\rho = \frac{c^2}{r}, \qquad \omega = -\theta.$$

In this case any circle inverts into a circle. The density of the membrane is then given by $\dfrac{D}{D_0} = \left(\dfrac{c}{r} \right)^4$. As this is infinite when r is zero, the centre of inversion must be external to the membrane.

654. *Example* 1.—The density of a membrane bounded by two concentric fixed circles of radii a and b at any point distant ρ from the centre is A/ρ^2. Let it vibrate symmetrically so that the nodal lines are concentric circles, then by Ex. 3, Art. 639, the possible periods of vibration are $2\pi (A/p^2 T)^{\frac{1}{2}}$, where p is such that $p (\log a - \log b) = i\pi$, and i is any integer.

Let us invert this with regard to an external point. We immediately have the following theorem.

A heterogeneous membrane is bounded by two fixed circles, centres C and C'. Let O be that point which has a common polar line in both circles, and let this polar line cut the straight line OCC' in the point R. Let the density at any point P be given by $D = A \cdot \left(\dfrac{OR}{OP \cdot RP} \right)^2$. Then the membrane can vibrate so that the nodal lines are circles, and the possible periods of vibration are $2\pi \left(\dfrac{A}{p^2 T} \right)^{\frac{1}{2}}$, where p is such that $p \log \dfrac{a \cdot OC'}{a' \cdot OC} = i\pi$, and a and a' are the radii of the circles whose centres are C and C'.

Ex. 2. A heterogeneous membrane is bounded by two rigid circles whose equations are respectively $\rho = \mu r$ and $\rho = \lambda r$, where ρ and r are the distances of any point from two fixed points S and R. The former is the outer circle and is fixed in space; the inner is free to move and is so loaded that its centre of gravity is at R. The surface density at any point P of the membrane is $4Ab^2/\rho^2 r^2$, where $2b$ is the distance between the fixed origins S and R. Prove that the membrane can oscillate so that the nodal lines are the circles $\rho = kr$, and that the periods P are given by

$$\tan \left[\frac{2\pi}{P} \left(\frac{A}{T} \right)^{\frac{1}{2}} \log \frac{\lambda}{\mu} \right] = \frac{P}{M} (AT)^{\frac{1}{2}}, \text{ where } T \text{ is the uniform tension of the membrane,}$$

and M the mass of the load.

655. *Example.*—The motion of a rectilinear membrane bounded by the axes of ξ and η and the straight lines $\xi = h$, $\eta = k$, is known to be given by the type

$$w = A \sin \frac{i\pi\xi}{h} \sin \frac{i'\pi\eta}{k} \cos pmt,$$

where i and i' are any integers which satisfy $\dfrac{i^2}{h^2} + \dfrac{i'^2}{k^2} = \dfrac{p^2}{\pi^2}$.

Let us invert this with regard to the origin, we see that—

The motion of an infinite membrane bounded by the axes of x and y, and the arcs of two circles whose diameters are h', k', and which touch the axes of x, y at the origin, is given by the type $w = A \sin \dfrac{i\pi h' \cos\theta}{r} \sin \dfrac{i'\pi k' \sin\theta}{r} \cos pmt,$

where the integers i and i' satisfy the equation $i^2 h'^2 + i'^2 k'^2 = \dfrac{p^2}{\pi^2} c^4,$

provided that its density is given by $D = \left(\dfrac{c}{r}\right)^4 \cdot \dfrac{T}{m^2},$

where $T = $ tension of the membrane.

656. *Example.*—If we transform the same theorem with $n = 2$, we see that—

The motion of a finite membrane bounded by two straight lines $OA = h'$, $OB = k'$, inclined at an angle $\pi/4$, and by two rectangular hyperbolas passing respectively through A and B, and having OB and OA for asymptotes, is given by the type

$$w = A \sin \frac{i\pi r^2 \cos 2\theta}{h'^2} \sin \frac{i'\pi r^2 \sin 2\theta}{k'^2} \cos pmt,$$

where i and i' are connected by $\dfrac{i^2}{h'^2} + \dfrac{i'^2}{k'^2} = \dfrac{p^2}{\pi^2} \dfrac{1}{c^2},$

provided that its density is given by $D = 4 \left(\dfrac{r}{c}\right)^2 \cdot \dfrac{T}{m^2}.$

657. Suppose, in an infinite homogeneous membrane, a very small circular area of radius c to become rigid, and to be constrained to move transversely with a motion given by $w = A \cos pmt$. Then waves will spread out equally in all directions, and, when the motion has become steady, the vibration at any point distant ρ from the centre of disturbance will be given by $w = J_0(p\rho) A \cos pmt$.

Here we have supposed c to be so small that $J_0(pc) = 1$. Such a small circular vibrating area may, for convenience, be called *a source of disturbance*, or more shortly *a source*.

If we transform this theorem by the method of conjugate functions, we see, for the reason to be given in Art. 653, that the infinitely small circle will transform into a similar figure, *i.e.*, into another circle.

658. *Example.*—The vibrations of an infinite homogeneous membrane bounded by a fixed straight line taken as the axis of x, and acted on by a *source* at some point (ξ_1, η_1), are given by $w = \{J_0(p\rho) - J_0(p\rho')\} A \cos pmt$,

where $\rho^2 = (\xi - \xi_1)^2 + (\eta - \eta_1)^2$,

and $\rho'^2 = (\xi - \xi_1)^2 + (\eta + \eta_1)^2$,

so that ρ, ρ' are the distances of the point (ξ, η) from the source, and its image on the other side of the axis of ξ.

Hence we infer that the vibrations of an infinite heterogeneous membrane bounded by two fixed radii vectores forming a corner of angle π/n, and acted on by a source at a point $r_1\theta_1$, are given by

$$w = \{J_0(pR) - J_0(pR')\} A \cos pmt,$$

where $c^{2n-2} R^2 = r^{2n} + r_1^{2n} - 2r^n r_1^n \cos n(\theta - \theta_1)$

$$c^{2n-2} R'^2 = r^{2n} + r_1^{2n} - 2r^n r_1^n \cos n(\theta + \theta_1),$$

provided that the density of the membrane is given by $\dfrac{D}{D_0} = n^2 \left(\dfrac{r}{c}\right)^{2(n-1)}$

Here r, θ are the running co-ordinates of any point of the medium, w is the transverse displacement at the point ρ, ω, and D_0 is a constant.

The method of deducing the motion of a heterogeneous from that of a homogeneous membrane was given by the author in the twelfth volume of the *Proceedings of the Mathematical Society*, 1881.

NOTES.

Art. 56. **Transformation to principal co-ordinates.** This method of transforming any co-ordinates θ, ϕ, &c. to the principal co-ordinates ξ, η, &c. may be presented in a purely Mathematical form. Let us first assume the transformation to be possible, so that we have

$$\left.\begin{array}{l} 2T = A_{11}\theta^2 + 2A_{12}\theta\phi + \ldots\ldots = a_{11}\xi^2 + a_{22}\eta^2 + \ldots\ldots \\ 2U = C_{11}\theta^2 + 2C_{12}\theta\phi + \ldots\ldots = c_{11}\xi^2 + c_{22}\eta^2 + \ldots\ldots \end{array}\right\} \quad \ldots\ldots\ldots\ldots (1),$$

where the accents have been dropped from the co-ordinates in $2T$ as being unnecessary for our present purpose. We have also omitted U_0 from the second equation for the sake of unity. Let the formulæ of transformation, which we have to find, be, as in Art. 69,

$$\left.\begin{array}{l} \theta = l_1\xi + l_2\eta + \ldots\ldots \\ \phi = m_1\xi + m_2\eta + \ldots \\ \&c. = \&c. \end{array}\right\} \quad \ldots\ldots\ldots\ldots\ldots\ldots\ldots (2).$$

Let us eliminate ξ^2 from the equations (1) and differentiate the result with regard to θ. Putting $p_1^2 = -c_{11}/a_{11}$ we have

$$\frac{d}{d\theta}(Tp_1^2 + U) = (a_{22}p_1^2 + c_{22})\,\eta\,\frac{d\eta}{d\theta} + (a_{33}p_1^2 + c_{33})\,\zeta\,\frac{d\zeta}{d\theta} + \&c. \quad \ldots\ldots\ldots(3).$$

This vanishes when we put $\eta = 0$, $\zeta = 0$, &c. whatever ξ may be. Hence if the transformation be possible we have after substitution from (2)

$$(A_{11}p_1^2 + C_{11})\,l_1 + (A_{12}p_1^2 + C_{12})\,m_1 + \ldots\ldots = 0 \quad \ldots\ldots\ldots\ldots (4).$$

In the same way by differentiating with regard to ϕ, we have, when $\eta = 0$, $\zeta = 0$, &c.

$$(A_{12}p_1^2 + C_{12})\,l_1 + (A_{22}p_1^2 + C_{22})\,m_1 + \ldots\ldots = 0.$$

Thus we see that p_1^2 is one value of p^2 obtained from Lagrange's determinantal equation as given in Art. 58, while the values of l_1, m_1, &c. are proportional to the minors of the determinant. Eliminating η^2, ζ^2, &c. in turn from the equations (1), the same argument applies to each of the other columns of coefficients in the formulæ of transformation (2). Thus we obtain the rule given in Arts. 53 and 56. The formulæ of transformation are written at length in Art. 56. We see that the coefficients of x, y, &c. are the values of the minors $I_{11}(p^2)$, &c.

If there were on the right-hand side of the equations (1) any term such as $\xi\eta$, this product would give on the right-hand side of (3) a term $(a_{12}p_1^2 + c_{12})\,\xi\,d\eta/d\theta$ when we eliminated ξ^2 and differentiated with regard to θ. It would give $(a_{12}p_2^2 + c_{12})\,\eta\,d\xi/d\theta$ when we eliminated η^2 and differentiated with regard to θ. Now the differential coefficients of ξ or η with regard to the co-ordinates θ, ϕ, ψ &c. cannot be all zero, for this would make ξ or η independent of all the co-ordinates. Also, if Lagrange's determinantal equation have all its roots unequal, the coefficients $a_{12}p_1^2 + c_{12}$ and $a_{12}p_2^2 + c_{12}$ cannot both vanish. Hence in this case, when the right-hand sides of (3) are made to vanish, there cannot be any products of co-ordinates in either of the expressions on the right-hand side of (1).

If Lagrange's equation have equal roots we know by Art. 61 that all the minors will be zero. The ratios of l, m, &c. found by the preceding rule will therefore be nugatory. To simplify the argument let us suppose that the equation has two equal roots and let these be p_1^2 and p_2^2. The ratios of the coefficients in the third and following columns of (2) may be found as before, because they depend on unequal roots in Lagrange's determinant. Since the first minors are zero for the equal roots, the equations (4) to determine the coefficients of either of the first two columns of (2) are not independent. Rejecting any one of these equations (as in Art. 273) we obtain by using the second minors all the letters in the first column in terms of any two, say l_1 and m_1. The letters in the second column are found in terms of l_2 and m_2 by the same formulæ. Thus we have two independent coefficients in each of these columns instead of one as before.

But if we use these formulæ of transformation without further limitation, we are not sure that terms containing the product $\xi\eta$ may not enter into the two right-hand sides of the expressions (1) provided they enter both with coefficients in the ratio $p_1^2 : 1$. To secure the absence of such terms, it will be sufficient to make the coefficient of $\xi\eta$ in *either* of the coefficients T or U equal to zero. If we choose T, we have by substituting from (2) in (1)

$$A_{11}l_1l_2 + A_{12}(l_1m_2 + l_2m_1) + \ldots \ldots = 0,$$

or as it is written in Art. 316

$$A\,(l_1l_2) = 0.$$

Regarding then l_1, m_1 and l_2 as arbitrary we have sufficient linear equations of the first order to find all the other coefficients of the two first columns in the formulæ of transformation. Thus we have three arbitrary constants instead of two.

Art. 60. **The conditions that a quadric should be one-signed.** The conditions briefly quoted from Williamson's *Differential Calculus* have reference to the quadric T, which is to be a positive one-signed function, and it is stated that the successive discriminants should all be positive.

If we assume that the sign of the discriminant is not altered by any linear transformation of the co-ordinates we may obtain an easy proof of this proposition. Let the quadric be

$$2T = A_{11}\theta^2 + 2A_{12}\theta\phi + A_{22}\phi^2 + \&c. \ldots\ldots\ldots\ldots\ldots\ldots(1),$$

and to simplify the argument let there be only four co-ordinates θ, ϕ, ψ, χ. Let D be the discriminant, D_1 the discriminant when any one co-ordinate, say χ, is put equal to zero, D_2 the discriminant when two co-ordinates, as χ and ψ, are both put equal to zero, D_3 the discriminant when three co-ordinates, χ, ψ and ϕ, are put equal to zero, and so on.

Collecting all the θ's together, then the ϕ's and so on, we may write T in the form

$$2T = B_1\,(\theta + a_1\phi + b_1\psi + c_1\chi)^2 + B_2\,(\phi + b_2\psi + c_2\chi)^2 + B_3\,(\psi + c_3\chi)^2 + B_4\chi^2,$$

where all the English letters on the right-hand side are rational functions of A_{11}, A_{12}, &c. and therefore are real.

We may now write this expression in the form

$$2T = B_1v^2 + B_2y^2 + B_3z^2 + B_4u^2 \ldots\ldots\ldots\ldots\ldots\ldots (2),$$

where $u = \chi$, $z = \psi + c_3\chi$, and so on.

Since (1) and (2) may be derived from each other by a linear transformation, their discriminants have the same sign. Hence the product $B_1B_2B_3B_4$ has the same sign as D. Again, putting $u = \chi = 0$ and repeating the argument, the product $B_1B_2B_3$ has the same sign as D_1. Similarly the product B_1B_2 has the same sign as D_2 and

B_1 has the same sign as D_3. Thus B_1, B_2, B_3, B_4 *are positive when the discriminants* D, D_1, D_2, D_3 *are all positive and not otherwise.*

The conditions that T should be a one-signed positive quadric follow immediately. The conditions that T should be a one-signed negative quadric may be deduced from these by changing the signs of all the coefficients A_{11}, A_{12}, &c. in the expression for T.

That the discriminants of (1) and (2) keep the same sign may be shown by the method indicated in Art. 71. Taking the second expression let us write

$$x = l_1\theta + l_2\phi + \dots \qquad y = m_1\theta + m_2\phi + \dots \qquad z = \&c. \dots\dots\dots (3).$$

Substituting in (2) we obtain a quadric expression whose discriminant is easily seen to be

$$\begin{vmatrix} B_1 l_1{}^2 + B_2 m_1{}^2 + \dots & B_1 l_1 l_2 + B_2 m_1 m_2 + \dots \&c. \\ B_1 l_1 l_2 + B_2 m_1 m_2 + \dots & B_1 l_2{}^2 + B_2 m_2{}^2 + \dots & \&c. \\ \&c. & \&c. & \&c. \end{vmatrix}$$

This is obviously the square of

$$\begin{vmatrix} \sqrt{B_1} l_1 & \sqrt{B_2} m_1 & \sqrt{B_3} n_1, & \&c. \\ \sqrt{B_1} l_2 & \sqrt{B_2} m_2 & \sqrt{B_3} n_2, & \&c. \\ \&c. & \&c. & \&c. & \&c. \end{vmatrix}$$

The discriminant of T when expressed as a function of θ, ϕ, &c. is therefore equal to

$$B_1 B_2 B_3 \dots \begin{vmatrix} l_1 & m_1, & \&c. \\ l_2 & m_2, & \&c. \\ \&c. & \&c. & \&c. \end{vmatrix}^2$$

The sign has therefore not been altered.

The determinant on the right-hand side is the Jacobian of x, y, &c. with regard to θ, ϕ, &c. We may therefore also immediately deduce from this result by a double transformation the theorem quoted in Art. 69.

Art. 631. **Stress in a curved flexible and extensible rod.** The statical theorems quoted in this article may be proved in the following manner. Let PQ be any element of the axis of the rod in its unstrained position, $P'Q'$ the same element in the strained rod. Let ds, ds' be the lengths of these elements, a, ρ the radii of curvature at P, P'. Then, since p, q are the proportional elongation and increase of curvature,

$$p = \frac{ds'}{ds} - 1, \qquad q = \left(\frac{1}{\rho} - \frac{1}{a}\right) a \dots\dots\dots\dots\dots(1).$$

Let a, θ be the co-ordinates of P, $a(1+u), \theta+\phi$ those of P'. Then since

$$ds = a\, d\theta, \qquad (ds')^2 = a^2 (du)^2 + a^2 (1+u)^2 (d\theta + d\phi)^2$$

we easily find that

$$p = u + d\phi/d\theta \dots\dots\dots\dots\dots\dots\dots\dots\dots (2).$$

Again, when we neglect the squares of small quantities we have

$$\frac{1}{\rho} = \frac{1}{r} + \frac{d^2}{d\theta^2}\left(\frac{1}{r}\right), \qquad r = a(1+u)$$

$$\therefore q = -\left(u + \frac{d^2 u}{d\theta^2}\right) \dots\dots\dots\dots\dots\dots\dots (3).$$

Let us refer the rod to the principal axes of the curved axis at P'. Let the normal measured inwards be the axis of z, let the tangent be the axis of x, and let y be perpendicular to the plane of the curve. We assume, as is usual in such problems, that the material particles of the rod which lie in a plane perpendicular to the axis continue to lie in a plane perpendicular to the axis when the rod is bent or stretched, and that their distances from the axis are not sensibly altered.

Drawing two planes normal to the axis at P', Q', let $R'S'$ be any elementary fibre of the rod parallel to the axis lying between these planes, and let RS be its unstretched length. Let y, z be the co-ordinates of R'; if ds be the unstretched length of $P'Q'$, the unstretched and stretched lengths of $R'S'$ are respectively

$$d\sigma = ds\left(1 - \frac{z}{a}\right), \qquad d\sigma' = ds'\left(1 - \frac{z}{\rho}\right) \dots\dots\dots\dots\dots (4).$$

The resultant tension of all the fibres which cross the elementary area $dydz$ is evidently $Edydz\left(\dfrac{d\sigma'}{d\sigma} - 1\right)$. Substituting for $d\sigma'$, $d\sigma$, $1/\rho$ their values given by (4) and (1) and rejecting all the powers of z/a above the second because the rod is thin, we find that the resultant tension of these fibres

$$= Edydz\left\{p - (1+p)q\left(\frac{z}{a} + \frac{z^2}{a^2}\right)\right\} \dots\dots\dots\dots\dots(5).$$

Let ω be the area of the section of the rod, ωk^2 its moment of inertia about the axis of y. Remembering that the centre of gravity of ω is the origin, we find by an obvious integration that the resultant tension T and couple L are given by

$$T = E\omega\left\{p - \frac{k^2}{a^2}(1+p)q\right\}, \qquad L = -E\omega\frac{k^2}{a}(1+p)q \dots\dots\dots(6).$$

Since the rod oscillates about its unstrained position we may neglect the squares and products of the small quantities p and q. These then reduce to the results used in Art. 631. Ex. 2.

The work of a fibre per unit area of section when pulled from its unstretched length $d\sigma$ to the length $d\sigma'$ is proved in Vol. I. Art. 343 to be $-\frac{1}{2}E(d\sigma' - d\sigma)^2/d\sigma$. Substituting as before for $d\sigma$, $d\sigma'$ and rejecting the cubes of z/a, we find that the work W is given by

$$Wds = -\frac{1}{2}E\omega ds\left\{p^2 + \frac{k^2}{a^2}q^2(1+p)^2\right\} \dots\dots\dots\dots\dots(7).$$

This reduces to the result given in Art. 631 when only the lowest powers of p and q are retained.

From the expression for W we may deduce the values of T and L. Keeping P' fixed, let the element $P'Q'$ be further stretched, without altering the curvature, so that its length becomes ds'', then $dp = (ds'' - ds')/ds$. The work done by the tension at the end Q' is $-T(ds'' - ds')$ and that done by the couple at Q' is $L(ds'' - ds')/\rho$. We therefore have

$$-T + \frac{L}{\rho} = \frac{dW}{dp}.$$

Next let the rod receive an increase of curvature without altering the length of the element. The tension at Q' does no work, while the work of the couple is $L(1/\rho' - 1/\rho)ds'$, where $1/\rho'$ is the new curvature. Since $dq = (1/\rho' - 1/\rho)a$, we see that

$$L = \frac{a}{1+p}\frac{dW}{dq}.$$

From these results we easily deduce the values of T and L given by (6).

The theorem quoted from statics in Art. 628, viz. that $L = \pm F/\rho$ where $F = k^2(E\omega + T)$, also follows easily from the equations (6). Remembering that the unstrained radius a is here infinite, and putting $q = a/\rho$ we have

$$T = E\omega p, \qquad L = -E\omega k^2(1+p)/\rho.$$

Eliminating p from the value of L, we have the result quoted.

Art. 639. Loaded Membranes. We may also deduce this result from the formulæ in Arts. 76 and 77. We shall begin by referring the unloaded membrane to principal co-ordinates. To effect this we write (see Art. 56) the complete expression for w given in Art. 636 in the form

$$w = \sin\frac{\pi i x}{a}\sin\frac{\pi i' y}{b}\,\xi + \sin\frac{\pi j x}{a}\sin\frac{\pi j' y}{b}\,\eta + \&c.,$$

then the quantities ξ, η, &c. are principal co-ordinates.

The vis viva of the membrane is easily seen to be

$$\iint (dw/dt)^2\,\rho\,dx\,dy = \tfrac{1}{4}\rho ab\,(\xi'^2 + \eta'^2 + \ldots)$$

where accents denote differential coefficients with regard to the time. If we now form Lagrange's determinant, every constituent will be zero except those in the leading diagonal. If $q_1{}^2$, $q_2{}^2$, &c. be the roots of the determinant and $M = \rho ab$, these constituents will be $\tfrac{1}{4}M(q^2 - q_1{}^2)$, $\tfrac{1}{4}M(q^2 - q_2{}^2)$, &c. Here q stands for the quantity represented by pm in Art. 636; the roots q_1, q_2 &c. are all found in that Article, and are expressed by giving i and i' all integer values.

Placing now a mass μ at the point (h, k), its displacement will be given by

$$W = \sin\frac{\pi i h}{a}\sin\frac{\pi i' k}{b}\,\xi + \&c.,$$

which we may abbreviate into

$$W = \alpha\xi + \beta\eta + \&c.$$

There will now be an additional term in the expression for the vis viva, while the force-function will be the same as before. This additional term will be

$$\mu\alpha^2\xi'^2 + 2\mu\alpha\beta\xi'\eta' + \&c.$$

There will therefore be an additional term to every constituent of Lagrange's determinant. The determinant will be

$$\begin{vmatrix} \tfrac{1}{4}M(q^2 - q_1{}^2) + \mu\alpha^2 q^2 & \mu\alpha\beta q^2 & \&c. \\ \mu\alpha\beta q^2 & \tfrac{1}{4}M(q^2 - q_2{}^2) + \mu\beta^2 q^2 & \&c. \\ \&c. & \&c. & \&c. \end{vmatrix} = 0.$$

Expanding this, and remembering that by Art. 76 only the first powers of μ can enter into the expansion, we have

$$(q^2 - q_1{}^2)(q^2 - q_2{}^2)\,\&c. + \frac{4\mu}{M}\,q^2\{\alpha^2(q^2 - q_2{}^2)\,\&c. + \beta^2(q^2 - q_1{}^2)\,\&c. + \&c.\} = 0.$$

Dividing by the first term we have

$$\frac{M}{4\mu q^2} = \frac{\alpha^2}{q_1{}^2 - q^2} + \frac{\beta^2}{q_2{}^2 - q^2} + \&c.$$

Substituting for α, β, &c. their values given above, and writing $q = pm$, we have the result given at length in Art. 639.

This method is clearly general, and will apply, when the proper values of α, β, &c. are substituted, to membranes of other forms.

Art. 641. Conjugate Functions. The application of the theory of conjugate functions to Hydrodynamics is probably well known to the student. By that theory the potential of a complicated fluid motion can sometimes be made to depend on that of some simpler motion. But this of course is beyond the scope of the present work. We may however notice some propositions which appear to be new.

When one fluid motion is changed into another by a method analogous to that described in Art. 642 for membranes, the kinetic energies of the two fluids which occupy corresponding elementary areas are equal. Thus *the whole kinetic energies of the two motions are equal, but differently distributed over the areas of motion.* This corresponds to the theorem proved in Art. 646 for membranes.

Suppose a vortex II of strength m to exist at any instant in one fluid at a point whose co-ordinates are (ξ, η). There will then be a vortex P of equal strength at the corresponding point (x, y) of the other fluid. These will not continue to move so as to occupy corresponding points, but we may sometimes, *without discussing the motion of the rest of the fluid*, deduce the motion of P from that of II by the following rule. *Let $\chi\,(\xi, \eta)$ be a current function (not the current function of the fluid) giving the motion of the vortex II, so that its velocities resolved parallel to the axes of ξ and η are respectively $\dfrac{d\chi}{d\eta}$ and $-\dfrac{d\chi}{d\xi}$. Then the instantaneous motion of P is given by a current function*

$$\chi'\,(x, y) = \chi\,(\xi, \eta) - \tfrac{1}{2} m \log \mu,$$

i.e. its velocities resolved parallel to the axes of x and y are respectively $\dfrac{d\chi'}{dy}$ and $-\dfrac{d\chi'}{dx}$, and its path is found by equating χ' to a constant. Here μ^2 is the quantity called D/D_0 in Art. 642. Generally we may say that *the current function of P is obtained from that of II by subtracting $\tfrac{1}{2} m \log \mu$, where*

$$\mu^2 = (d\xi/dx)^2 + (d\xi/dy)^2 = (d\eta/dy)^2 + (d\eta/dx)^2.$$

In using this rule the strength m of a vortex is to be considered positive when the vortex rotates in the direction opposite to the hands of a watch, that is from the positive direction of ξ to the positive direction of η.

To prove this theorem we notice that the current function at any point (ξ_1, η_1) in one fluid, or at the point (x_1, y_1) in the other is

$$\psi = -\tfrac{1}{2} m \log \{(\xi_1 - \xi)^2 + (\eta_1 - \eta)^2\} + R,$$

where in the latter fluid the Greek letters are regarded as known functions of the English ones. Here R represents a series of terms, similar to the first, due to the presence of other vortices. Since the vortex P does not move itself, we can deduce its motion from that of the neighbouring points by superimposing on the latter the reversed motion due to the vortex. This relative motion is given by the current function,

$$\psi = -\tfrac{1}{2} m \log \{(\xi_1 - \xi)^2 + (\eta_1 - \eta)^2\} + \tfrac{1}{2} m \log \{(x_1 - x)^2 + (y_1 - y)^2\} + R.$$

Let $\xi_1 = \xi + \xi'$, $\eta_1 = \eta + \eta'$, $x_1 = x + x'$, $y_1 = y + y'$. Let us expand the expression for ψ in powers of x', y' by substituting

$$\xi_1 - \xi = \xi_x x' + \xi_y y' + \tfrac{1}{2}\,(\xi_{xx} x'^2 + 2\xi_{xy} x' y' + \xi_{yy} y'^2) + \&c.$$

with a similar expression for $\eta_1 - \eta$. Here the suffixes x, y &c. denote differentiations. We find, after retaining the cubes of the small quantities, that the factor $x'^2 + y'^2$ divides out. Expanding the logarithm we have

$$\psi = -\frac{m}{2} \left\{ \frac{x'\, d \log \mu}{dx} + y'\, \frac{d \log \mu}{dy} \right\} - m \log \mu + R,$$

where $\mu^2 = \xi_x^2 + \xi_y^2$. The effect of the first term of this series is to give P resolved velocities equal to $-\tfrac{1}{2} m d \log \mu/dy$ and $\tfrac{1}{2} m d \log \mu/dx$ parallel to the axes of x and y.

Consider next any term of R due to the presence of a vortex at (ξ_0, η_0), say

$$R = -\tfrac{1}{2} m \log \{(\xi - \xi_0)^2 + (\eta - \eta_0)^2\}.$$

The resolved velocities of a point of the fluid at II are found by differentiating this with regard to η, ξ, and changing the sign in the latter case; let these be u, v. The resolved velocities of a point at P are similarly found to be $u\eta_y - v\xi_y$ and $-u\eta_x + v\xi_x$. If there be only one independent vortex, the vortices included in R are images of Π and their positions are determined by that of Π. Let the conditions of the question be such that the resolved instantaneous velocities of Π are $u = \chi_\eta$, $v = -\chi_\xi$, then the resolved velocities of P due to the same terms are χ_y, $-\chi_x$. Taking therefore all the terms of ψ, the resolved velocities of P are $\chi_y - \frac{1}{2}md \log \mu/dy$ and $-\chi_x + \frac{1}{2}md \log \mu/dx$.

As an example of this rule, let us investigate the path of a vortex P swimming in the corner formed by two straight lines inclined at an angle equal to π/n. This problem is discussed by Prof. Greenhill in the *Quarterly Journal*, Vol. xv. Let us first suppose a vortex Π to swim in the infinite space bounded by the axis of ξ. Placing an image on the negative side of this axis, we see that the vortex Π moves parallel to the axis of ξ with a velocity $m/2\eta$. Its stream function is therefore $\frac{1}{2}m\log\eta$. Taking any point on the axis of ξ as origin, we shall turn the negative side of the axis round the origin until it makes an angle equal to π/n with the positive side. To express this we use the formulæ of transformation given in Art. 653. We thus have $\eta = c\,(r/c)^n \sin n\theta$. The value of μ is therefore $n\,(r/c)^{n-1}$. According to the rule the stream function which gives the motion of the vortex P in the corner is

$$\chi' = \frac{1}{2}m\log\eta - \frac{1}{2}\log\mu$$
$$= \frac{1}{2}m\log(r\sin n\theta).$$

The path is therefore given by $r\sin n\theta = c$ where c is a constant. It may be noticed that n need not be an integer.

If two circles intersect in A and B, we may find, by inverting this result, the *motion of a vortex V in the space between the circular boundaries*. Let θ be the angle the circle through A, B and the vortex V makes with either circular boundary, and let α be the angle between the circular boundaries. Then the current function of the vortex V is found by subtracting $\frac{1}{2}m\log\mu$ from the value of χ' given above, where $\mu = \left(\dfrac{c}{r}\right)^2$, as shown in Art. 653. The current function of the vortex V is therefore $\chi = \dfrac{m}{2}\log\left(AV.BV.\sin\dfrac{\pi\theta}{\alpha}\right)$.

The path of the vortex is given by the equation $AV.BV.\sin\dfrac{\pi\theta}{\alpha} = C$, where C is a constant.

The chief objection to using the method of conjugate functions in Hydrodynamical problems is the difficulty of finding the proper formulæ of transformation. But to discover these we have a convenient rule, viz. that *if we know the motion of a fluid within the space bounded by one or two infinite curves, we can in general find the motion with the same boundaries when complicated by the presence of sources and vortices.* To prove this, let ξ and η be the velocity and stream potentials of this motion. Then η is constant along the boundaries. If we use ξ, η as our formulæ of transformation, the given boundaries will transform into straight lines parallel to the axis of ξ. The motion due to vortices and sources in this space has already been investigated. Hence the motions in the more general spaces may be deduced.

We may regard any closed curve, such as an ellipse, as a section of an infinite cylinder. If we know its potential at any external point when charged with a given quantity of electricity, we may immediately deduce the motion of a fluid with vortices and sources outside this curve from the corresponding motion round a circle.

For these theorems we refer the reader to a paper by the author published in the twelfth volume of the *Proceedings of the Mathematical Society*, 1881.

𝕮𝖆𝖒𝖇𝖗𝖎𝖉𝖌𝖊:

PRINTED BY C. J. CLAY, M.A. AND SONS,

AT THE UNIVERSITY PRESS.

For EU product safety concerns, contact us at Calle de José Abascal, 56–1°,
28003 Madrid, Spain or eugpsr@cambridge.org.

www.ingramcontent.com/pod-product-compliance
Ingram Content Group UK Ltd.
Pitfield, Milton Keynes, MK11 3LW, UK
UKHW010853090126
466816UK00011B/201